Coastal and Estuarine Studies

Coastal and Estuarine Studies

formerly Lecture Notes on Coastal and Estuarine Studies

36

W. Michaelis (Ed.)

Estuarine Water Quality Management

Monitoring, Modelling and Research

Springer-Verlag
Berlin Heidelberg New York London Paris Tokyo Hong Kong

Editors

W. Michaelis
Institute of Physics, GKSS Research Centre Geesthacht
P.O. Box 1160, 2054 Geesthacht, FRG

ISBN-13:978-3-642-75415-9 e-ISBN-13:978-3-642-75413-5
DOI: 10.1007/978-3-642-75413-5

Softcover reprint of the hardcover 1st edition 1990

2131/3140-543210 – Printed on acid-free paper

LUDWIG PRANDTL
GKSS

Photograph on page V:

RV LUDWIG PRANDTL

The ship, operated by GKSS, is a special development for estuarine research.

Dimensions:	Overall length 23 m. Beam 6 m. Draught 1 m.
Speed:	10 kn.
Propulsion:	Two air-cooled diesel engines, 180 kW each, with fixed propellers; stem and stern pumpjet, 88 kW each.
Scientific facilities:	Electronic and data processing laboratory for hydrographic measurements; chemical laboratory for sample preparation.
Hoist:	5 mt stern crane; A-frame; single-conductor cable slip-ring winch.
Equipment:	Hydrographic measuring system (HYDRA) with two sensor packages, one at a bow cantilever for horizontal profiling and another one in a lowering probe operated from a starboard A-frame for vertical profiling. Precision navigation systems (RALOG and SYLEDIS) for application of the moving-boat technique.

PREFACE

The pollution of rivers, estuaries and the sea, the associated impact on these ecosystems and the effect on organisms, food-chains, water-supply and finally on man himself are becoming more and more recognized all over the world. Estuaries are often surrounded by highly industrialized and densely populated regions and, consequently, are particularly endangered by anthropogenic polluton. They act in a sense as a link between the limnetic and marine environments and are thus characterized by a variety of complex, mutually interacting physical, chemical and biological processes. A lot of the phenomena are not yet sufficiently understood. This makes efficient water quality management in estuaries a difficult task. The knowledge on the pollution loads that estuaries discharge into the sea is quite fragmentary. According to a recent compilation only 25 % of the world-wide 260 major rivers discharging into the oceans are regularly monitored for water quality.

In view of this background the International Symposium "Estuarine Water Quality Management - Monitoring, Modelling and Research" was held between 19th and 23rd June, 1989, at the Sachsenwald Congress Centre at Reinbek, adjacent to the City of Hamburg. It was organized by the GKSS Research Centre Geesthacht jointly with the National Committee of the International Hydrological Programme (IHP) and the Operational Hydrological Programme (OHP) of the Federal Republic of Germany, and the United Nations Educational, Scientific and Cultural Organization (UNESCO). The conference was a contribution to the International Hydrological Programme (IHP) and it was devoted to two main objectives:

(i) to serve as a forum for the exchange of information on current estuarine research and

(ii) to elaborate fundamentals and criteria for planners and decision-makers in water quality management.

The meeting brought together work from a broad spectrum of disciplines. It was attended by more than two hundred participants from many countries all over the world. A review on estuarine water quality management, 7 keynote lectures on the most important topics, 41 contributed papers describing original research as well as specific case-studies and field evaluations, and 25 posters were presented. This volume summarizes the work reported at the Symposium. Numerous authors have contributed to achieving the objects envisaged.

As the editor, I express my hope that the volume will provide a useful aid for scientists, engineers, planners and administrators in universities, research laboratories, regulatory agencies, water authorities, ministries and industries which are in a broad sense concerned with estuarine water quality.

W. Michaelis

Acknowledgments

The financial support by the following private sponsors is gratefully acknowledged:

Verein der Freunde und Förderer des GKSS-Forschungszentrums Geesthacht e.V.
IBM Deutschland GmbH, Hamburg
Dresdner Bank AG, Hamburg
Drägerwerk AG, Lübeck
Deutsche Bank AG, Hamburg
Rotring-Werke Riepe KG, Hamburg
Fürst Bismarck-Quelle, Aumühle

The editor is indebted to Dr. Peter Beaven, Dr. Desmond Murphy and Mrs. Karen Wiltshire for their linguistic assistance in revising the manuscripts. I also feel obliged to Mrs. Wiebke Jansen and Mrs. Karin Rahn for retyping most of the contributions.

CONTENTS

Chapter I

REVIEW ON ESTUARINE WATER QUALITY MANAGEMENT

Chapter II

ESTUARINE MODELLING

Chapter III
INTEGRATED CONCEPTS, STRAGEGIES

Chapter IV
WATER DYNAMICS AND TRANSPORT PROCESSES

Chapter V
ESTUARINE MEASUREMENT TECHNIQUES

CHAPTER VI
SEDIMENT-WATER INTERFACE

Chapter VII
MONITORING

Chapter VIII
NUTRIENTS AND OXYGEN BUDGET

Chapter IX
BIOLOGICAL PROCESSES, ENVIRONMENTAL IMPACT

Chapter X
WATER CHEMISTRY AND SPECIATION

CHAPTER I

REVIEW ON ESTUARINE WATER QUALITY MANAGEMENT

ESTUARINE WATER QUALITY MANAGEMENT: PLANNING, ORGANIZING, MONITORING AND MODELLING

Neil S. Grigg
International School for Water Resources
Colorado State University
Ft Collins CO 80523 USA

1. OBJECTIVE OF PAPER

The value of estuaries is well known, but managing their water quality faces obstacles due to scientific and political complexity. Unique action plans, based on good science, are required for each estuary. Effective water quality monitoring and modelling programs can help.

IHP-II included a project to identify research needs in coastal waters, and one recommendation was for an IHP-III project on "the effectiveness of estuarine water quality models in policy analysis and decisionmaking" (IHP-III 10.3). This paper results from the writer's participation in both the IHP-II and IHP-III studies; it reports on principles and developments in planning, organizing, monitoring and modelling to improve the effectiveness of estuarine water quality management.

2. PROBLEM DESCRIPTION

Much of the world's population lives close to and depends on estuaries for income, food and shipping, and water quality problems are the result. Little is known about the global pollution loads that estuaries bring to the oceans. UNESCO (1977) compiled a world register of rivers discharging into the oceans. They showed that of 260 major rivers discharging into the oceans, only 25 % were regularly monitored for water quality.

There are about 850 estuaries in the US alone (National Academy of Science, 1983). The US National Oceanic and Atmospheric Administration (NOAA) is compiling a National Estuarine Inventory (NOAA, 1985). Common problems with US estuaries include: shellfish bed closures, lost and altered wetlands, disappearance of submerged aquatic vegetation, threats to livng resources from toxics, diseased fish and shifts of fish species, and nuisance blooms of algae (Davies, 1985).

Similar threats have spurred pollution control action in other regions such as the Mediterranean and the North Sea. The United Nations Environment Programme (UNEP) recently assessed the health of the world's oceans, seas and coastal waters, and issued a report showing improvements in some situations, such as oil pollution, but not others, such as radioactivity (United Nations Environment Programme, 1988). UNEP reports progress in the Mediterranean, but needed improvements in other regions such as the Caribbean and South Asia which suffer from chemical fertilizer and pesticide pollution from farming.

3. MANAGING WATER QUALITY IN ESTUARIES

Meeting the challenge of marine pollution will require innovation and initiative in pollution control. General tools for water pollution include standards, discharge permits and regulatory actions, but estuary problems require more complex approaches. Authorities have turned to the concept of "management plans" for estuaries. In the US steps are underway to develop them for a number of key estuaries through the National Estuary Program, which uses experiences with the Chesapeake and Great Lakes programs to seek three goals: living resource management, water resource management and pollutant load reduction (Davies 1988).

Management principles applied in the National Estuary Program are based on a phased program approach and collaborative problem solving. The phased approach identifies priority problems, establishes priorities and develops strategies to solve them. This requires monitoring and research, and may involve modelling. It involves assessment of laws and regulations, water uses and impacts, and identification of alternative management strategies and development of specific action plans with implementation schedules. In short, it is the application of the general problem solving process to the complex problem of estuary management.

The collaborative approach involves concerned parties in problem solving and secures commitments for needed actions. The National Estuary Program requires the organization of a "Management Conference" which is intended to be a forum for open discussion, cooperation and compromise that will result in consensus and action. The Management Conference is a formal technique that seeks to provide effective collaboration for the phased problem solving approach.

Management planning for estuaries requires a complex mixture of science and intergovernmental public-private cooperation. Its status will bear watching. It tests the ability of modern governments to solve common resource management problems. Management planning needs to bring current knowledge in many areas to bear on solving complex pollution problems involving many different interest groups. These management plans will take years to implement, and will require sustained commitments of effort and scientific resources.

4. MONITORING OF WATER QUALITY

A monitoring program is necessary to assess estuarine water quality and to make sure all phases of a management program are working. The terms "monitoring" and "assessment" are sometimes used to mean different processes. The US National Academy of Sciences defined them this way: monitoring is the "... repetitive collection of water quality data for some specific purpose, e.g., compliance and enforcement, or establishment of a management strategy." Assessment, on the other hand, uses monitoring data and other information to evaluate or interpret data in terms of ambient conditions, identification of water quality problems, sources of pollutants and their impact, trends and effectiveness of control programs (National Academy of Sciences, 1986).

The US Environmental Protection Agency (1984, 85) has published guidance for the design of monitoring programs. Three basic purposes of monitoring are listed: to conduct water quality assessments, to develop water quality-based controls and assess compliance with and effectiveness of controls. Data uses are given as: national water quality assessment, statewide water quality assessment, regional oversight, program management and wasteload allocations, construction grants design and to check water quality standards. Aspects of estuaries that need measurement are listed: "... tides and currents, stratification, substrate characteristics, importance of salinity, dissolved oxygen and nutrient enrichment, species diversity, plant and animal populations, and physiological adaptations which permit freshwater or marine organisms to survive in the estuary."

A partial list of characterization parameters, according to Davies (1985), would include: physical parameters (land use, hydrology, shoreline development, erosion rates and storm events); chemical parameters (nutrients, dissolved oxygen, phosphate, total nitrogen, inorganic nitrogen, nitrate, ammonium, organic nitrogen, toxic metals, pesticides and organics); and biological parameters (landings, catch per unit, nursery areas juvenile index, spawning areas, and plant data). The complexity involved in designing monitoring programs for these is evident. The Potomac Estuary study by the US Geological Survey shows this, with water quality problems indexed to sampling activities (USGS, 1984). For example, sedimentation as a water quality problem requires studies of the following processes: nutrient transport, hydrodynamics, nonpoint sources, submerged aquatic vegetation and sedimentation. This data requires a major investment of scientific resources, and shows the tight linkage between monitoring and research necessary to establish action plans.

The state of the art of monitoring will continue to evolve due to the complexity of the processes to be monitored and studied. It will be important that monitoring be designed to complement the use of mathematical models in the future. Monitoring needs can be so expensive and extensive that they may in the future need to go beyond governmental programs and include citizen volunteers (Rhode Island Sea Grant Program, 1988).

5. MATHEMATICAL MODELS FOR ESTUARIES

Models simulate what occurs in actual systems. For estuaries, this is a tall order. First, the model must simulate complex hydraulics. Then the model must be able to handle mass transport phenomena. Next there are complex biological and chemical changes to be handled, such as algal sequencies and oxygen transfer. Then there is the relationship between the benthic layer and water column, and finally, living things in the food chain must be considered.

Why model estuaries? Models help our understanding of complex relationships that we cannot measure and provide information for decisionmaking. The first objective, to improve understanding, is rather forgiving; if the model ist not completely valid, we can excuse that because it is a research tool. The second reason, decisionmaking, is not as forgiving, however, because decisions have consequencies, such as imposing costs on businesses, cities and farmers. With this observation it is not surprising that there are many estuary research models, but not so many that have been used for decisionmaking.

Since there are so many models, comprehensive inventories and guideline reports have been prepared. Examples include: Roesner, Walton and Hartigan (1986) who cover practical considerations; McAnally (1987) who presents a course handbook; and Hinwood & Wallis (1975 a, b) who analyzed 108 models.

A literature review of reported model applications (Grigg, 1987) showed that models are in regular use to predict how river segments respond to waste discharges, but actual applications for estuary management are limited. The writer attempted to use the results from several models for decisionmaking in a tributary to Albemarle Sound, located on the US East Coast. Due intractable problems a scientific review panel was convened to examine all the Chowan River evidence. They concluded that "existing mathematical models developed in past Chowan research will not be directly useful in making management decisions for nutrient control, because limitations in their composition precludes their being used to evaluate relevant consequencies of management strategies" (Chowan River Review Committee, 1980).

The survey and literature review by the writer found diverse applications. They included: a Dynamic Estuary Model which had been applied to evaluate management alternatives on the Delaware and Potomac Estuaries; a screening model by US-NOAA, giving a framework for preliminary estuarine assessment; the "Rational Method" for stream analysis, applied to Lake Maracaibo in Venezuela; a number of one- and two-dimensional models to evaluate situations such as the effect of a proposed levee on adjacent marshes, the effect of reservoir regulations and consumption on the salinity of the Delaware estuary, the DO und BOD of the Tyne estuary (England) resulting from new waste water treatment works, the costs and options in the Tees estuary (England), material transport and load in the Elbe estuary (Germany), predicting the salinity intrusion effects of deepening the Mississippi navigation channel, studying the circulation, sedimentation and salinity intrusion in Atchafalaya

Bay, a tributary to the Mississippi, evaluating the effect of warm cooling water on the heat budget in a bay in Japan, and studying contaminants, sediment uptake, eutrophication and toxicity in the North Sea near the Durch coast.

Wallis (1987) provided a list of model applications in Australia which shows similar applications: Darwin Harbor, 1-D model of Disposal site plan; Mooloolaba Estuary, 1-D model of effluent limits; Botany Bay, 2-D, 1-D models of effluent diversion plan; Hanns Inlet, 1-D model for problem identification; Port Phillip Bay, 2-D model of effluent diversion plan; and Hawkesbury Estuary, 1-D model of gravel extraction impacts.

Although use of models for decisionmaking has proceeded slowly, some are in actual use for management. Claims about the use of models for management need to be carefully evaluated, since many claims refer to research, and may never be used in management.

6. CURRENT ESTUARY MANAGEMENT ACITIVITIES

In the US the best known programs are on Chesapeake Bay and the Great Lakes. In the Great Lakes, recommendations are being implemented for point and nonpoint source controls, and a detergent ban has been placed in effect in many of the states around the Great Lakes. The Chesapeake Study aimed at land use controls near the shoreline, development of nonpoint source controls, control of point sources, and strengthening of wetlands protection laws. As a result of the program the riparian states have taken action (Davies, 1985).

Examples of management plans can be found for other estuaries. The Puget Sound Water Quality Authority's plan (1987) includes the following elements: nonpoint source pollution, shellfish protection, municipal and industrial discharges, contaminated sediments and dredging, stormwater and combined sewer overflows, laboratory support, wetland protection, oil spill response planning, monitoring, research, education and public involvement, household hazardous waste, and legal support. An action plan was developed in 1979 for the Chowan River, a key tributary of the Albermarle-Pamilico estuary. The river suffered from outbreaks of blue-green algae and fishing problems (Grigg, 1982). Results of action planning have been merged into an action plan by the state and USEPA, and the estuary has now been included in the National Estuary Program.

The Mediterranean Action Plan (MAP) is an example of a complex, multinational program. MAP emerged in 1975 at a conference called by the United Nations Environment Programme in Barcelona when 16 countries approved an action plan (United Nations Environment Programme, 1985). The plan called for a series of treaties, the creation of a pollution monitoring and research network and a socio-economic program that would reconcile economic and environmental priorities. After a declaration at Genoa of a second decade of MAP, priority targets for action have been identified. The future of the Mediterranean Plan will be tough, since there are many pollution problems to solve, and much political will and public education are needed.

The North Sea is another example of an international water system needing management action. The Dutch have prepared a management analysis that calls for development of a data base and modelling system and for policy studies. This illustrates the need to base decisions on scientific information developed through research (Rijswaterstaat, 1988). Another example is the Gulf of Mexico program launched in 1988 by the US Environmental Protection Agency. EPA lists six areas of concern: nutrients, toxics, habitat loss, shellfish pollution, freshwater flow alteration and public health concerns. The action plan for the Gulf will build on Great Lakes and Chesapeake lessons (USEPA, 1988).

Transnational or interstate aspects of a cleanup program, such as with MAP, are the rule, not the exception. Chesapeake Bay is an interstate waterway, being surrounded by Virginia, Maryland, Pennsylvania and the District of Columbia. This complicating factor in estuary water quality management shows the importance of multi-state action.

Ecologists, as well as engineers, participate in management planning. The US established an estuarine reserve system as part of the Coastal Zone Management Act of 1972, and it provides for estuarine reserves that are managed for long term environmental monitoring and scientific research. By 1989 there were 18 such reserves that had been designated.

7. CONCLUSIONS

The difficulties in estuary management are scientific and political. Success requires unique action plans based on good science, with realistic implemention goals supported by effective collaboration.

To manage estuaries, authorities have turned to the concept of "management plans". In the US, principles for management plans are based on a phased approach which requires monitoring and research, and may involve modelling. The phased approach must involve collaborative problem solving with hard commitments for needed actions. Management plans take years to implement, and require sustained commitments of effort and scientific resources. Much is happening in management planning for estuaries. The US National Estuary Program and the Mediterranean Action Plan are two programs to learn from.

An effective monitoring program is necessary to support the research needed for action planning and to assess compliance with controls. Monitoring technology and practice are evolving, and it is necessary to adapt them to support management programs.

Modelling can improve understanding of estuaries and provide information for decisionmaking about management actions. Modelling has proceeded slowly, but some models are in limited use for management. Others are used for research and planning.

Estuary management presents a good example of what the Earth faces in its future environmental struggles: the need to blend scientific understanding with effective political cooperation to solve critical natural resource issues.

ACKNOWLEDGEMENTS

The writer acknowledges the support of UNESCO which was provided through IHP-II and III leading to ideas and organization of the studies behind this paper. The US Geological Survey also provided support. Dr. K. Sivakumaran assisted in the resarch leading to the completion of his doctoral dissertation at Colorado State University in 1989.

REFERENCES

Chowan River Review Committee (1980) An assesment of algal bloom and related problems of the Chowan and recommendations towards its recovery, Raleigh

Davies, Tudor (1985) Mangagement principles for estuaries, Environmental Protection Agency, unpublished speech, Washington

Davies, Tudor (1988) Institutional structures to deal with regional water problems: the Chesapeake Bay example, 22nd Water for Texas Conference, Houston

Grigg, Neil S (1982) Experience in estuary water quality management, Water International, No. 7, Urbana

Grigg, Neil S (1987) Effectiveness of models and monitoring in estuary water quality management, working paper for IHP-III Project 10.3, Colorado State University, Ft Collins

Hinwood JB, Wallis IG (1975) Classification of models for tidal waters, in Proceedings ASCE, Journal of Hydraulics Division, Vol.101, No: HY10

Hinwood JB, Wallis IG (1975) Review of models of tidal waters, in Proceedings ASCE, Journal of Hydraulics Division, Vol.101, No:HY11

McAnally, WH Jr (1987) Numerical modeling overview (An introduction), in Course Handbook, Numerical Analysis (2D) of Sediment Transport (TABS-2), USAE Waterways Experiment Station, Vicksburg

National Academy of Science (1983) Fundamental research on estuaries: the importance of an interdisciplinary approach, National Academy Press, Washington

National Academy of Sciences (1986) Water Science and Technology Board, national water quality monitoring and assessment, report of a Colloquium, Washington

National Oceanic and Atmospheric Administration (1985) National estuarine inventory: Data Atlas, Washington

Puget Sound Water Quality Authority (1987) Puget Sound water quality management plan, Seattle

Rhode Island Sea Grant Program (1988) Citizen volunteers in environmental monitoring, USEPA Report 503/9-89-001, Washington

Rijswaterstaat (1988) North Sea Directorate, management analysis North Sea, Hollanb

Roesner, Larry A, Raymond Walton, John P. Hartigan (1986) Realistic water quality modeling, in urban runoff pollution, NATO ASI Series, Vol G10, eds. H.C. Torno, J. Marselek and M. Desbordes, Springer Verlag, Berlin

United Nations Environment Programme (1985) Mediterranean coordinating unit, Mediterranean Action Plan, September

United Nations Environment Programme (1988) UNEP News, The state of the marine environment 1988, Nairobi

UNESCO (1977) World register of rivers discharging into the oceans, Paris

US Environmental Protection Agency (1985) Guidance for State water monitoring and wasteload allocation programs, EPA 440/4-85-031, Washington

US Environmental Protection Agency (1984) Technical support manual: water body surveys and assessments for conductintg use attainability analyses, Volume 1, General, Volume 2, Estuarine Systems, Volume 3, Lake Systems, Washington

US Environmental Protection Agency (1988) The gulf initiative, Washington

US Geological Survey (1984) A water-quality study of the tidal Potomac River and estuary - An Overview, ed. by Edward Callender, Virginia Carter, D.C. Hahl and Barbara Schulz, USGS water supply paper 2233, Washington

Wallis, Ian G. (1987) Personal communication

CHAPTER II

ESTUARINE MODELLING

PROGRESS IN ESTUARINE WATER QUALITY MODELLING

Wim van Leussen and Job Dronkers
Rijkswaterstaat, Tidal Waters Division
P.O. Box 20907, 2500 EX The Hague, The Netherlands

ABSTRACT

Starting from the needs of estuarine water quality management, the paper opens with an introduction into the developments in estuarine modelling, showing the trend of building frameworks of models, in which the various interdisciplinary building blocks are incorporated. Next a review is given of the progress in estuarine water quality modelling from a physical point of view. The paper concludes with a discussion on the main points which are to be expected to need much attention in the development of estuarine models in the coming years.

1. INTRODUCTION

Estuaries are unique water systems; they are the interface between fresh river water and saline coastal water. They have a high biological productivity, and are generally situated in densely populated areas, where often many pollution problems exist. To preserve the water resources in these systems, where complex interactions of physical, chemical and biological factors occur, decisions have to be made in the midst of many conflicting interests.

It is the goal of estuarine water quality management to optimize these decisisons in such a way that the quality of a specific estuary will be maintained or improved, weighting out the costs of diminishing discharges and the conflicting interests from recreation, fishery, shipping, industry, etc. In fact it is an optimization of the nonparallel commercial, recreational and ecological values of the system. Making these decisions there will always be incomplete and inadequate information, so that in estuarine management we have to learn to handle with environmental risks. Also cost-benefit analysis is generally incapable of properly assessing these environmental risks, because some risks are per se unacceptable, many costs cannot be adequately quantified and we are unable to predict all the environmental consequences of alterations in the ecosystem (Kamlet, 1978). To make well-considered decisions in a more quantitative way Environmental Quality Standards (EQS) have been defined. A considerable amount of work has been done to define these standards. Although such standards are always somewhat arbitrary, they form an important guidance in the judgement of the control of discharges. In the past decades much research was concentrated on the effects of domestic sewage discharges; more recently attention has focussed upon more persistent pollutants, such as organohalogens, and certain toxic trace metals. The response of the flora and fauna to these contaminants is not well known and there is considerable difficulty in selecting appropriate standards (Wilson and Halcrow, 1985). At this moment there is a trend to give more attention to a quantification of effects in the ecosystem, than starting from arbitrary criteria for allowable concentrations of toxic substances. For long-term planning of water quality management decisions it is recommendable to define a Environmental Quality Objective (EQO). The Rhine Action Program decided by the rhine riparian states can be cited as an example. The target of this program is the recovery of the Salmon in the River Rhine (Rijn Aktie Plan, 1987).

Control and improvement of the water quality in an estuarine system, requires insight in relevant phenomena in order to predict the consequences of environmental actions. Examples of such phenomena are:

- eutrophication due to excessive algal blooms in nutrient-rich waters;
- depressed dissolved oxygen conditions as a result of waste water discharges;
- accumulation of chemicals in the estuarine mud deposits;
- accumulation of toxic substances in the food chains;
- ecological effects of these toxins, resulting in disease, mortality or reproduction problems of organisms and possible risks for human health.

There is an increasing trend of applying mathematical models in estuarine water quality management. This trend is parallel to the rapid developments in computing facilities during the last decades. Models have been shown to be valuable tools in the evaluation of alternatives. They also often drew attention to some aspects of a problem that were unrealized in the management decisions. It has to be understood that models are always a more or less schematized version of the real system, based on some hypotheses and descriptions of processes, wherein several aspects are yet unknown. Although this means that the accuracy of quantitative predictions is limited, especially when biological processes are included, the model remains a very useful instrument in evaluating management decisions and decisions on future directions in estuarine research.

Because of the complexity of nature all the details can never be quantified in a model. However, for management decisions not all these details have to be known. Therefore, each problem needs its own specific model. Although there is a trend to increase the complexity of the models, it should be emphasized that the model to be applied should be no more complicated than is required to give an accuracy that is acceptable to the decision-maker. In addition to models, due to the inherent complexity of nature, accurate measurements in the real estuary are of great value. Application of both sources of information in an optimum way to minimize uncertainties is the art of excellent estuarine water quality management. To understand the characteristics of an estuarine system holistic models are developed, incorporating all the physical, chemical and biological processes which are thought to be of importance in the given system. Knowledge of the complex interactions of these processes is fundamental to our understanding of estuarine behaviour and requires interdisciplinary studies. Gordon et al. (1986) presented the philosophy that in such cases the process of model development is more important than the model created, because just this development phase was experienced as a period to get a much better understanding of the specific estuary from a whole ecosystem perspective.

Interdisciplinary estuarine modelling collects the knowledge and experience of the various disciplines in forming a framework, in which the various disciplinary building blocks are brought together, as is schematically shown in Fig. 1. Thereby special attention should be given to the interactions between these blocks. The - transport model, predicting the circulation and mixing in an estuary, forms the basis of the water quality models. The suspended sediment transport is of special importance, because the fine sediment particles are important carriers for substances such as trace metals, radionuclides and organic micro-pollutants, owing to their adsorptive capacity.

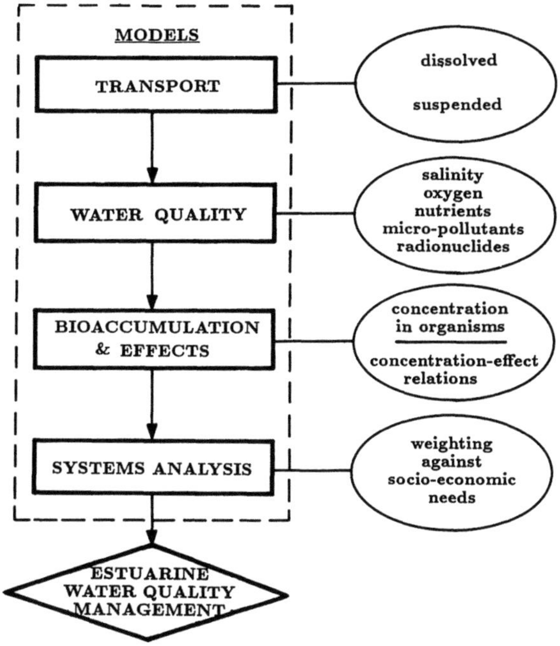

Figure 1: Framework of models, containing the interdisciplinary building blocks of estuarine modelling

Often water quality is defined in terms of concentrations of the various dissolved and suspended substances in the water, as for example salinity, temperature, dissolved oxygen, turbitity, nutrients, phytoplankton, bacteria, heavy metals, organic micro-pollutants, radionuclides, etc. Starting from the results of the transport model, the distribution of these substances in a specific estuary under specified conditions has to be calculated by the water quality model. Generally, a variety of processes is included in the model, as there are hydrolysis, oxidation, photolysis, volatilization, biological activities, interactions between the sediment and contaminants (adsorption and desorption), etc. The models can vary from a relatively simple model, including a few dominant processes in a simple way (cf. the dissolved oxygen model of Streeter and Phelps, 1925) to models with a large number of processes and parameters (cf. Baretta and Ruardij, 1988), as well as very complex holistic ecosystem models with various trophic levels, trying to simulate the whole food web.

From the distribution of the various substances the uptake by organisms and especially the accumulation of toxic substances as a function of time is calculated by the bioaccumulation models. A tedious point is the chemical selectivity of the organisms (bioavailability). The effect models should predict the effects of these accumulated toxins, both on individual organisms and on communities.

The aim of systems analysis models is not in the first place to simulate the physical, chemical or biological behaviour of estuaries, but to support directly management decisions. Estuary simulation models are coupled to social and economic models, in order to predict the impact of measures not only on the ecosystem itself, but also on other aspects relevant to society. Often such models have a structure similar to those used in economics: input/output analysis, optimization techniques, cost/benefit analysis, etc. Mostly they are simple models, in which much attention is given to clear presentations, using modern computer graphics techniques. They meet the needs to come to a well-considered decision between alternatives and to elucidate the consequences of them.

2. PROGRESS IN ESTUARINE WATER QUALITY MODELLING

As has been concluded by Mann (1982) it is impossible to understand biological and chemical processes in coastal waters and estuaries without understanding the physical oceanography of the system under consideration. The complex interaction of the local tides, river flows, winds and topography gives the estuary a highly dynamic character, while also the density difference between the fresh river water and the salt sea water can affect the circulation in the estuary enormously. For adequate water quality modelling a good insight is needed into the material fluxes in the estuary, both from advective and dispersive structure.

The first attempt to model water quality in estuaries was based on the turn-over concept: the estuarine basin is a well-mixed reservoir exchanging water and substances with the adjacent sea at a limited rate. The time scale at which a fraction e^{-1} remains in the basin is called turn-over time (Ketchum, 1951; Bolin and Rohde, 1973). After extension of this concept to multiple-box models, a considerable improvement was achieved with the introduction of the 1D-longitudinal dispersion concept. The longitudinal distribution of steadily released dissolved substances can be predicted without detailed knowledge of the current behaviour. The dispersion coefficient is derived from the salinity distribution. This approach has been further improved, firstly by showing that it can also be used for non-steady problems, within certain limits (Dronkers, 1982); secondly, by identifying the major physical processes which determine the dispersion coefficient (Fischer et al., 1979; Smith, 1980; Van de Kreeke, 1988). This even allows application to water quality predictions in estuaries where projected engineering works affect the current regime. For wide shallow estuaries the relation $D \propto u_0^2$ applies, where u_0 is the amplitude of the tidal velocity (Dronkers et al., 1981).

In wide shallow estuaries density driven currents are small compared to geometry-induced residual currents. Therefore, the dependence of the dispersion coefficient on the fresh water discharge is small compared to the influence of the tidal discharge. Field data do not entirely support this assumption - an increase of the dispersion coefficient with fresh water discharge Q is generally observed in the range $D \propto Q^{0.1}$ to $Q^{0.5}$ (Eggink, 1965; Helder and Ruardij, 1982).

For the initial spreading of a dissolved substance, for example after a calamitous release, the 1D approach is not appropriate. As long as the contaminant is not spread out over the whole cross-section of the estuary it is necessary to use models which take into account the local and instantaneous current structure. In shallow estuaries one may employ a 2D depth-averaged model. A very convenient way to simulate the contaminant is to introduce a bunch of particles at the point of release, which move with the local instantaneous current velocity (Maier-Reimer, 1973).

During the past decades 2D depth-averaged tidal models have been so far improved that even the complicated flow pattern in shallow multi-channel estuaries is reproduced to a good accuracy (Langerak and Leendertse, 1986). It has also been shown that these models may simulate the spreading and transport of dissolved substances, without the introduction of a numerically large dispersion coefficient. The drifter-particle method described above is the most appropriate means for the simulation of spreading and transport (Van Dam, 1982).

The prediction of contaminant-concentrations after a calamitous ship-release requires a fast operating simulation-instrument. The spreading in an estuary is very rapid in the first tidal cycles after the release, warning times therefore being short. In spite of this fast spreading, chaos theory has learned that high concentration patches may subsist a long time in the estuary. Sufficiently detailed 2D flow models are able to reproduce this patchy behaviour (Pasmanter, 1988).

In stratified estuaries the flow simulation with present mathematical models is still less accurate. For narrow estuaries a 2D width-averaged approach is most indicated, otherwise a complete 3D flow simulation is necessary. Especially in the latter case the spatial resolution puts a limit on the accuracy. In the former case it is the physical description of vertical momentum exchange under stratified conditions which still poses some fundamental problems. For most practical questions concerning the behaviour of dissolved substances in estuaries present models provide sufficient accuracy, however.

In practice, one has to deal often with substances which adsorb strongly to fine sedimentary particles or are incorporated in flocs of sedimentary particles. In the adsorbed state the bio-chemical activity of contaminants is strongly reduced. However, the contaminant is not definitively eliminated from the system. During transport only few releases of contaminants occur, most of the adsorbed contaminant is stored in the bottom at places where fine sediment accumulates. From there a slow release may take place, depending on the chemical condition of the bottom (O_2, pH, organic matter). This presents a prolonged danger for the aquatic ecosystem. Little insight exists into the fate of the enormous quantities of contaminants which have been and which still are released directly (also by the atmosphere) into coastal waters. Following recent estimates a considerable fraction is stored in estuarine systems.

Modelling the pathway of these adsorbed contaminants requires knowledge of settling, deposition, consolidation and resuspension processes of the fine sediments. No reliable and predictive physical description exists at present. In practice empirical formulations are used which require calibration for each field situation. Systematic investigations are necessary, in order to achieve a better understanding of the above-mentioned processes. The important role ascribed to organisms (bacteria, diatoms, worms) and organic matter makes these investigations very complex. Besides, an appropriate description of the near bottom current structure requires further improvement of present estuarine tidal models.

In mathematical ecosystem studies the scales of interest, both in time and space, can differ significantly from that of the underlying transport models. In hydrodynamical and transport modelling the continuity equations are generally solved by finite difference or finite element techniques, with, dependent on the problem and the geometric situation, a grid size of some hundreds of meters to a few kilometers, and time steps of a few minutes. However, the time scales of the chemical and biological processes are often much larger than that of the hydrodynamic and fine sediment transport model. From the biological standpoint one is primarily interested in seasonal variations, while for example for fine sediment transport the deposition and resuspension processes within a tidal cycle should be considered.

In water quality modelling also the space scale will be chosen much larger than in transport modelling. Because of the patchy character it is only worthwhile to consider the mean values in a specific area. Therefore, in complex models with chemical and biological interactions between a large number of parameters, often the estuary is schematized into a limited number of compartments. The results of the transport models should be averaged both in space and time to give the long-term transport across element boundaries within the water quality or ecosystem model.

3. COMPENDIOUS DISCUSSION ON MAIN POINTS OF ESTUARINE WATER QUALITY MODELLING

3.1 Estuarine modelling

Estuarine modelling has been shown to be an important tool in estuarine water quality management. Although the accuracy of the quantitative predictions is still in discussion, they are very helpful in analysing the complex ecosystem with a large number of interacting processes. Because of the highly dynamic character of an estuary and its complex structure it would be impossible to understand the main characteristics of the system and its reactions on various forcing functions. In addition, the development of such models has proved to be important in stimulating the interdisciplinary approach to estuarine research.

3.2 Validation of the model

The process of validation involves comparison between computed values of the various variables and measured field data. When this comparison is satisfactory, the model is said to be validated. What constitutes "satisfactory" depends on the nature of the problem, the structure of the model, and the extent of available data (O'Connor et al., 1977). Besides a check of the structure of the model algorithms, three steps can be distinguished in this process (Di Toro et al., 1982): calibration, verification and post-audit. In the calibration process some coefficients are optimized in the model to get a best fit to the observations in the estuary. In the verification phase the model predictions are compared to field measurements that were not used in the calibration, while the coefficients are held unchanged. In the post-audit phase the adequacy of the model can be established by evaluating decisions, based on model predictions, from comparing the model output with field measurements collected after implementation. However, very few ecosystem models have been validated against adequate field data that are also completely independent of calibration data (Radford and Ruardy, 1987). Testing the predictive power of ecosystem models requires continuation of the research effort during some five to ten years after implementation of measures.

3.3 Uncertainty in forecasting water quality

The accuracy of model predictions is influenced by a number of uncertainties, as there are uncertainties and errors in the field data, uncertainty in the process formulations, and uncertainty in the process parameters. The available field data may also be inadequate because of too low sample frequency or too short records of time-series, especially from the fact that estuaries are inherently variable in time and space. Temporally they respond to a combination of forcing functions in a variable fashion over a broad spectrum of frequencies (Wolfe, 1986). Therefore, sampling densities and frequencies for the validation of water quality models must be carefully fitted to the variability of the estuarine system. There is a strong need for reliable long-term measurement series.

The last decade has seen an increasing interest in uncertainty analysis in ecosystem modelling (Beck et al., 1983; Di Toro, 1984; Loehle, 1987), treating the model parameters as random variables and investigating the effect of these variabilities on model results for example by Monte Carlo techniques, i.e. repeated simulations of the model with random selected parameter values. Also parameter estimation techniques have been developed for assessing the model parameters in the case of poor field data (cf. Keesman and Van Straten, 1987).

3.4 Framework of the present-day simulation models

Notwithstanding the complexity of the various processes in transporting, transferring and transforming chemicals, significant progress has been made by modelling the fate of toxic substances in the aquatic environment. Generally, linear relations are applied in the equations. Validation of the existing models as well as improvements of the process descriptions, especially of the sediment/chemical interactions, should acquire the highest priority. Accurate data sets both from the field and laboratory are needed. In large areas, for example coastal seas, the effect of atmospheric deposition should also be incorporated in the calculations.

In water quality-ecosystem models, simulating the nutrient distribution and the plankton growth, accurate predictions are much more difficult. According to Mann (1982) two reasons can be stated: firstly the fact that organisms are inherently variable in their physiological properties, and secondly the hierarchical structure of ecosystems (communities - populations - individuals). From the last point he concludes that it is virtually impossible to predict the behaviour of the integrated system from a study of lower levels of organization. Therefore, one should look for an ecological theory that will provide the connection between the dynamics of populations and the behaviour of ecosystems. From the non-linear interactions between nutrients and organisms a complex system of equations arises, which induce unstable solutions, corresponding to explosive growth. This puts a limit on the temporal and spatial accuracy of predictions. In spite of these fundamental questions concerning the predictive capabilities of the models, it may be concluded that ecosystem models have strongly contributed to the insight into the estuarine ecosystems.

3.5 Sources of nutrients and toxic substances

For an accurate modelling of the distribution of the various substances a sufficient knowledge of the major inputs is a prerequisite. These inputs can be divided in point sources and diffuse sources. The point sources, such as sewage treatment plants and industrial facilities, are generally well known. The diffuse sources, such as agricultural and urban runoff, dredged material disposal, groundwater infiltration and atmospheric deposition, are much more difficult to compile. Their relative significance becomes increasingly important after an improved control of the loadings from point sources. The study of modelling a specific area should start with thoroughly inventorying both the point and non-point sources of the relevant substances as well as their temporal variations. Good examples of such studies are given by Tippie (1984) for the Chesapeake Bay and by Van Meerendonk et al. (1988) for the Dutch Wadden Sea and Ems Estuary. It should be realized that estuaries generally act as filters between land and sea, essentially trapping and recycling significant quantities of nutritious and toxic substances (Kennedy, 1983). These phenomena should be modelled in a satisfactory way.

3.6 Spatial and temporal scales in modelling

Physical, chemical and biological processes vary over a broad spectrum, both in time and space. Spatial variations depend on the topography of the estuary and gradients in salinity and fine sediment concentration, while with respect to temporal variability long-term (yearly), seasonal (monthly), diurnal (hourly) and short-term (minutes) time scales can be distinguished. Starting from a series of coupled models (transport - water quality - bioaccumulation - effects - systems analysis) different time scales can be applied. From a standpoint of transport modelling it could be needed, especially in the case of fine sediment transport with deposition and resuspension cycles during the tide, to simulate the processes over relatively short periods, while for the water quality model generally a tide-averaged model will be preferred. The bioaccumulation and effect models will work over much longer time scales. Systems analysis has to deal with a long-term interest from a management perspective. Therefore, a coupled set of models should be developed each with its own spatial and temporal scales.

If necessary the results will be averaged in space and/or time to pass them to the next model. Coupling of the models needs sufficient attention.

3.7 Bioaccumulation and effects

Bioaccumulation models received increasing attention during the past years, because the concentrations of contaminants in organisms could be considered as a measure of environmental quality. Generally, a specific pelagic and/or benthic organism is chosen as representative. The models are strongly empirical, ranging from fully black box to more physiological models with descriptions for uptake, growth and biological excretion (Van Haren et al., 1989). Important points for attention are chemical speciation, physiology of the model organisms and metabolism of organic contaminants. The development of bioaccumulation models is still in its infancy. The modelling programmes should be accompanied by biomonitoring programmes.

Basically not the bioaccumulation itself is relevant, but the effects, being the relations between the contaminant concentrations in the organisms and the biological effects such as disease, mortality and reproduction, and the possible effects on human health. Much effort is being given to the development of dose/effect relations by conducting experiments in micro- and mesocosms. Effect models may be developed at various levels: organism-level, population-level and ecosystem-level. Models on the organism-level are in progress and are statistical in conception. Effect models on higher levels may be expected to become available only in the far future.

3.8 Systems analysis

Systems analysis has been defined by Thomann (1987) as "the engineering art of integrating and synthesizing the physical, chemical, biological and mathematical sciences with the social and economic sciences to construct frameworks that elucidate the consequences of alternative water quality and water use objectives". One of the characteristics is the module-structure of the framework which makes it easy to replace elements when better data or more accurate descriptions of processes become available. Models to optimize both the economic, social and ecological benefits of management decisions have been developed in the past decade for many purposes. Examples are the Netherlands Policy Analysis Water Management model (PAWN, 1982), the Policy Analysis Oosterschelde (POLANO, 1977) and the BARrier CONtrol model (BARCON, 1985). Modern presentation techniques, that elucidate the mutual links in the estuarine system, are an essential part of the systems analysis models.

One of the requirements of a systems analysis model is the possibility to investigate a large number of alternative measures. Therefore, the models which are included in the systems analysis framework are simplified as much as possible. Certain constituting models are just simple input-output relationships with tabulated coefficients derived from complex dynamic models. The acquired accuracy is often determined by the constituting model which introduces the largest uncertainty.

3.9 Integrated modelling/monitoring

One should endeavor to integrate the framework of models with a monitoring network in the estuary. From the monitored parameters the model will be in a continuously improving process, while otherwise from the co-operation with the model an optimum monitoring system can be developed, in which sampling frequencies, space density, etc. are optimized (cf. Radford, 1986). In this way an optimum system originates for operational water quality management.

REFERENCES

BARCON (1985) Controlling the Oosterschelde Storm-Surge Barrier. Reports Rijkswaterstaat. Zeeland Directorate, Middelburg

Baretta J, Ruardij P (eds) (1988) Tidal Flat Estuaries. Simulation and Analysis of the Ems Estuary. Springer-Verlag, Berlin Heidelberg New York

Beck MB, Van Straten G (eds) (1983) Uncertainty and Forecasting of Water Quality. Springer-Verlag, Berlin Heidelberg New York

Bolin B, Rohde H (1973) A note on the concepts of age distribution and transit time in natural reservoirs. Tellus 25: 58-62

Di Torro DM, Donigean AS, Games LM, Hern S, Lassiter RR, Matsuoka Y (1982) Validation and application testing, synopsis of discussion session. In: Dickson KL, Maki AW, Cairns J (eds) Modelling of the Fate of Chemicals in the Aquatic Environment. Ann Arbor Science, pp 387-395

Di Torro DM (1984) Statistical methods for estimation and evaluating of water quality parameters and predictors. Delft Hydraulics Report R1310-13

Dronkers J (1982) Conditions gradient-type dispersive transport in one-dimensional, tidally averaged transport models. Est Coast Shelf Sci 14: 599-621

Dronkers J, van Os G, Leendertse JJ (1981) Predictive salinity modelling of the Oosterschelde with hydraulic and mathematical models. In: Fischer HB (ed) Transport Models for Inland and Coastal Waters. Academic Press, New York, pp 451-482

Eggink HJ (1965) The estuary as a receiving water system of large quantities of wastes. Dissertation. Staatsdrukkerij, The Hague (in Dutch with English summary)

Fischer HB, List EJ, Koh RCY, Imberger J, Brooks NH (1979) Mixing in Inland and Coastal Waters. Academic Press, New York

Gordon DC, Keizer PD, Dahorn GR, Schwinghamer P, Silvest WL (1986) Adventures in holistic ecosystem modelling: The Cumberland Basin ecosystem model. Neth J Sea Res 20(2/3): 325-335

Helder W, Ruardij P (1982) A one-dimensional mixing and flushing model of the Ems-Dollard estuary. Calculation of time scales at different river discharges. Neth J Sea Res 17: 293-312

Kamlet KS (1978) An environmental lawyer's uncertain quest for legal and scientific certainty. In: Wiley ML (ed). Estuarine Interactions. Academic Press, New York London, pp 17-29

Keesman KJ, Van Straten G (1987) Modified set theoretic identification of an ill-defined water quality system from poor data. In: Beck MB (ed) Systems Analysis in Water Quality Management. Pergamon Press, pp 297-308

Kennedy VS (ed) (1984) The Estuary as a Filter. Academic Press, New York London

Ketchum BH (1951) The exchanges of fresh and salt water in tidal estuaries. J Mar Res 10: 18-38

Langerak A, Leendertse JJ (1986) Predictive ability of two-dimensional models for mixing in estuaries. In: Radojkovic M, Maksimovic C, Brebbia CA (eds) Hydrosoft-Hydraulic Engineering Software. Springer-Verlag, Berlin Heidelberg, pp 135-154

Loehle C (1987) Errors of construction, evaluation, and inference: a classification of sources of error in ecological models. Ecol Modelling 36: 297-314

Maier-Reimer E (1973) Hydrodynamisch-numerische Untersuchungen zu horizontalen Ausbreitungs- und Transportvorgängen in der Nordsee. Dissertation. Mitt für Meereskunde der Universität Hamburg, No 21

Mann KH (1982) Ecology of coastal waters: a systems approach. Studies in Ecology. Vol 8. Blackwell, Oxford

O'Connor DJ, Thomann RV, Di Toro DM (1977) Water-quality analyses of estuarine systems. In: Estuaries, Geophysics and the Environment, National Academy of Sciences, Washington DC, pp 71-83

Pasmanter RA (1988) Deterministic diffusion, effective shear and patchiness in shallow tidal flows. In: Dronkers J, Van Leussen W (eds) Physical Processes in Estuaries. Springer-Verlag, Berlin Heidelberg New York, pp 42-52

PAWN (1982) Blumenthal KP (ed) Policy Analysis for the National Water Management of the Netherlands, Rijkswaterstaat Communications No 31

POLANO (1977) Protecting an Estuary from Floods - a Policy Analysis of the Oosterschelde. Goeller BF, Abrahamse AF, Bigelow JH, Bolten JG, De Ferranti DM, De Haven JC, Kirkwood TF, Petruschell RL. RAND Report R-2121/1-NETH

Radford PJ, West J (1986) Models to minimize monitoring. Wat Res 20 (8): 1059-1066

Radford PJ, Ruardij P (1987) The validation of ecosystem models of turbid estuaries. Cont Shelf Res 7 (11/12): 1483-1487

Rijn Aktie Plan (1987) Arbeitsplan für die Durchführung des Aktionsprogramms "Rhein". Internationale Kommission zum Schutze des Rheins gegen Verunreinigung

Smith R (1980) Buoyancy effects upon longitudinal dispersion in wide well-mixed estuaries. Philosophical Transactions of the Royal Society London, A296: 467-496

Streeter HW, Phelps EB (1925) A Study of the Pollution and Natural Purification of the Ohio River. Bulletin no 146, US Public Health Service

Thomann RV (1972) Systems Analysis and Water Quality Management. Mc Graw-Hill Book Co, New York

Thomann RV (1987) Systems analysis in water quality management - a 25 year retrospect. In: Beck MB (ed) Systems Analysis in Water Quality Management. Pergamon Press, pp 1-14

Tippie VK (1984) An environmental characterization of Chesapeake Bay and a framework for action. In: Kennedy VS (ed) The Estuary as a Filter. Academic Press, pp 467-486

Van Dam GC (1982) Distinct particle simulations. Section VIII of chapter 2. In: Kullenberg G (ed) Pollutants Transfer and Transport in the Sea. Vol I, CRC Press, Boca Raton, Florida

Van Haren R, De Vries MB (1989) UPTIMUM: a dynamic model of the uptake of heavy metals by Mytilus Edulis. Report on development, calibration and validation of bioaccumulation models. Rijkswaterstaat, Tidal Waters Division and Delft Hydraulics (in Dutch)

Van de Kreeke J (1988) Dispersion in shallow estuaries. In: Kjerfve B (ed) Hydrodynamics of Estuaries Vol I. Estuarine Physics, CRC Press, pp 27-39

Van Meerendonk JH, Janssen GM, Frederiks B (1988) De aanvoer van voedingsstoffen en microverontreinigingen naar de Waddenzee en Eems-Dollard. Rijkswaterstaat, Report GWWS-88.002

Wilson JG, Halcrow W (eds) (1985) Estuarine Management and Quality Assessment. Plenum Press, New York London

Wolfe DA (1986) Estuarine Variability. Academic Press, New York London

STRATEGY OF CURRENT AND TRANSPORT MODELLING IN DIFFERENT ESTUARIES

Jürgen Sündermann
Institut für Meereskunde, Universität Hamburg (IfM)
Troplowitzstr. 7, D-2000 Hamburg, Federal Republic of Germany
Kurt C. Duwe
HYDROMOD GbR
Bahnhofstr. 34, D-2000 Wedel, Federal Republic of Germany
Ingeborg Nöhren and Klaus D. Pfeiffer
IfM and HYDROMOD
Ye Longfei
South China Sea Institute
Guangzhou, Peoples Republic of China

ABSTRACT

This paper summarizes an oral and two poster presentations by the authors. The close thematic relation between the research and the applications can be presented in a more structured and logical way in one paper. Cross–referencing in this volume can also be avoided.

Synoptical and high resolution data of the waterlevel, current, and mass field which basically govern estuarine transport processes, as well as prognostic data, can be only acquired by numerical models carefully verified with consistent field data. The numerical schemes and the modelling strategy presented here show some prospects for almost universal application in treating a broad band of estuarine research problems. If the models resolve the topographical features properly the parameterization presented only adapted by scaling arguments to the respective estuary, seems to cope with the subgrid processes without further 'tuning' of the models. This is shown by verifications and hindcasts of field experiments in estuaries of different shape and hydrography with significant variations in the boundary and forcing conditions.

INTRODUCTION

Research in estuaries is based on 'classical' hydrographic measurements (including modern methods of remote sensing), laboratory experiments (including hydraulic models) and numerical model simulations. The high variability in time and space of all state quantities describing an estuary and the large logistic effort required in field measurements makes the development of numerical models more and more important. This is supported by advanced computer facilities.

From the original application to hydrodynamic (tidal waves and storm surges, fresh water discharge) and thermodynamic (heat transport and salt intrusion) problems modelling more and more tries to tackle transport problems (dissolved and suspended material) and the simulation of ecosystems.

The complexity of a natural estuary can only be considered systematically with the help of a model concept. Although it is not yet possible to describe all processes involved (especially biological, chemical and sedimentological ones), sensitivity analysis by means of numerical models gives an objective tool for an iterative improvement in the formulation of causalities. It also provides help in assessing the relative importance of processes and interactions in an estuary, and therefore allows economic research strategies to be developed.

In the following some experience with different estuarine models is shown and a brief outlook to future needs is given.

BASIC QUESTIONS IN MODELLING

Starting with modelling there are always two different subjects to consider, fundamental (e.g. sensitivity, response, and process studies) or applications oriented (for example coastal protection, navigation, environmental protection, water quality management). The question is to plan and design the respective model in such a way that this subject is well treated with low expense. Moreover there is no unique strategy as this will depend firstly on the actual state of hard- and software in the institution and secondly on the available time and data set. Sometimes it may be of some advantage to use a somewhat overcomplicated model or an oversized computer for a fast solution if both are available.

But there are a number of scientific points of view which have to be considered in each case independent of special situations. Besides questions concerning the physics and numerical mathematics there are two other most important factors:

1. Space and time scales

It has to be said that very little is known about this subject. With growing accuracy and frequency in measuring it is now known that in natural estuaries the so called mean conditions (mean annual values, cross-section averages or even mean flood and ebb conditions) do not represent estuarine processes at all. In general the fluctuations themselves take place in a broad band of time and space dependent scales, which range from climatic variations (interannual and seasonal scale) to short events and patches (hours, local scale). Coherency scales, those space and time intervals, where structures and events are determined, are presumably of the order of hours to minutes, of kilometres to metres, depending on the topographic conditions.

The usual discretizations in estuarine modelling generally lie above the coherency scales, e.g. the processes involved are not resolved by the model. A deterministic formulation of the natural conditions presumes the parameterization of the subscale processes. The system parameters have to be determined by the adaptation of the model to the data (calibration, inverse modelling). Often the parameterization depends on the data base, so it depends on the area, on the time scale considered, and so, on a specific event. In this case the model generated can neither be transferred to other areas nor to other time intervals (for example for forecasting). It is then necessary to include the most important subscale processes by using a higher resolution (for example the generation of fronts and eddies etc.).

Integral values such as water level or mass transport are more easily reproduced by a large scale model than differential ones such as velocity or concentrations, which show large gradients in time and space. If the dissipation of energy in a locally strong current cannot be simulated due to a lack of resolution, other terms in a large scale model take over this part of the energy balance (for example the turbulent viscosity) - deviating from the natural conditions.

2. The data base

Because of the large variability in natural estuaries, field data are only representative of a single event, for a single measuring campaign. So they are only comparable with model solutions which resolve on the same scale. This may cause a high digree of inconsistency. Already the synoptic interpretation of measured data ('typical summer situation') from different campaigns gives a spatial distribution, which in that way never exists (and therefore cannot be reproduced by a deterministic model). Most of the measured data resolve far above the coherency scale, therefore the values are statistically independent and can be analyzed only by statistical methods. Experience shows that the (small scaled) topography in estuarine dynamics is very important. Many model calculations indicate that using a consistent bathymetry (that means valid in the moment of the hydrographic measurements and sufficiently resolved by the grid net) is the decisive step for verification. System parameters turn out to be most universal if the topography is well recorded. Data sets without matching topography are more or less useless for model verification.

3. Model selection and parameterization

There is no doubt that a model can only be an idealized view of nature. Its quality (e.g. the similarity to natural processes, prognostic ability) strictly depends on the physical processes introduced, resolved and parameterized. There is always a compromise between effort (e.g. resolution, numerical schemes, data adaptation and acquisition, computer time) and quality of model results. A slight quality increase in model results may yield a large increase in data acquisition, adaptation, man power, and running costs.

On the other hand, if sophisticated processes in water quality management are to be treated with models or by using model results, high quality and high resolution data of the governing parameters (basically the current and mass field) are an essential requirement. This leads to some standards which estuarine models have to cope with, and significant restrictions on model simplifications, resolution, parameterization, and dimension reduction possibilities which strictly depend on the nature of the processes investigated. Therefore the following requirements should be fulfilled by a model used for such purposes:

- three, and only in limited cases twodimensional,
- fully prognostic,
- baroclinic,
- low numerically diffusive and non–damping,
- able to cope with a broad band of boundary and forcing conditions.

As a single model is only particulary able to fulfil the aforementioned requirements on all time and length scales to be resolved in estuarine dynamics, problem depending model approaches are commonly used. This generally yields a great loss in information and a high parameterization effort which usually still will not give satisfactory results. Therefore the authors propose the use of a model family consisting of Eulerian and Lagrangian model schemes which are modularily combined to the problem oriented model system. They are generally based on a semi–implicit scheme for the computation of the Navier–Stokes equation, and flux–corrected transport or Lagrangian higher order schemes for the simulation of both dynamically passive and active water ingredients. Details concerning the physical processes included in the model equations and of their numerical treatment have already been presented in various publications by some of the authors (e.g. Duwe and Sündermann 1986, Pfeiffer, Duwe, and Sündermann, 1988, Duwe 1988) and will not be outlined here.

Basically all model segments contain the same physics and therefore the quality of the results – providing the physical processes included are formulated correctly – depends on the scales of the model, and the availability and quality of input data, as well as on the selection of a problem–suited numerical scheme. The scales necessary to resolve complex estuaries require time steps from approx. 1 to 20 minutes, horizontal grid spacings from some 10 to some 100 metres, and a vertical spacing in the range of some decimetres to some metres. Then a suitable parameterization has to be adapted according to scaling arguments which refer to the grid spacing only. Free parameters in the model system mentioned are:

- the bottom friction coefficient to be constant in time and space for all simulations (0.0025),
- the drag friction coefficient (an empirical function of the wind speed),
- the vertical exchange coefficients for momentum, mass, and water ingredients (empirical functions of the current and mass field itself),
- the horizontal exchange coefficients for momentum, mass, and water ingredients (constant in time and space, and chosen according to scaling arguments only),
- in Lagrangian schemes the exchange coefficents are transformed to related random velocities or random displacements of the particles with referring bandwidth.

A strict application of the aforementioned parameterization was done for all model applications described below and no additional calibration or tuning of the respective models for field data comparisons was necessary. The models were usually verified by hindcasts of at least two independent and consistent data sets containing boundary and forcing conditions as well as data of various parameters in the model area itself. Probability studies, qualitative verification approaches and comparisons with other theoretical research were done in cases where insufficient data were available.

By this procedure the models showed their prognostic ability and application possibility to estuaries and coastal waters with different characters and hydrographical features in which the dynamics are driven by different boundary conditions and forcing. Limitations on accurate hindcasting and with respect to longterm prediction occur due to the lack of consistent and highly accurate data with the resolution required, as well as due to non–predictable and stochastic effects (e.g. storm surges) and longterm changes (climate, topography, human impact). This easily extends to inconsistencies in model input data and to recent data acquisition requirements. The introduction of statistical,

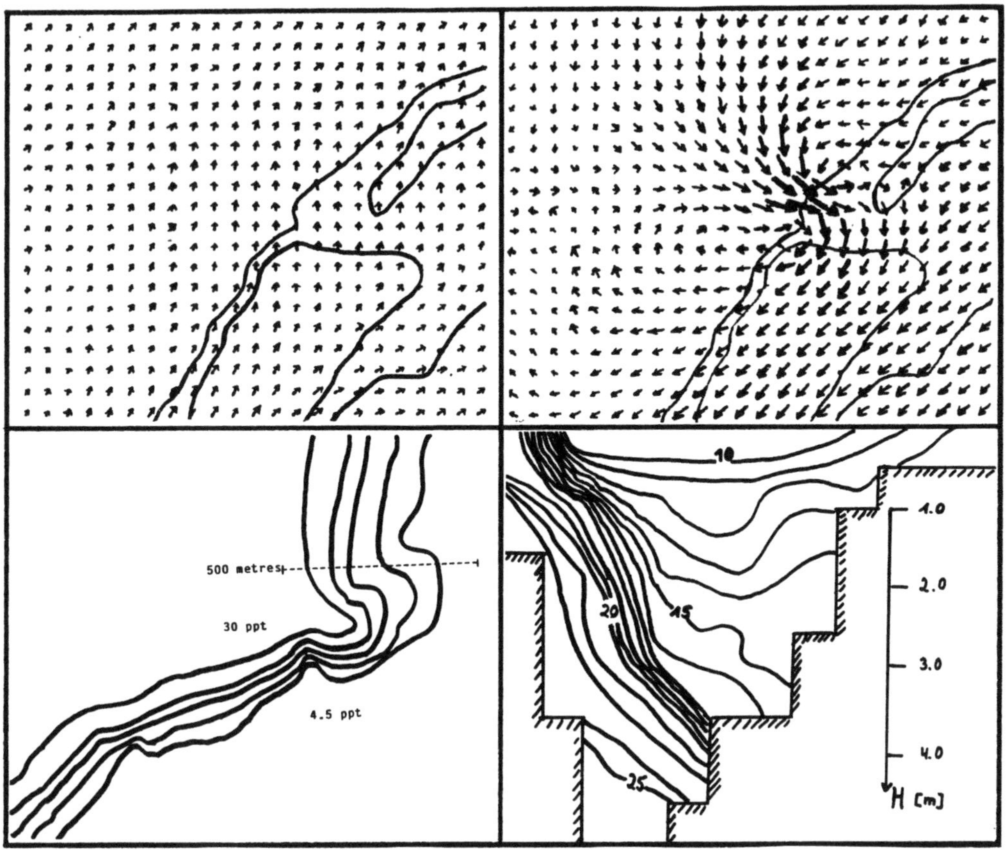

Figure 1: Current and salinity distribution in a part of the brackish water zone of the Elbe estuary around highwater. The upper left picture shows the current field for a barotropic computation, the upper right one the circulation patterns when the salinity and temperature transports were included in the calculation using an active fourth order Lagrangian scheme. The horizontal and vertical density gradients force a completely different circulation including eddies and cross–isobathic flows (the 2 and 5 meter isobathes are indicated). The lower left picture shows the corresponding isohalines (5 ppt space between isolines) clearly indicating the formation of an intensive front. The lower right picture shows the corresponding vertical salinity distribution in a cross section normal to the front.

variational, and inverse modelling techniques in estuarine modelling might therefore bring significant progress as well.

SELECTED APPLICATIONS

In the following, results of model simulations of the lower Elbe (FRG), Odra(GDR, Poland) and Pearl river estuaries (China) and of a lagoon near Venice (Italy), are shown to demonstrate the statements above. There are more detailed experiences with the estuaries of the Shannon (Ireland), Schlei (FRG), and Queule river (Chile), as well as in the coastal waters Frisches Haff (Poland) and Jiaozhou Bay (China).

1. The Elbe estuary

In the Elbe estuary, (see also Duwe and Sündermann 1986, Duwe, 1988) a partially mixed estuary with dominant tidal forcing and large tidal flat regions extensive numerical investigations concerning the estuarine dynamics and transport processes have been carried out. Some of these are described in this symposium (see Krohn and Duwe in this volume, related poster presentations). The model system applied to this estuary was verified by various hindcasts of measurement campaigns with respect to waterlevels, current velocities, salinity, temperature, and density. Therefore only two experiences might be mentioned here:

- The spreading of tracers on the scales applied is dominated by advection: there is much less diffusion in nature than in most numerical schemes. The finer the model resolution, the smaller and less sophisticated to treat are the diffusion coefficients which describe the spreading of water masses. If the model resolution reaches the dominant topographic scales, the diffusion left is a stochastic scattering of deterministic paths by practically accidental stimulation due to wind or fine scale topography.
- At some locations the baroclinic pressure gradient has a strong influence on the local dynamics (see fig. 1) which determines the lateral entrainment of the general estuarine circulation. These topographically induced, density driven currents are in no way reproducable by means of coarse grid models, diffusive schemes or dimension reduced models.

2. The Odra estuary

A three-dimensional baroclinic numerical model was adapted to the Odra estuary for the description of water level changes as well as velocity and salinity distributions in that area. This estuary, shaped as an inverse delta, shows a very sensitive response to the local wind field, the density gradient between the Baltic Sea and the inner basin, and to the fresh water discharge. Whereas tides are almost negligible, seiches in the Baltic and wind induced rising and lowering of the waterlevel in the adjacent bay have a strong influence on the current and mass field in the estuary. During calm conditions even very small changes in the waterlevel elevation of the Baltic Sea yield completely different flow patterns and change the total water transport rapidly from in– to outflow and vice versa.

As a first approach, despite the above mentioned parameterization, the vertical exchange coefficients were taken to be constant; their values were taken from model calculations of the Elbe estuary. The longitudinal distributions of velocity and salinity along the main waterway show a good accordance between measured and calculated values. The change of flow directions in different layers, which is characteristic in the Świna strait, is also quite well reproduced. A higher resolving model and variable exchange coefficients may give better results (see fig. 2, 3).

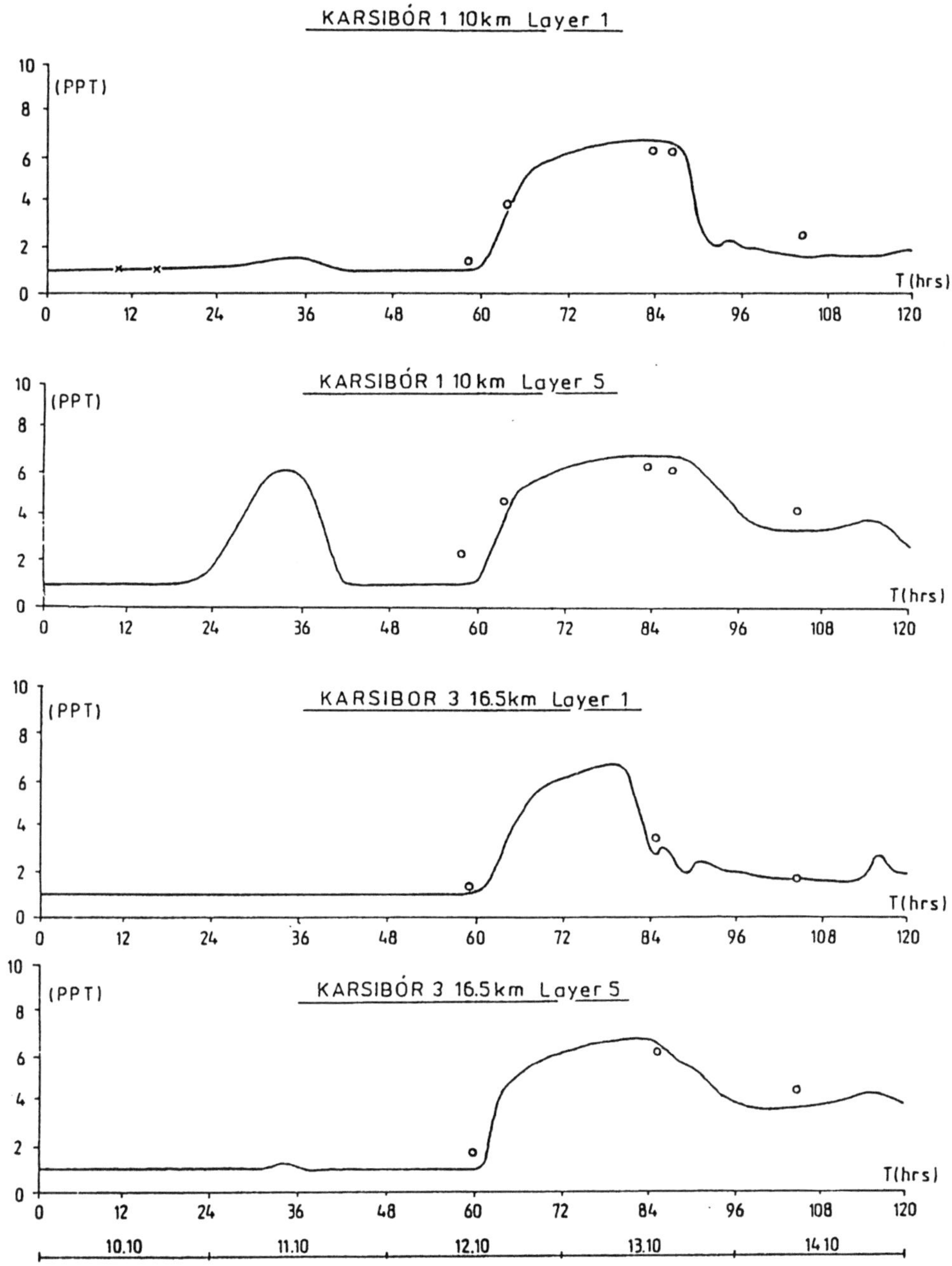

Figure 2: Comparisons between computed and measured (circles) salinity at two positions in the Świna Channel (Odra) in 1 and 10 metres depth for a hindcast of an experiment done in October 1983.

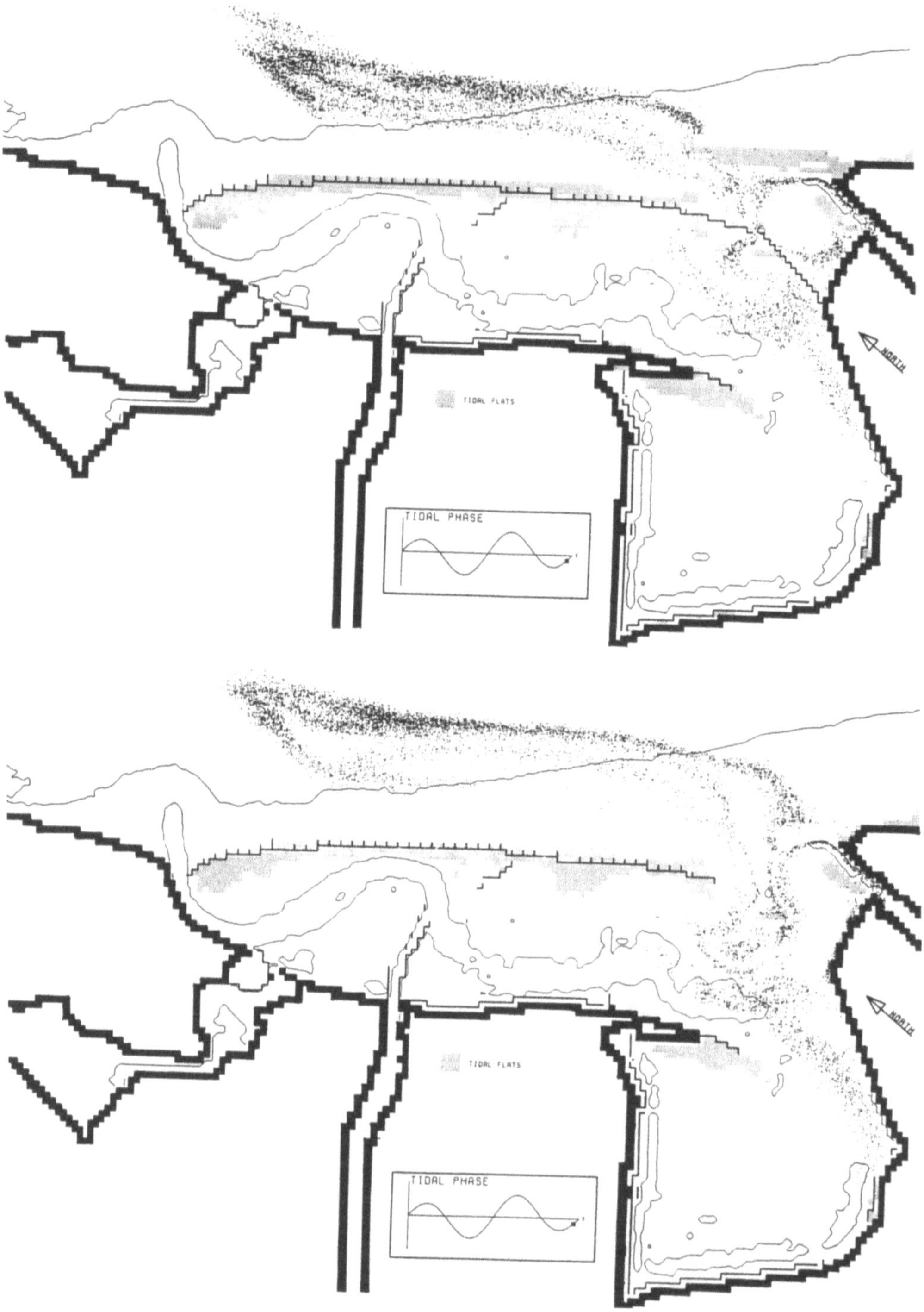

Figure 3: Distribution of passive tracers marking water of the river Po for the present (upper) and past (lower picture) situation 24 hours after continuous release of 1 tracer per $100m^3$ runoff.

3. The lagoon Valle Vallona

The shallow lagoon of Valle Vallona near the city of Venice was modelled using a two–dimensional approach. This model was very carefully verified with respect to waterlevels and current velocities (Pfeiffer, Duwe, and Sündermann, 1988).

Simple water quality estimations were done by using mark and trace techniques with a Lagrangian model driven with currents separately calculated with the current model. Although no quantitative verification was possible due to lack of data, the calculated concentrations of river water spreading into the lagoon correspond well to reported

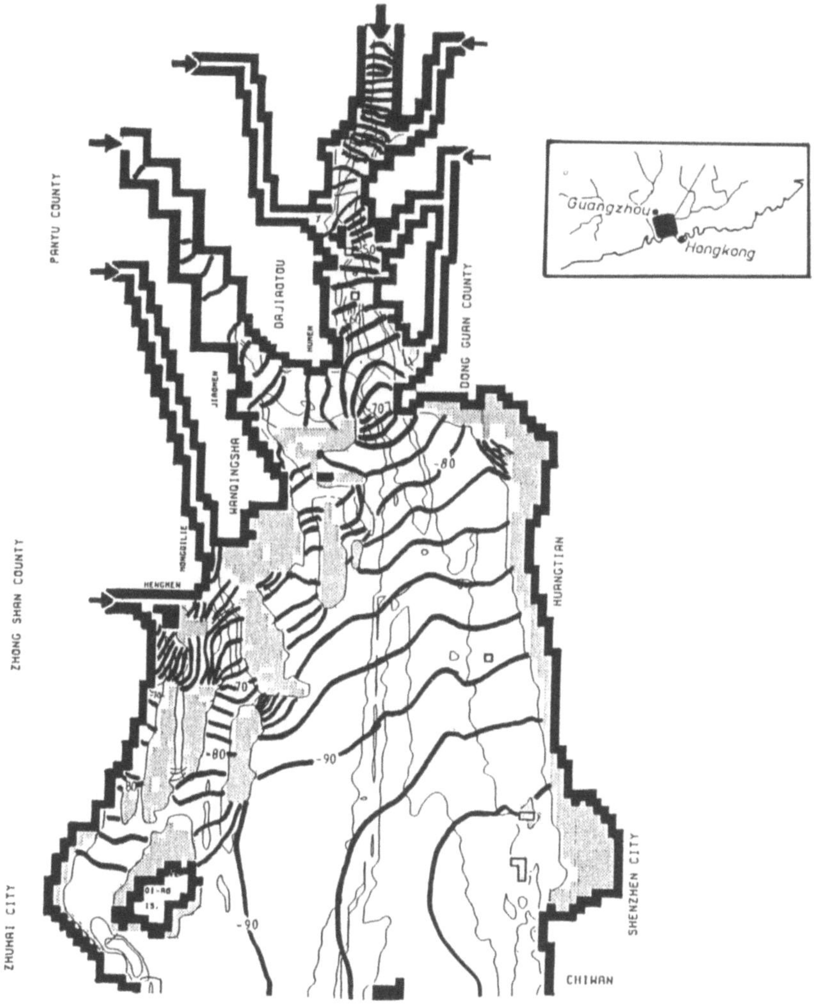

Figure 4: Water level distribution in the Lingdingyang region of the Pearl river estuary at ebb tide at the southern boundary. Grey shaded areas indicate tidal flats; arrows indicate river runoffs applied as boundary conditions. The river branches in the model are further extended to give a nature–like retention area (not shown in the figure). The mean river discharge is approx. 15.000m^3/s reaching more than double during the monsoon season.

features from local authorities, fishfarmers, fishermen, and other institutions located in the area. The results coincide well with a deemed reduction of water quality and changes in the local sedimentation and erosion tendencies due to the erection of a dam. This now prevents a major water exchange to the Adriatic Sea and locally changes the current patterns significantly (fig. 3).

4. The Pearl river estuary

The Lingdingyang region of the Pearl river estuary has a funnel shape and its dynamics are significantly influenced by the large river run–off as well as by tidal forcing. A crude description of the dynamical features can be done by some analytical estimates concerning the superposition of incoming and partially reflected Kelvin waves. The model investigations showed a fairly good agreement with field data available. The general circulation was close to the analytical results but the variable topography and the time dependent density gradients yield local deficiencies to the analytical solution. Since analytical results are very seldom available in estuarine research this application gave an opportunity to prove the models' abilities from this point of view.

OUTLOOK

The more and more advanced computers allow in an economic way the general modelling of estuarine dynamics in the mesoscale reaching up to the coherency scales. The governing importance of parameterizations is hereby limited, the scattering of system parameters (for example bottom friction coefficient) becomes smaller. Lower resolution models, although still useful for integral values (water level in storm surges, total transports), can be calibrated by comparison with more high resolution models, and processes can be parameterizised objectively. Hydrographic measurement campaigns and model runs have to be done consistently. Techniques of inverse modelling, as they are known in oceanography, will become more and more important in estuarine research. They start from fundamentally similar treatments of hydrographic and numerical data, but allow a different weighting according to reliability, and minimize the total error. Therefore variational methods become important in hydrodynamic modelling.

REFERENCES

Duwe KC, Sündermann J (1986) Currents and salinity transport in the Lower Elbe estuary: Some experiences from measurements and observations. In: v. d. Kreeke J (ed), Lecture notes on coastal and estuarine studies, 16, 30-39, Springer, Berlin.

Duwe KC (1988) Modellierung der Brackwasserdynamik eines Tideästuars am Beispiel der Unterelbe. HYDROMOD Publikation Nr.1, Wedel

Krohn J, Duwe KC (1989) Mathematical modelling of hydrodynamics in the Elbe estuary. Same volume.

Pfeiffer KD, Duwe KC, Sündermann J (1988) A highresolving Eulerian current and Lagrangian transport model of the lagoon Valle Vallona near Venice (Italy). In: Schrefler BA, Zienkiewicz OC (eds) Computer modelling in ocean engineering. Balkema, Rotterdam Brookfield, p 393–400

MATHEMATICAL MODELLING OF HYDRODYNAMICS IN THE ELBE ESTUARY

Joachim Krohn and Kurt C. Duwe
Institut für Physik, GKSS Forschungszentrum Geesthacht GmbH
Postfach 1160, D – 2054 Geesthacht, Federal Republic of Germany

ABSTRACT

Experiences in modelling the hydrodynamics of the tidally dominated Elbe estuary are presented. Because of the high spatial and temporal variability in the momentum and mass field, three-dimensional baroclinic current models were used together with low-diffusive transport algorithms (FCT, Lagrangian Tracers). The results show a good approximation to both the topographical and forcing scales, and that a nonlinear baroclinic three-dimensional modelling approach is necessary.

INTRODUCTION

An important prerequisite for modelling water quality in an estuary is an in depth knowledge of the underlying current, transport and mixing processes occurring in such a variable physical regime. The Elbe estuary has a pronounced tidal range of up to 4 m, has strong tidal currents, extensive tidal flats, a large seasonal fresh water input variation and large topographical gradients especially near river banks.

The dynamics are furthermore influenced by density gradients in the brackish water zone which is usually coupled with a turbidity maximum, the position of both depending on atmospheric parameters, the actual tidal range and river discharge. Due to strong currents, bottom gradients, and shallow water depths, nonlinear effects may increase to a similar magnitude as that of the linear part of the motion.

This means that a satisfactory simulation of estuarine processes can only be achieved by means of high-resolution three-dimensional baroclinic models, and for this reason a model family of these properties has been developed (Tab. 1). The algorithms used are based on the nonlinear shallow-water equations and the transport equations of temperature and salinity. A very economic semi-implicit numerical scheme is used to solve the system by a well-established finite difference scheme (Duwe, Hewer and Backhaus, 1983).

DESCRIPTION OF THE MODEL SYSTEM

The model family (Fig. 1) consists of a basic model for the entire estuary, and is used (1) to study processes where the oscillation system of the estuary is involved and (2) to provide boundary values for driving local, more finely resolved segment models. Those segment models complete the system and are applied, for example, to the brackish water zone, a very highly resolved small tidal flat area, and a section of the fresh water zone of the tidal river (for details see Krohn et al., 1986).

For calculating the transport of salinity and temperature either a low-diffusive FCT-algorithm is used or a Lagrangian approach with active tracers. In the case of other conservative substances with point release, passive tracers are used. Details of these approaches may be found in Duwe and Pfeiffer, 1988.

Model area	Δs (m)	Δz (m)	Δt (sec)	CFL factor	grid points	CPU (h)	p.g.	d.t.
Entire estuary	250	4	150	10	18000	2	c	FCT
Brackish water zone	250	2	150	10	16000	1.5	c	FCT
Neufeld wadden area	50	1	30	10	9000	2.5	c	tracer
Fresh water zone	100	1.5	60	5	9000	1	t	none

Tab. 1 Characteristic data of the submodels: Δs - horizontal, Δz - vertical grid size, Δt - time step, CFL factor - quotient implicit to explicit time step, grid points give only those involved in the computations (i.e 'wet' points), p.g. - pressure gradient indicating baroclinic processes included (c) or excluded (t), d.t. - density transport algorithms (FCT - Fast Fourier Transform)

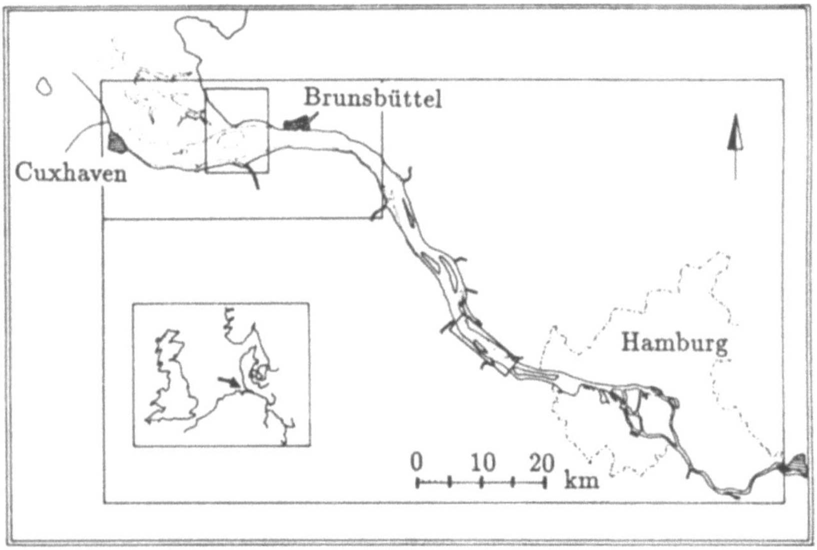

Fig. 1 Location of the Elbe estuary and boundaries of the above mentioned current and transport models

RESULTS AND CONCLUSIONS

Generally a good agreement between modelled and observed velocity fields is achieved; an example for surface and near bottom flow in a small tidal channel of the very highly resolved small scale model is given in Fig. 2.

Whereas the parametrization of the horizontal momentum and mass exchange has been found to be of minor importance, the vertical dimension proved to be the crucial factor in this respect. Accordingly, the vertical momentum and mass exchange is simulated by a temporally and spatially variable approach depending on the density gradient and current shear.

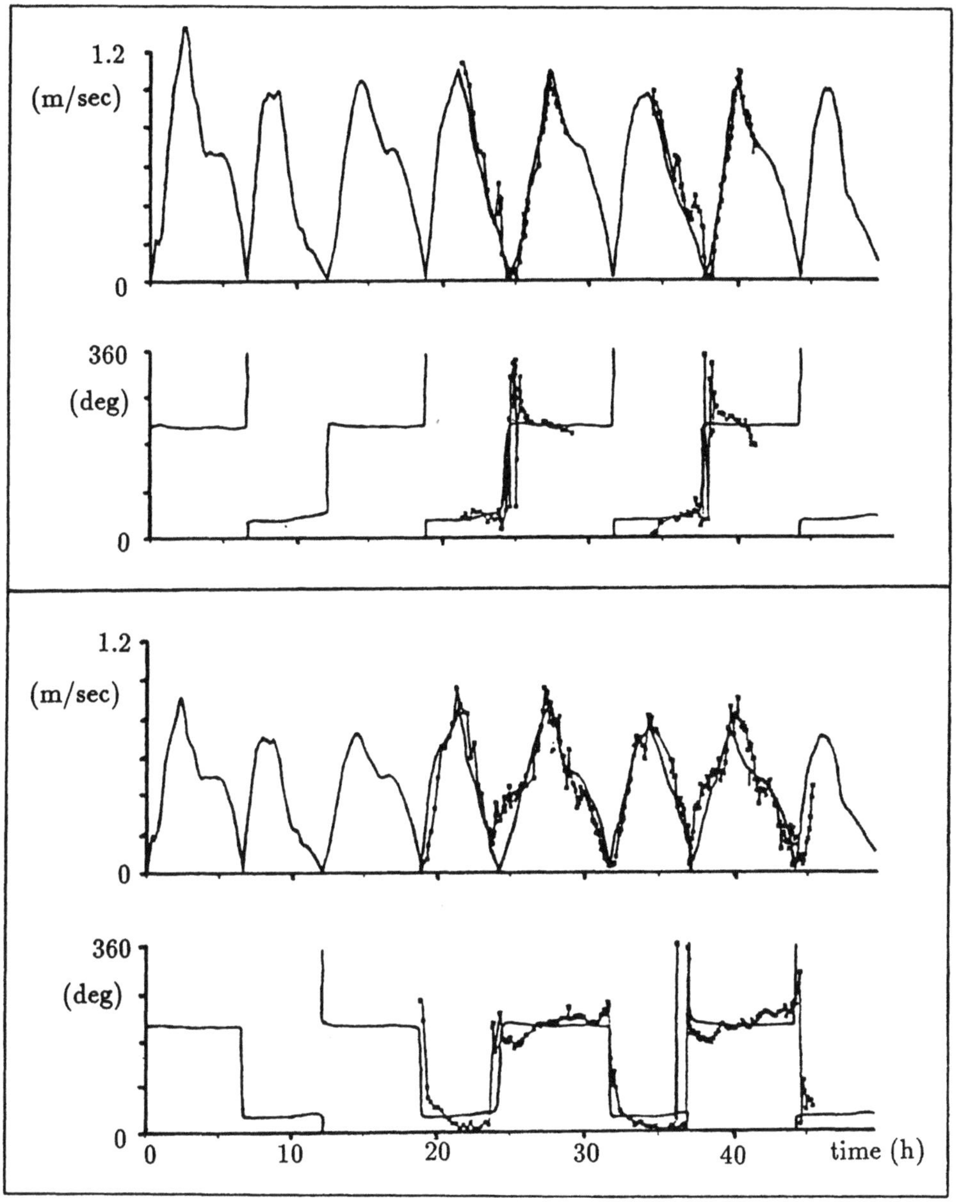

Fig. 2 Current speed and direction in a small tidal channel: the phase lag between bottom and surface currents is due to baroclinic effects. Corresponding salinity variations of between 5 and 32×10^{-3} are simulated by an active Lagrangian tracer technique. Upper panel: surface, lower panel: near bottom; solid line: simulation, marked line: observation.

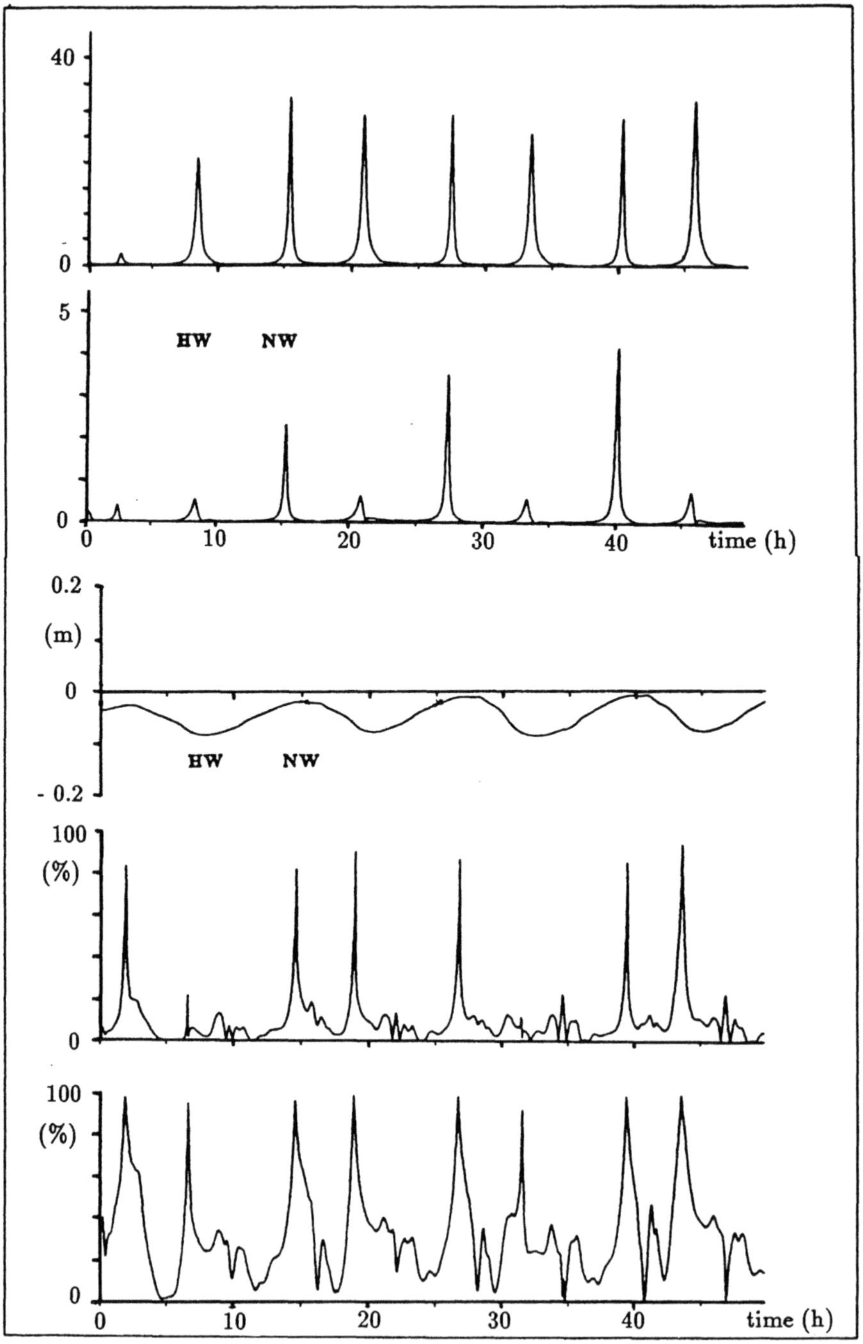

Fig. 3 Local Richardson number and baroclinic forcing (in percentage of total forcing) in the main waterway: the strong temporal changes show the impact of both vertical and horizontal density gradients on vertical momentum and horizontal currents. From top to bottom: local Richardson number near surface, same near bottom, baroclinic water level anomaly, baroclinic forcing near surface, same near bottom; HW - high water, NW - low water

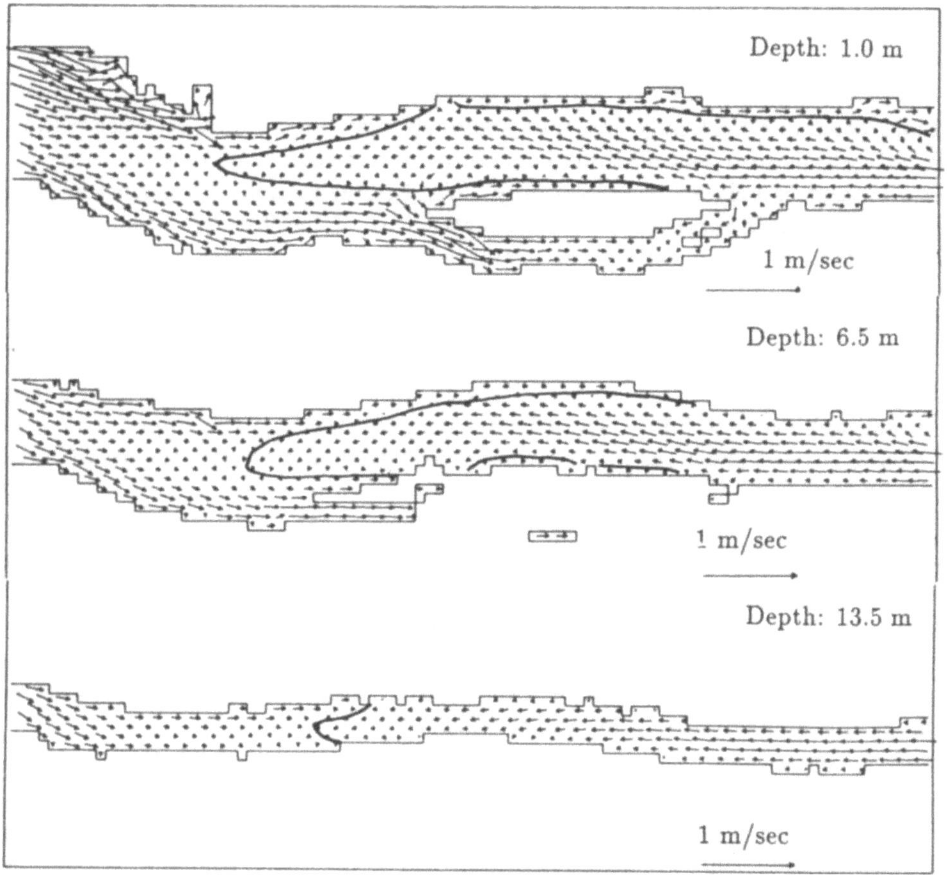

Fig. 4 Current velocities near slack water time in the fresh water zone: even in a barotropic environment the spatial variability of currents is evident. Solid line indicates reversal of flow direction.

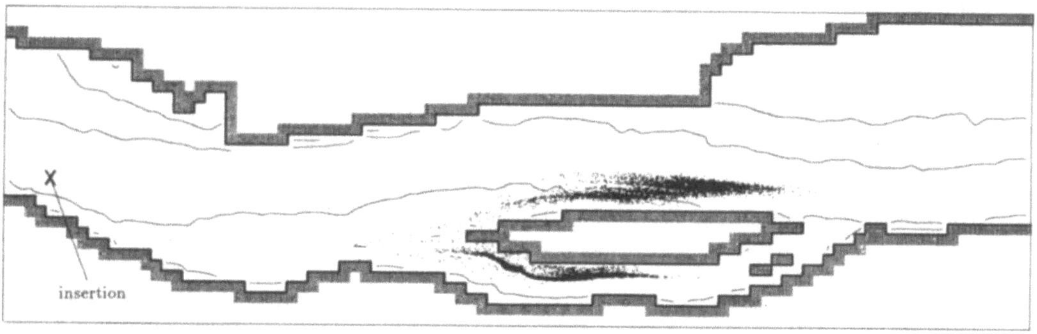

Fig. 5 Dispersion of a conservative substance 3 hours after release in the fresh water zone (point marked by 'X'): the horizontal dispersion is dominantly due to lateral topographic gradients and longitudinal curent shear.

In Fig. 3 the variance of the local Richardson-number shows the impact of stratification near slack water time. Not accounting for this effect would have resulted in total homogeniety for the water column during the whole tidal cycle.

More economic vertically integrated models may be applied to fresh water areas if investigations aim at questions concerning water level and total barotropic water transport and associated processes only. Wind influence may be neglected (Krohn & Lobmeyr, 1988). If, however, the vertical structure of the flow has implications for the processes that are in question, for example sediment transport or suspended matter transport, three - dimensional models should be applied even in fresh water areas. The vertical variation of tidal flow reversal in Fig. 4 demonstrates the significance of the vertical current profile.

Dispersion in the estuary is generally very high due to the strong tidal energy and the topographical variability. Even in barotropic regions of the upper river (Fig. 5) the strong current shear results in pronounced dispersion effects (for details see Duwe et al., 1986).

The impact of salinity fronts and stratification in the brackish water zone is due mainly to advection processes. Their spatial variation is a result of horizontal and vertical current shear. In certain locations, strong tidal mixing allows for the development of salinity fronts, vertical stratification generally occurs during during slack water only but occasionally during ebb tide. Flood tide enhances vertical mixing. In this instance, advection of denser water over lighter water takes place by vertical current shear, resulting inevitably in buoyancy effects.

Results of computations and comparison with observations have shown the importance of obtaining very exact topographic and boundary values to properly resolve the currents and corresponding short-term transports in a complex estuary like the Elbe. Treatment of more sophisticated problems, such as the generation of a turbidity zone or the transport of suspended matter including its biochemical processes is based on these results.

REFERENCES

Duwe KC, Hewer RR, Backhaus JO (1983) Results of a semi-implicit two step method for the simulation of markedly nonlinear flows in coastal seas. Cont Shelf Res 2(4):255-274

Duwe KC, Krohn J, Pfeiffer K, Riedel-Lorjé J, Soetje KC (1987) Ausbreitung von wassergefährdenden Stoffen in der südlichen Deutschen Bucht und im Elbeästuar nach Freisetzung durch Schiffe. Report GKSS 87/E/18

Duwe KC, Pfeiffer KD (1988) Three-dimensional modelling of transport processes in a tidal estuary and its implications for water quality management. In: Schrefler BA, Zienkiewicz OC (eds) Computer modelling in ocean engineering. Balkema, Rotterdam Brookfield, p 419 - 426

Krohn J, Duwe KC, Pfeiffer KD (1987) A high resolving 3D model system for baroclinic estuarine dynamics and passive pollutant dispersion. In: Nihoul JCJ, Jamart BM (eds) Threedimensional models of marine and estuarine dynamics. Elsevier, Amsterdam Oxford New York Tokyo, p 555 - 571 (Elsevier Oceanographic Series, vol 45)

Krohn J, Lobmeyr M (1988) Application of two- and three - dimensional high resolving models to a small section of the Elbe estuary - a comparative study. Report GKSS 88/E/52

Modeling Wind-Induced Mixing and Transport
in Estuaries and Lakes

by

Y. Peter Sheng[1]

ABSTRACT

Wind-induced mixing and transport in estuaries and lakes is modeled by means of a time-dependent three-dimensional numerical model of estuarine and lake circulation, which allows the use of generalized curvilinear grids to resolve the complex shoreline geometry and bathymetry. Model formulation is briefly presented, followed by model applications to Chesapeake Bay and Lake Okeechobee.

INTRODUCTION

Numerical models are routinely used for scientific investigations and management studies of circulation and transport processes in estuaries and lakes. However, a number of complexities are associated with the numerical modeling of estuarine and lake circulation and transport. First of all, the geometry and bathymetry of estuaries and lakes are quite complex. Secondly, circulation and transport in estuaries and lakes waters are generally turbulent as affected by wind, tide, wave, and density stratification. This paper focuses on the simulation of wind-induced mixing and transport in estuaries and lakes with complex geometries. In particular, wind-induced de-stratification and restratification in Chesapeake Bay and wind-driven circulation and sediment transport in Lake Okeechobee are studied.

A generalized curvilinear-grid model for circulation and transport in estuaries, lakes, and coastal waters has been developed. The model resolves the contravariant components of horizontal velocity vectors, thus simplifies the resulting equations of motion and boundary conditions in the transformed grid. In the limiting cases of conformal grids or orthogonal grids, the model equations become substantially simplified.

MODEL FORMULATION

The basic circulation and transport model is developed in terms of the contravariant velocity components (u^1 and u^2) in the horizontal directions (x^1 and x^2) and the vertical velocity (ω) in the vertical direction of a σ-stretched curvilinear grid (Sheng, et.al., 1978). The basic equations of motion in dimensionless and tensor-invariant forms can be written as (Sheng, 1986):

$$\zeta_t + \frac{\beta}{\sqrt{g_o}} \frac{\partial}{\partial x^k} \left(\sqrt{g_o} H u^k \right) + H \frac{\partial \omega}{\partial \sigma} = 0 \tag{1}$$

[1]Professor, Coastal and Oceanographic Engineering Department, University of Florida, Gainesville, Florida 32611

$$\frac{1}{H}\frac{\partial Hu^k}{\partial t} = - \varsigma !^k - g_{nj}\varepsilon^{kj}u^n \tag{2}$$
$$- \frac{Ro}{H}\left[(Hu^\ell u^k)_{,\ell} + \frac{\partial Hu^k\omega}{\partial\sigma}\right]$$
$$+ \frac{E_v}{H^2}\frac{\partial}{\partial\sigma}\left(A_v\frac{\partial u^k}{\partial\sigma}\right) + E_H A_H u^k_{,m}!^m$$
$$- \frac{Ro}{Fr_d^2}\left[H\int_\sigma^0 \rho !^k d\sigma + H!^k\left(\int_\sigma^0 \rho d\sigma + \sigma\rho\right)\right]$$

where $k = 1$ and 2, $(x^1, x^2) = (\xi, \eta)$ represent the coordinates in the computational domain which is always consisted of uniformly spaced rectangular grid, $\partial/\partial x_k$ is the partial derivative, $g_{\ell n}$ is the metric tensor while g_o is the determinant of the metric tensor, ς is the free surface displacement, $H = h + \varsigma$ is the total depth, u^k is the contravariant velocity component, $\sigma = (z - \varsigma)/(\varsigma + h)$, $(\,)_{,\ell}$ represents the covariant spatial derivative, $!^k$ represents the contravariant spatial derivative, and ε^{kj} is the permutation tensor, $Ro = u_r/fXr$ is the Rossby number, $E_v = A_{vo}/fH^2$ is the vertical Ekman number, $E_H = A_{Ho}/fL^2$ is the lateral Ekman number, $Fr_d = u_r(gD\Delta\rho_o/\rho_o)^{0.5}$ is the densimetric Froude number, $\beta = (Ro/Fr)^2 = gD/f^2L^2$, f is the Coriolis parameter, and (u_r, D, L) represent the reference (velocity, depth, horizontal length).

The above Eq. (2) can be expanded into 2 equations for u^1 and u^2. For example, denoting (u^1, u^2) by (u, v), the following equation represents the momentum equation in the ξ direction:

$$\frac{1}{H}\frac{\partial Hu}{\partial t} = - \left(g^{11}\frac{\partial\varsigma}{\partial\xi} + g^{12}\frac{\partial\varsigma}{\partial\eta}\right) + \frac{g_{12}}{\sqrt{g_o}}u + \frac{g_{22}}{\sqrt{g_o}}v \tag{3}$$
$$- \frac{R_o}{g_o H}\left\{x_\eta\left[\frac{\partial}{\partial\xi}(y_\xi\sqrt{g_o}Huu + y_\eta\sqrt{g_o}Huv) - \frac{\partial}{\partial\eta}(y_\xi\sqrt{g_o}Huv + y_\eta\sqrt{g_o}Hvv)\right]\right.$$
$$- y_\eta\left[\frac{\partial}{\partial\xi}(x_\xi\sqrt{g_o}Huu + x_\eta\sqrt{g_o}Huv) - \frac{\partial}{\partial\eta}(x_\xi\sqrt{g_o}Huv + x_\eta\sqrt{g_o}Hvv)\right]$$
$$\left. + g_o\frac{\partial Huw}{\partial\sigma}\right\} + \frac{E_v}{H^2}\frac{\partial}{\partial\sigma}\left(A_v\frac{\partial u}{\partial\sigma}\right)$$
$$- \frac{Ro}{Fr_d^2}\left[H\int_\sigma^0\left(g^{11}\frac{\partial\rho}{\partial\xi} + g^{12}\frac{\partial\rho}{\partial\eta}\right)d\sigma\right.$$
$$\left. + \left(g^{11}\frac{\partial H}{\partial\xi} + g^{12}\frac{\partial H}{\partial\eta}\right)\left(\int_\sigma^0 \rho d\sigma + \sigma\rho\right)\right]$$
$$+ E_H A_H \cdot (\text{Horizontal Diffusion of } u)$$

The nonlinear inertia terms in Eq. (3), have been derived by means of the finite-volume method and allow better satisfaction of global momentum conservation than the earlier non-conservative nonlinear terms presented in Sheng et. al. (1988).

The transport equations for temperature, salinity, and sediment concentration are much simpler and can be written as:

$$\begin{aligned}
\frac{1}{H}\frac{\partial H\phi}{\partial t} &= \frac{E_v}{Pr_v}\frac{\partial}{H^2\partial\sigma}\left(K_v\frac{\partial\phi}{\partial\sigma}\right) \\
&- \frac{Ro}{H\sqrt{g_o}}\left(\frac{\partial\sqrt{g_o}Hu\phi}{\partial\xi}+\frac{\partial\sqrt{g_o}Hv\phi}{\partial\eta}+\frac{\partial\sqrt{g_o}H\omega\phi}{\partial\sigma}\right) \\
&+ \frac{E_H K_H}{Pr_H}\left(g^{11}\frac{\partial^2\phi}{\partial\xi^2}+2g^{12}\frac{\partial^2\phi}{\partial\xi\partial\eta}+g^{22}\frac{\partial^2\phi}{\partial\eta^2}\right)
\end{aligned} \tag{4}$$

where ϕ can be temperature (T), salinity (S), or suspended sediment concentration (C), (Pr_v, Pr_H) represent the vertical and lateral turbulent Prandtl numbers, and (K_v, K_H) are the dimensionless vertical and horizontal turbulent diffusivities.

The above system of equations are solved in conjunction with the proper boundary conditions at the free surface, the bottom, and the lateral boundaries. At the bottom, the velocity profile asymptotically approaches that of the law-of-the-wall, with a drag coefficient which depends on the bottom roughness.

Turbulence parameterization used in the present model is similar to the one described in Sheng (1987). While the horizontal turbulent eddy coefficients are prescribed as uniform values which decrease with the grid spacing, the vertical eddy coefficients are determined from simplified versions of a second-order closure model of turbulent transport (Sheng ,1984; Sheng and Villaret, 1989).

WIND-INDUCED MIXING

Chesapeake Bay (Fig. 1), the largest estuary in the U.S., is generally stratified in the vertical direction. As a storm or a front passes over the bay, the water column can become destratified due to wind-induced mixing and later restratified as the wind decreases. Such an event was observed in September 1983 (Fig.2). The three-dimensional curvilinear-grid model described above was used to simulate the one-month event. Using the curvilinear grid shown in Fig. 3 and 7 vertical layers, the model was able to simulate the destratification and restratification cycle as evidenced in Fig. 4. The same basic model is being used to the upper Chesapeake Bay (Johnson, 1988).

Lake Okeechobee (Fig. 5) is a large shallow lake with an average depth of 3m and extensive vegetation over the western portion of the lake. Recent studies (Sheng, et. al. , 1989) have measured the currents at six locations in the lake at 2 Hz frequency over extended time period. As shown in Fig. 6, wind data showed significant temporal variation, which is also found in the current data at all stations. As shown in Fig. 7, the presence of vegetation apparently lead to rather strong currents parallel to the vegetation boundary. Model-computed vertically-averaged circulation pattern (Fig. 8) due to a steady wind from the east also showed similar trend. Although wind-driven circulation can affect the sediment transport, resuspension of sediments is primarily due to wind waves. A shallow water wave model and a turbulent boundary layer model coupled with a sediment bed model have been developed for the sediment study.

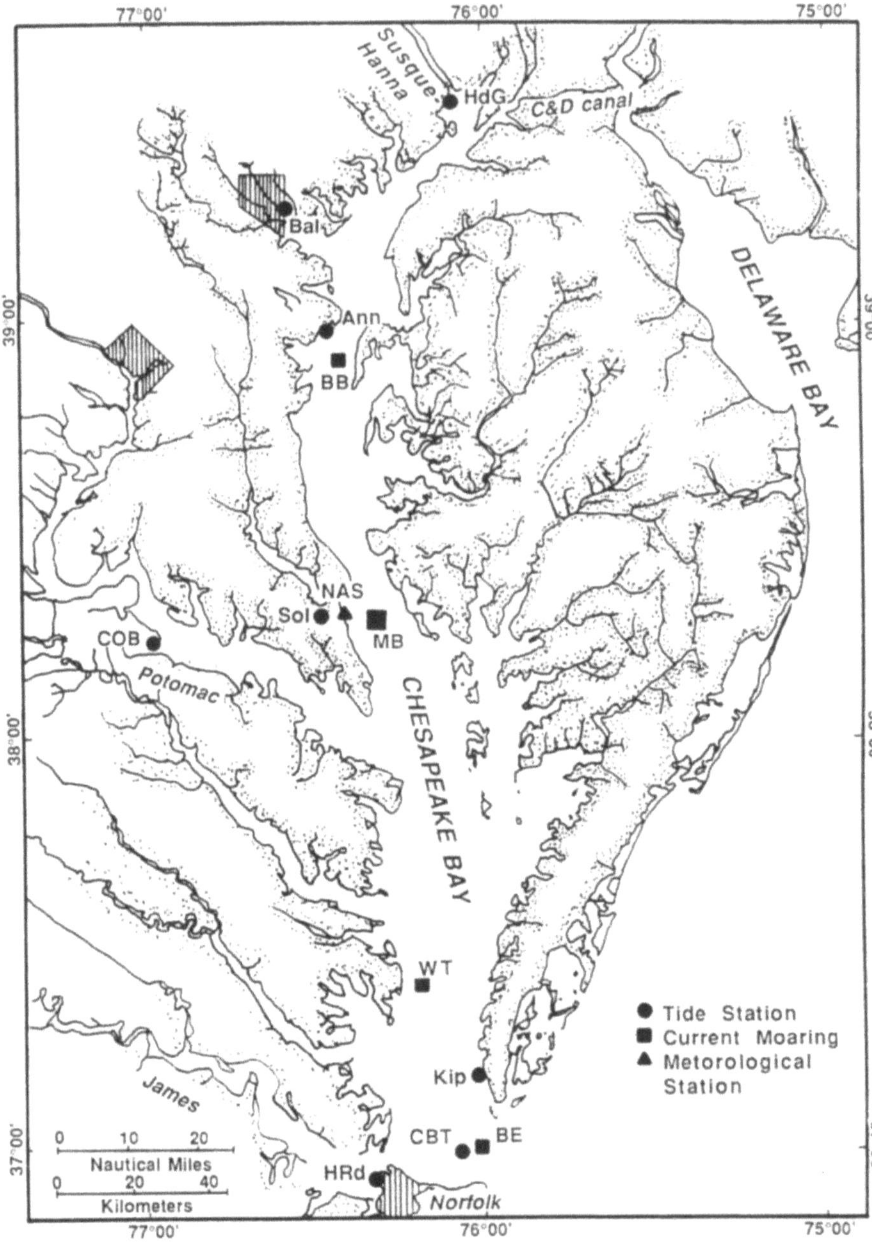

Fig.1 Chesapeake Bay sampling stations.

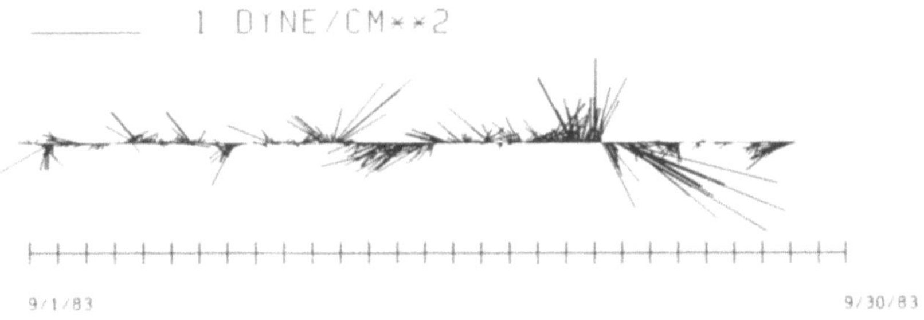

Fig.2 Wind stress used for the September 1983 simulation.

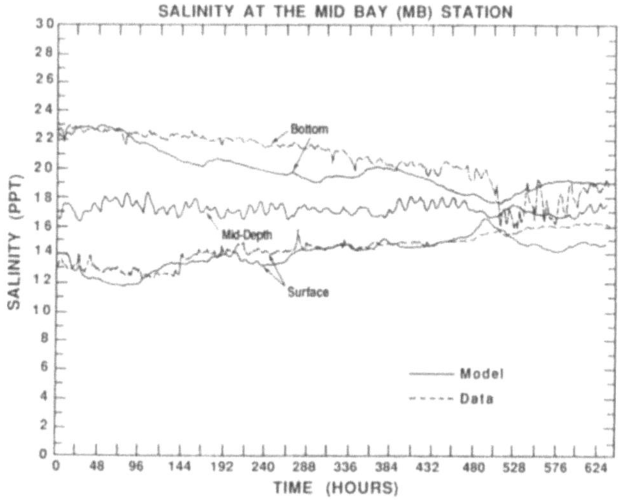

Fig.4 Temporal variation of salinity at the mid-bay (MB) station during September 1983.

Fig.3 Chesapeake Bay numerical grid.

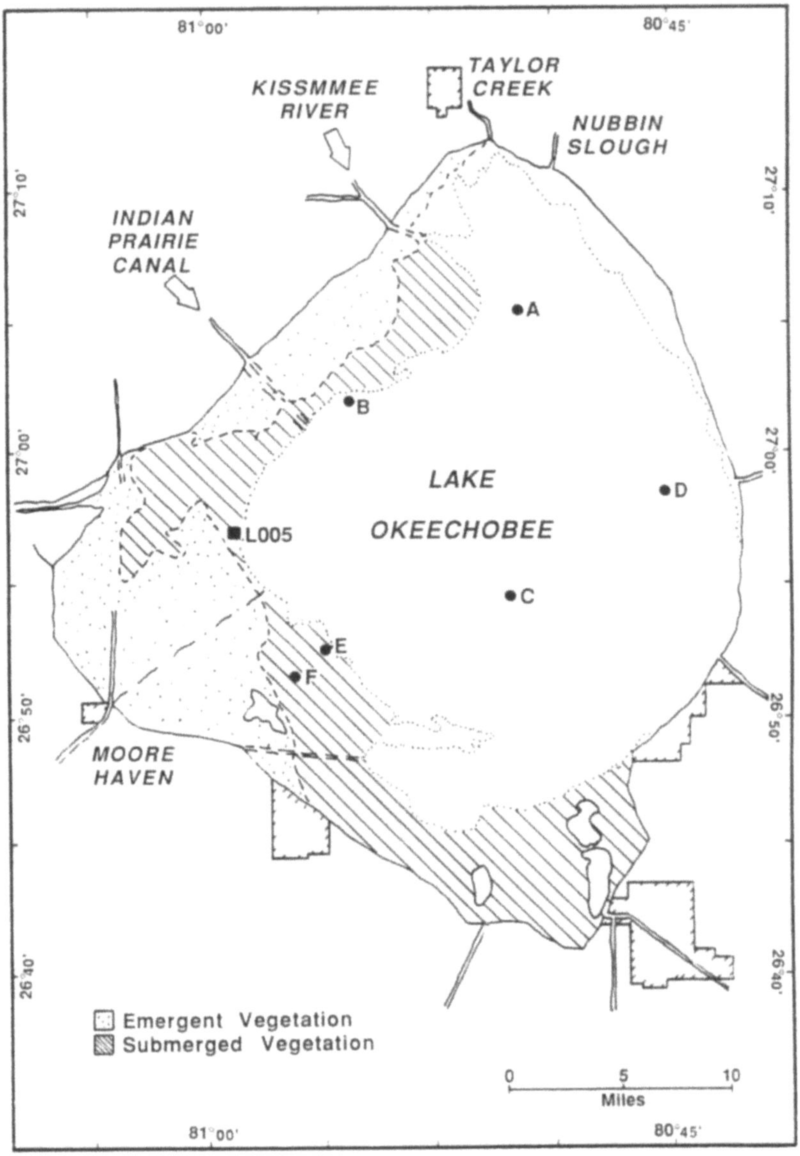

Fig.5 Lake Okeechobee with platform locations.

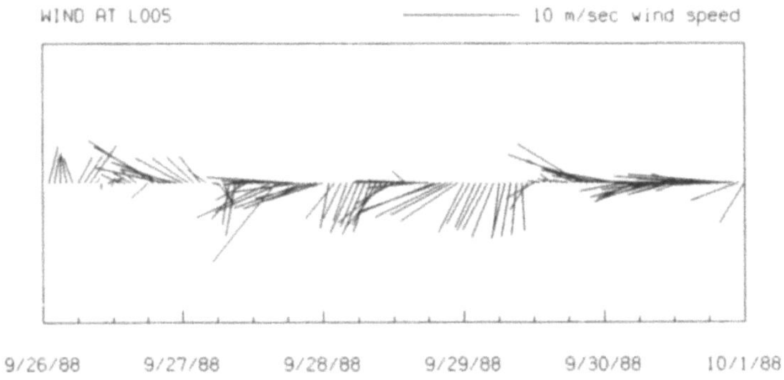

Fig.6 Measured wind velocity at L005 in Lake Okeechobee.

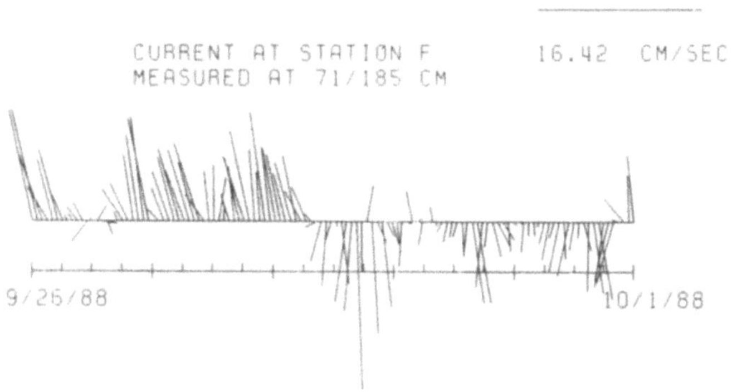

Fig.7 Measured current at platform F in Lake Okeechobee.

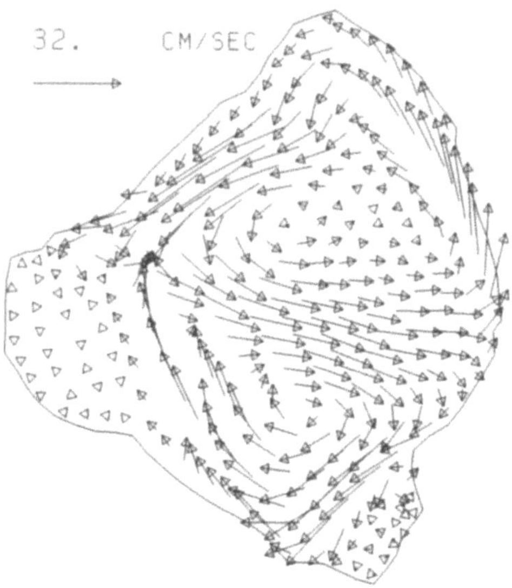

Fig.8 Circulation gyres in Lake Okeechobee due to a 2 dyne/cm^2 wind from the east.

ACKNOWLEDGEMENT

Supports from U.S. Army Engineer Waterways Experiment Station, Virginia Institute of Marine Science, South Florida Water Management District, U.S. Environmental Protection Agency and Pittsburgh Supercomputing Center are acknowledged.

REFERENCES

Johnson B H (1988) An Approach for Modeling the Upper Chesapeake Bay , Proc of the 1988 National Conference on Hydraulic Engineering, ASCE, pp. 588-93.

Sheng Y P (1984) A Turbulent Transport Model of Coastal Processes , Proc of the 19th ICCE, ASCE, pp. 2380- 2396.

Sheng Y P (1986) A Three-Dimensional Mathematical Model of Coastal, Estuarine and Lake Currents Using Boundary-Fitted Grid , ARAP Technical Report No. 585, Aeronautical Research Associates of Princeton/California Research and Technology/Titan Systems, Princeton, N.J.

Sheng Y P (1987) On Modeling Three-Dimensional Estuarine and Marine Hydrodynamics , in Three-Dimensional Models of Marine and Estuarine Dynamics (J C J Nihoul and B M Jamart, eds), Elsevier Oceanography Series, Elsevier, pp. 35-54.

Sheng Y P W Lick, R T Gedney, and F B Molls (1978) Numerical Computation of Three-Dimensional Circulation in Lake Erie: A Comparison of a Free-Surface Model and a Rigid-Lid Model , J Phys Oceanog Vol 1, pp. 713-27.

Sheng Y P , T S Wu, and P F Wang (1988) Coastal and Estuarine Hydrodynamic Modeling in Curvilinear Grids , Proc of the 21st ICCE, ASCE, pp. 2655-2665.

Sheng Y P and C Villaret (1989) Modeling the Effect of Suspended Sediment Stratification on Bottom Exchange Processes , Journal of Geophysical Research, In Press.

Sheng Y P , S Peene, V Cook, P F Wang, K M Ahn, J Capitao, and G Chapalain: Hydrodynamics and Transport of Suspended Sediments and Phosphorus in Lake Okeechobee, Technical Report, Coastal and Oceanographic Engineering Department, University of Florida.

DEVELOPMENT OF A WATER QUALITY MODEL FOR THE PATUXENT ESTUARY

Wu-Seng Lung
Department of Civil Engineering
University of Virginia
Charlottesville, VA 22901

Introduction

The Patuxent River (Figure 1) has been a major focal point for water quality management in Maryland for well over a decade. As the Baltimore and Washington, D.C. metropolitan areas encroached on the Patuxent watershed, long-time residents made policy-makers increasingly aware of the river's declining vitality and called for a reversal of declining aquatic resources and negative water quality trends. Substantial controversy concerning the choice of the best water quality protection strategy for the Patuxent prevented much progress until a consensus was reached in December, 1981.

With the 1981 consensus as a foundation for management actions, the 208 Water Quality Management Plan adopted in 1983 for the Patuxent River Basin outlined a broad and ambitious approach for improving water quality in the river by:

1. formally adopting specific goals for nutrient control from point and nonpoint sources, specifically nitrogen and phosphorus
2. proposing that efforts to protect the estuary must proceed based on the best available information at that time
3. calling for an intensive effort to improve the understanding of the estuary
4. providing for management strategy adjustment over time as new information becomes available

Following adoption of the Water Quality Management Plan, the Maryland Department of the Environment (MDE) has implemented the following:

1. point source control focusing on upgrading major sewage treatment plants for phosphorus and nitrogen removal
2. nonpoint source control centering on demonstration of best management practices and providing aid for their installation
3. a Patuxent River monitoring, research, and modeling strategy has been carefully designed to monitor the state of the river over time, provide a much clearer understanding of the factors influencing water quality, and improve our ability to predict the consequences of various water quality management decisions.

At the present time, the broad efforts to clean up the Patuxent are substantially underway with measurable progress in implementation of the policies set forth in the 1983 Water Quality Management Plan. Based on a modeling study in the 1983 water quality management plan, phosphorus removal has been installed at major wastewater treatment plants with flows greater than 1 mgd in the Patuxent River basin. Additional removal of nutrients (i.e., nitrogen) has been considered for the Patuxent River basin. In 1985, federal funds for advanced treatment (e.g., nitrogen removal) facilities at four treatment plants in the basin were requested by the State of Maryland.

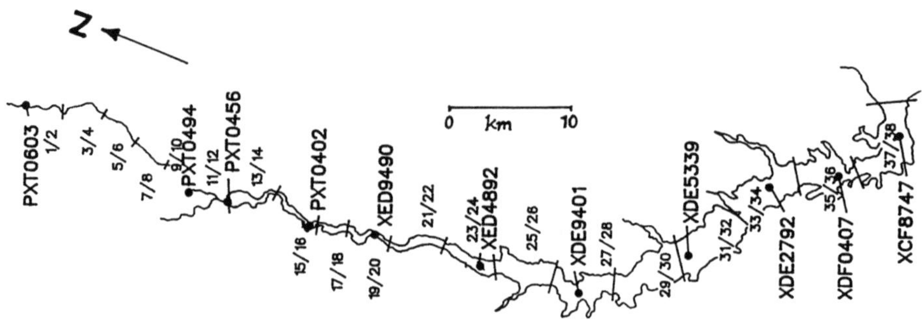

Figure 1. The Patuxent Estuary (model segmentation and water quality sampling stations)

To better address nutrient controls from point and nonpoint sources, a more comprehensive model is needed. That is, additional factors associated with nutrient limitations and phytoplankton growth dynamic should be incorporated. For example, D'Elia and Bishop (1985) showed that large doses of phosphorus in the Patuxent River caused almost no algal growth in the warmer months and only mild, delayed-action blooms in the colder months. Conversely, nitrogen stimulated large blooms of algae during the summer. During the high-flow season in late winter and early spring, nitrogen inputs and abundance exceed phosphorus. Phosphorus, therefore, is less available and more limiting during this time of year. During the low-flow season in late summer and early fall, nitrogen may become the limiting factor. The previous model of the Patuxent Estuary by Hydroscience (1981) only addressed steady-state conditions in the Patuxent Estuary. Thus, a new model with the capability to quantify seasonal variations is needed.

The Patuxent Estuary Water Quality Model

The newly developed model is a time-variable model to account for seasonal dynamics of phytoplankton and nutrient uptake/recycle. Spatially, the water column is sliced into two layers, each of which is subdivided into 19 segments in the longitudinal direction. Additional features of the model include multiple algal functional groups and a sediment layer. The algal groups are diatoms, green algae, non-nitrogen fixing and nitrogen fixing blue-green algae. They are included in the model to characterize seasonal succession of algal species. The sediment plays a major role in eutrophication of the Patuxent. Spring and summer algal growth in the upper estuary results in subsequent deposition of algal biomass to the sediment where decomposition consumes dissolved oxygen. Depressed oxygen levels have been observed in the bottom water of the lower estuary in the summer. Nutrient (orthophosphate and ammonia) releases from the sediment under anaerobic condition represent a significant nutrient source in the estuary. Such sediment-water interactions are incorporated into the model by including a sediment layer in addition to the two layers in the water column.

Water quality constituents modelled include chlorophyll *a* for the four algal groups, zooplankton carbon, nitrogen components (organic nitrogen, ammonia, and nitrite/nitrate), phosphorus components (non-living dissolved and particulate organic phosphorus, dissolved and particulate inorganic phosphorus), dissolved oxygen, and total suspended solids in the water column and sediments. Sediment nutrient fluxes are calculated in terms of concentration gradients of the nutrients between the water column and interstitial water. Sediment oxygen demand is calculated in a similar manner.

A comprehensive water quality monitoring program has been conducted by MDE since 1982 and provides a significant data base to support the model development effort. Sediment nutrient release fluxes and oxygen demand were measured in 1983 (Lantrip *et al.*, 1987). The model is calibrated and verified using the data from 1983, 1984, and 1985. The hydrologic conditions in these three years differ significantly and provide a challenge for the model to reproduce the nutrient and phytoplankton growth dynamics, sediment oxygen demand, and nutrient release fluxes in the Patuxent.

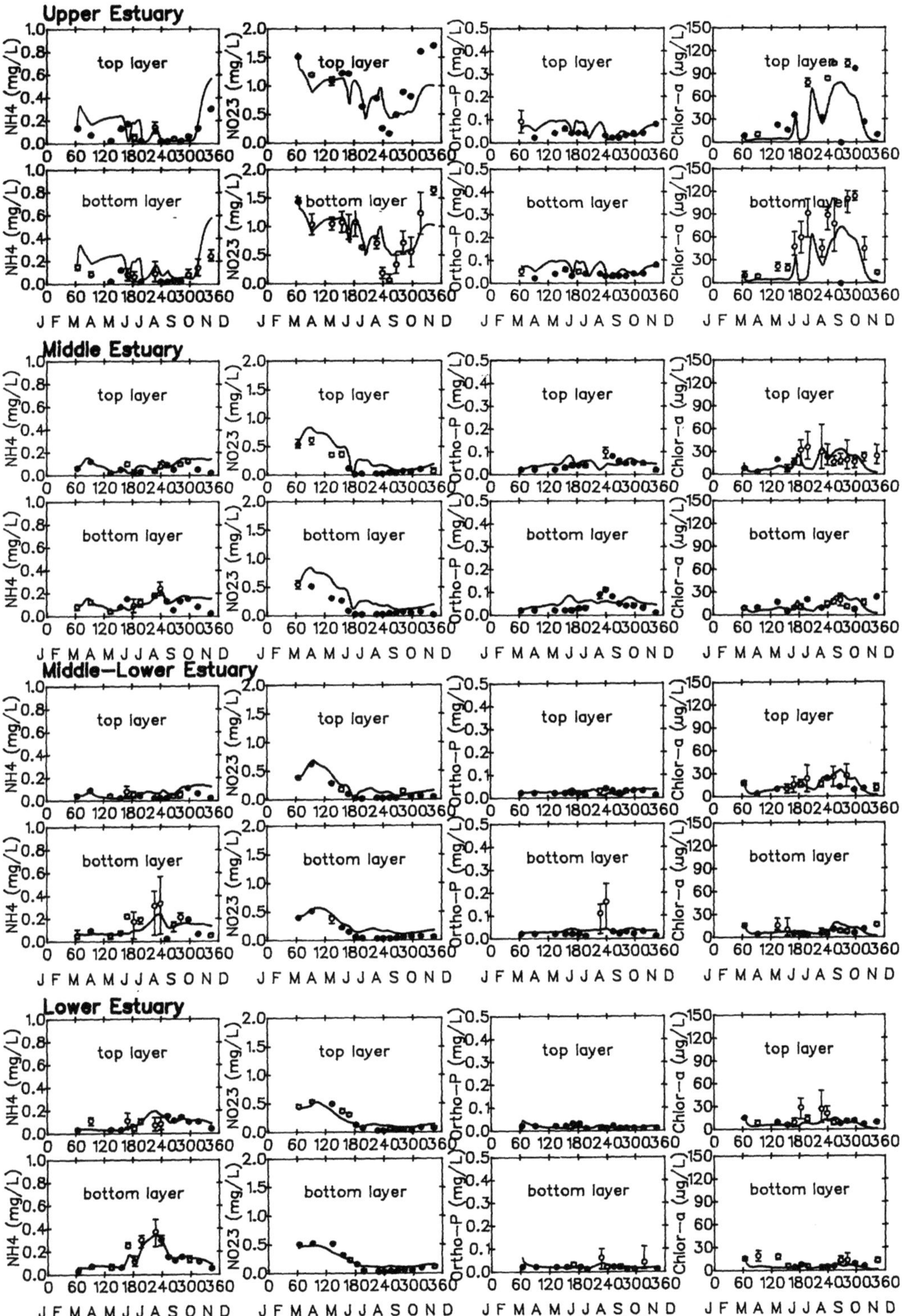

Figure 2. Model calibration results using 1984 data

Model Calibration and Verification

In model calibration, the time-variable mass transport pattern in the estuary was first determined by reproducing the salinity distribution throughout the year in a two-layer fashion. The calibrated mass transport is then integrated with the kinetics to calculate the concentrations of the water quality constituents in the water column without the sediment layer. Field measured sediment nutrient release fluxes and oxygen demand are initially input to the model. Such a step is needed to properly calibrate the kinetic coefficients in the **water column** with known sediment-water fluxes. The next step is to attach the sediment layer to the water column. As such, the model calculates sediment release fluxes and oxygen demand which are compared directly with measured fluxes. Subsequent adjustments of the diffusion coefficients across the sediment-water interface and kinetic coefficients in the sediment layer further fine tune the model. The model calibration and verification results for three years of data are extensive and have been fully documented in another report (Lung, 1989). Due to limited space, only the salient features of the calibration results using the 1984 data are presented in this paper.

Salinity distributions measured at different times of 1984 for the entire estuary at 12 locations with one-meter resolution in the vertical direction are used to derive the first estimates of the mass transport coefficients: longitudinal and vertical velocities, and vertical diffusion coefficients using the methodology developed by Lung and O'Connor (1984). Additional refinement of these coefficients is made to reproduce the salinity distributions throughout the year.

The calibrated mass transport coefficients are then used for water quality calculations. Figure 2 shows the model results and observed data at four locations representing the upper, middle, middle-lower, and lower estuary for key water quality constituents: ammonia, nitrite/nitrate, orthophosphate, and chlorophyll *a* in the top and bottom layers of the Patuxent throughout 1984. As illustrated, the model mimics the seasonal phytoplankton growth and nutrient dynamics in the water column very well. Low ammonia levels in August and September are calculated by the model, matching the observed data (Figure 2). Modest ammonia release rates from the sediment prevents a nitrogen limiting condition in the **upper** estuary. Nitrite and nitrate concentrations in the upper estuary are dominated by the upstream input to the estuary at the fall line. Inorganic nitrogen concentrations gradually decrease in the downstream direction, approaching a nitrogen limiting condition for algae in the **lower** estuary. Phosphorus does not appear to be a limiting factor for algal growth in 1984 as the orthophosphate levels are consistently higher than the Michaelis-Menton coefficient.

Figure 3 shows calculated and measured inorganic nitrogen to orthophosphate ratios in the top layer of the water column in 1984. Model calculated ratios follow a seasonal trend in the middle and lower estuary, indicating a *potential* of phosphorus limitation in the spring and nitrogen limitation in the summer - a confirmation of the finding by D'Elia and Bishop (1985). Model results generally match the data except in the upper estuary. Model calculations and observed data show that the summer ratios in the middle estuary are consistently below 10, a threshold suggested by Thomann and Mueller (1987) for nitrogen limitation.

The model reproduces the dissolved oxygen concentrations in the surface and bottom layers very well (Figure 4). The dissolved oxygen concentrations in the bottom layer almost reached zero in August 1984 in the lower estuary. Seasonal variations of dissolved oxygen correlate closely with those of algal chlorophyll *a* in the surface layer (Figure 2). Dissolved oxygen concentrations in the bottom layer are strongly controlled by the decomposition of organic matters in the water column and sediment oxygen demand (Lung, 1989).

Model Applications

The verified model is now used to address an important management question: what is the incremental water quality benefit following nutrient reductions into the Patuxent Estuary? Reductions of nutrient input would come from both nonpoint and point sources. While best management practices for agricultural areas are recommended to control nonpoint nutrient loads, phosphorus and nitrogen removals are needed to reduce point source input. Phosphorus loads from wastewater treatment plants in the Patuxent basin have steadily reduced their phosphorus loads from 840 kg/day in 1980 to 150 kg/day in

1986 (Lung, 1986). [Eight of the nine major treatment plants in the basin are located above the fall line.] With more plants scheduled to have phosphorus removal in operation, the major phosphorus source for the estuary would be the nonpoint runoff. In the mean time, nitrogen removal is being installed at the Western Branch plant (located below the fall line) and considered for other plants. It is therefore essential to determine the water quality improvement resulting from phosphorus control and further to quantify the incremental benefit due to nitrogen removal. The 1983 condition is used as the basis to evaluate these water quality management scenarios as a drought condition existed in the summer months of that year, creating a stressed condition. Thus, the model is used to project what would have happened in the estuary if a series of nutrient control measures had been implemented in 1983. It should be pointed out that model projections of this nature are by no means **actual** predictions of the water quality conditions in the Patuxent. Rather, the model results should be viewed to determine incremental benefits over the range of nutrient control measures.

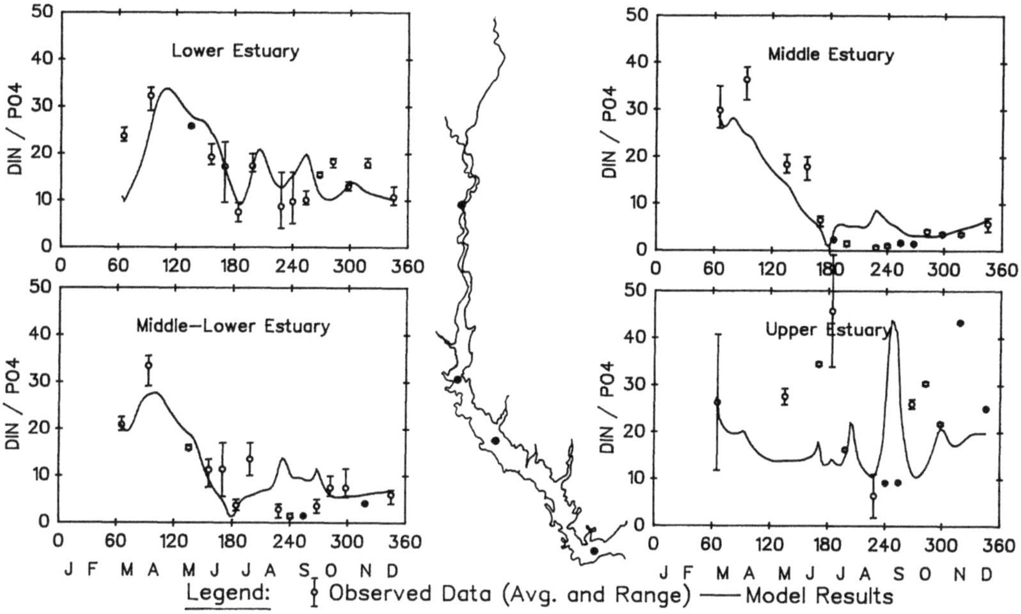

Figure 3. Dissolved inorganic nitrogen to orthophosphate ratios (1984 condition)

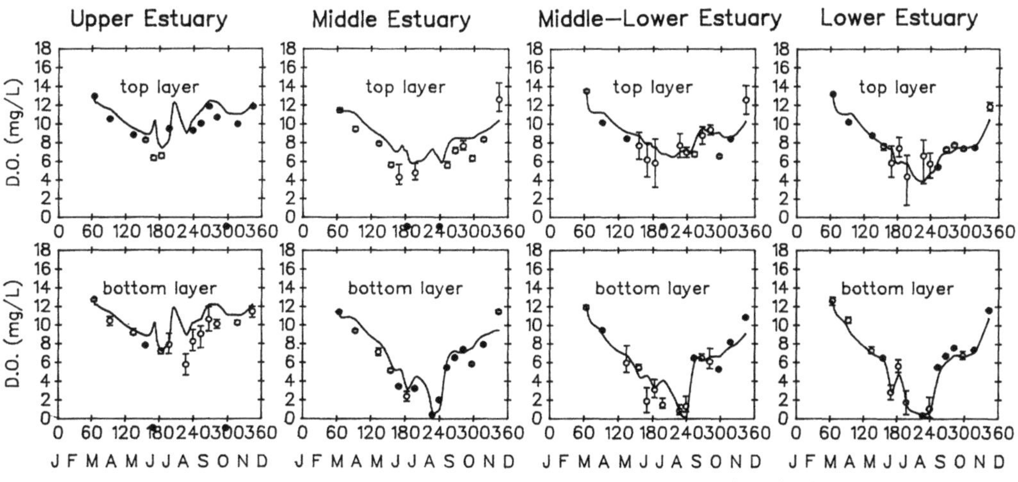

Figure 4. Model calculated and observed dissolved oxygen concentrations in the water column

The model results are summarized in Figure 5 showing the reduction in algal biomass in the surface layer of the upper estuary and increase in dissolved oxygen in the bottom layer of the lower-middle estuary. Phosphorus removal at all major wastewater plans (total phosphorus concentration of 1 mg/L in the effluent) and nonpoint phosphorus reductions from best management practices would reduce the peak chlorophyll *a* concentration from about 70 μg/L to 54 μg/L in the upper estuary. Nitrogen control would further reduce it to 49 μg/L. While the dissolved oxygen concentrations in the bottom waters near the lower estuary would slightly increase by 0.3 mg/L due to phosphorus control and additional 0.4 mg/L due to nitrogen control.

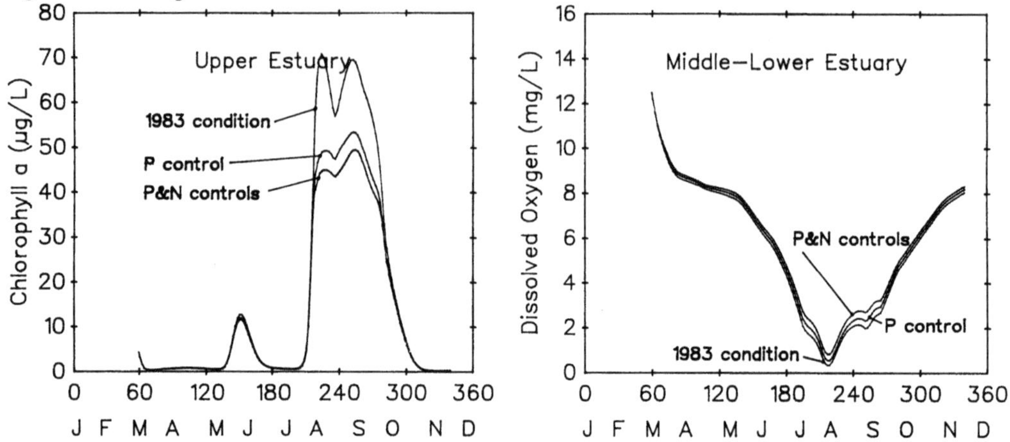

Figure 5. Projected water quality improvement due to nutrient controls

Summary and Conclusions

A mathematical model has been developed to assist the water quality management of the Patuxent Estuary. The model has been verified with a comprehensive database (data from 1983, 1984, and 1985) and is reproducing the algal growth and nutrient dynamics in the estuary on a time-variable basis. The verified model has been used to evaluate nutrient control scenarios. The model results indicate water quality improvement from phosphorus control and appreciable incremental water quality benefit from nitrogen removal at point sources.

Acknowledgement

This modeling study has been sponsored by Maryland Department of the Environment since 1986. Their continuing interests and support are greatly appreciated.

References

1. D'Elia C F, Bishop D J (1985) Growing green: nutrient enrichment of Chesapeake Bay. The Chesapeake Citizen Report 5 (Fall 1985)
2. Hydroscience (1981) Water quality analysis of the Patuxent River. Report prepared for the State of Maryland
3. Lantrip B M, Summers R M, Phelan D J, Andrle W (1987) Sediment/water-column flux of nutrients and oxygen in the tidal Patuxent River and Estuary, Maryland. U.S. Geological Survey Water-Supply Paper 2296, Washington, D.C.
4. Lung W S (1986) Phosphorus loads to the Chesapeake Bay: a perspective. *J. Water Pollution Control Fed.* 58:749-756
5. Lung W S (1989) A water quality model for the Patuxent Estuary. Report under preparation for Maryland Department of the Environment
6. Lung W S, O'Connor D J (1984) Two-dimensional mass transport in estuaries. *J. Environ. Eng.*, ASCE 110:1340-1357
7. Thomann R V, Mueller J A (1987) Principles of surface water quality modeling and control. Harper & Row Publisher, Inc., New York, NY

NUMERICAL MODELING OF SUSPENDED SEDIMENT DYNAMICS IN THE WESER ESTUARY

Günther Lang and Mark Markofsky
Institut für Strömungsmechanik und Elektronisches Rechnen im Bauwesen
Universität Hannover, Appelstraße 9a, D-3000 Hannover 1

1. INTRODUCTION

The accumulation and dynamics of suspended sediments in estuaries and also the development of an estuarine turbidity maximum are influenced by a wide variety of physical, chemical and/or biological mechanisms. In general the following processes are known to be important: Baroclinic circulation, transport due to tidal pumping, resuspension and deposition of suspended sediments, salt-flocculation and floc-breakup, the death of microorganisms in the brackish water zone, internal waves as well as mixing due to structures like bridges etc. The relative significance of the different processes mentioned is mainly dependent on hydrodynamic factors (tidal range, fresh water flow-rate, etc.), the density-stratification (gradients of salinity, temperature, etc.) and the presence of organic material.

Due to the numerous processes involved, the complexity of their mutual interactions and the characteristics of the estuary considered, the application of a numerical model in close connection with data interpretation is a valuable tool for deepening the insight of the hydrodynamics as well as the behaviour of the turbidity maximum of an estuary.

In the following some results of a field survey (the Mud and Suspended Sediment Experiment MASEX '85) and numerical simulations of the short-term behaviour of suspended sediment dynamics in the Weser estuary are presented. A comprehensive description of the strategy and the realization of the field survey as well as an overview over the measured data was presented in a separate paper (Riethmüller, R. et al., 1988).

2. THE NUMERICAL MODEL TISAT-S

TISAT-S (Tide, Salinity, Temperature and Suspended Sediment) is a coupled, explicit, three-dimensional finite-difference numerical model, which has been used for the simulation of MASEX '85. TISAT-S is mainly based on classic ideas about three-dimensional models presented in the early seventies (Leendertse, J.J. et al., 1973). More specific information about TISAT-S, e.g. turbulence model, parametrization of deposition and resuspension etc., can be found elsewhere (Markofsky, M. et al., 1986a; Markofsky, M. et al., 1986b; Lehfeldt, R. and Bloß, S., 1988; Lang, G. et al., 1989).

A comparison between measured and calculated suspended sediment concentrations has lead to an improved parametrization of the bottom shear stress, which takes the near bottom density stratification and the velocity gradient into acount (Lang, G. et al., 1989). This model assumes that the presence of a (stable) stratification (due to the combined effects of temperature, salinity and suspended sediment) yields to a reduction of the bottom shear stress.

3. VERTICAL PROFILES OF SALINITY AND SUSPENDED SEDIMENT

It has been shown that TISAT-S is able to reproduce some of the observed essential processes related to the turbidity maximum of the Weser estuary (e.g. deposition, resuspension and advection of suspended material) with respect to their spatial distribution and time-dependency (Lang, G. et al., 1989). Measured values of salinity (Figure 1a) vary between 3 and 9 ppt close to the bottom; the maximum stratification was about 1.5 ppt over the water depth. It can be seen from this figure that the numerical model reproduces the (time-dependent) mean value of salinity with discrepancies in the correct form of the vertical stratification. The measurements frequently show an s-shaped profile - well mixed close to the bottom and close to the surface, stratification at mean depths - while the model does not reflect the well mixed near surface layer (see Section 4).

A comparison between measured and calculated suspended sediment concentrations (Figure 1b) shows larger differences than for salinity. From the measurements three phases with different shapes of the vertical suspended sediment distribution can be distinguished: phase 1 (7:21 h, 8:45 h and 10:06 h) shows a well mixed layer in the upper half of the water column and a strongly stratified one in the lower half of the water column; phase 2 (11:31 h, 12:01 h and 13:33 h) is characterized by almost constant values over the whole depth; during phase 3 (15:07 h, 16:52 h and 18:21 h) the profiles are very similar to exponential (equilibrium) profiles.

4. DISCUSSION AND CONCLUSIONS

The measured well mixed layer of near surface salinity may be in part due to the strong winds present during the experiment. In the model, the vertical eddy-diffusivity is calculated from a mixing-length approach which uses the Richardson number as a parameter for vertical stability. The calculation of the Richardson number assumes equilibrium between production and dissipation of turbulent kinetic

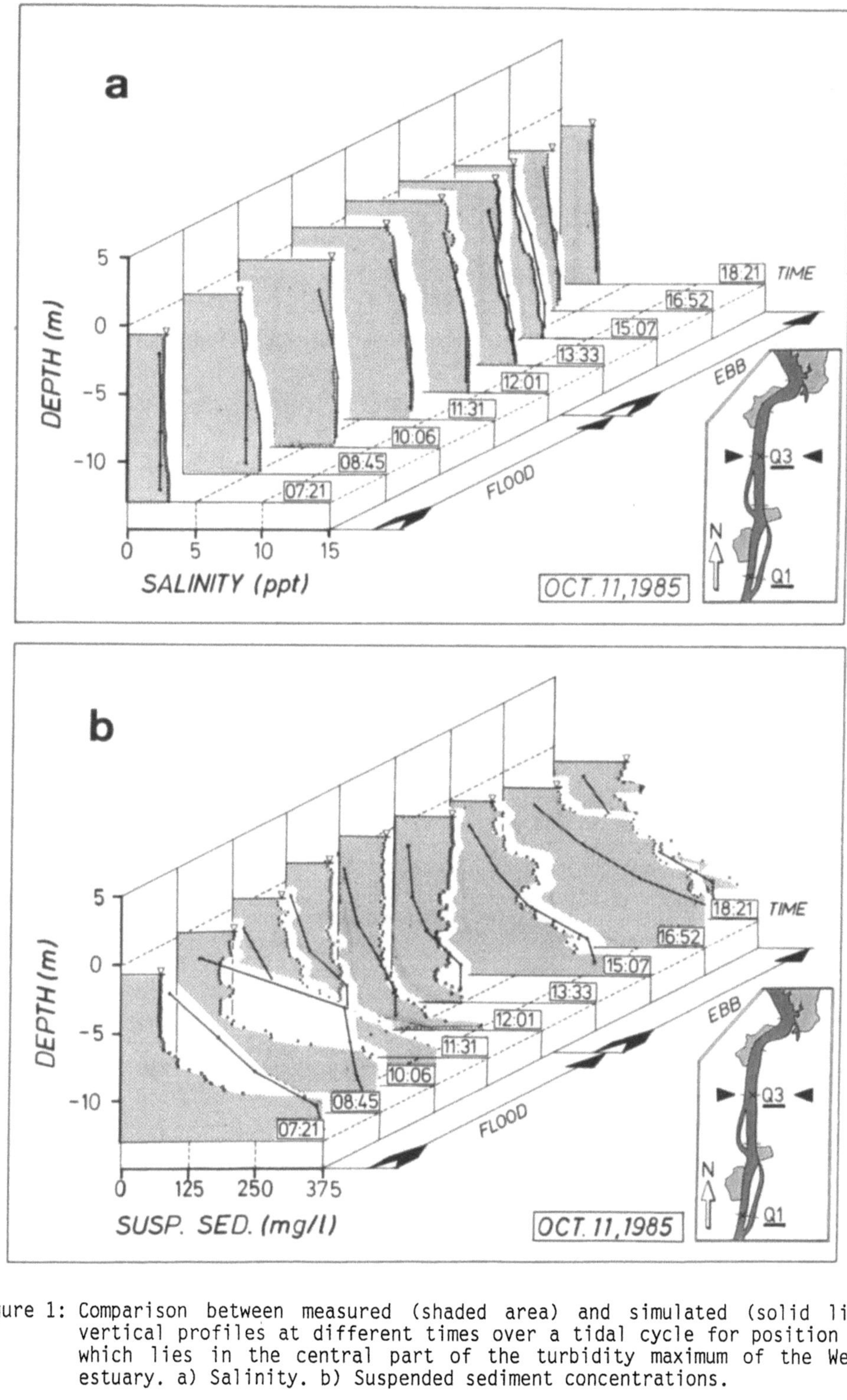

Figure 1: Comparison between measured (shaded area) and simulated (solid line) vertical profiles at different times over a tidal cycle for position Q_3, which lies in the central part of the turbidity maximum of the Weser estuary. a) Salinity. b) Suspended sediment concentrations.

energy. Therefore, in the near surface layer, the input of wind induced turbulent kinetic energy (wind-mixing) should also be considered in the calculation of the Richardson number.

The same is valid for the near surface mixed layer of suspended sediment. On the other hand, the observed stratification near the bottom (phase 1), which might be closely related to the fine-structure of the salinity stratification, cannot be reproduced with a coarse vertical grid resolution (spacing about 2 m) used. Therefore, future simulations should be made with a much better vertical resolution (about 0.5 to 1 m). The vertical profiles observed during phase 2 indicate a very low value of the particle settling velocity. The actual simulations were performed using a constant settling velocity. Whether an empirically determined, concentration dependent, formulation is sufficient or a more complex flocculation model should be used is still an open question. During phase 3 the calculated profiles are in quite good agreement with the measured ones.

The vertical profiles presented offer an impression of the pronounced variability (shape of the profile, mean value) of suspended sediment concentrations in the central part of the turbidity maximum of the Weser estuary. Analysis with the numerical model has provided new insight into this complicated phenomenon.

REFERENCES

Lang G, Schubert R, Markofsky M, Fanger H-U, Grabemann I, Krasemann HL, Neumann LJR and Riethmüller R (1989) Data interpretation and numerical modeling of the Mud and Suspended Sediment Experiment 1985, Journal of Geophysical Research. C-Oceans (in press)

Leendertse JJ, Alexander RC and Liu S-K (1973) A Three-Dimensional Model for Estuaries and Coastal Seas: Vol 1, Principles of Computation, Technical Report R-1417-OWRR, The Rand Corporation, Santa Monica, CA 90406, USA

Lehfeldt R and Bloß S (1988) Algebraic turbulence model for stratified tidal flow, in Physical Processes in Estuaries, edited by Dronkers J and van Leussen W, pp. 278-291, Springer-Verlag, Berlin Heidelberg New York London Paris Tokyo

Markofsky M, Lang G and Schubert R (1986a) Numerische Simulation des Schwebstofftransportes auf der Basis der Meßkampagne MASEX '83. Die Küste 44: 171-189

Markofsky M, Lang G and Schubert R (1986b) Suspended sediment transport in rivers and estuaries, in Lecture Notes on Coastal and Estuarine Studies: Physics of Shallow Estuaries and Bays, edited by van de Kreeke, pp. 210-227, Springer-Verlag, Berlin Heidelberg New York Tokyo

Riethmüller R, Fanger H-U, Grabemann I, Krasemann HL, Ohm K, Böning J, Neumann LJR, Lang G, Markofsky M and Schubert R (1988) Hydrographic measurements in the turbidity zone of the Weser estuary, in Physical Processes in Estuaries, edited by Dronkers J and van Leussen W, pp. 332-344, Springer-Verlag, Berlin Heidelberg New York London Paris Tokyo

ANALYSIS OF THE DISTRIBUTION OF SUSPENDED PARTICULATE MATTER, BACTERIA, CHLOROPHYLL _a_ AND PO_4 IN THE UPPER ST. LAWRENCE ESTUARY, USING A TWO-DIMENSIONAL BOX MODEL

Jean Painchaud, Denis Lefaivre, Gilles-H. Tremblay, Jean-Claude Therriault
Institut Maurice-Lamontagne, Ministère des Pêches et des Océans
CP 1000, Mont-Joli, Québec, G5H 3Z4, Canada

ABSTRACT

A box model was used to study the importance of biogeochemical vs physical processes on suspended particulate matter (SPM), attached and free bacteria, chlorophyll _a_ and dissolved PO_4 in the Upper St. Lawrence Estuary (USLE). Flux calculation revealed that SPM distribution was in a steady state and that the dynamics of attached bacteria were closely related to that of SPM. The estuary was a sink for free bacteria and chl _a_, whereas PO_4 was essentially controlled by estuarine hydrodynamics. Turn-over time values showed that the upstream portion of the estuary was the most biogeochemically active.

INTRODUCTION

In estuaries, the distribution of non-conservative variables is controlled by estuarine hydrodynamics and by biogeochemical processes. To interpret correctly the distribution of a variable, both sets of processes must be taken into account, which is not possible with distribution data and dilution plots alone. A simple way to address this problem is to use a box model approach and to calculate fluxes. Variables controlled solely by estuarine circulation will show null net fluxes, but any biogeochemical process at work on a given variable will result in positive or negative net fluxes, thus allowing to identify zones of source or sink. Moreover, knowing box volumes, estimation of rates of production or loss, and of turn-over time, is possible. The objective of this study was to use this approach on SPM, attached and free bacteria, chlorophyll _a_ and PO_4 in order to assess the relative importance of biogeochemical and physical processes in the control of their respective distribution in the USLE.

MATERIALS & METHODS

The samples were collected hourly over 12-hour periods from Aug 17 to 27, 1987 at

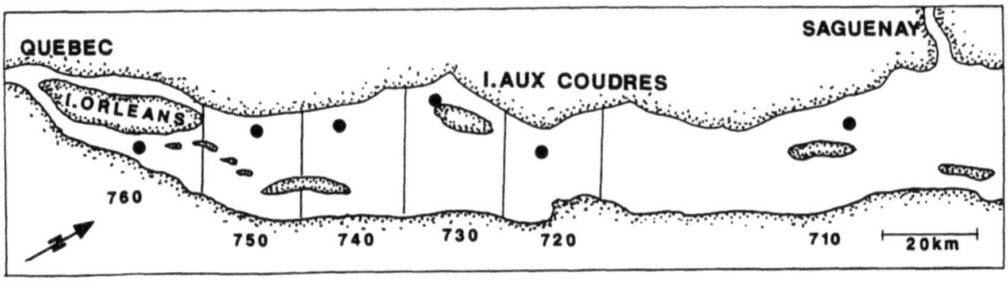

Fig.1. Station location and box model boundaries

6 anchor stations in the USLE (Fig.1), using Niskin bottles lowered at 2 depths: above and below the pycnocline. Sampling depths were determined from vertical profiles of temperature and salinity, obtained using a Guildline CTD probe, model 8750. Bacteria were preserved with glutaraldehyde and counted using epifluorescence microscopy (Hobbie et al. 1977). Chl a samples were determined fluorometrically (Parsons et al. 1984). Suspended particulate matter (SPM) was collected on preweighed 0.45 µm, 47 mm Nuclepore filters, later dried and weighed again. Dissolved PO_4 samples were analyzed on board with a Technicon AutoAnalyzer II after filtration with 0.45 µm Nuclepore membranes.

Flux calculation was based on the methodology by Officer (1980) for a mixed one and two dimensional box model. The upstream boundary of box 750 was set at the limit of salt intrusion (eastern tip of I. d'Orléans), below which four 20-km boxes were delimited (Fig.1). Box 760 was one dimensional and a two dimensional model was used for downstream boxes. Boundaries between upper and lower boxes were set at the average depth of the pycnocline. Following Officer's notation, variables or numbers relating with lower boxes are denoted with ' .

Horizontal and vertical exchange coefficients (E_{ij}), based on salinity (S_{ij}) distribution and freshwater discharge at the time of sampling ($R = 10000\ m^3\ s^{-1}$, Ministère de l'Environnement, Québec), were calculated (see Fig.2) using Officer's equations 101 to 104 (box 2) and equations 49-52 (boxes 3-5).

BOX i		BOX j
S_i, C_i, V_i	$E_i \rightarrow$	S_j, C_j, V_j
S'_i, C'_i, V'_i		S'_j, C'_j, V'_j

$$E_i = \frac{S'_j}{S'_j - S_i} \cdot R \qquad (m^3\ s^{-1})$$

$$Flux_i = E_i \cdot C_i \qquad (qty\ s^{-1})$$

$$Net\ flux_i = \Sigma flux_i \qquad (qty\ s^{-1})$$

$$Pro/los_i = \frac{Net\ flux_i}{V_i} \qquad (qty\ l^{-1}\ s^{-1})$$

$$Turn\text{-}over_i = \frac{C_i}{Pro/los_i} \qquad (h)$$

Fig.2. Schematic representation of data and calculations required to obtain fluxes, production/loss rates and turn-over time

These exchange rates have units of m^3 s^{-1} and are essentially fluxes of water; they can be used to calculate water residence time (Officer's equation 54). The product of these fluxes with tidally-averaged concentrations (C_{ij}) of the studied variables yielded fluxes of these variables in units of mass or number s^{-1} (Figs. 3-8a). Summation of all fluxes in a box gave the net flux of the variable (inside ellipses, Figs. 3-8a), which was subsequently divided by the volume (V_{ij}) of the box to obtain the rate of production or loss (mass or number l^{-1} h^{-1}, Figs. 3-8b). Turnover time (h) of the variables was estimated by dividing tidally-averaged concentrations by rates of production or loss (Figs 3-8c). Turn-over time was taken as an index of biogeochemical transformations and was compared to water residence time. The length of turn-over time with respect to water residence time allows to interpret the importance of these biogeochemical transformations vs physical transport.

RESULTS AND DISCUSSION

Tidally-averaged values of each variable are presented in Table 1. Salinity distribution showed that sta 760 was vertically homogeneous, whereas increasingly strong stratification was observed from sta 750 to 710. SPM peaked at sta 750, which is the usual summer location of the core of the turbidity maximum and decreased seaward to values one order of magnitude lower; freshwater values were intermediate. Large vertical gradients were observed everywhere, except at sta 760. Attached bacteria had the same pattern than SPM, but became virtually absent at sta 710. Free bacteria were most abundant in fresh water and decreased rapidly from sta 760 to 740, below which their abundance levelled off; vertical gradients were weak. Chl a followed similar horizontal and vertical trends. Dissolved PO_4 showed a typical estuarine gradient, increasing from fresh to marine waters and had little vertical structure.

Fluxes of water are shown in Fig.3a. Vertical fluxes were large in box 750; elsewhere, horizontal fluxes were largely dominant. Water residence time was short at the head of the estuary and increased seaward. SPM fluxes (Fig.4a) were negative in the upper layer and positive in the lower layer; vertical fluxes were upward in all boxes and were particularly large in box 750. This supports earlier observations that sediment resuspension, particularly near the head of the estuary, is the main process contributing SPM to the turbidity maximum (Silverberg & Sundby 1979). The removal of SPM from the upper layer is also consistent with observations of sediment exchange and deposition on tidal flats during summer (Serodes & Troude 1984, Lucotte & d'Anglejan 1986). The larger rates of loss in

Table 1. Tidally-averaged values of variables in upper and lower (') boxes

VARIABLES	BOX					
	760 760'	750 750'	740 740'	730 730'	720 720'	710 710'
S °/oo	0.1 0.1	8.1 9.3	12.5 15.1	15.5 21.5	20.1 25.9	23.7 30.6
Chl <u>a</u> (μg l^{-1})	2.9 2.9	2.6 2.0	1.0 1.5	0.8 0.8	0.6 0.5	0.6 0.4
SPM (mg l^{-1})	11.3 11.3	30.4 41.8	8.9 15.3	4.2 8.5	2.7 4.1	1.9 0.8
PO_4 (μM)	0.4 0.4	0.9 0.9	1.0 1.1	1.0 1.3	1.2 1.4	1.4 1.5
Att. bacteria (10^6 ml^{-1})	0.8 0.8	1.3 1.7	0.2 0.2	0.1 0.1	0.04 0.05	0 0
Free bacteria (10^6 ml^{-1})	3.9 3.9	1.5 1.5	0.9 0.8	0.7 0.5	0.5 0.3	0.4 0.3
Volume (10^6 m^3)	- -	1300 1300	2100 2100	2400 2400	4100 4100	- -

upstream boxes are probably due to the higher ratio of marsh area to water volume at the head of the estuary. It is significant that the sum of negative fluxes in the upper layer is almost exactly balanced by the positive fluxes of the deeper layer (121.5 vs 116.7 x 10^7 mg s^{-1}), which implies that in summer the turbidity maximum is in a steady state and is maintained by a dynamic equilibrium between bottom sediment resuspension and sedimentation on tidal flats. Turn-over time (Fig.4c) was very short for SPM, particularly in boxes 740 & 730 where it nearly equalled water residence time. These very short turn-over time values show that SPM is a very dynamic variable in the USLE.

Fluxes of attached bacteria closely parallel those of SPM (Fig.5a,b), which supports the hypothesis that these bacteria are essentially controlled by suspended sediment dynamics (Painchaud & Therriault in prep.). Thus it implies that bottom sediment is the source of attached bacteria which are then retained by intertidal flats. Relatively short turn-over times (Fig.5c) suggest that attached bacteria are also a dynamic variable in the USLE. Free bacteria, on the other

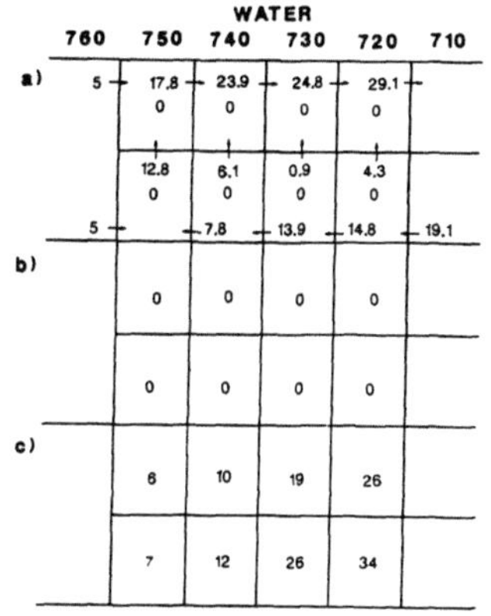

Fig.3. Water. a) Fluxes (10^3 m^3 s^{-1}) b) rates of production/loss, c) turn-over time (h)

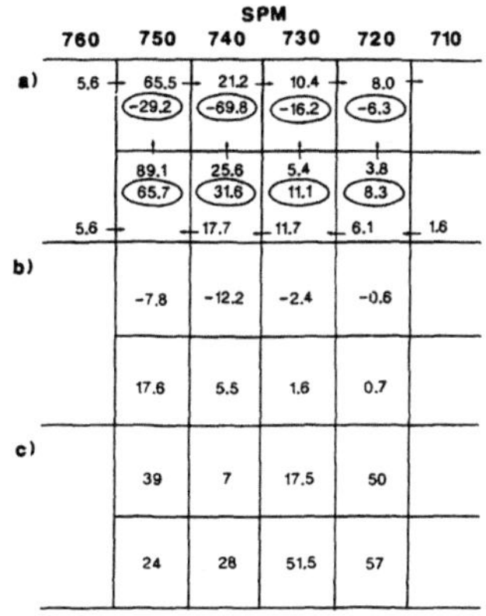

Fig.4. SPM. a) Fluxes (10^7 mg s^{-1}), b) rates of production/loss (10^{-1} mg l^{-1} h^{-1}), c) turn-over time (h)

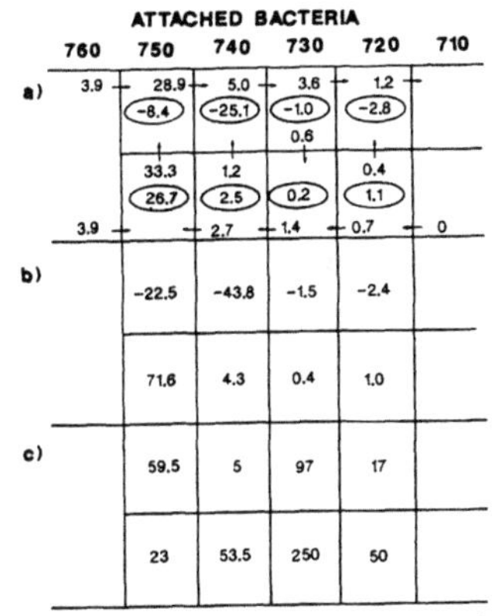

Fig.5. Attached bacteria. a) Fluxes (10^{15} cells s^{-1}), rates of production /loss (10^6 cells l^{-1} h^{-1}), c) turn-over time (h)

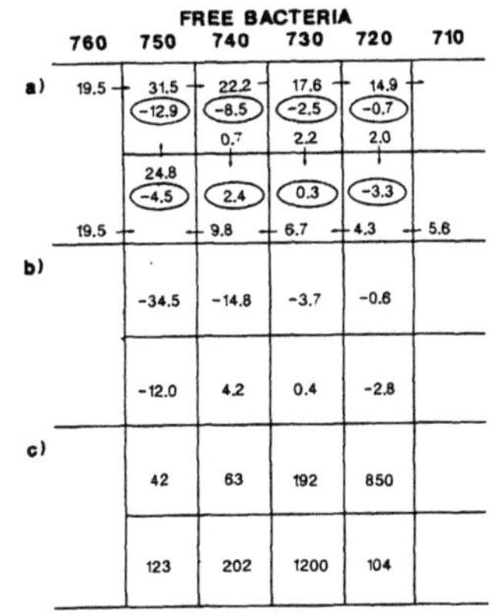

Fig.6. Free bacteria. a) Fluxes (10^{15} cells s^{-1}), b) rates of production/loss (10^6 cells l^{-1} h^{-1}), c) turn-over time (h)

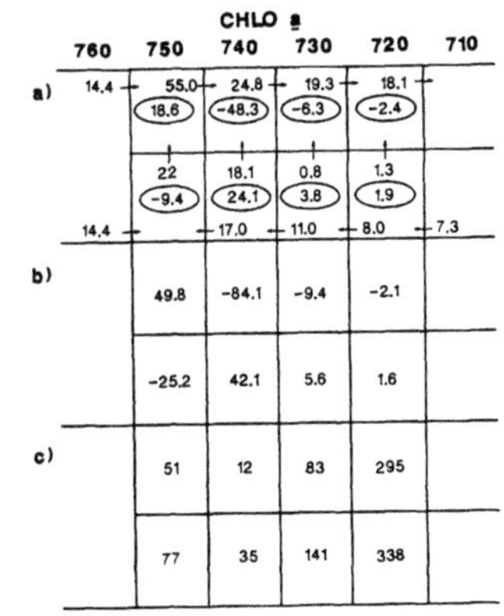

Fig.7. Chl a. a) Fluxes (10^3 mg s^{-1}) b) rates of production/loss (10^{-3} mg m^{-3} h^{-1}), c) turn-over time (h)

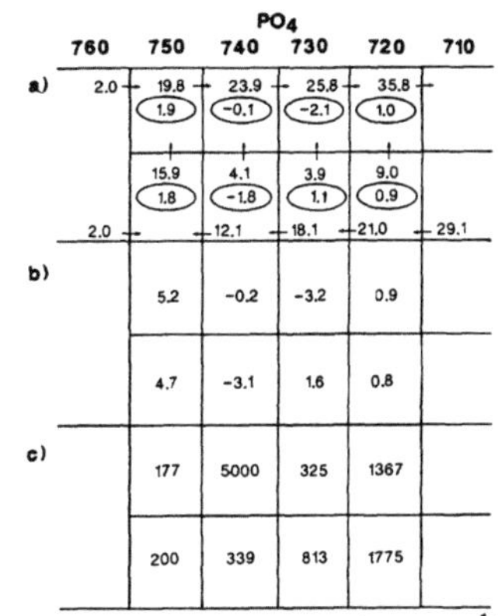

Fig.8. PO_4. a) Fluxes (mol s^{-1}), b) rates of production/loss (10^{-3} μM h^{-1}), c) turn-over time (h)

hand, showed negative fluxes in all boxes, except 740' and 730' (Fig.6a,b), which suggests that the USLE is largely a sink for free bacteria. A large upward flux is observed at station 750 and much smaller, downward fluxes are seen elsewhere. Boxes 750, 750' and 740 are sites of large rates of loss. An earlier hypothesis (Painchaud et al. 1987) interpreted similar losses as indicative of mortality of freshwater bacteria due to osmotic stress. However, experiments with diffusion chambers showed freshwater bacteria to be unaffected by exposure to water of salinity up to 10 °/oo, which is in the range observed at station 750 (unpubl. data). An alternative hypothesis is that there is excessive predation in most of the estuary, particularly at the head. Growth of estuarine bacteria and predation by microzooplankton have been reported to be in balance (Wright & Coffin 1984), which is probably also the case in the USLE. However, predation by benthic organisms, particularly in the intertidal zone would exert an additional pressure, mainly on bacteria of the upper layer, and could account for the observed loss of bacteria. Turnover time (Fig.6c) was long relative to water residence time, which indicates that biological processes did not dominate their dynamics.

Chl a fluxes are paradoxical: the upper layer was largely a sink and the lower layer, a source (Fig.7a,b). All vertical fluxes were upward. Two processes can account for the negative fluxes observed in the upper boxes: mortality of freshwater phytoplankton and zero net productivity of estuarine phytoplankton. In the Tamar estuary, freshwater phytoplankton appeared to die at salinities > 8 °/oo (Jackson et al. 1987); lysis, decay and predation on the dying cells would thus account for the losses observed seaward of station 750. Moreover, net productivity of estuarine phytoplankton approaches zero (respiration > photosynthesis) when the ratio between photic depth (Z_p) and depth of mixing (Z_m) is between 0.1 and 0.5 (Cloern 1987). In the study area, Z_p/Z_m ranges from 0.2 to 0.5, thus the water column tends to be a sink for phytoplankton production. In these circumstances, the observation that lower boxes are a source of chlorophyll is intriguing. Chl a is a very dynamic variable in the USLE: turnover time (Fig. 7c) was rapid in upstream boxes, particularly in box 740, where it equalled water residence time. Turn-over time lengthened away from the head of the estuary.

Net fluxes of dissolved PO_4 (Fig.8a,b) were small relative to vertical and horizontal fluxes, and turn-over time was long (Fig.8c). These observations indicate that PO_4 distribution was largely controlled by the estuarine circulation. In boxes 750 and 750', small sources of PO_4 were observed. In this area, Lucotte & d'Anglejan (1988) observed PO_4 enrichment of the water column, possibly due to P desorption and diffusion from intertidal marsh sediments. However, this source is relatively small, since PO_4 turnover time in boxes 750 and

750' was much longer than water residence time.

CONCLUSION

The use of a box model has revealed significant biogeochemical processes in the USLE, particularly in the upstream boxes of the study area. This indicates that the zone of salinity transition (0-15°/oo), between Ile d'Orléans and Ile aux Coudres, is very bio- and geochemically dynamic. The observed fluxes have also allowed to propose hypotheses implying exchanges between water column, sediment and intertidal marshes, suggesting that such interactions should be studied in this zone.

REFERENCES

Cloern JE (1987) Turbidity as a control on phytoplankton biomass and productivity in estuaries. Continent Shelf Res 7:1367-1381

Hobbie JE, Daley RJ, Jasper S (1977) Use of Nuclepore filters for counting bacteria by epifluorescence microscopy. Appl environ Microbiol 33:1225-1228

Jackson RH, Williams PJleB, Joint IR (1987) Freshwater phytoplankton in the low salinity region of the River Tamar estuary. Estuar Coast Shelf Sci 25:299-311

Lucotte M, d'Anglejan B (1986) Seasonal control of the Saint-Lawrence maximum turbidity zone by tidal-flat sedimentation. Estuaries 9:84-94

Lucotte M, d'Anglejan B (1988) Seasonal changes in the phosphorus-iron geochemistry of the St. Lawrence Estuary. J Coast Res 4:339-349

Officer CB (1980) Box models revisited. In: Estuarine and wetland processes with emphasis on modelling, p.65. Plenum New York

Painchaud J, Lefaivre D, Therriault J-C (1987) Box model analysis of bacterial fluxes in the St.Lawrence Estuary. Mar Ecol Prog Ser 41:241-252

Painchaud J, Therriault (in prep) Uncoupling between bacteria and phytoplankton in the Upper St. Lawrence Estuary: dominance of bacterial biomass

Parsons TR, Maita Y, Lalli CM (1984) A manual of chemical and biological methods for seawater analysis, 1st edn. Pergamon, Oxford New York Toronto

Serodes J-B, Troude J-P (1984) Sedimentation cycle of a freshwater tidal flat in the St. Lawrence Estuary. Estuaries 7:119-127

Silverberg N, Sundby B (1979) Observations in the turbidity maximum of the St.Lawrence Estuary. Can J Earth Sci 16:939-950

Wright RT, Coffin RB (1984) Measuring microzooplankton grazing on planktonic marine bacteria by its impact on bacterial production. Microb Ecol 10:137-149

CALCULATION OF ESTUARINE CONTAMINANT TRANSPORT WITH STRONG PARTICULATE-DISSOLVED PARTITIONING

R J Uncles and J A Stephens
Plymouth Marine Laboratory, West Hoe, Plymouth PL1 3DH, UK.

Theoretical calculations of the distributions of dissolved, particulate and bed-sediment concentrations of a contaminant within a muddy estuary are presented for the case of strong partitioning between dissolved and particulate phases (K_d = 200 m^3 kg^{-1} (200 l g^{-1})). Results for both continuous fluvial and marine contaminant inputs are averaged over winter and summer periods and are plotted as contaminant-salinity mixing diagrams. Water-column concentrations (contaminant mass per unit volume of water) are dominated by particulate-contaminant levels. For a fluvial input of contaminant there is a marked enhancement of dissolved levels above conservative values at higher salinity, although maximum water column (dissolved plus particulate) concentrations occur in the turbidity maximum region. For a marine input of contaminant, maximum water-column concentrations during summer also occur in the turbidity maximum region and are roughly twice those in coastal waters, contrary to intuition.

INTRODUCTION

The turbidity maximum within some macrotidal estuaries migrates axially in response to changing river flows (Allen *et al.* 1977, 1980). There is an accompanying shift in the sediment pattern, which consists of sediment accumulation in the upper estuary during summer and in the lower estuary in winter and spring (Bale *et al.* 1985). Theoretical studies of these sediment transport processes and their consequences for the transport of a contaminant which partitions between dissolved and particulate forms have been given by Uncles *et al.* (1987, 1988). A typical value of the partition coefficient, K_d , considered was 5 m^3 kg^{-1} (5 l g^{-1}), which is similar to that measured for several trace metals within the estuarine environment (zinc, caesium and cadmium; IMER, 1986/1987; Salomons, 1980).

The objective of this paper is to investigate the seasonal transport and behaviour of a contaminant which has the much higher partition coefficient of 200 m^3 kg^{-1} (typical of, say, Cobalt in the coastal zone; IAEA, 1985). The chemistry of the contaminant is not taken into account. We are concerned only with physical transport processes in a macrotidal estuary. The model estuary is topographically similar to the Tamar, UK. Chemical transformations within the sediment are excluded. Speciation is not treated explicitly. Total dissolved and total particulate states only are considered, and equilibrium exists everywhere between these states, with $C_p = K_d C_d$, where C_p is the particulate concentration (kg kg^{-1}), C_d is the dissolved concentration (kg m^{-3}) and $K_d = 200\ m^3\ kg^{-1}$.

BASIC MODEL AND RESULTS

Details of the model are given in Uncles *et al.* (1987, 1988) and Harris *et al.* (1984). Currents, tidal elevations and sediment transport are determined from a one-dimensional, hydrodynamical model of the Tamar Estuary. A tidally-averaged, one-dimensional model is used to compute the dispersal of a contaminant (Harris *et al.* 1984). Tidally-averaged, longitudinal and vertical fluxes of sediment are used. Within-tide resuspension and deposition are treated as a vertical mixing process superimposed on a tidally-averaged vertical transport.

We consider two types of contaminant input to the estuary. First, a continuous fluvial input at the head, where the dissolved contaminant concentration (C_d) is constant at 10^{-6}kg m^{-3} (1 μg l^{-1}). The coastal value is zero when salinity reaches 34 ppt. Second, a continuous marine input of contaminant through the mouth, where the dissolved contaminant concentration (C_d) is 10^{-6} kg m^{-3} (1 μg l^{-1}) when the salinity reaches 34 ppt. The fluvial concentration (zero salinity) is zero.

Dissolved Concentrations

Figure 1(A) shows the steady-state, conservative (zero partitioning, $K_d = 0$) mixing line (C_d against salinity, line(1)) for a fluvial input of contaminant. A typical summer, spring-tide distribution of suspended particulate matter (SPM) concentration is shown as line (2) in Figure 1(A), on a normalized scale of 0 - 1 (corresponding to 0 - 0.34 kg m^{-3}). The turbidity maximum is a very pronounced feature of the SPM concentrations at low salinity. If the partitioning is now 'switched-on', with $K_d = 200\ m^3\ kg^{-1}$, then the dissolved concentration, C_d, is immediately reduced to the low levels shown in line (3) of Figure 1(A). Almost complete removal of dissolved contaminant occurs in the turbidity maximum region at very low salinity.

The dynamical situation for a fluvial input, depicted in Figure 1(B), is very different. Averaging winter (line 2) and summer (line 3) periods over a four-year simulation period shows that dissolved levels exceed conservative values for most of the salinity range, but that depletion still occurs in the turbidity maximum region.

This depletion is particularly marked in summer owing to the very strong turbidity maxima which can develop then.

The enchancement of dissolved levels at higher salinity is a consequence of particle transport. If SPM concentrations were very small the dissolved contaminant concentrations, C_d, would be approximately conservative (line (1) in Figure 1(B)). Because the partition coefficient, K_d, is constant, the particulate concentrations of instantaneously resuspended SPM (C_p, kg kg^{-1}) would also follow this conservative line, according to $C_p = K_d C_d$. Subsequent down-estuary movement of these suspended

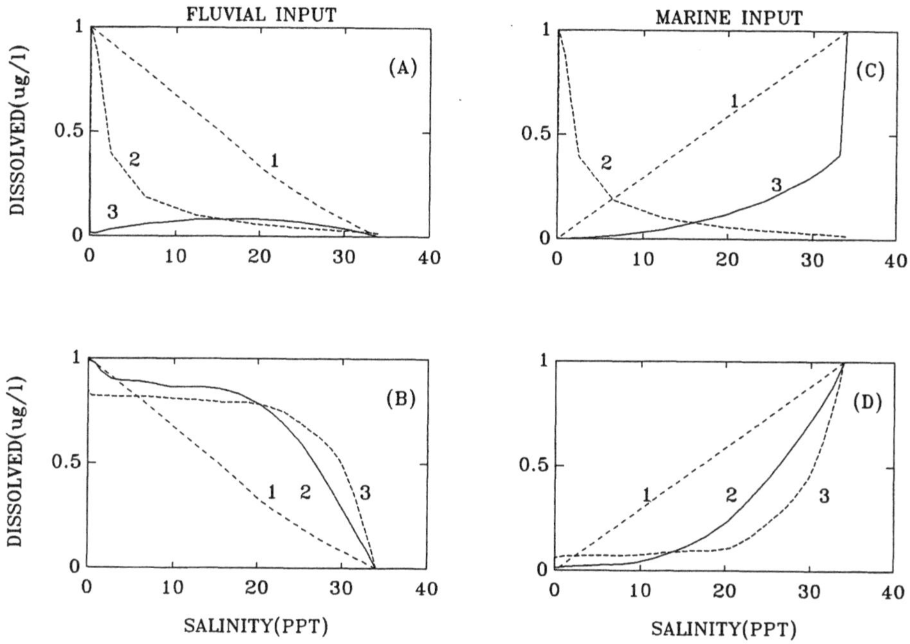

FIGURE 1. Dissolved levels of contaminant against salinity for marine and fluvial inputs. Lines(1) are conservative levels. Lines(2) in (A,C) are normalized SPM levels. Lines(3) in (A,C) are dissolved levels after partitioning. Lines(2) and (3) in (B,D) are winter and summer dissolved levels, respectively.

particles would carry them into higher salinity waters, with a smaller particulate-contaminant loading (smaller C_d implies smaller C_p). The excess contaminant is therefore desorbed to the dissolved phase, which causes an elevation of the dissolved levels. Conversely, up-estuary particulate transport in the turbidity maximum region must lead to adsorption and a lowering of dissolved levels.

Figure 1(C) shows the steady-state, conservative ($K_d = 0$) mixing line for a marine input of contaminant (line (1)). Line (2) again shows typical summer, spring-tide SPM concentrations on a normalized scale of 0 - 1 (corresponding to 0 - 0.34 kg m^{-3}). If the partitioning is now 'switched-on', with $K_d = 200\ m^3\ kg^{-1}$, then the dissolved concentrations, C_d, are immediately reduced to the low levels shown in line (3) of Figure 1(A).

The dynamical situation for a marine input is depicted in Figure 1(D). Averaging winter (line 2) and summer (line 3) periods over a four-year simulation period shows that dissolved levels are less than conservative values for most of the salinity range. In the turbidity maximum region, at low salinity, dissolved levels

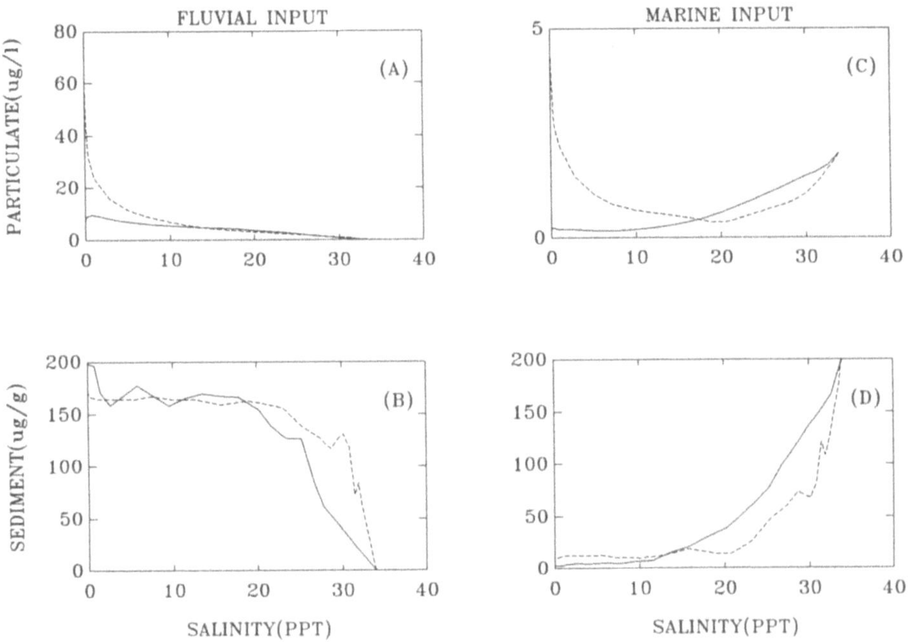

FIGURE 2. Particulate and bed-sediment contaminant levels for marine and fluvial inputs. Continuous and dashed lines are winter and summer levels, respectively.

are elevated above conservative values, the effect being most pronounced in summer owing to the stronger turbidity maxima which can develop then.

The depletion of dissolved levels at higher salinity is a consequence of particle transport. If SPM concentrations were exceedingly small the dissolved contaminant concentration, C_d, would be approximately conservative (line (1) in Figure 1(D)). Because the partition coefficient , K_d, is constant, the particulate concentrations of instantaneously resuspended SPM (C_p, kg kg^{-1}) would also follow this conservative line, according to $C_p = K_d C_d$. Subsequent down-estuary transport of these suspended particles would carry them into higher salinity waters, with a higher particulate-contaminant loading (larger C_d implies larger C_p). The deficit contaminant is therefore adsorbed from the dissolved phase, which causes a depletion of the dissolved levels. Conversely, up-estuary particulate transport in the turbidity maximum region must lead to desorption and an enhancement of dissolved levels.

Particulate and sediment levels

Particulate and bed-sediment concentrations for a fluvial-contaminant input are shown in Figures 2 (A,B) for winter (continuous lines) and summer (dashed lines) conditions. Particulate-contaminant concentrations are shown as contaminant mass per unit water-volume (units of 10^{-6} kg m^{-3} ($\mu g\ l^{-1}$)). Levels rapidly decrease progressing down-estuary (Figure 2(A)), but are generally much higher than dissolved levels. Summer values are much higher than winter values in the turbidity maximum region owing to higher SPM concentrations. Bed-sediment contaminant concentrations (C_b, Figure 2(B), kg kg^{-1}) in the mobile sediment essentially follow dissolved levels (Figure 1(B)) and are of order K_dC_d.

Particulate and bed-sediment concentrations for a marine-contaminant input are shown in Figures 2 (C,D) for winter (continuous lines) and summer (dashed lines) conditions. For the particulate-contaminant concentrations (shown in Figure 2(C) as contaminant mass per unit water volume), summer levels (dashed line) increase progressing towards the head. The total water column concentrations (sum of dissolved and particulate values) maximize in the turbidity maximum region, owing to the adsorption of contaminant on the high concentrations of suspended particles there and the presence of the bed-sediment shoal in the shallow upper reaches. The bed-sediment contaminant concentrations decrease towards the head (Figure 2(D)) and essentially follow dissolved levels (Figure 1(D)), satisfying $C_p = K_dC_d$.

CONCLUSIONS

For a continuous input of contaminant which has a relatively high K_d of 200 m^3 kg^{-1} (200 l g^{-1}) the calculations show that water-column concentrations are dominated by particulate-contaminant levels. For a fluvial input, we conclude that:

(1) Dissolved levels greatly exceed conservative (zero decay and partitioning) values in the higher salinity range. Thus, suspended and deposited fine sediment act here as contaminant sources to the dissolved phase.

(2) Dissolved levels are depleted in the low salinity, turbidity maximum region, especially in summer when a strong turbidity maximum develops.

(3) Particulate-contaminant concentrations (mass per unit volume of water) decrease towards the mouth and have highest values in the turbidity maximum region during summer.

For a continuous marine input of contaminant with the same K_d:

(1) Dissolved levels are greatly depleted below conservative values in the higher salinity range, with suspended and deposited sediment acting as sinks to the dissolved phase.

(2) Dissolved levels are elevated in the low salinity region, especially in summer when strong up-estuary transport of particles into the turbidity maximum region occurs.

(3) Particulate-contaminant concentrations (mass per unit volume of water) increase markedly in the upper reaches during summer and greatly exceed input concentrations. Maximum values occur in the turbidity maximum region.

REFERENCES

Allen, G P, Sauzay G, Castaing P, Jouanneau J M (1977) Transport and deposition of suspended sediment in the Gironde Estuary, France. In: Wiley M (ed) Estuarine Processes. Academic Press, New York, p 63

Allen G P, Salomon J C, Bassoullet P, Du Penhoat Y, De Grandpre C (1980) Effects of tides on mixing and suspended sediment transport in macrotidal estuaries. Sedimentary Geology 26: 69-90

Bale A J, Morris A W, Howland R J M (1985) Seasonal sediment movement in the Tamar Estuary. Oceanologica Acta 8: 1-6

Harris J R W, Bale A J, Bayne B L, Mantoura R F C, Morris A W, Nelson L A, Radford P J, Uncles R J , Weston S A, Widdows J (1984) A preliminary model of the dispersal and biological effect of toxins in the Tamar Estuary, England. Ecological Modelling 22: 253-284

IAEA (1985) Sediment K_ds and concentration factors for radionuclides in the marine environment. International Atomic Energy Agency, Technical Reports Series No. 247, p14

IMER (1986/7) Report for 1986/87 of the Institute for Marine Environmental Research. Plymouth Marine Laboratory, Plymouth PL1 3DH, UK, p41

Salomons W (1980) Adsorption processes and hydrodynamic conditions in estuaries. Environ Technol Lett 1: 356-365

Uncles R J, Woodrow T Y, Stephens J A (1987) Influence of long-term sediment transport on contaminant dispersal in a turbid estuary. Continental Shelf Research 7: 1489-1493

Uncles R J, Stephens J A, Woodrow T Y (1988) Seasonal cycling of estuarine sediment and contaminant transport. Estuaries 11: 108-116.

THE SPECIFICATION OF SHEAR STRESSES IN ESTUARY MODELS

Roy E Lewis
ICI Brixham Laboratory
Brixham Devon TQ5 8BA, UK

SYNOPSIS

The widespread use of a coefficient of eddy viscosity, Nz, to describe the vertical distribution of shear stress in estuary models may not be justified. The concept of shear stress has a more sound physical basis than Nz and offers more scope since stresses can be determined directly from measurements in estuaries. Above the constant stress layer at the bed, the stress can be considered to be made up of a boundary layer drag which decreases linearly with height and a variable internal stress associated with accelerations and density structure in the flow.

THEORETICAL APPROACHES

Shear stress, T, at any point in an estuary is related to the local gradient of velocity by the coefficient of eddy viscosity:

$$T = p \, Nz \, du/dz \qquad (1)$$

where the stress is defined as being positive in the direction of mean flow.

Bowden and Hamilton (1975) have given a summary of formulae commonly used to describe the variation of Nz over depth, these being derived from considerations of mixing length or from observed velocity profiles. These expressions have the general form:

$$Nz = k \, u^* \, z \, f(z) \qquad (2)$$

where k is Von Karman's constant (0.41), u* is the friction velocity and z is the height up from the bed. The function f(z) may be constant, linearly decreasing or a power law of z.

LIMITATIONS OF THEORETICAL FORMS FOR Nz

There are a number of weaknesses in the physical concept of Nz and in the way its functional form is deduced. Equation (1) assumes that momentum is a conservative transferable property which implies that pressure fluctuations do not affect the mean transfer of momentum. In a turbulent flow the pressure fluctuations are far from negligible as is demonstrated in Fig 1 which shows that, in the Tees estuary, the topography causes pressure variations normal to the bed which result in matching internal waves on a density interface. When du/dz=0, equation (1) implies that there is no net transfer of momentum. A zero gradient of velocity is known to be a common occurrence in estuaries, particularly at intermediate depths on flood tides, and the equation must be regarded as untenable under these circumstances.

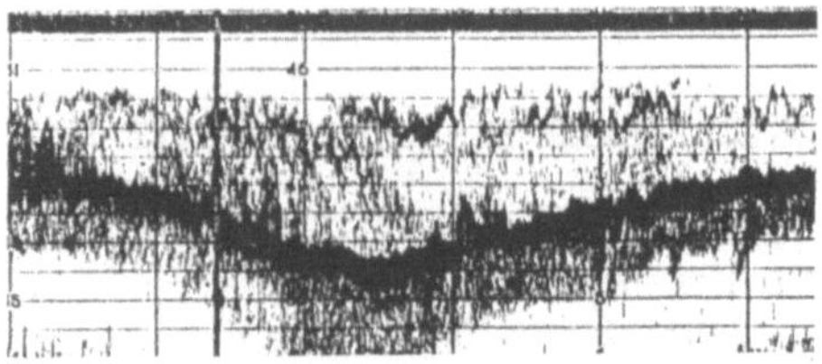

Fig 1 Interfacial waves matching topographic variation - depth lines correspond to 2m intervals, vertical lines correspond to approximately 170 m intervals.

Although the formulae for Nz as expressed in equation (2) have proved of use, they are based on such assumptions as constant stress or logarithmic velocity profiles over the whole depth of water and steady state conditions of flow. Even assuming that an approximate steady state exists at times of maximum flood or ebb, estuary profiles rarely display a logarithmic profile over depth (corresponding to an approximate 1/7 th power law). Velocity profiles in the central reach of the Tees estuary correspond to approximate 1/8 th and 1/3 power laws on flood and ebb tides respectively (Lewis & Lewis, 1987). Furthermore, there is considerable evidence that at times of maximum current, shear stress profiles increase almost linearly with depth (Bowden et al, 1959). Only near the bed are shear stresses nearly constant over depth and these are associated with logarithmic profiles, even under strong pressure gradients (Duncan et al ,1970 :p337).

INTERNAL AND EXTERNAL STRESSES

Eddy viscosity has proved to be a characteristic of the flow itself rather than the fluid and is subject to wide variations in space and time. Unlike Nz values, shear stresses within a flow can be estimated directly, either from measurements of the turbulent components of velocity or from surface slopes using corresponding observations of currents and density structure. If the causes of these stresses can be understood, there is a better prospect of correctly modelling velocity shears and the associated dispersion processes.

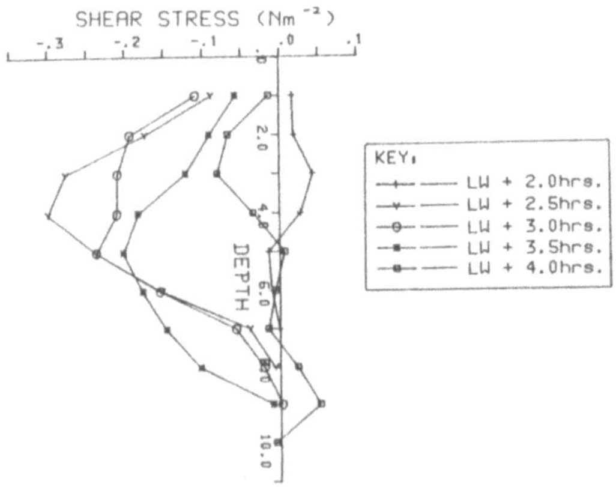

Fig 2 Internal shear stresses on an ebb tide - Tees estuary at Smith's Dock

By splitting shear stress into terms corresponding to internal and external components, the causes of stress can be investigated in more detail (Abraham, 1980). Data for the Tees estuary were analysed on the assumption that the bed generated external stress, Tb, was constant within a near bed layer and then decreased linearly up to the surface. The residual internal stresses determined by this approach (Fig 2) approximate to a parabolic profile of so that,

$$T = Tb\ (1-z/h) + 4Ti\ z/h(1-z/h) \qquad (3)$$

where Ti is the maximum internal stress which generally occurs at about mid-depth. Values of Ti derived from records at five stations in the Tees ranged between 0.21 and 0.31 Nm^{-2} on a flood tide and between 0.05 and 0.34 Nm^{-2} on an ebb. No clear pattern for the temporal variation in Ti could be established from the existing data, although high values of Ti at times of maximum current did appear to be associated with spatial accelerations.

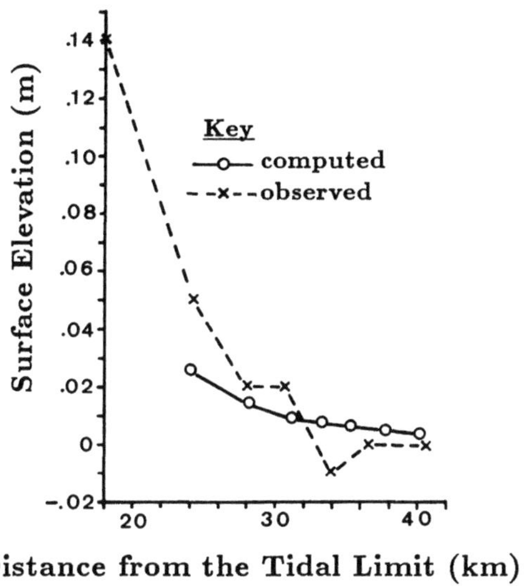

Fig 3 Observed and predicted tidal mean surface slopes in the Tees.

Such accelerations give rise to pressure gradients which apparently control the boundary drag and these processes can explain the observed asymmetry in drag coefficients determined on flood and ebb tides (Lewis & Lewis, 1987). For example, a tidal mean can be assumed to approximate to steady state flow and this concept has been used to derive a theoretical relationship between surface slope and density gradient, assuming equation (1) with a constant Nz over depth (Officer, 1976: p116). Fig 3 shows that the actual tidal mean slope along the Tees is greater than theory predicts, indicating that the theoretical relationship is an oversimplification of the stress variation. More detailed analyses have revealed that the difference is largely due to the asymmetry in the drag coefficients.

REFERENCES

Abraham, G (1980) On internally generated estuarine turbulence Second International Symposium on Stratified Flows, Norwegian Institute of Technology, Trondheim, Norway.

Bowden,K.F., Fairbairn,L.A. & Hughes,P. (1959) The distribution of shearing stresses in a tidal current Geophysical Journal of Royal Astronomical Society, 2, N4, 288-305

Bowden,K.F. & Hamilton,P. (1975) Some experiments with a numerical model of circulation and mixing in a tidal estuary Estuarine and Coastal Marine Science, 3, 281-301

Duncan,W.J.,Thom,A.S. & Young,A.D. (1970) Mechanics of Fluids Edward Arnold Great Britain

Lewis,R.E. & Lewis,J.O. (1987) Shear stress variations in an estuary Estuarine, Coastal and Shelf Science, 25, 621-635

Officer,C.B. (1976) Physical Oceanography of Estuaries (and Associated Coastal Waters) John Wiley, New York

CHAPTER III

INTEGRATED CONCEPTS, STRATEGIES

THE BILEX CONCEPT
- RESEARCH SUPPORTING PUBLIC WATER QUALITY SURVEILLANCE IN ESTUARIES -

W. Michaelis
Institute of Physics, GKSS Research Centre Geesthacht
D-2054 Geesthacht, FRG

ABSTRACT

Estuaries are characterized by pronounced spatial heterogeneities and temporal variabilities in the composition of the water body. These features make water quality assessment and flux measurements a difficult task. Well-aimed research may effectively support the functions of the public authorities. Field campaigns performed by the GKSS Research Centre are understood along these lines. The concept is described and typical applications are discussed: investigations of the representativeness of stationary data recording stations, the validation of numerical transport models, studies of the fluxes of particulate matter and pollutants during intensive field campaigns, and the determination of the long-term discharge over selected cross-sections.

1. INTRODUCTION

In contrast to the nearly stationary conditions in the upper course of a river, strongly varying effects determine the transport of pollutants in the estuary (Fig. 1). The amount and sign of the flow, and thus also the water level and the river width, vary periodically with the tides. Turbulent diffusion is superimposed on the purely advective transport. Sedimentation of suspended particulate matter, resuspension of this material, and possible erosion of consolidated sediments change with the flow and influence the particulate matter concentration in the water column and, consequently the pollutant load, since significant amounts of heavy metals, chlorinated hydrocarbons and other harmful substances are bound to the particulate phase. Flocculation, sorption, desorption, and disintegration processes control the interaction between suspended matter and the liquid phase. Uptake, release, and detritus characterize the interrelation between organisms and the water body. Trace substances may be remobilized from the sediment as a function of time. The concurrence of fresh and sea water leads to stratification and mixing phenomena which produce vertical and horizontal salinity gradients, which in turn are subject to the influence of the tides. The freshwater plankton dies off with increasing salinity, whilst the marine plankton dies with the growing influence of fresh water. A characteristic feature of estuaries is the occurrence of a turbidity zone with a high content of dead organic material. Gradients in the chemical environment, e.g. in the salt or oxygen content, may change the speciation of trace elements with the possible consequence that, for instance, the transport mechanisms vary with time and location. Furthermore, dredging activities and atmospheric deposition processes may influence the composition of the water body.

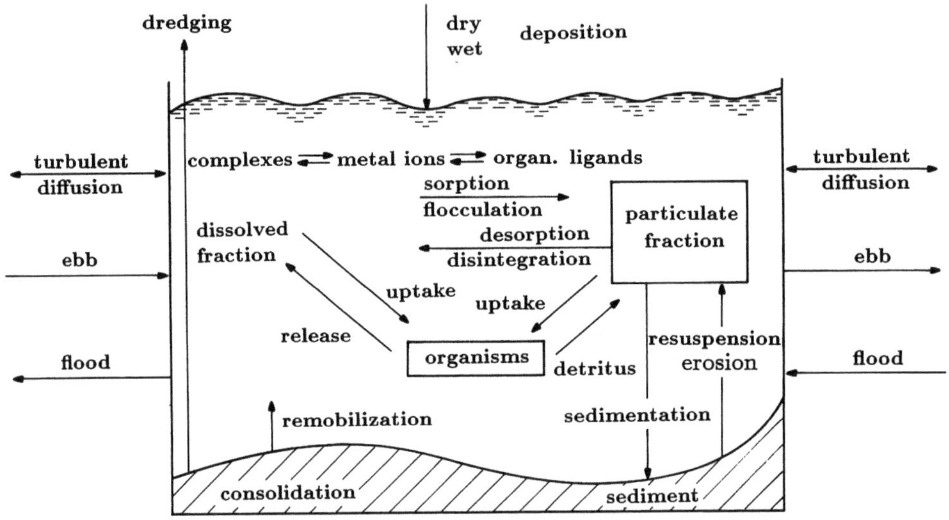

Figure 1: Transport-relevant processes in the box of a tidal river

These complex phenomena result in pronounced spatial heterogeneities and temporal variabilities which lead to severe difficulties in assessing the water quality and in measuring fluxes in tidal rivers. On the other hand, a knowledge of the transport processes operating is both of ecological and economic importance. The transport of large amounts of suspended matter and sediment causes considerable costs for dredging out the fairways and harbour basins. Ecologically, the ability to describe the relevant processes with the final goal of establishing a balance is a prerequisite for locating sinks and sources, for predicting pollution trends, for assessing the self-purification power, for decision-making in water quality management, and for determining the net discharge into the sea.

2. THE BILEX CONCEPT

The public authorities have to provide solutions to these various problems on the basis of economical strategies and easy-to-handle data recording systems. It is a challenge to estuarine research to support these functions by establishing the necessary fundamentals for optimum instrumentation and strategies. Examples of such activities are: (i) investigations of the representativeness of moored stationary platforms, (ii) validation of numerical transport models, (iii) studies of the fluxes of suspended particulate matter and trace substances during intensive field campaigns, and (iv) investigations of the long-term discharge over selected cross-sections.

For performing research of this kind, a methodology has been developed at the GKSS Research Centre which combines in an effective way hydrographic measurements using the moving boat technique and moored platforms with numerical transport models and trace analytical procedures [1 - 3].

Field data are obtained by means of the hydrographic measuring and sampling system HYDRA [4]. Operated on board a ship, it comprises two main components. A sensor package mounted below the water surface on a cantilever at the bow of the ship is used for horizontal profiling of flow velocity, optical attenuation, temperature, conductivity, oxygen content and water depth (Fig. 2). Current determination is accomplished by combining a two-dimensional ultrasonic flow meter based on the entrainment effect with a radio position finder which simultaneously measures the distance variations relative to two land-based stations by utilizing the Doppler effect. A nearly identical sensor arrangement, with an additional pressure gauge, is mounted on a probe which can be lowered to allow vertical profiles to be measured with the dynamically positioned ship. Similar equipment is available for two stationary platforms.

Suspended particulate matter concentrations are derived from continuous light attenuation measurements. Repeated calibration is performed by taking water samples during selected vertical profiling cycles, and by separating the particulate material immediately using 0.45 µm pressure filtration on board the ship. Settling velocities of the suspended particles are determined in the field by means of the Owen tube technique [5, 6]. The loaded filters and the filtrates are subsequently analysed for trace substances at the Geesthacht laboratories. For heavy-metal analysis several complementary methods are used in order to ensure the required precision and accuracy [7]: total-reflection X-ray fluorescence analysis [8], neutron activation analysis, atomic absorption spectrometry, and anodic stripping voltammetry. Organic pollutants are analysed by gas chromatography/mass spectrometry [9].

The measuring campaigns are performed according to the BILEX concept. The German acronym BILEX stands for "Bilanzierungs-Experiment" which may be interpreted as pollutant inflow-outflow balancing for a selected river section (Fig. 2). Hydrographic and analytical data are determined with high spatial and temporal resolution on the open boundaries of the reaches to be studied. These data then serve as boundary conditions for the application of numerical models which simulate the transport processes along the river section concerned. Usually the measurements extend over several tidal periods. They are repeated under quite different hydrographic conditions and thus provide a growing insight into the processes involved. Interpolation between these intensive investigations is achieved by long-term measurements using automatic stationary platforms. Of course, the representativeness of the data obtained at such stations requires careful examination.

For the simulation of the transport of water, particulate matter and trace substances, models of varying complexity are the subject of current research. The one-dimensional model FLUSS [10] has been applied with great success to the fluxes of suspended particulates, heavy metals, and chlorinated hydrocarbons both in the Elbe and the Weser estuaries [2, 11]. Two- and three-dimensional models are currently under development.

In the series of investigations carried out according to the BILEX concept, a strategy is being pursued in which the river sections under study are gradually shifted downstream corresponding to increasing complexity (Fig. 3). The first experiment was performed in 1982 on the Lower Elbe River [1] with boundaries only 1.2 km apart and with rather easy-to-survey sources, sinks and topography. The

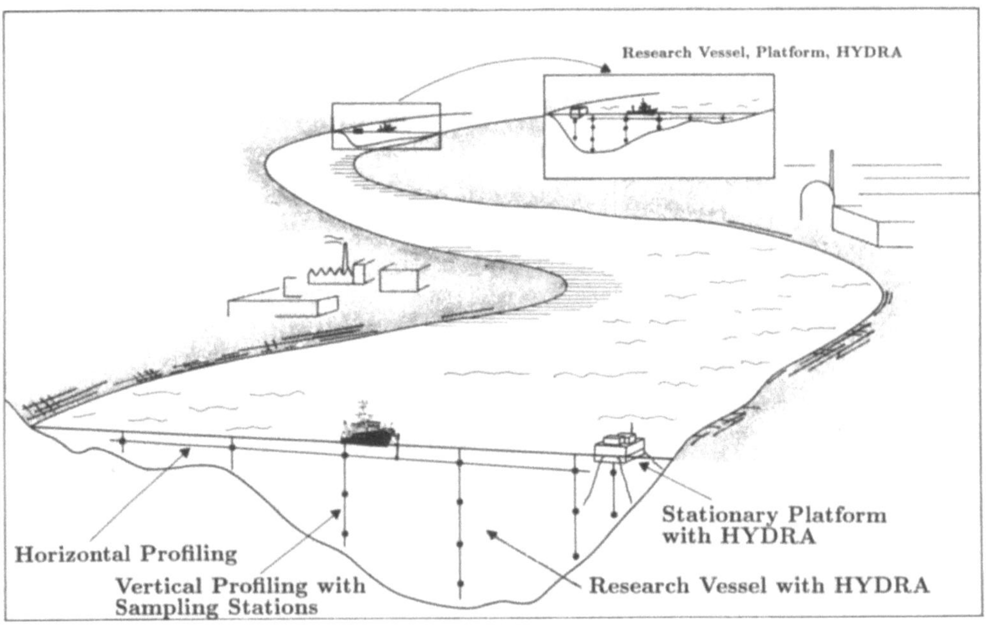

Figure 2: Schematic representation of field campaigns according to the BILEX concept

goal of this campaign was essentially to test the experimental and theoretical tools developed up to that time. In principle, they stood the test and the experience gained gave hints for possible improvements and how to proceed in further investigations. The next studies, BILEX '84 and BILEX '85 [2], took place on the 22 km long river section (A to D in Fig. 3) between the weir at Geesthacht and Oortkaten, which is located close to the entrance to the Hamburg Harbour. A study on the transport of particulate and dissolved matter through the harbour was performed in 1986 in close cooperation with several public authorities [3]. The investigation area of BILEX '89 and '90 is on the Lower Elbe River about 40 to 60 km downstream from Hamburg.

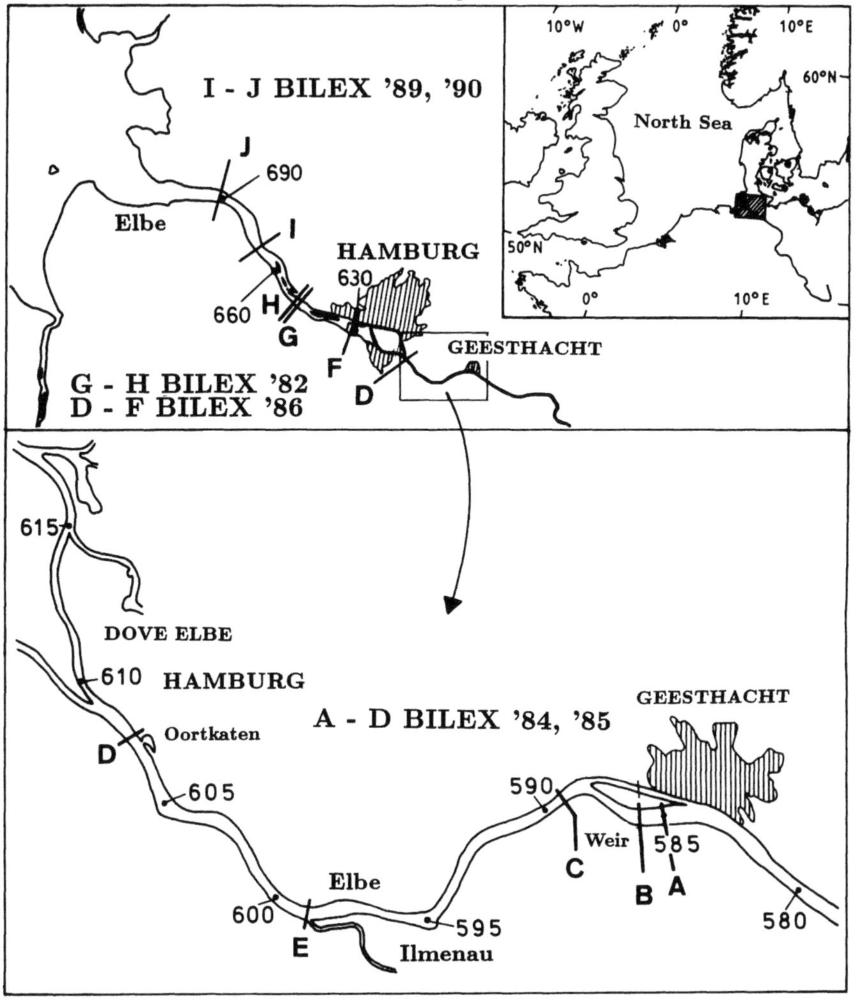

Figure 3: Investigation sites of the BILEX campaigns

3. TESTING THE VALIDITY OF DATA OBTAINED AT STATIONARY RECORDING STATIONS

On cross section D, at the northern border of the Elbe fairway (Fig. 3), a permanent automatic measuring platform (AMO) has been operated for several years by the Technical Department of the Hamburg Harbour authorities [12]. One of the goals of BILEX '84 and '85 was to find out to what extent the platform yields data that are representative for the cross-section. The timing of these experiments ensured that the studies were performed under quite different hydrographic conditions (water discharge 342 and 1165 m^3/s, respectively).

The results for the cross-section-averaged values of residual flow velocity, v_r, and net flux of suspended particulate matter, q, are summarized in Table 1 and compared with the local data at the site of the platform. It can be seen that AMO overestimates these quantities by approximately 10 to 25 %, irrespective of the discharge conditions. Similar studies have been performed on the Lower Elbe River [13].

Table 1: Residual flow velocity, v_r, and net flux of suspended particulate matter, q, at cross-section D (Fig. 3): Comparison of cross-sectional averages (Q) with the local data at the site of AMO.

	v_r[m/s]		q[g/m²s]	
	BILEX '84	BILEX '85	BILEX '84	BILEX '85
AMO	0.22	0.72	4.0	27.5
Q	0.17	0.61	3.6	25.5
Q/AMO	0.77	0.85	0.90	0.93

A crucial check of the moving boat technique as an essential tool in such investigations is the comparison of the water discharge as obtained by HYDRA and the official water-gauges. The results for the campaigns in 1982, 1984 and 1985 are as follows (official values in brackets): 355 (362), 310 (345) and 1100 (1165) m^3/s. Hence one may conclude that the experimental errors are about 10 % at the most.

4. VALIDATION OF PARTICULATE MATTER AND POLLUTANT TRANSPORT MODELS

From the hydrographic data and the trace analytical results for the liquid and the particulate phase, the transport rates during a BILEX campaign can be calculated. Moreover, the experimental data obtained on one of the open boundaries may be used to simulate the fluxes at the other cross-section by means of numerical models [2]. For instance, the data from boundary A in Fig. 3 allow the cross-section-averaged particulate matter and total heavy metal concentrations at D, 22 km downstream to be calculated. Examples are given in Figs. 4 and 5 in which the calculat-

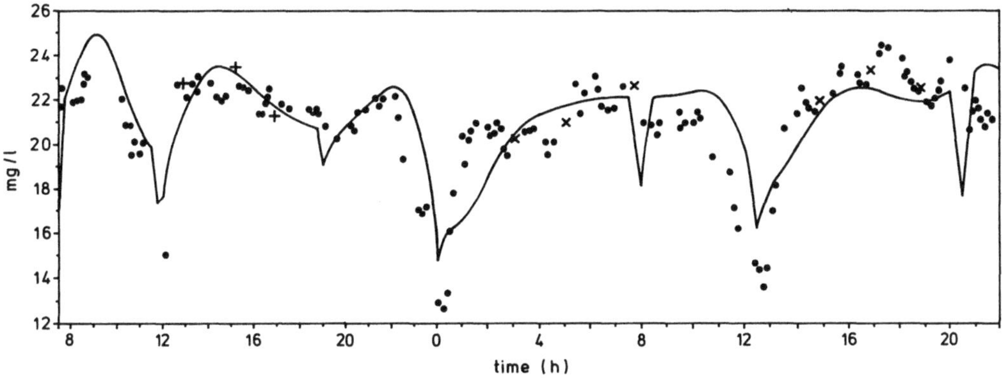

Figure 4: Modelling the transport of suspended solids: Comparison of experimental data (crosses: water sampling; full circles: light attenuation measurements) and model predictions for the content of suspended particulate matter at cross-section D during BILEX '84

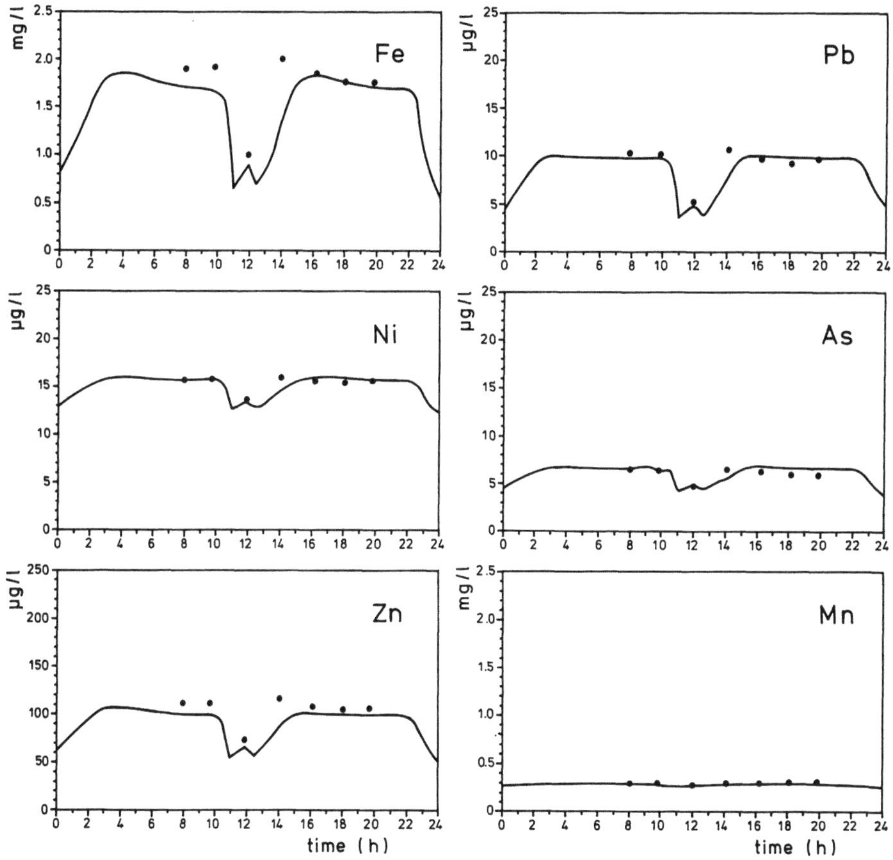

Figure 5: Temporal variation of the cross-section-averaged total trace element concentration at D (Fig. 3): Comparison of analytical data with model calculations (BILEX '85)

ed temporal variations of these parameters are compared with the measured data. The agreement is quite promising in view of the length of the reach and the complexity of the transport processes involved. It is found that the results for the suspended matter are strongly influenced by interactions at the sediment-water interface, and by the non-consolidated sediment covering on the river bottom [10, 11]. The temporal variation of the trace metal concentration strongly depends on the partitioning of the element considered between the liquid and the particulate phase (Fig. 5).

5. TRACE METAL DISCHARGE DURING BILEX CAMPAIGNS

Since the campaigns in 1984 and 1985 were performed during different seasons (September and April), the results provide information on the influence of the seasons and thus the water discharge on the transport processes (cf. section 3). It was found that the specific trace element load of the particulate matter is significantly enhanced in spring for Sc, Ti, V, Cu, Zn, Zr, Cd and Sb. The values for Mn and As, however, are lower than in autumn. The concentrations of Mn, Fe and As in the filtrates are obviously reduced at high water discharge by dilution effects.

In Table 2 daily discharge values during the experiments are summarized for a sample of a total of 20 elements investigated [2]. The data reflect the different levels both in water discharge, suspended matter concentration, and element content in the two phases. For comparison, model predictions have been included in the table (cf. section 4).

Table 2: Daily trace metal discharge through the Elbe cross-section at Oortkaten (D, Fig. 3) during BILEX '84 and '85 in kg/d. DL = detection limit.

Element	BILEX '84				BILEX '85			
	Experiment			Model	Experiment			Model
	Particulate	Dissolved	Total	Total	Particulate	Dissolved	Total	Total
Ni	50	360	410	511	425	1 005	1 430	1 550
Zn	1 000	700	1 700	1 720	8 200	2 500	10 700	10 000
As	45	131	176	240	296	250	546	650
Sb	1.7	30	32	-	16	80	96	-
Hg	15	1	16	-	125	2.5	128	-
Pb	150	< DL	150	150	1 096	< DL	1 096	1 000

6. LONG-TERM DISCHARGE

A knowledge of time-integrated fluxes is an indispensable basis for environmental policy. This refers, for instance, to transboundary fluxes, to the discharge of cohesive sediments into harbour areas, or to the discharge of pollutants into the sea. Typical hydrographic situations are considered during the BILEX campaigns which give a detailed insight into the relevant processes and their interactions. Of course, such experiments can only be performed over limited periods. Therefore, temporal interpolation is achieved on the basis of properly contrived single measurements and sampling, preferably by making use of stationary platforms at carefully chosen locations. As an example, Table 3 summarizes a few data derived in this way on the Elbe cross-section D (cf. Fig. 3) over a one-year period which includes BILEX '84 and '85.

The data listed in the table approach very well the input to the Hamburg Harbour. Since during the 1984 experiment a considerable part of the water body from the harbour passed the cross-section with the flood tide, it could be shown that influxes in this area only contribute an additional 5 to 15 % to the heavy metal load of the river, depending on the element considered. This result confirms conclusions drawn by the Hamburg Environmental Authority on the basis of the official emission inventory. It is expected that the data in Table 3 are also not excessively far from those representing the input into the North Sea. A thorough examination of this aspect is the subject of the BILEX campaigns in 1989 and 1990. The results will be of considerable importance for the next North Sea Conferences.

Table 3: Annual discharge [t/a] in the tidal Elbe River on cross-section D (Fig. 3) during the period September 1, 1984, to August 31, 1985. SPM = suspended particulate matter (dry substance).

SPM	493 000	Cu	170
Al	18 000	Zn	1 530
Sc	3.6	As	110
Ti	1 120	Se	9
Cr	202	Cd	9
Mn	6 100	Sb	19
Fe	24 000	Hg	17
Ni	252	Pb	134

REFERENCES

[1] Michaelis W, Knauth H-D (eds) (1985) Das Bilanzierungsexperiment 1982 (BILEX '82) auf der Unterelbe. GKSS 85/E/3

[2] Michaelis W, Fanger H-U, Müller A (eds) (1988) Die Bilanzierungsexperimente 1984 und 1985 (BILEX '84 und BILEX '85) auf der oberen Tideelbe. GKSS 88/E/22

[3] Fanger H-U, Kappenberg J, Männing V (1989) A study on the transport of dissolved and particulate matter through the Hamburg Harbour. This volume, chapter IV

[4] Fanger H-U, Kuhn H, Maixner U, Milferstädt D (1989) The hydrographic measuring system HYDRA. This volume, chapter V

[5] Owen MW (1976) Determination of the settling velocities of cohesive muds. Hydraulics Research Report IT 61, Wallingford, UK

[6] Puls W, Kühl H, Lobmeyr M, Müller A, Schünemann M (1989) Investigations on suspended matter transport processes in estuarine and coastal waters. This volume, chapter IV

[7] Prange A, Niedergesäß R, Schnier C (1989) Multielement determinaton of trace elements in estuarine waters. This volume, chapter X

[8] Michaelis W, Prange A (1988) Trace analysis of geological and environmental samples by total-reflection x-ray fluorescence spectrometry. Nucl Geophys 2: 231-245

[9] Sturm R, Knauth H-D (1989) Use of nonvolatile chlorinated hydrocarbons in suspended particulate matter as anthropogenic tracers for estimating the contribution of the rivers Weser and Elbe on the pollution of the German Bight. This volume, chapter X

[10] Kunze B, Müller A (1988) in Ref. [2]

[11] Müller A, Grodd M, Weigel P (1988) Lower Weser monitoring and modelling. This volume, chapter VI

[12] Neumann LJR (1985) AMO - Die automatische Meßstation Oortkarten - Meßsystem zur Sammlung von Schwebstoffproben und Messung von Begleitparametern. Die Küste 42: 151-161

[13] Kappenberg J, Fanger H-U, Männing V, Prange A (1989) Suspended matter and heavy metal transport in the lower Elbe River under different flow conditions. This volume, chapter IV

WASTE DISPOSAL AND THE ESTUARINE ENVIRONMENT

J Orr, D T E Hunt and T J Lack Water Research Centre, Henley Road, Medmenham, PO Box 16 Marlow, Buckinghamshire. SL7 2HD

ABSTRACT

Waste discharges to UK estuaries and coastal waters are controlled by the application of the Environmental Quality Objective/Environmental Quality Standard (EQO/EQS) philosophy, which takes into account the dispersive capacity of the receiving waters when Discharge Consent Conditions are set. This paper describes the rationale of the approach and the mechanism whereby it is employed, illustrated by reference to a case study. Current research aimed at providing robust yet sensitive procedures for assessing the environmental impact of complex discharges is also described.

INTRODUCTION

UK estuaries are used extensively for the disposal of industrial effluents and for the marine treatment of sewage through sea outfalls. To avoid possible conflict with other uses of estuaries - such as fisheries, shell fisheries, recreation and amenity - and to ensure that the requirements of legislation are met, these disposal practices must be carefully controlled and monitored. The cornerstone of the approach adopted in the UK to manage discharges to tidal waters is the Environmental Quality Objective/Environmental Quality Standard (EQO/EQS) philosophy. Under this philosophy the ability of the receiving waters to dilute and assimilate contaminant loads is taken into account when Discharge Consent Conditions are set; however, the application of the principle of "Best Available Technology Not Entailing Excessive Cost (BATNEEC)" ensures also that the industrial community does not have the freedom to contaminate the estuary to the limit of its assimilative capacity.

THE CONTROL OF DISCHARGES TO COASTAL WATERS BY THE EQO/EQS APPROACH

The basis of the EQO/EQS approach is that a Use-related EQO (eg "The protection of marine life") is established for the receiving waters and a related EQS defines the permitted concentration of the contaminant in those waters, providing a yardstick for pollution control [1]. An EQS may be set as a maximum average concentration over any period from a tidal cycle to a year. If an area of water has more than one designated use - and therefore more than one EQO - then the most stringent EQS relevant to those EQOs will apply. The general requirement is that the receiving waters should support healthy biological systems compatible with the characteristics of the water body concerned, in line with the EC Dangerous Substances Directive, which requires that pollution be eliminated or reduced. The EQO/EQS approach aims to improve water quality progressively, exploiting technical advances where practical. Discharge control must not be relaxed simply because the EQS is easily met.

For all discharges whose concentration of the contaminant exceeds the EQS there will exist a region in the receiving water where the latter cannot be met. This region is known as the Mixing Zone (MZ), and it is of fundamental importance for the EQO/EQS philosophy. (Although a formal MZ will not exist unless an EQS is in operation, the underlying concept of assessing and controlling environmental concentrations and impact should also apply when Uniform Emission Standards are adopted).

THE MIXING ZONE

The MZ is simply that volume of the receiving water in which excedence of the EQS is acceptable to the regulatory authority - in practice, it is treated as an area (Figure 1). If the EQS is set as an annual average concentration, the MZ is defined as the area outside which the annual average concentration must be below the EQS.

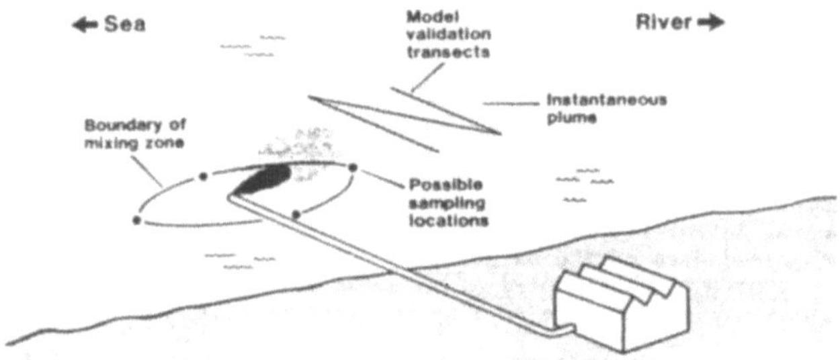

Figure 1. The Mixing Zone Concept

In general, the setting of the MZ is a local decision, inevitably involving a degree of judgement. Costs, public reaction, the views of industry and relevant Government guidance must all be considered. There should be no adverse effect on the general and visible amenity, and any area of acute toxicity near the the discharge must be minimised - within "reasonable cost"- by using the best available diffuser technology and by limiting the concentration, as well as the load, of the contaminant in the discharge. Moreover, consideration must also be given to the possibilities of build-up of potentially toxic substances in organisms, of "overlapping" effects from adjacent discharges and of future requests to discharge to the same water body.

The MZ should not normally be allowed to impinge on a shoreline with public access. It should not produce a public health hazard or an aesthetically objectionable effect, impinge upon a Site of Special Scientific Interest or interfere with any other recognised Uses of the water body. MZs should not occupy more than one third of an estuary's width at any point, to allow the passage of fish and other mobile organisms.

APPLICATION OF THE EQO/EQS APPROACH TO THE CONTROL OF DISCHARGES TO TIDAL WATERS.

In estuaries, monitoring the MZ boundary for compliance is impractical, because tidal movement of the contaminant plume causes the concentration at the boundary to fluctuate wildly; an enormous number of samples would have to be taken to define the mean concentration at the boundary with useful accuracy. Figure 2 outlines a practical alternative which is described in detail below.

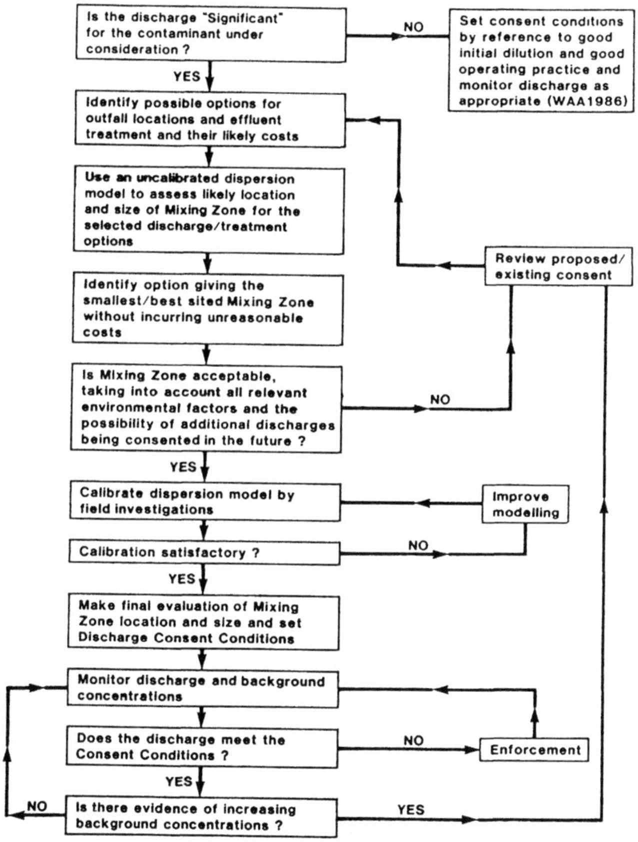

Figure 2. Strategy for the application of the EQO/EQS approach to pollution control in marine waters

MZs will need to be set only for the relatively small number of discharges with an important potential impact, as judged by an agreed screening procedure [2]. What represents an acceptable MZ for an important discharge is a local decision, made within national guidelines under the principle of BATNEEC.

In applying BATNEEC, a balance must be struck between the wish to minimise the MZ, by outfall design and/or effluent treatment, and the associated costs. The regulatory authority needs to consider the nature of the contaminant and the possible impact of the MZ on the receiving waters, and have information on the costs of achieving MZs of different sizes. As shown in Figure 2, a dispersion model will help to identify the best option.

The EQO/EQS/BATNEEC approach does not sanction increasing discharges to tidal waters until their diluting and assimilating capacity is exhausted. Rather, it combines responsible use of that capacity with a requirement to minimise, within reasonable cost, the environmental consequences of discharges.

SETTING THE MIXING ZONE AND CONSENT CONDITIONS

In view of the large uncertainties in model predictions, substantial "safety factors" must be incorporated, and model calibration carried out, when MZs and Consent Conditions are set. Calibration involves determining the contaminant concentration along transects across the plume, or using a tracer as a surrogate - see Figure 1. The proposed Consent Conditions are then adjusted to allow for any consistent bias in the model predictions. The calibration need be repeated only if a subsequent change in hydrographic conditions would be likely to cause significant changes to the MZ.

Consent Conditions are then set - in terms of both a daily average and a daily maximum load - from the average load estimated by modelling to be consistent with the allowable MZ. As noted previously, the Consent Conditions must also specify a concentration limit, to minimise any region of acute toxicity.

MONITORING

Because there is no prospect of making an annual assessment of compliance by routine monitoring of the contaminant concentration at the MZ boundary, day-to-day pollution control is exercised by monitoring the discharge itself. Environmental monitoring is, however, also essential to ensure that the background levels of the contaminant do not increase beyond those prevailing when the MZ and Consent Conditions were set. Obviously, this monitoring must be performed at points distant from the MZ.

CASE STUDIES

A case study is currently being undertaken to assess the practicality of the approach described above, involving a mathematical dispersion model produced for an outfall discharging effluent from an industrial plant on the Severn estuary. A 1250m long outfall continually discharges the plant effluent, at an average rate of 14.8Ml/d. This effluent has a pH range of 3.4 to 10.1 (depending on the operation of different chemical processes within the works), an average BOD of 207mg/l and contains a number of List 1 and 2 metals and significant concentrations of ammonia.

A mathematical dispersion model was constructed for the discharge, using a 125m x 125m grid based on a 500m x 500m grid hydrodynamic model. This model was used to predict concentrations of the effluent over the estuary for spring tide and neap tide conditions which were then compared with field observations of the dispersion of a tracer selected to mimic the dispersion and dilution of the effluent.

A radiotracer (bromine-82) was used for the tracer surveys. On two occasions (one spring tide and one neap tide) 200-300 GBq of bromine-82 were added to the effluent over a tidal cycle while two vessels, each deploying in-situ scintillation detectors, plotted the distribution of the label in the estuary. In addition, a third detector was deployed from a fixed platform positioned in the estuary, at a point where the model predicted tracer would be detected, to see how the predictions at one point varied with time.

Data from these field exercises is currently being compared with the model predictions, both for spatial dispersion and changes with time at a fixed point.

THE FUTURE - TOXICITY BASED CONSENTS

The control of industrial discharges by means of EQS's for specific chemicals is a sound approach where the effluent contains a limited number of known toxins, but is less satisfactory with complex discharges containing mixtures of many chemicals. In such cases, there is great potential for the use of toxicity tests for the direct control of effluent quality. WRc is undertaking research into such techniques, which should ideally be sensitive, simple, rapid, reproducible and relevant to UK coastal conditions. A number are already capable of routine deployment, whereas others are still being developed; they exploit genetic, developmental, biochemical and bioenergetic responses to pollution stresses. Figure 3 shows the sensitivity of several novel techniques to one contaminant, copper, and compares their sensitivities with those of conventional acute toxicity tests and with the current EQS.

One of these tests currently in use exploits the fact that development of mussel and oyster larvae to the shelled stage is extremely sensitive to disturbance by pollutants. This test has been applied to sewage discharged from outfalls at Weymouth and Tenby - with the results shown in Figure 4, which suggest the presence of a potential toxin or toxins in the Weymouth sewage, but not in the (weaker) Tenby sewage. (A control of distilled water in seawater was used because the lowering of salinity by sewage itself has a deleterious effect on development.)

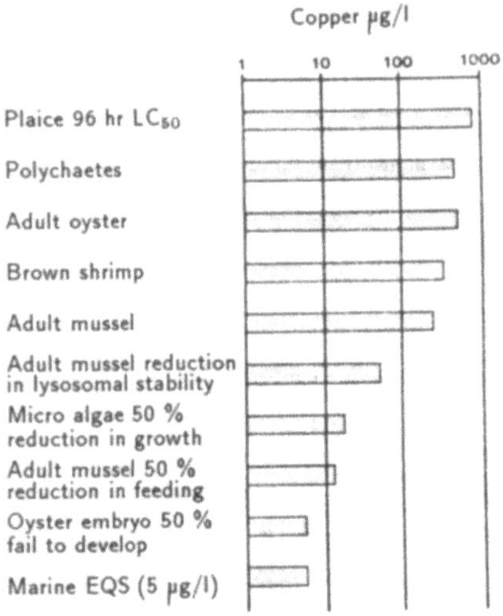

Figure 3. The sensitivity to copper of some marine lethal and sub-lethal toxicity tests

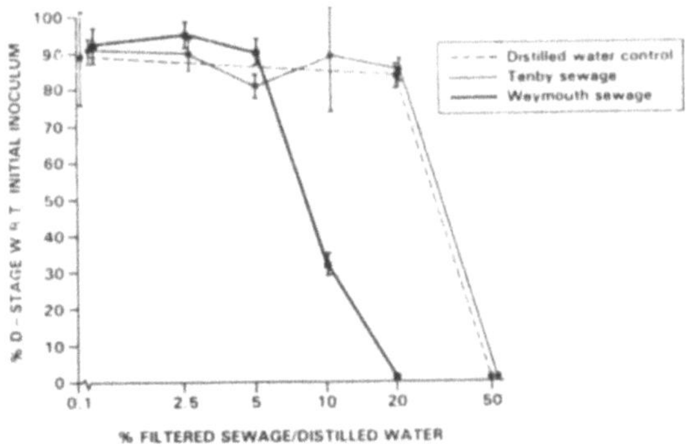

Figure 4. Application of the bivalve larval toxicity test to sewage discharges

The bivalve larval test applies to the water itself, but new tests are also needed for assessing the impact of wastes on benthic organisms. WRc is currently investigating stress tests on such organisms but, in the absence of a routine procedure, has applied a commercial bacterial toxicity testing system ("Microtox") to organic extracts of sediments in the vicinity of the outfall at Weymouth. The results showed that toxicity increased with proximity to the outfall, and was significantly correlated with the concentration of the faecal sterol, coprostanol [3], in the sediments -indicating that the toxicity derived from the sewage.

REFERENCES

1) Gardner J, Mance G (1984) Proposed Environmental Quality Standards for List II Substances in Water. Introduction. TR 206, Water Research Centre, Medmenham, UK

2) Anon (1986) Mixing Zones. Guidelines for Definition and Monitoring. Water Authorities Association, London

3) Walker RW, Wun LK, Litsky W, Dutka BU (1982) Coprostanol as an indicator of faecal pollution. Critical Reviews in Environmental Control 112:91 - 112

DEVELOPMENT OF A WATER QUALITY AND BIO-ACCUMULATION MODEL FOR THE SCHELDT ESTUARY

G.Th.M. van Eck[1] and N.M. de Rooij[2]

[1]Ministry of Transport and Public Works, Tidal Waters Division
P.O. Box 8039, 4330 EA Middelburg, The Netherlands
[2]Delft Hydraulics, P.O. Box 177, 2600 MH Delft, The Netherlands

INTRODUCTION

Estuaries are one of the most important coastal features for mankind. Large civilizations have developed on the shores of many estuaries and most of the present day harbours and large industries are located within their sphere. This often results in water quality problems in many estuaries, including the Scheldt estuary. This estuary is faced with two major problems:

1. Extensive dredging and dumping operations of (contaminated) sediments, necessary to keep the port of Antwerp accessible and
2. large domestic and industrial waste-water discharges. This results, for example, in low or even zero oxygen concentrations in the upper estuary and high concentrations of micro-pollutants in the biota and abiota.

A few years ago Rijkswaterstaat, the authority responsible for the management of the Western Scheldt, decided to built a management oriented model related to the second problem: the influence of the large waste-water discharges on the ecosystem of the Western Scheldt. The objective of the project was to develop a coupled transport, chemical/biochemical and bio-accumulation model for pollutants/micro-pollutants for the Scheldt estuary. Pollutants/micro-pollutants in this context does not only mean heavy metals and/or organics, but also BOD, COD and nutrients (N, P, Si). The model attempts to relate the pollutant discharges into the estuary to the concentrations found in organisms (Tidal Waters Division, 1987).

The project has three stages, which are related to the formulation of an integrated policy and management plan for the Western Scheldt. Stage one (1987), already completed, consists of the development of a O_2-N-(pH-CO_2-alkalinity) model for the water-phase (Delft Hydraulics Laboratory and Tidal Waters Division, 1988). Stage two (1988 - 1989) consists of the development of a prototype model and stage three (1990 - 1991) consists of the development of the final model. In this paper the developed prototype model is presented.

THE SCHELDT ESTUARY

The Scheldt estuary (Figure 1) is situated in the SW part of the Netherlands and the NW part of Belgium. The Scheldt estuary is usually divided into three

1. The lower estuary from Vlissingen to the Dutch/Belgium border, 55 km long and commonly called Western Scheldt;
2. the upper estuary from the Dutch/Belgium border to Rupelmonde, 40 km long and commonly called Zeeschelde, and
3. the fluvial estuary from Rupelmonde to Gent, 65 km long. The fluvial estuary contains only freshwater.

The Scheldt estuary is well-mixed and has a water residence time of two to three months. The freshwater input is about 150 m³/s, the tidal amplitude is 5 m and the average water depth is 10 m.

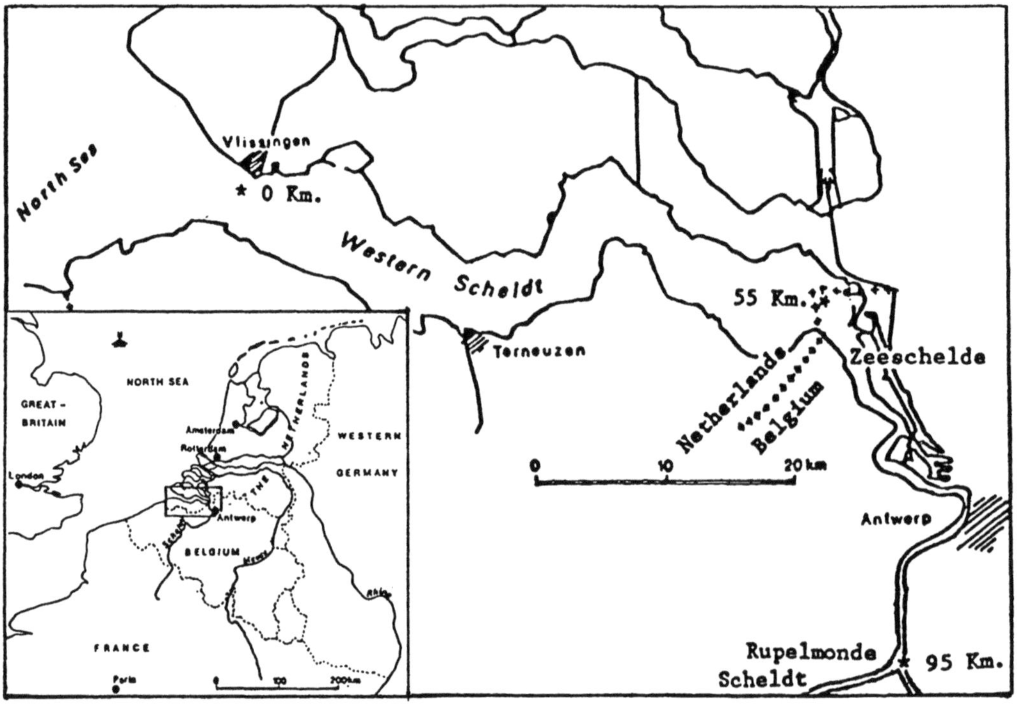

Figure 1: The Scheldt estuary

MODELLING APPROACH

Figure 2 shows the detailed modelling approach for the prototype model.

Firstly, the chloride and suspended sediment concentrations are calculated with the water, chloride and sediment discharges and a water and sediment transport model (DELWAQ, Postma, 1984).

Secondly, pH, eH, O_2, BOD, COD, DOC, POC, chlorophyll and nutrient (NH_4-N, NO_2-N, NO_3-N, DON, particulate-N, PO_4-P, DOP, particulate-P and dissolved silicate) concentrations are calculated with the BOD, COD, nutrient etc. discharges and a water quality model (CHARON; de Rooij, 1980, Los et al., 1984, de Rooij, 1988) and a separate plankton model (DYNAMO; Glas, 1989) which contain the process formulations for nutrients, primary production and other variables (such as O_2). The general water quality and suspended sediment concentrations are calculated first, because the sediment transport largely determines the transport of micro-pollutants, and the general water quality (Cl, pH, O_2) largely determines their behaviour.

Thirdly, the concentrations of dissolved and particulate matter as well as the speciation in the water-phase are calculated from the micro-pollutant discharges with a micro-pollutant model (CHARON/IMPAQT; de Vries, 1987).

Finally, the concentrations of micro-pollutants in organisms are calculated with respect to the abiotic compartments using a bio-accumulation model.

With the calculated concentrations in organisms and the results of related laboratory research presently being carried out we hope that something can be said about the effects of the bio-accumulated micro-pollutants on the selected organisms. It is obvious that not all the existing micro-pollutants can be modelled. The micro-pollutants selected are the heavy metals Cd, Cr, Cu and Zn, the PCB congeners PCB-52 and 153, the PAHs B(a)P and fluoranthene, pentachlorophenol (PCP) and γ-HCH

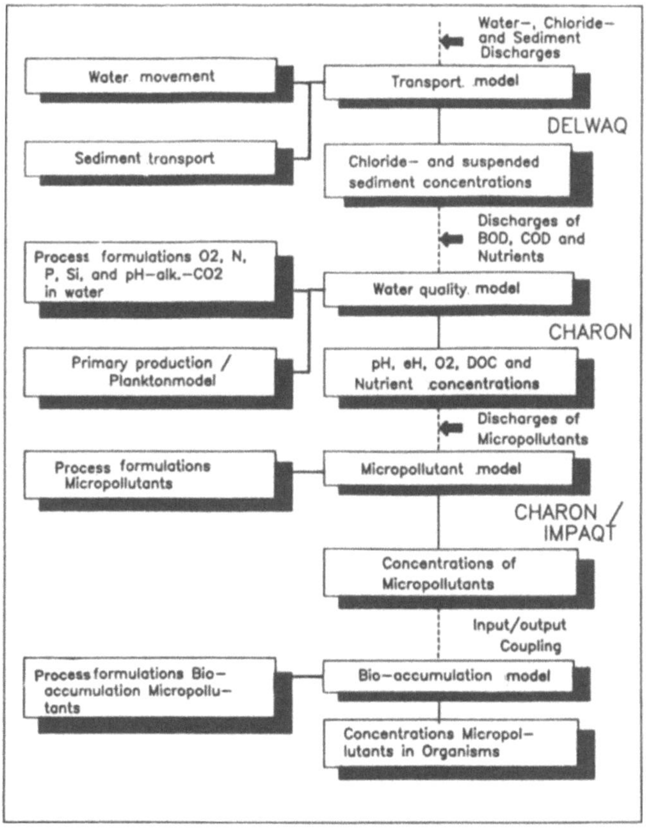

Figure 2: Modelling-approach, prototype model

(lindane). The selection is based on which of the measured micro-pollutants in the Western Scheldt are thought to have the greatest possible impact on the ecosystem. The modelling approach described is based on two hypotheses. The model approach is only valid if these hypotheses need not be rejected. The hypotheses are such that the water quality, micro-pollutant and bio-accumulaton sub-models can be considered as three separate sub-models without feedback. The two hypotheses are:

1. that the micro-pollutant concentrations do not affect the processes in the water quality and plankton sub-models and
2. that the amounts of micro-pollutants present in organisms are negligible compared with the quantities contained in the abiotic compartments.

MODEL CHARACTERISTICS

The developed model is a box model. The Scheldt estuary is divided into 14 compartments (Figure 3). The model considers transport between the compartments and processes within a compartment. The transport between the compartments is divided into advective, for the discharge of fresh water, and dispersive transport. The dispersive transport is characterized by the dispersion coefficients, which are calculated with the measured chloride concentrations in the compartments. The processes in the compartments are divided into those which attain equilibrium instantaneously, an example is the pH-CO_2-alkalinity system, and those which are kinetically controlled processes. Examples are the degradation of organic matter, the exchange of O_2 and CO_2 with the atmosphere and nitrification/denitrification. The kinetically controlled processes are described as zero-, first- or second-order differential equations.

The developed model is a 1D, dynamical, tidal average model which runs on a Personal Computer. More details about the various models, such as background, process formulations, mathematics and numerical schemes, can be found elsewhere (Delft Hydraulics Laboratory and Tidal Waters Division, 1989). A 1D approach has been chosen because the Scheldt estuary is well-mixed and the lateral gradients are small compared to the longitudinal gradients.

The model requires a large amount of input data, even if monthly averages are used, because the number of discharge points (i.e. 60) is high and the model requires quite a few input variables (i.e. 28). For one year about 20,000 input data are therefore required. The systematically collected input data already contains very valuable information. Figure 4, for example, shows the cadmium balance for the Scheldt estuary in 1988. Figure 4 also shows that the inflow at Rupelmonde (46 % of the total load of 21 ton) and the industrial discharges in Belgium (39 %) almost completely determine the cadmium balance.

RESULTS

Because the prototype model is not yet fully developed and calibrated, only some preliminary model results are given below.

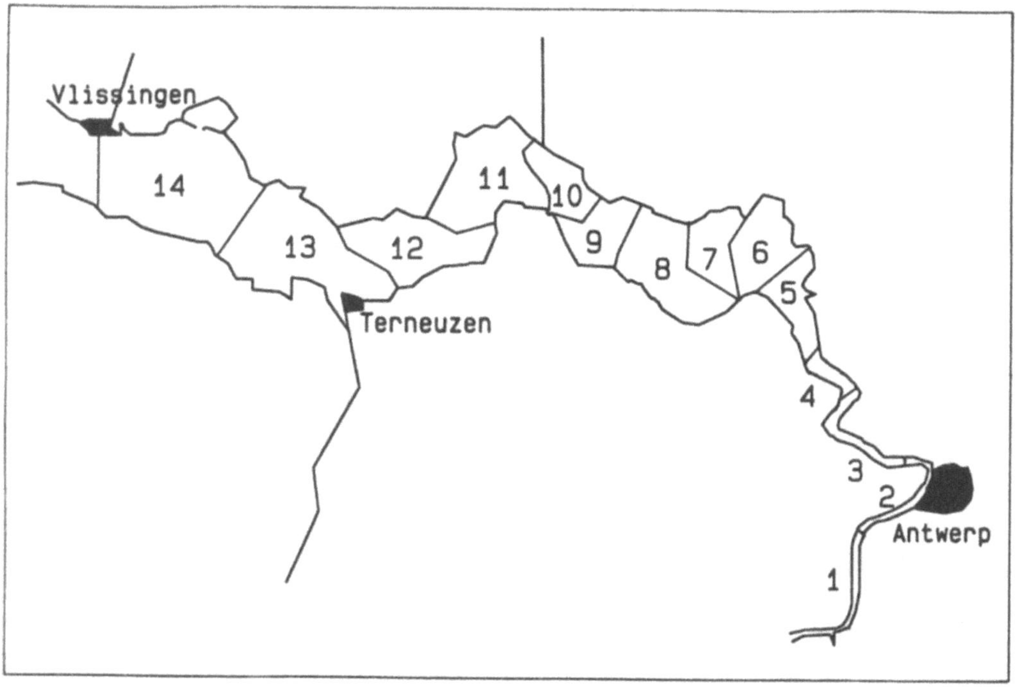

Figure 3: Model compartments

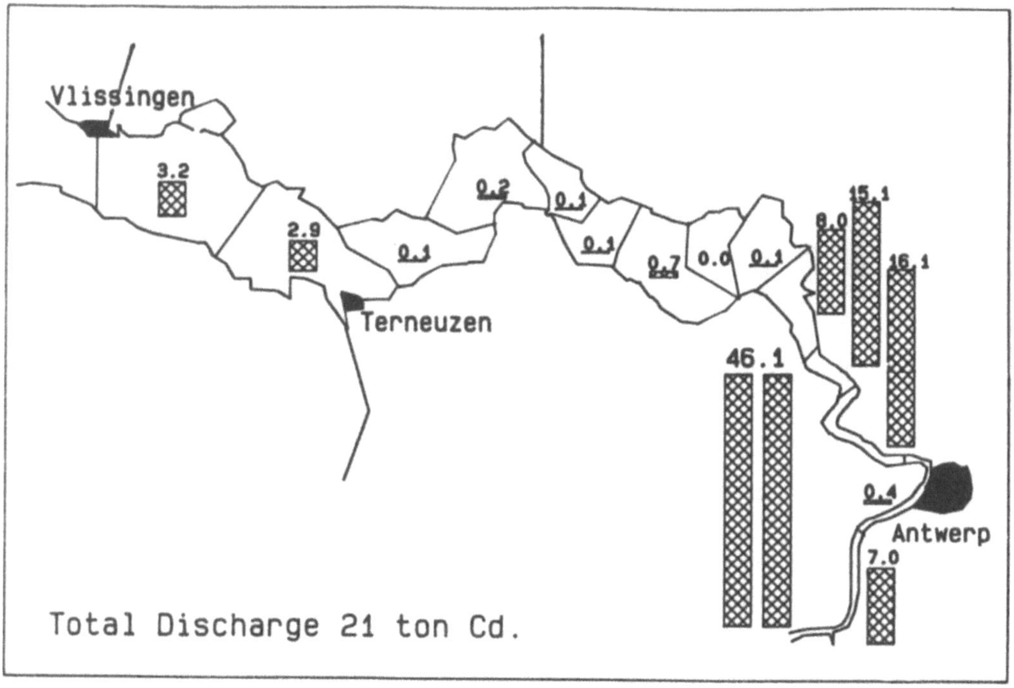

Figure 4: Cadmium discharges (%) in the Scheldt estuary 1980

Silt transport

Figure 5 shows the resulting transport of marine and fluviatile silt across the 15 compartment boundaries. Sand is not accounted for in the model. The transports in Figure 5 are calculated from:

1. the (calculated and measured) annual import of fluviatile silt at Rupelmonde of 320,000 ton;
2. $\delta^{13}C$ measurements in the bottom sediments of each compartment to determine the ratio of marine to fluviatile silt, and
3. the change in the amounts of silt in each compartment between 1980 and 1985 determined by soundings.

In the final model, the silt transports will be calculated with a 2D horizontal silt transport model. Figure 5 also shows that large amounts of silt settle in the eastern part of the Western Scheldt around the Dutch/Belgian border.

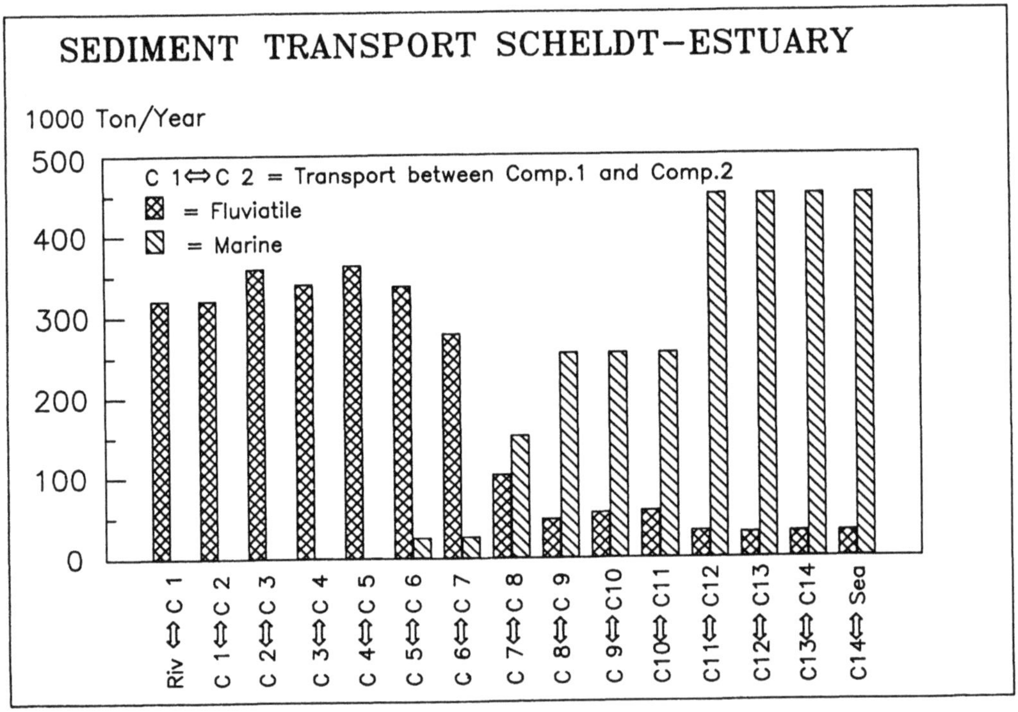

Figure 5: Transport of fluviatile and marine silt between the model-compartments

Echo-soundings and side-scan sonar recordings indicate that the bottom of the Western Scheldt typically consists of sand ripples of 1 to 25 m in length and 0.1 to 1 m in height. The displacement of the sand ripples is high. In the model it is therefore assumed that a sediment layer of 0.5 m erodes and re-settles twice a year.

Chloride

Figure 6a shows the calculated and measured longitudinal chloride profile between Vlissingen and Rupelmonde for two days in 1983. It is demonstrated that the water transport is correctly modelled.

Oxygen

For oxygen the following processes are incorporated into the water quality model:

1. degradation of organic matter (modelled as $CH_2O + O_2 \rightarrow CO_2 + H_2O$);
2. primary production ($CO_2 + H_2O \rightarrow CH_2O + O_2$);
3. re-aeration;
4. chemical oxidation of reduced compounds (Mn^{2+}; Fe^{2+}; S^{2-}), and
5. nitrification ($NH_4^+ + 2O_2 \rightarrow NO_3^- + 2H^+ + H_2O$).

Figure 6b shows the calculated and measured oxygen profiles between Vlissingen and Rupelmonde for two days in 1983. The agreement is quite good.

Micro-pollutants

The heavy metals are not yet fully modelled. The behaviour of heavy metals in the Scheldt estuary is discussed by Zwolsman and van Eck, 1989 (this volume). The processes described there will be incorporated in the model.

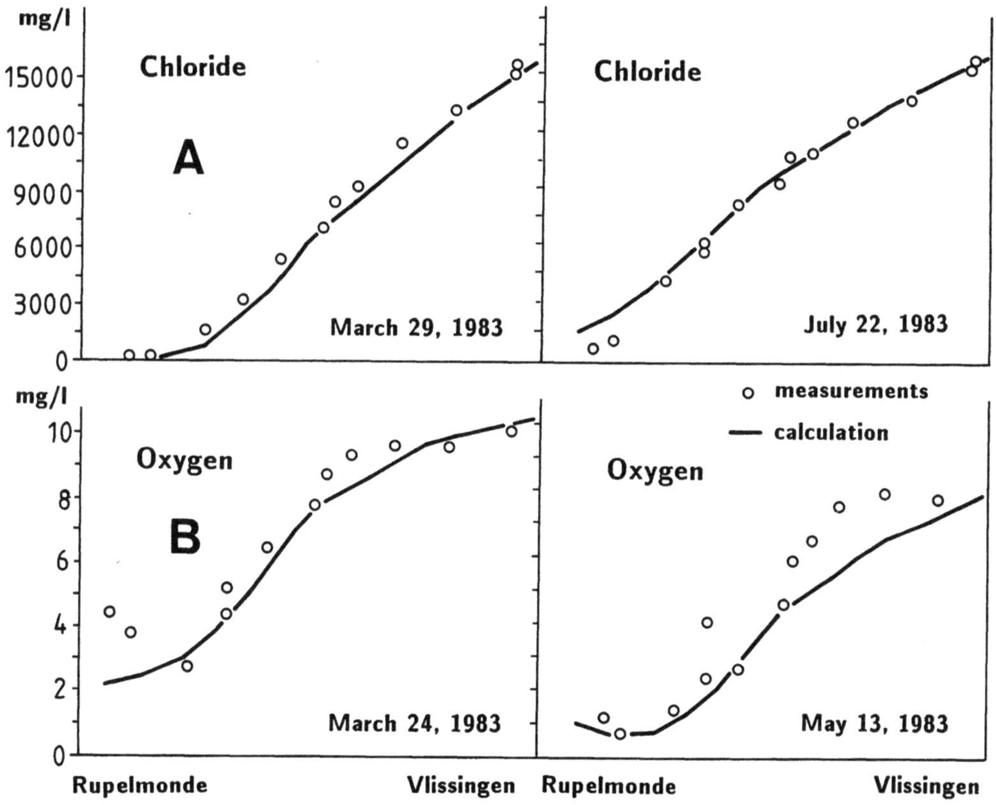

Figure 6: Measured and calculated chloride and oxygen concentrations

Figure 7 illustrates the processes in the organic micro-pollutant sub-model. The model considers precipitation, volatilization, biodegradation, photolysis, sorption onto POC and DOC, sedimentation, resuspension, bioturbation and diffusion. The figure also shows the PCB-52 balance for the entire estuary calculated by the model after calibration. Of the load of 5 kg PCB-52 per year, 10 % comes from precipitation and 90 % from discharges. The outflow to the North Sea is 60 %, the volatilization 15 % and the net sedimentation 25 %. Of the concentration in water, 15 % is in the dissolved state and 85 % bound to either POC or DOC. On the bottom, 100 % is bound to POC. Resuspension and sedimentation are high, diffusion is negligible.

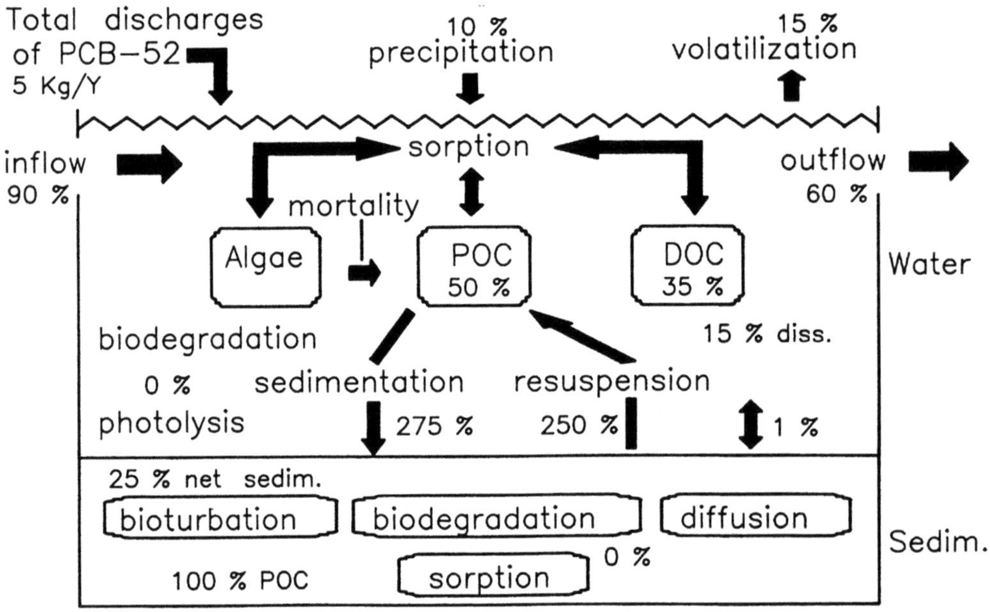

Figure 7: Organic-pollutants model and model results for PCB-52

Bio-accumulation

The bio-accumulation sub-model of the prototype model contains only one organism, the common mussel Mytilus edulis. In the final model, flounder (Platichthys flesus) and lugworm (Arenicola marina) will also be incorporated. The calculation of the micro-pollutant concentrations in organisms is further restricted to three segments in the Western Scheldt. For the prototype model a very simple bio-accumulation model for heavy metals and organic micro-pollutants has been developed. In the final version a more deterministic model, based on available ligands inside the organism, will be used. The present model with results for heavy metals is described in van Haren (1989).

The developed model states that the relationship between the PCB concentration in the mussel and the particulate PCB concentration is a straight line through the origin. The equation can be derived simply if an equilibrium between the PCB concentration in water, in particulate organic carbon and in the mussel is assumed, and that the uptake by the mussel is considered to be linearly related to the dissolved concentration and that the release by the mussel is linearly related to the concentration in the mussel.

In Figure 8, the relationship is given for data from the Western Scheldt and North Sea coastal area. The figure also shows that the very simple bio-accumulation model which has been developed can be used in the prototype model.

CONCLUSION

The preliminary results of the prototype model show that the model under devel opment can become a valuable management tool for the Scheldt estuary.

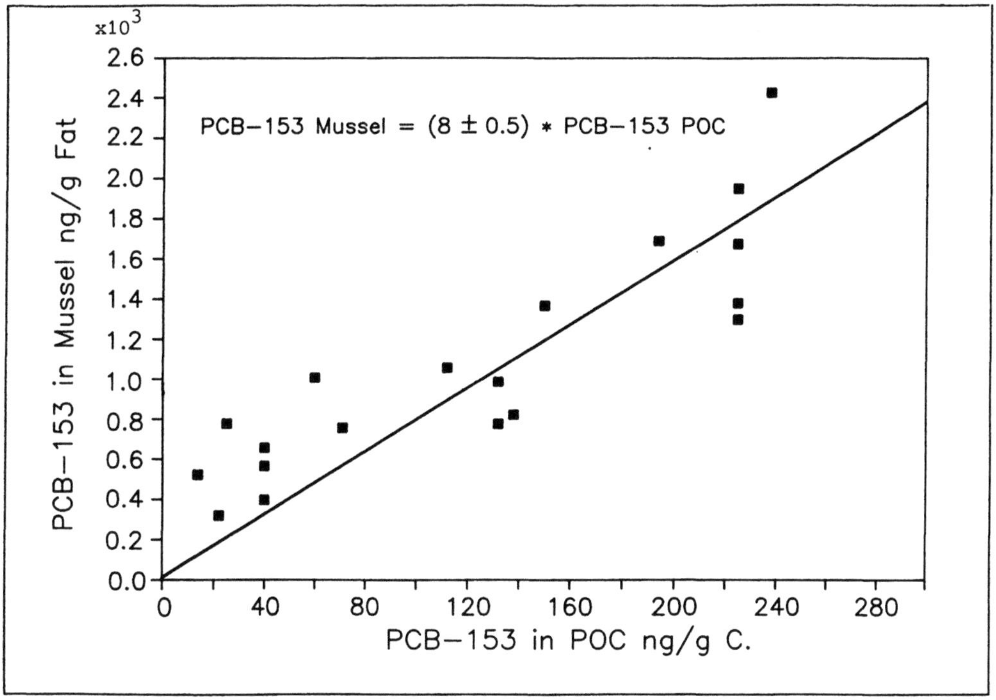

Figure 8: PCB-153 in POC vs PCB-153 in mussel

REFERENCES

Delft Hydraulics Laboratory and Tidal Waters Division (1988) Oxygen model Scheldt estuary. Delft Hydraulics, Delft, The Netherlands, report T0257 (in Dutch)

Delft Hydraulics Laboratory and Tidal Waters Division (1989) SAWES: a water quality and bio-accumulation model for the Scheldt estuary. Delft Hydraulics, Delft, The Netherlands

De Rooij NM (1980) A chemical model to describe nutrient dynamics in lakes. In: Barica J and Mur LR (eds.). Developments in Hydrobiology, Vol 2. Junk, The Hague, The Netherlands, 139-149

De Rooij NM (1988) Mathematical simulation of biochemical processes in natural waters by the chemical model CHARON. Delft Hydraulics, Delft, The Netherlands, report R1310-10

De Vries DJ (1987) IMPAQT: a physico-chemical model for the fate of hydrophobic organic micro-pollutants in aquatic systems. Delft Hydraulics Laboratory, Delft, The Netherlands

Glas PCG, Nauta TA (1989) A North Sea computational framework for environmental and management studies: an application for eutrophication and nutrient cycles. Proceedings International Symposium on Integrated Approaches to Water Pollution Problems (SISI PPA 89), Lissabon, June 1989

Los FJ, De Rooij NM, Smits JGC (1984) Modelling eutrophication in shallow Dutch lakes. Verh Internat Verein Limnol 22: 917-923

Postma L (1984) A two-dimensional water quality model application for Hongkong coastal waters. Water Sci Technol 16: 643-652

Tidal Waters Division (1987) Project plan SAWES. Report GWAO-87.103. Tidal Waters Division, Middelburg, The Netherlands

Van Haren RJF, Van der Meer J, De Vries MB (1989) Cadmium and copper accumulation in the common mussel Mytilus edulis in the Western Scheldt estuary: a model approach. Submitted to Hydrobiologia

Zwolsman JJG, Van Eck GThM (1989) The behaviour of dissolved Cd, Cu and Zn in the Scheldt estuary. This volume

WATER QUALITY MANAGEMENT FOR THE HOOGHLY ESTUARY

S.K. Bose
Consulting Engineering Services (India) Pvt. Ltd.
H-31, C.I.T. Buildings, Christopher Road, Calcutta-700014, INDIA

P. Ray and B.K. Dutta
Department of Chemical Engineering, Calcutta University,
92, A.P.C. Road, Calcutta-700009, INDIA

INTRODUCTION

The pollution problem of the Hooghly Estuary within the Calcutta Metropolitan District (CMD) has been considered in this study. Hydrodynamic, water quality as well as water quality management models have been formulated and an in-depth study of major pollution parameters has been considered to attain an optimum solution for effective management of the water pollution of the Hooghly Estuary.

THE STUDY ZONE

This study deals with the 95 km stretch of the Hooghly Estuary flowing through the Calcutta Metropolitan District (area 1370 km^2), the importance of which lies mainly in dense population (presently 10 million), industrialization and urbanization. This estuary is also unique in terms of river hydraulics as well as from the pollution point of view (Bose and Dutta, 1986).

WATER QUALITY MANAGEMENT

The principal problems of concern in water quality management in rivers and estuaries, can be classified into three groups of pollution beginning with pathogenic pollution followed by gross pollution and lastly chemical pollution. In this study on water quality management, hydrodynamic and water quality models (BOD-DO) were formulated for finding out estuarine water quality due to untreated waste discharge. Subsequently with the help of output parameters a linear programming model was formulated. These models help in finding out the least cost solution for water pollution control and management.

EXISTING STATUS WITHIN THE CMD

It is unfortunate that only a small fraction of the CMD population enjoys proper sewerage facilities. Even for the city of Calcutta there is no proper sewage treatment plant. Sewage treatment facilities are at Kalyani, Bhatpara, Titagarh, Howrah, Salt Lake and Calcutta, but most of the plants are non-operational. The Hooghly Estuary is also extensively used as raw water source for drinking purposes. Water treatment plants are at Palta, Hooghly-Chinsura, Serampore, Kamarhati, Howrah and Garden Reach (Bose et al., 1988b).

WASTE WATER AND ESTUARINE STATUS

The National Environmental Engineering Research Institute (NEERI) has conducted surveys both for the waste water discharged into the Hooghly Estuary as well as estuarine water quality and found the estuary to be grossly polluted in terms of BOD and bacteriological content (Bose et al., 1988b).

WATER QUALITY STANDARD

The Indian Standard Institute (ISI) has laid down standards for different parameters discharged into the water bodies as well as tolerance limits for raw water for public water supply and bathing ghats. These standards have been considered as the guideline for the present pollution control and management programme.

MATHEMATICAL MODEL

In the formulation of a mathematical model for water quality, normally the physical system is translated into a mathematical problem on the basis of suitable assumptions by application of the principle of conservation of mass. The resulting advection-diffusion equation can be expressed as

$$\frac{\partial \vec{C}}{\partial t} + \text{div}\,(\vec{U}\,\vec{C}) = \text{div}\,(\vec{E}\,\text{grad}\,\vec{C}) + P \qquad (1)$$

Appropriate initial and boundary conditions should be provided to make the statement of the mathematical problem complete. Equation (1) in a three-dimensional system is a formidable task. The equation can be reduced to the following two-dimensional form:

$$\frac{\partial \vec{C}}{\partial t} + u\,\frac{\partial \vec{C}}{\partial x} + v\,\frac{\partial \vec{C}}{\partial y} = E_x\,\frac{\partial^2 \vec{C}}{\partial x^2} + E_y\,\frac{\partial^2 \vec{C}}{\partial y^2} + P\,(x,y,t,\vec{C})$$

The initial and boundary conditions are

I.C. $t=0, \vec{C}(x,y,t) = \vec{C}_0(x,y)$ (t = time) (3)

B.C. $x=0, \vec{C} = \vec{C}_0' (y,t)$ (4)

$x = \pm \infty, \vec{C} = \vec{C}_\infty$ (5)

$y=0, \frac{\partial \vec{C}}{\partial y} = 0$ (6)

$y=y_p, \frac{\partial \vec{C}}{\partial y} = 0$ (7)

Equation (3) is the initial distribution of BOD or DO, Equation (4) the transient transverse distribution at the outfalls. Equation (5), constant concentration (zero for BOD and saturation concentration for DO) assumed at a distance from an outfall. Equation (6), lateral symmetry of the concentration field. Equation (7) implies zero flux at the wall.

$P = -(k+k')L + W$ (L = BOD) (8)

$= r(C_s - C) - kL - D_B - R + p$ (C = DO) (r = reaeration-rate coefficient) (9)

C_s = solubility of oxygen, $\vec{C}$ = BOD or DO, $\vec{C}_0$ = initial distribution of BOD or DO at x = 0, $\vec{C}_0'$ = transient distribution of BOD or DO at x = 0, $\vec{C}_\infty$ = BOD or DO at large distance from the origin, D_B = benthal oxygen demand, E = effective dispersion coefficient, E_x = axial dispersion coefficient, E_y = transverse dispersion coefficient, k = rate constant for biochemical decay, k' = rate coefficient for BOD removal by sedimentation, p = oxygen generation rate due to photosynthesis, P = source/sink term, R = oxygen generation rate due to plant respiration, u, v = velocity component, W = waste load and x,y = co-ordinate of a point.

Modelling studies of the Hooghly Estuary have been made considering one-dimensional steady state and time-varying (Bose and Dutta, 1986) and two-dimensional steady state situations. A two-dimensional steady state solution was also tried for a short length within the study zone. It may be mentioned that the velocity term is in general a function of space and time. To know the velocity as well as water level data a suitable hydrodynamic model (one and two-dimensional) was also formulated and solved. The other parameters are taken from different correlations and also from experimental study. A linear programming model was formulated for the water quality management of the estuary and the output of the one-dimensional steady state model was utilized for the constraint. The modelling procedure and solution technique have been shown in (Bose et al., 1988a).

MODEL OUTPUT

The mathematical model has been solved to find out the BOD and DO profiles for both one and two-dimensional cases. Model parameters depend on flow conditions and other physical factors, and BOD and DO computations have been carried out under a range of parameters such as discharge rate, tidal exchange coefficient, effective longitudinal dispersion coefficient, transverse mixing coefficient, reaeration rate coefficient, BOD decay rate, solubility of oxygen, source and sink terms etc.

RESULTS

A steady state model (one-dimensional) has been computed for all the stations within the study zone (i.e. Tribeni to Mayapur) considering the waste load in terms of BOD at different stations. The effect of different parameters has been critically studied. A time-varying one-dimensional model has also been computed within the same study zone. Some of the typical results show the range of BOD for stations Palta, Kamarhati and Garden Reach as (0.85 - 4.10), (2.00 - 3.10) and (1.50 - 2.50) mg/lit. These data are in agreement with the only reported data (Bose and Dutta, 1986).

For the two-dimensional approach a 20 km stretch (Kamarhati to Howrah) has been taken to see how the model output tallies with the predicted values in the one-dimensional model. The computed values of the BOD and DO are rather low as compared to the one-dimensional model output. The main reason for this discrepancy lies in the basic data on the velocity field calculated by using the two-dimensional hydrodynamic model which failed to yield satisfactory results. Rectification of this measurement and correct description of the channel geometry will yield correct results. The steady state model indicates a gradual DO increase downstream. An important factor for decision making is the linear programming model (Bose et al., 1988a). However, as per waste loading even with the specified constraints and calculated treatments, the DO level cannot be desirably improved up to station Bhatpara because of the rather low DO upstream. Improvement can be attained if control measures are taken upstream. Beyond this station the water quality conforms to the specified level increasing up to a value of 6 mg/l.

POSSIBLE RESTORATION STEPS

It is expected that unless proper action is taken right now conditions will deteriorate further. Therefore an appropriate and economically viable solution is required. Given below are some of the specific points which may be considered for water quality management of the estuary: low cost sanitation schemes, renovation of existing plants and pipe lines, construction of new treatment plants for sewage and industrial wastes utilizing gas for power generation, utilization of treated effluent for agricultural use, minimum preliminary treatment or discharge of untreated waste at downstream points considering mixing and dispersion phenomena (Bose et al. 1987), water intake stations should be properly located, laboratories for constant vigilation, use of mathematical model for effective pollution control and management and lastly installation of computers for constant monitoring and data processing.

THE PROGRAMME

Attention has been focussed on the cost of waste removal within the study zone (Bose et al., 1988). For proper control and management of the estuarine pollution a phase-wise implementation based on population density will be beneficial. There are 17 class I (population more than 0.1 million) towns. These towns should be started in the first phase. In the second phase the 15 class II towns (population less than 0.1 million) should be taken up. The population of class I towns covers 84% of the total population living in urban settlements on the river bank and constitutes 87% of the total sewage production. Therefore a major portion of the total expenditure ought to be utilized here. The total expenditure should again be subdivided phase-wise for orderly distribution of the work load. In the earlier modelling studies cost involvement was developed considering the waste load at particular points. If the assumptions are slightly shifted to the removal of waste loads considering all the towns by the sides of the estuary, a slightly higher estimated value can be seen. It is estimated that for controlling the total waste load removal there will be a 44% increase in expenditure as compared to the previous one i.e. amounting to 2300 Million Rupees which is approximately 157 Million U.S. Dollars (Bose et al., 1988b).

IMPLEMENTATION

An authority with a directorate and executive bodies is necessary for proper implementation. The detailed schemes must be economically viable. Proper coordination is required at every stage. Fund allocation, execution and monitoring are important factors to be considered. Physical and financial monitoring is necessary for proper utilization, implementation and timely completion of the projects as delay in completion may increase the ultimate project cost. Lastly greater emphasis should be given to operational management of the executed schemes in order to obtain the correct results.

CONCLUSION

The present paper attempts a detailed study on water quality management of the Hooghly Estuary. In the course of this study details regarding water quality, hydrodynamic and water quality modelling, management modelling, cost involvement and lastly programme implementation have been discussed. However, more detailed studies are required for effective implementation.

REFERENCES

Bose, SK, Dutta BK (1986) Steady state and time-varying water quality models for the Hooghly Estuary, India. Wat Sc Tech 18: 257-265.

Bose SK; Ray P, Dutta BK (1987) Mathematical models for mixing and dispersion in forecasting and management of estuarine water quality. Wat Sc Tech 199:183-193.

Bose SK, Ray P, Dutta BK (1988a) Water quality management of the Hooghly Estuary - a linear programming model. Wat Sc Tech 20 6/7: 235-242.

Bose SK, Ray P, Dutta BK (1988b) Long term plan for pollution control for the Hooghly Estuary. In: Panswad T, Polprasert C, Yamamoto K (ed) Water pollution control in Asia, Second IAWPRC Asian Conf on Water Pollution Control, Bangkok, 9-11 Nov 1988, Pergamon Press, p 21.

STRATEGY FOR ESTUARINE WATER QUALITY MANAGEMENT IN DEVELOPING COUNTRIES: THE EXAMPLE OF NIGERIA (WEST AFRICA)

O. Ojo
Department of Geography, University of Lagos, Nigeria

INTRODUCTION

The history of the petroleum industry in Nigeria started in the estuarine areas of the country and dated back to 1937 when the first explorations were made. However, it was not until 1958 when oil was first produced in commercial quantities at Oloibiri. Since 1958, there had been a phenomenal increase in the quantity of crude oil produced, from an initial average of only 5,000 barrels a day, in 1958, at Oloibiri to an all time peak of about 2.3 million barrels a day in 1979.

Consequent to the rapid development of the petroleum industry and the resulting increase in foreign exchange earnings in Nigeria, the last three decades have witnessed a rapid industrial development and their adverse consequences on the environment. Probably the worst affected is the water environment. For example, the effects are considerable on the estuarine water environment, which has long been a source of living for a considerable number of people particularly along the coast of Nigeria. The industrial wastes - solid, liquid and gaseous - as well as the other accidental and non-accidental pollutants have thus had grave adverse consequences on the estuarine water quality, causing considerable impairment to it. The wastes have also caused a lot of damage and considerable losses to man, animals and plants. For example, considerable hazards have been caused to marine life, human health and vast destruction of the environment as a whole, while a lot of estuarine and marine activities have been hindered by the pollution effects.

A lot of awareness and concern for the need to formulate and effectively implement policy decisions relating to protection and control of the water environment have been created among the people and governments of Nigeria. Yet, not much has been done to effectively solve the pollution problems, particularly those related to the estuarine water quality management. The inability to effectively solve the problems therefore raises the question as to what strategy should be used to effectively tackle this problem. The present paper examines and discusses this question and solicits an integrated management strategy.

STUDY AREA

The study areas consist of the estuarine zones which border the Atlantic Ocean and occur at the southern end of Nigeria. The areas stretch from the boundary of Nigeria with the Republic of Benin on the west to the boundary of the Cameroun

Republic to the east. These areas are composed of coastal marshes and include the maze of creeks, rivers and lagoons which provide a cheap and in some localities the only means of transportation. Like estuarine areas in many other parts of the world, the estuaries of Nigeria contain life of all kinds, and by their nature, are very difficult to clean if polluted by oil or other contaminants. Probably the most commonly polluted areas are mainly in the environment of the Niger Delta, which extends from the Benin river to the west to the Bonny river to the east (Fig. 1). Inland, the Niger Delta begins a few kilometers below the village of Aboh at a point where the Niger forks into the Nun and Forcados rivers. The Delta is a low lying region with an intricate system of natural water channels through which the Niger finds the ways to the sea.

POLLUTION IN THE ESTUARINE AREAS: PROBLEMS AND SOLUTIONS

As already noted, the pollution problems arise mainly because of the impact of the petroleum industry. All the various activities, in all phases of the industry, are sources of emission of pollution and therefore are causes of the deterioration of the estuarine water quality. However, production activities contribute more to the deterioration of the water quality in the Niger Delta than either refining or marketing. Of the various production operations, by far the greatest impact comes from accidental discharge of oil (e.g. leakages and blow out) which may occur at the various stages of the petroleum production. In March and May 1979, for example, about 9718 m^3 and 9062 m^3 of oil, respectively, were recorded as oil spills in the Niger Delta. Many of these spills occurred because of improper handling or mishaps such as burst pipes or hose failures.

The issues related to the problems of the estuarine water quality should undoubtedly be of concern to several institutions including the governments, the industrialists and the public. Thus, the effective management and implementation of any measures designed to solve the problem require the involvement of and cooperation from these institutions. This strategy of "integrated management" will provide a mechanism necessary for coordinating any plans or measures designed at finding solutions to the problem of the estuarine water quality in the study area.

PREVENTIVE MEASURES AND INTEGRATED MANAGEMENT PLANS

For the strategy of integrated management advocated in this paper, two types of measures must be considered, namely the preventive and the remedial measures. It is normally not possible to completely prevent pollution of the estuarine water; consequently, there must be the general policy of no "harmful" pollution (i.e. no harmful discharge) which must aim at using preventive measures. This policy will minimize discharge when pollution occurs, and any preventive measures taken in the management plans would represent the backbone of effective long-term protection.

Among the aspects of the management issues related to preventive measures are good operation practices, adequate supervision of performance by the industry,

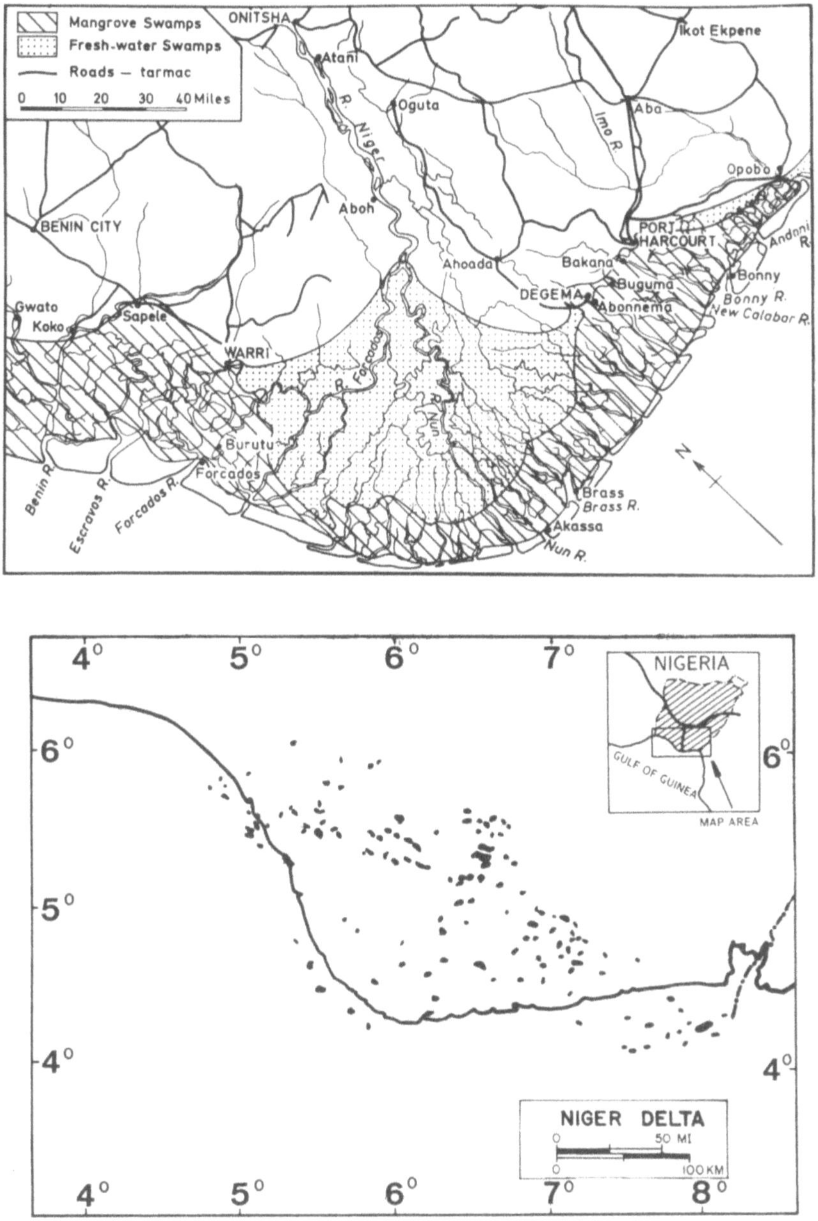

Figure 1: The Niger delta and the distribution of the major oil fields

legislation and education, training and research. Good operation practices concern both the government and the industries. It is imporant that the industrial establishments should realize the need to draw up and operate a "good housekeeping" practice based on international standards. It is also important to note that any regulations on the "good operation practices" must be strictly enforced. The responsibility of supervision of performance by the industry also concerns both the governments and the industry. In this case, there must be teams of professionals, in the area of the industries responsible for the pollution, who will be charged with the responsibility of supervising the activities of the oil companies.

Probably the most important aspect of government activity as far as preventive measures are concerned, is in the area of legislation for setting up standards in the performance of the industries which are responsible for the pollution of the estuarine environment. It is, however, important to ensure that in the procedure for making legislation, all the necessary institutions are fully involved and consulted and necessary advice and contributions obtained from them.

REMEDIAL MEASURES

While preventive measures prevent water pollution, remedial measures are actions such as clean-up operations which are meant for treatment of the polluted water environment. Of course, clean-up operations of oil on water surfaces in estuarine areas can be a very difficult task particularly because of the easiness with which the oil spreads into the surrounding areas and the problems related to the physical cleaning operations between mangrove roots. Moreover, the operations require the employment of well trained personnel, the acquisition of clean-up materials and an effective management to organize the operations. All these require enormous capital investment in addition to proper coordination of the activities.

There are also measures relating to responsibilities and liabilities for damages caused by oil spillage or leakages. Such liabilities should cover the cost of previous existing state. In general, it is probably reasonable to recommend that the liability for a spill whether intentional or unintentional should fall on the industry or its insurers. This is regardless of who is to blame. Industrial operations should normally include this liability as one of their risks.

Education and training is an aspect of the broader management responsibility of human resource development. As far as Nigeria is concerned this measure is needed both as a preventive and as a remedial measure. In general, societal perception and ignorance have strong influence on the success or failure of efforts which are put into combating and controlling the causes of environmental pollution. Consequently, there is need for societal education in order to break down barriers which are due to ignorance, malice, suspicion and lack of understanding of problems and efforts put into solving them. The concept of "community involvement" in finding solutions to the management problems of the estuarine water quality must therefore be emphasized.

Results of research activities on the estuarine pollution issues is another significant approach for identifying and solving problems of estuarine pollution. Such research activities are particularly significant in contributing and exchanging ideas on environmental issues relating to the estuarine areas and the measures for improving the water quality management.

WATER QUALITY INCLUDING THE ECOSYSTEM

Wulf Greve
Biologische Anstalt Helgoland, Hamburg

ABSTRACT

Estuaries are transient zones impacted from and impacting upon the freshwater and the marine ecosystem. The transition zone is thus of special importance. Water quality management, originally handling the boundary conditions of limnetic ecosystems, has to consider the complexity of this part of the biosphere. The far reaching effects, the system properties, and the multiple resource utilization may be respected using advanced decision support methods.

1. INTRODUCTION

The management of natural resources is confronted with the general dilemma of decision-finding in an uncompletely specified world. While resource utilization usually has a clear focus, the protection of the resource potential of the system concerned is less clear. Values have to be given to either goal. Traditionally, the value of the resource under consideration was more highly priced than the ecological potential of the whole system. Only conflicting resource utilizations were sufficiently valued to lawfully limit the desire for resources. "Water quality" standards were one way to manage resource utilization by industrial, domestic and other users. The classification of water quality follows various definitions with respect to the content of oxygen, the biochemical oxygen demand, the toxicity of the chemical composition, and the organisms present within the hydrosphere. These parameters are regarded as indicators for that part of the biosphere to be managed by a resource decision.

This volume is concerned with the management or decision-finding in estuaries, the transition zone from the terrestrial and limnetic to the marine parts of the biosphere. Estuaries are thus gateways out of and into either system and act as a link between them. Therefore, decision-making in resource-management requires high sophistication. Some examples may help to clarify the necessary framework for estuarine water quality management as seen by a marine ecologist.

2. SIGNIFICANT ECOLOGICAL ESTUARINE PROCESSES

Estuaries are the transition zone from freshwater to marine habitats. This transition induces changes in the physical, chemical and biological conditions. For the ecosystem this change stands for a phase transition from the freshwater stability regime to the marine stability regime. This phase transition along the salinity gradient may include various sub-regimes, vertically as well as horizontally. Organisms and ecosystems may depend on this phase-transition as an obligatory structure for life-cycle strategies.

2.1 Basic estuarine structures

Dissolved salt increases the weight of seawater as compared to freshwater. The volume to surface ratio of seawater generally exceeds that of flowing freshwater, thus indicating a shift in the air exchange ratio, so as to be registered in temperature and oxygen exchanges. The sea, generally, is subdued to the tidal regime, which induces mixing, and to the greater impact of storms and storm-driven currents and waves. Rivers generally supply the sea with nutrients, dissolved by the runoff of rainwater or groundwater. To-day the nutrient-load is enhanced by human action, as is the pollution load. Further anthropogenically induced changes result from hydraulic engineering activities, creating deeper navigable rivers and reducing flood plains.

Ecologically, the transition from freshwater to seawater results in a change in species composition. Remane (1934) first described this shift in populations. The species-specific ecological niche of the populations concerned determines this composition. This is to be seen in the composition of bacteria, for example (Gunkel, pers. com.), in tertiary consumers (Greve and Reiners, 1981) and in organisms which live in the estuary proper as well as in the outer river plume. Estuaries thus provide good examples for the study of community formation and ecological system behaviour.

2.2 Estuarine variability

Estuaries vary with respect to physical structure, runoff quantity and regime, climatic zone, intensity of anthropogenic impact and other parameters. Accordingly, they provide a variety of cases for the study of comparative systems ecology. Such-studies have been descriptive in the past (Kühl and Mann, 1983), analytical in the present for single estuaries (Kremer and Nixon, 1978; Baretta and Ruardij, 1988), and will hopefully incorporate comparative systems ecology in the future.

Thereby, it will be possible to define the relative stability of each ecosystem (Grümm, 1975) and its resilience (Holling, 1973) to potentially hazardous resource utilizations. These system parameters may then be used as management criteria (Fig. 1).

2.3 Transient processes

Estuaries are thresholds in either way. The ecological niche of many species requires the ontogenetic change of the biotope between the marine and freshwater habitat. Anadromous (e.g. salmon) and catadromous (e.g. eel) fish cross the brackish water zone on their way from the spawning area to the feeding grounds of the adults. The spawning ground may be positioned just inside the estuary, as in some herring populations, or near the spring of a river. If any part of the ontogenetically changing niche of such a fish is not met, populations may become extinct (e.g. sturgeon and salmon in the Elbe). Often salt-marshes, sea grass communities and/or tidal flats are related to the rich nutritional supplies from rivers within the estuary. A rich avifauna and mussel-beds are connected with such conditions.

The populations have adapted themselves to the specific local floods and the regional vegetation pattern. The organisms are part of the local ecosystem, even if they just live in the estuary for a few weeks during their migration, e.g., birds gathering energy for longer flights to their nesting grounds in order to play their role in a distant ecosystem which is coupled to the estuary by just this periodic visit.

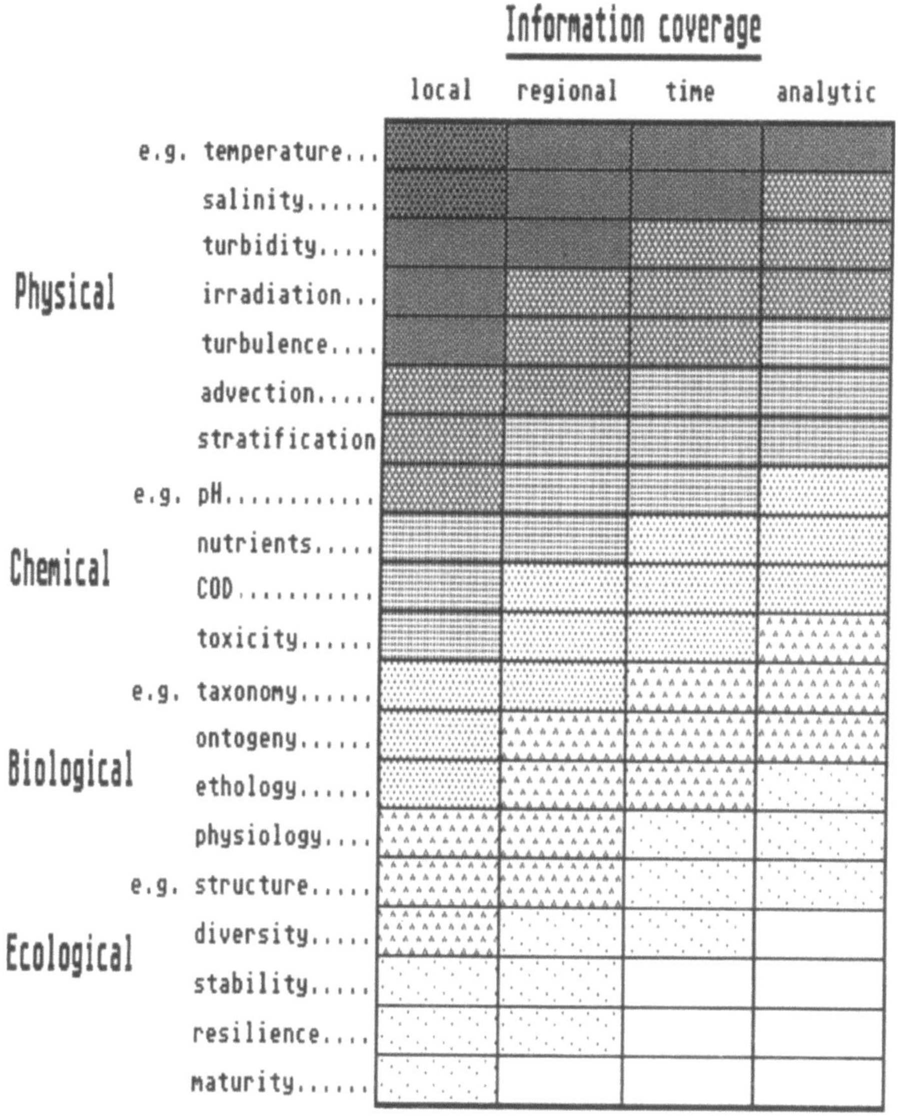

Figure 1: Ecological water quality parameters. The degree of information coverage is indicated in estimated levels of darkness.

2.4 Utilization of estuaries

Men have always liked to settle close to estuaries. The richness of the sea and the availability of freshwater, the nutrient supply by floods for agriculture, and the transport facilities for ships have a long cultural background. The management of such estuarine ecosystems is a cultural heritage endeavour which has to prove our cultural and technological status.

So far, rivers have been misused as a dump site for the catchment area. Industry has been placed close to the sea to make use of the large water body as it was assumed that "the solution to pollution is dilution". We know better now. But enormous loads of anthropogenic materials, nutrients and toxic materials have to be dealt with by estuaries, reduced in area by diking and in surface-to-volume ratio by dredging. The variability of the brackish water ecosystem supported by the sea and by the freshwater ecosystem has not shown the drastic signs of distress, visible on either side of it in oxygen deficiency events. Yet, the vital importance of the estuaries for other parts of the biosphere requires wise decision-making.

3. THE MANAGERIAL OPTIONS

Decision-finding in estuaries has to take into account the accepted rules in terrestrial systems as well as the "state-of-the-art" in marine systems. Both fields undergo permanent changes and so far - perhaps in principle - there is no valid general rule of decision-finding in estuaries. In each case the above given ecological processes have to be respected in any such procedure.

3.1 Single resource management

The traditional approach in the management of natural resources is the partitioning of nature according to the potential users requirements. Finn (1982) in his study on "managing the ocean resources of the United States" suggests this strategy. It conforms with regional planning in Germany (Hanke, 1981). Current-influenced systems cannot, as yet, be separated within the waterbody.

3.2 Conflicting resource management

Recently Richardson (1985) and Koudstaal (1987) introduced integrated marine resource management in marine decision-finding. These approaches, based on simulation models for the functional description of (i) the management rules, (ii) the ecosystem, and (iii) the control values, come close to the methodology developed for the "Adaptive Environmental Assessment and Management" (AEAM) by Holling and his co-workers (1978). The possibility of assigning changing values to ecosystem processes and single resources makes simulation models a valuable tool for decision-makers. This procedure has been applied in several coastal environmental impact studies (Environmental and Social Systems Analysts Ltd., 1982).

3.3 Ecosystem management

Men are organisms. Our illnesses are treated on the basis of experiences gathered from animals. Our survival within the outer natural support system - the ecosystem - has just become obvious to many of us in the last decades. Estuarine management on the basis of the ecosystem requires adequate information on the structure and functioning of the ecosystems concerned. This includes many far reaching effects, as ecosystems are open systems with interactions on various scales with other systems. It is questionable whether our present knowledge is adequate enough.

3.3.1 Knowledge basis

Ecosystems consist of physical, chemical and biological components. These vary in time and space. This four-dimensional set of data has to be defined in value and variance for each parameter for a complete knowledge basis. Further, the functional relationships between the parameters should be known. This includes deterministic, chaotic and catastrophic functional relationships.

The information required here is only partially available. Even if measurements have been made, detailed investigations are necessary to describe the functional relationships within the system. These relationships are complex. Simplifications are permitted only in those cases where changes in the system behaviour can be excluded. Some of the simplifications are long-lived. The "marine food chain" was useful in the sixties. In the seventies it was stated that 'food chains ... are rare in nature' (Kinne, 1977). It was further found that an ecosystem has to be defined in its "efficiency, maturity, resistance to deformation, stability, healthiness and diversity". Resistance to deformation may be understood as being synonymous to "resilience" (Holling, 1973). Our knowledge of these components of the water quality in estuaries is generally very low.

3.3.2 Knowledge engineering

The handling of the available information, the definition of research requirements, and the extraction of general rules from the rudimentary knowledge of the ecosystem, requires a flexible data handling system which supports the development of hypotheses, the testing of them against independent data-sets and the identification of control functions and indicators for successful management. Ecosystem parameters such as stability, resilience, and maturity are routinely investigated by the diverse functional representations of the multi-disciplinary approaches. Expert systems make it possible to use external expertise in the investigation of specific problems. Technologies as used by Fedra (1989) support the decision-finding in estuarine resource management.

3.3.3 Second-generation Adaptive Environmental Assessment and Management (AEAM)

Estuaries are living systems; they respond to managerial decisions. Resource values thus also vary according to commercial or political conditions. Decision-finding must be adapted to such changes. The decision scenario must have the possibility to include further resource utilization, changes in the ecosystem or other altering components. Based on a multi-disciplinary cooperation during the construction of a functional description of the system and on detailed research and numeric analysis, expert systems will be developed for a hopefully wise decision-making in future water quality management in estuaries.

OUTLOOK

Growing concern over the environmental behaviour of men, and the progress in computer application, are the boundary conditions for future estuarine water quality management. Physical and chemical parameters become increasingly surveyable by automatic analytical procedures. The biotic measurements still require the greatest effort. It is tempting to restrict the measurements to some boundary conditions of the biotic system, pretending that all ecosystems will respond equally.

It may be better to include a set of organisms to be monitored such as in the Dutch "amoeba system" (Colijn, pers. comm.). Even this attempt at defining indicators of the health of the ecosystem will not provide the necessary information on the ecosystem. Using taxonomic identification, statistical treatment of data, the functional description of the niches occupied, the analysis of the system characteristics (resilience, stability, maturation, etc) in management scenarios for optimal decision-finding is possible and has to be a prerequisite for preserving and developing the genetic resources and ecosystem identity in estuaries.

REFERENCES

Baretta J and Ruardij P (1988) Tidal Flat Estuaries. Springer, Berlin

Cairns J and Dickson KL (Eds) (1973) Biological Methods for the Assessment of Water Quality. American Society for Testing and Materials, Philadelphia, 256 pp

Environmental and Social System Analysis Ltd (1982) Review and Evaluation of Adaptive Environmental Assessment and Management. Environment Canada, 116 pp

Fedra K (1989) Modelling hazardous chemicals: software tools for environmental risk analysis. Proc Envirotec Vienna (in print)

Finn DP (1982) Managing the Ocean Resources of the United States. Springer, Berlin

Greve W and Reiners F (1980) The impact of prey-predator waves from estuaries on the planktonic marine ecosystem. Estuarine Perspectives, Academic Press: 405-421

Grümm HR (1975) Analysis and Computation of Equilibria and Regions of Stability. IIASA, Laxenburg CP-75-8, 283 pp

Hanke H (1981) Handbuch zur ökologischen Planung. Schmidt, Berlin, 310 pp

Holling CS (1973) Resilience and Stability of ecological systems. Ann Rev Ecol Syst 4; 1-23

Holling CS (ED) (1978) Adaptive Environmental Assessment and Management. Wiley, Chichester 377 pp

Kinne O (1977) International Helgoland Symposium "Ecosystem research". Helgoländer wiss. Meeresunters. 30, 709-727

Koudstaal R (1987) Water quality menagement plan North Sea: Framework for Analysis Balkema, Rotterdam 132 pp

Kremer JN and Nixon SW (1978) A Coastal Marine Ecosystem. Springer, Berlin, 217 pp

Kühl H and Mann H (1983) Main characteristics of the environment and zooplankton of Eider, Elbe, Weser and Ems. In Wolf (Ed) Ecology of the Wadden Sea Balkema, Rotterdam 143-145

Liebmann H (1969) Der Wassergüteatlas. München/Wien

Remane A (1934) Die Brackwasserfauna. Verh Deutsch Zool Ges

Richardson JG (Ed) (1985) Managing the Ocean, Resources, Research, Law Lomond Pub, Mt Airy 407 pp

THE SENSITIVITY-SURVEY
- A USEFUL DECISION-AID FOR THE PROTECTION OF THE MUD-FLATS IN THE GERMAN NORTH SEA COASTAL REGION -

Heinrich Reincke
Wassergütestelle Elbe, Focksweg 32a, 2103 Hamburg 95

1. INTRODUCTION

The world's largest area of mud-flats is situated on the North Sea coast lying between the Netherlands and Denmark. There are approximately 1.300 different species of living creatures and organisms existing in this unique coastal landscape which contributes to the world's most productive nature reserve. It is characterized by small islands, holms, bights, estuaries, tidal creeks, open mud-flats and shallows protected against the open sea. The manifold and nourishing sources of food, typical of mud-flats and salt-marshes, attract many different species of migrating birds from the most distant parts of the world, such as for example, the Siberian Bar-Tailed Gotwit (whose home is more than 4.000 kilometres away), the Brent Goose from the Russian Taimyr-Peninsula and the Redshank from Northern Canada. Furthermore, the mud-flats serve as a rest-site and breeding ground for millions of wild birds.

The increasing marine pollution caused by effluents from rivers, creeping oil pollution and the dumping of ships' waste all threaten to destroy this environment. All these problems require particular attention; and what is especially urgent is the immediate protection of these coastal areas.

The questions arise: (i) What can be done to protect these regions against a disaster caused by an oil-tanker accident, and (ii) whether there is any possibility, whatsoever, to take preventative measures in protecting the flora and fauna in such a case.

2. RECOGNIZING THE PROBLEMS AND FINDING THE SOLUTIONS

In 1983 a project team was brought together by the Stade Water Authority with the aim of developing a classification and sensitivity rating of damage caused by the detrimental effects of pollution, particularly oil. The first district to be selected was the Wurster mud-flats, situated between Cuxhaven and Bremerhaven, an area of 430 km². Previous sensitivity drafts were based mainly on geological and sedimentary representations, and were evolved for areas quite different from those mud-flats on the German coast. It is essential, particularly regarding these mud-flats, that special attention be given to the environmental aspects. In this case, considerable modifications are required if the original concepts are to be used. Particular consideration should also be devoted to aspects such as infrastructure, economy and ecology, as well as protecting nature. On this basis of acquired information, the authority should be able to classify the damage, analyse it and then be able to give detailed instructions as to how the cleaning and restoration programme should be carried out. The results should be simply documented when possible, without relinquishing the finer details.

3. THE RESULTS

3.1 Environment

While checking the information for a classification, a list of the ecological data has been compiled for each benthic species according to the correlation between biotic and abiotic factors. The meaning of these factors, involving their sensitivity to oil, their importance for food, their variability, the diffusibility and their time of reproduction, is expressed for each species by a number of valuations. These valuations enter into the relative weightedness of the benthic stations with respect to diversity, abundance and structure of the sediment.

The differences in sensitivity between the districts and their bird and fish-life, as well as the salt-marshes, have to be treated with equal importance. These differences can be catalogued under their particular regions or superimposed. To simplify the matter, the results of the environmental investigations can be distributed under 4 stages of sensitivity (Fig. 1). From May to October the need for protection is paramount. When working on the system of the 4 stages, it must be remembered that, in principle the whole area under examination is susceptible to pollution. The lowest sensitivity area (shaded blue) means that regeneration is possible after contamination, but the highest sensitivity area (shaded red) cannot be regenerated and will, therefore, be permanently damaged.

3.2 Economy

The results of economical research show that depending on the time of year of an oil-tanker disaster (either in December or March of one year) the economical losses, calculated over a 3-year period, are about 210 million DM (in December) or 235 million DM (in March). The most outstanding economical loss is, quite understandably, to tourism. The fishery and the connecting harbour industries are greatly affected by the loss of approximately 17 million DM and the least affected is the agricultural industry with a loss of 5 million DM approximately.

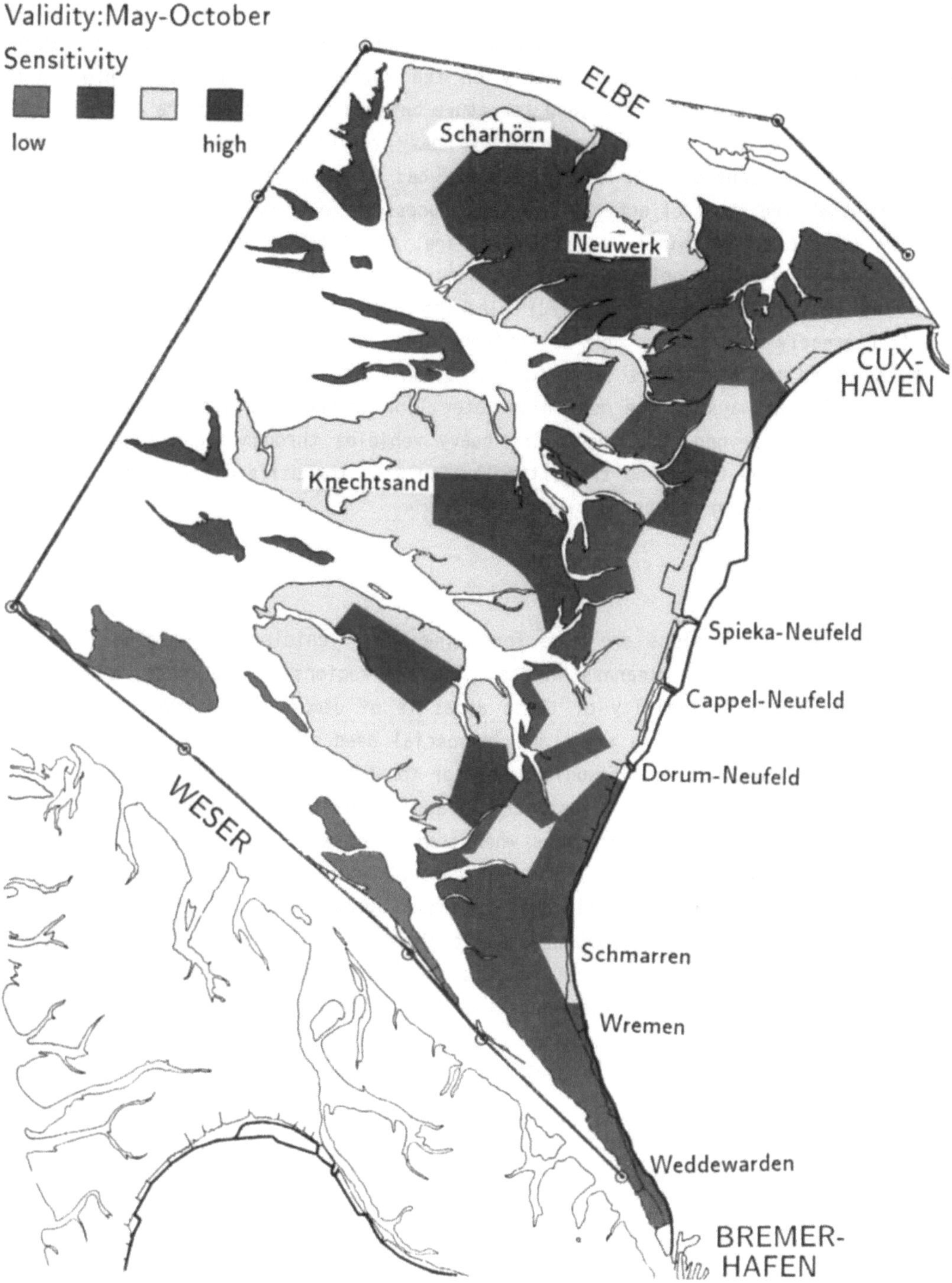

Figure 1: Wurster mud-flats sensitivity-survey

3.3 Infrastructure

An effective cleaning process on the foreshore (from the landward side - mean tidal high-water area) depends mainly on the regional circumstances, e.g., the structure of the sediment, the land structure between the dike and the open sea and whether or not this area is passable by heavy vehicles or machinery. The basic groundwork in connection to the foreshore, has already been calculated with the result that the most effective cleaning process is extremely limited when using heavy vehicles or machinery on the foreshore. An effective cleaning process from the landward side is only possible on sand-flats in the Cuxhaven-Duhnen region. In the remaining area a successful cleaning process from the landward side is either greatly hindered, or due to environmental reasons, is not advisable because of the mixed mud structures (e.g. soft silt and mud). Another drawback is the lack of suitable pathways to the mean high-water line through the wide foreland and to the marshlands beyond. By maneuvering heavy vehicles through these regions, more damage is created by mixing the oil pollution with the different types of mud and silt than by leaving nature to clean itself.

4. OUTLOOK

Before any decisions are made for using heavy vehicles and machines for the employment of chemical cleaners in the different regions of the coast, a thorough knowledge of the sensitivity of these areas is of utmost importance. It must be completely understood which areas are in special need of protection, at which time of year and in which cases it would be better to abandon a cleaning operation.

With these preliminary investigations and evaluations an essential foundation is laid for assisting in decisions when confronted with future oil pollution or evaluating and documenting the resultant damage. From this study an estimation of the possible damage in the individual regions, taking into consideration the time and the place, can be made. Furthermore, it gives environmental, economical, nature preservation and infra-structural aspects. Depending upon the nature of the oil pollution, a detailed or a less detailed classification can be made in order to carry out a still further and more precise evaluation of the resulting damage. Acting on this information, a decision can be made whether to use alternative methods or not.

CHAPTER IV

WATER DYNAMICS AND TRANSPORT PROCESSES

A STUDY ON THE TRANSPORT OF DISSOLVED AND PARTICULATE MATTER THROUGH THE HAMBURG HARBOUR

H.-U. Fanger, J. Kappenberg, V. Männing
Institute of Physics, GKSS Research Centre Geesthacht GmbH
P.O.Box 1160, D-2054 Geesthacht, Federal Republic of Germany

INTRODUCTION

In the early eighties, the GKSS Research Centre Geesthacht developed the 'BILEX' strategy [1] in which hydrographic measuring techniques (including the moving boat method), trace-analytical procedures and numerical modelling are combined in an effective way to investigate pollutant load and transport processes in tidal rivers. Three experiments following this approach were performed on the tidal Elbe river in the years 1982, 1984, and 1985 [2, 3]. Thereafter, a more comprehensive experiment (BILEX'86), which is described presently, was carried out in the Hamburg Harbour over a period of ten days and nights in September 1986. The hydrological situation of BILEX'86 can be seen from Figure 1 which displays the course of river water discharge for the years 1985 through 1987. In this diagram, the previous BILEX'85 campaign as well as subsequent investigations (described in another contribution to this conference [4]) and the September 1986 experiment are indicated by vertical bars. The fresh water discharge for BILEX '86 was obviously at the lower end of the dynamic range.

In this campaign, a fluorescent dye tracer (Rhodamine B) was also used to study transport and residence times of river water and dissolved constituents. This part of the investigation was carried out in close cooperation with the German Hydro-

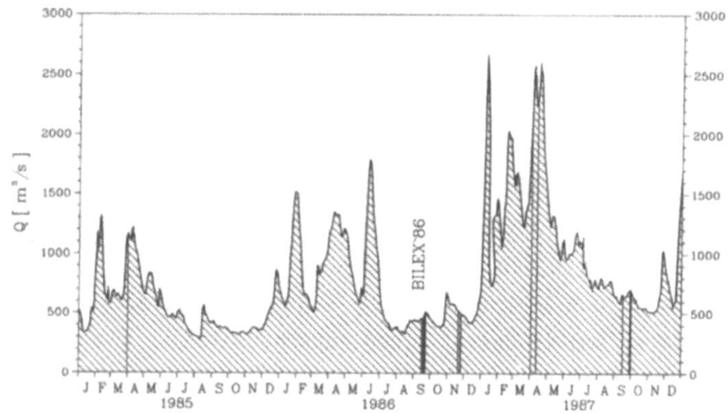

Figure 1: Course of the Elbe river discharge for 1985, 1986 and 1987 with marks for GKSS experiments

graphic Institute (DHI), the Hamburg Harbour Authority (Strom- und Hafenbau, S+H) and the Ministry for the Environment, Hamburg. Details are given in a separate contribution to this conference [5].

THE BILEX'86 EXPERIMENT

At the beginning of the experiment, 120 kg of Rhodamine-B was released into the Elbe river some 10 km upstream from the bifurcation near to the harbour area (see map in [1]). The movement and spreading of the dye patch was then recorded for 5 days by the S+H measuring boat REINHARD WOLTMAN which cruised along the river branches and in some harbour basins. Simultaneously, the GKSS research vessel, LUDWIG PRANDTL, continuously surveyed, in detail, a cross-section downstream from Hamburg (Nienstedten, Elbe km 631.5), at a distance of about 30 km from the release position.

The main activities of the LUDWIG PRANDTL on this cross-section were the uninterrupted operation of the hydrographic measuring system HYDRA [6], water sampling and onboard filtration. Current velocity, turbidity, temperature, conductivity, oxygen, and water depth were also measured in a continuous sequence of horizontal and vertical profiles over more than six coherent tidal cycles (Figure 2). Water sampling was carried out jointly with vertical profiling, but with reduced frequency, i.e. four times per tidal period at three locations and at two depths.

A cross-section at Oortkaten (Elbe km 607.5) was selected as the boundary upstream from Hamburg. At this site, the automated measuring station AMO had been running the last of, totally, three years [1, 7]. There, besides accumulative water sampling in half-tidal intervals, a set of hydrographic parameters, including current velocity, was measured continuously. Additional water sampling was performed during this experiment to obtain a better resolution in time.

RESULTS AND DISCUSSION

For a 1.0 to 1.5 m sub-surface position in the navigational channel of the Elbe river at Nienstedten, the variation in time of some of the parameters measured is shown in Figure 2. The amplitudes of the current velocity are close to 1.5 m/s for both ebb (positive sign) and flood stream. The beginning of flood tide is characterized by a steep slope which is accompanied by a sudden increase of suspended particulate matter (s.p.m.) concentration which far exceeds the maximum values recorded for ebb tide. Amplitudes of flood current velocity and seston content grow with the last tidal cycles as changing weather conditions with strong westerly winds reinforced the flood stream. The similar temporal structure of temperature and oxygen profiles reveals the existence of a gradient: with the end of flood

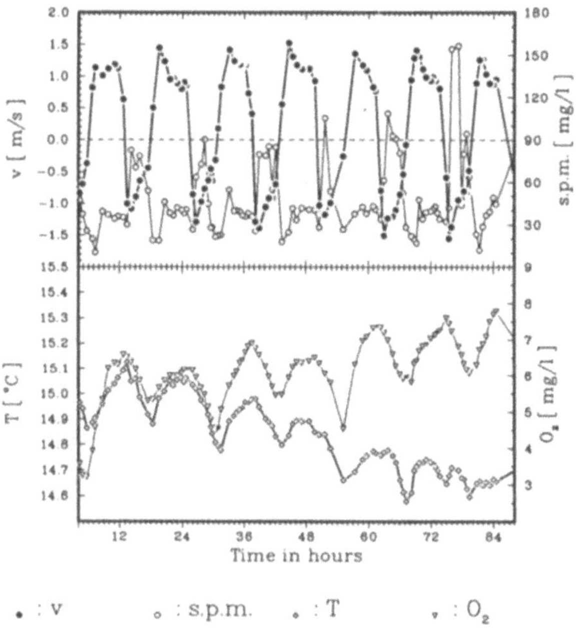

Figure 2: Tidal cycles of current velocity, s.p.m., oxygen, and temperature for the Elbe river at Nienstedten (see text). Time zero: Sept 18, 1986, 04:00

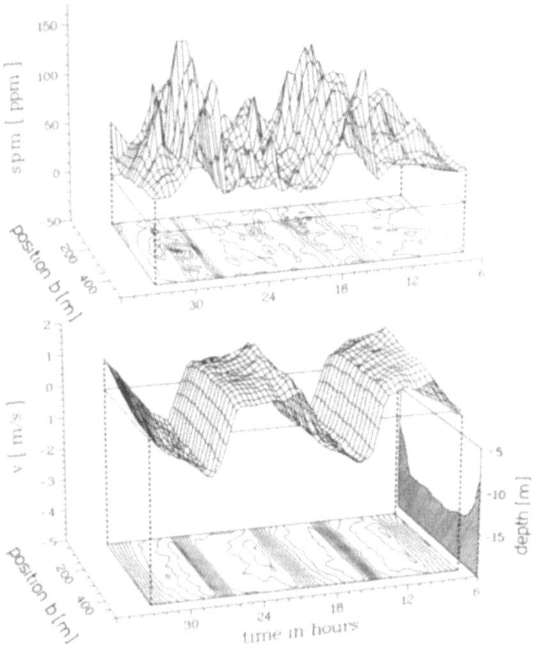

Figure 3: Near-surface current velocity (bottom) and s.p.m. concentration (top) across the Elbe river at Nienstedten over two tidal cycles on Sept 19. Position zero: Northern bank

tide, temperature and oxygen concentrations are regularly at their lowest. Both effects can be understood as (i) the harbour area upstream from this position is a heat source and (ii) the well-known oxygen minimum of the Elbe river develops downstream in summer-time.

The inhomogeneity of the lateral s.p.m. distribution as a function of time is demonstrated in Figure 3 for two of the tidal cycles described above. The values have been extracted from horizontal near-surface profiles measured (roughly) every half hour. A similar variability exists at greater depths, as can be seen from vertical profiles (not shown here). The data reveal the difficulty of operating a measuring platform for the purpose of determining cross-sectional transport rates.

It is interesting to compare measured hydrographic data with the predictions of a hydrodynamical 1-D model of the Danish Hydraulic Institute (dhi) [8]. The input data used for running the model were tidal gauge values from Cuxhaven and river discharge data from Neu Darchau. Evidently, the calculated ebb current velocities are generally 10 to 30 cm/s lower than the experimental ones (Figure 4, top), whereas the flood data agree more or less if it is considered that the very pronounced maxima at the beginning of the high tide might sometimes be missed in the half-hourly measuring cycles. The experimental cross-sectional areas (Figure 4, bottom) determined by echo sounding during the traversing cruises, show higher values for both high and low water, the tidal range being roughly the same. Since it is extremely unlikely that the measured current velocity data are excessively high for ebb tide only (there is also no doubt on the correctness of the calibration), it may be concluded that the model does not reproduce the local effects exactly, although the general agreement is not bad.

A most interesting tidal feature became evident by trace element analyses of both s.p.m. and filtrates [9]. The behaviour of the chromium content in seston, as shown in Figure 5, having maximum values at low-tide slack water and vice versa, is typical of many other heavy metals and trace elements like Ni, Cu, Zn, As, Se, Sr, Ag, Cd, Sb, Ba, Hg, Pb and U. In accordance with this, the corresponding concentrations at the entrance of the harbour (Oortkaten) are higher (Table 1). A similar effect has been observed for a smaller group of elements (Cr, Mn, Fe, Co, and Ag) in the dissolved fraction. Both phenomena might be explained by the admixture with less contaminated material (seston and water) downstream from Hamburg. One exception to this with just the opposite behaviour (Figure 5), is bromine in s.p.m. the quantity of which was small compared to the regularly occurring dissolved bromine. In this case a source of bromine downstream from Hamburg has to be assumed.

Since both hydrological and chemical parameters at the harbour entrance did not change significantly during the experiment, the transport rates at Oortkaten are easily derived from the average values for the river discharge of 550 m^3/s [10], the seston concentration of 24 mg/l (or the resulting s.p.m. flow rate of 13.2 kg/s) and the elemental data of Table 1. The daily rates for these elements are compiled in Table 2. The values correspond very well to those obtained during the BILEX'84 and BILEX'85 campaigns [1, 2].

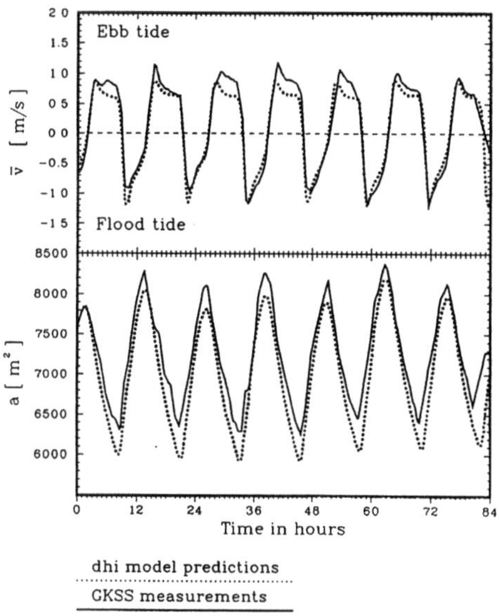

Figure 4: Temporal variation of measured current velocity (top) and cross-section area (bottom) in comparison with corresponding 1-D model predictions due to [8]. Dotted line: dhi model predictions. Full line: GKSS measurements

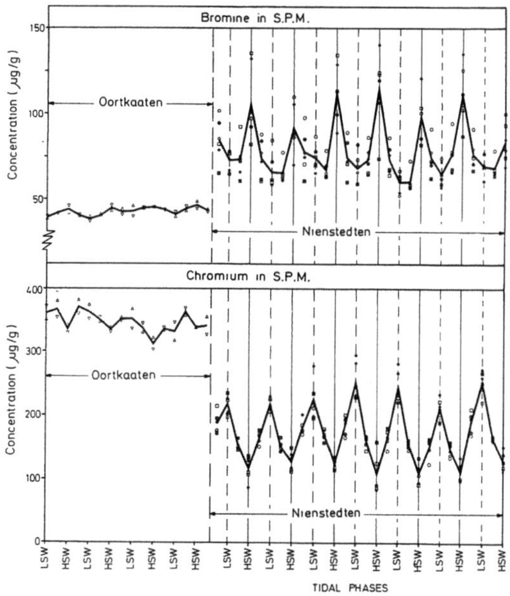

Figure 5: Tidal variations of cross-sectionally averaged Cr and Br concentration in s.p.m. upstream and downstream from Hamburg. The symbols refer to different depths (near bottom and near surface) and locations across the river. LSW/HSW: Low-tide/high-tide slack water.

Table 1: Trace element concentrations in µg/g seston and µg/l filtrate

Element	Oortkaten (upstream) part.		diss. (average)		Nienstedten (downstream) part.		diss. (max/min or average)	
Cr	350	(a)	2	(a)	300/90	(a)	2.5/1	(a)
Ni	88	(a)	11	(b)	77/33	(a)	10	(b)
Cu	290	(b)	4	(b)	215/65	(b)	4	(b)
Zn	1750	(a)	33	(b)	1350/500	(a)	27	(b)
As	71	(a)	6	(b)	67/33	(a)	6	(b)
Br	42	(a)	160	(a)	138/55	(a)	185	(a)
Cd	11	(a)	.06	(c)	7/2	(a)	.05	(c)
Hg	24	(a)	.13	(c)	15/4	(a)	.07	(c)
Pb	210	(b)	-	-	185/85	(b)	-	-

'max/min' values are only given for pronounced tidal variations of data.
Methods of analysis are indicated by a) INAA, b) TXRF, c) AAS.

Table 2: Daily transport of selected trace elements in kg/d through a cross-section upstream from Hamburg Harbour (Oortkaten)

Element	particulate	dissolved	total
Cr	400	95	495
Ni	100	525	625
Cu	330	190	520
Zn	1995	1570	3565
As	80	285	365
Br	50	7600	7650
Cd	13	3	16
Hg	(27) *)	(6) *)	33
Pb	240	-	240

*) The total rather than the fractional transport should be considered due to potential sorption effects during filtration of water samples.

Because of the pronounced temporal variabilities and local heterogeneities, the evaluation of data with respect to water, s.p.m. and heavy metal balances at the cross-section near Nienstedten is more complex. In short, the procedure may be outlined as follows: At 25 m intervals across the river, the half-hourly taken horizontal profile data are interpolated in time and transformed into water-column averaged values by means of the hourly taken vertical profiles at five (sometimes: three) locations. By considering the associated local water depths measured simultaneously by echo sounding, the time dependent cross-sectional water and seston transport rates, or, by a further step of summation over tidal cycles, the corresponding net transports are calculated. Additionally, prior to combining the three-hourly determined element-analytical data with the cross-sectional water or seston transport rates, continuous functions of time have been derived by interpolation. Vertical and lateral dependencies of element concentrations are not treated explicitly, as they are small compared to the temporal variation.

The results are compiled in Table 3 and, partially, presented in Figure 6. The tidal averages in Table 3 have been arranged according to the tidal cycles as can be identified in Figures 2 or 6, and starting with the first high-tide slack water at 6:00 h on September 18, 1986. The river discharge rates show systematic fluctuations corresponding to the daily inequalities in tide elevations (Figure 6); the minimum of 15 m^3/s during the 5th cycle is due to a wind induced increase at high tide. The balance of seston is rather critical; as the concentrations observed are highest in the flood phases, low water discharge tends to result in negative, i.e. upstream directed seston transport. The average over the six tidal cycles considered (last column) is 845 m^3/s water discharge and -16 kg/s seston transport; i.e. the mean s.p.m. input from the lower Elbe river into the harbour at Nienstedten approximately equals the input from the upper Elbe river at Oortkaten (13 kg/s).

Table 3: Tidally averaged transport of water (m^3/s), seston (kg/s), and selected trace elements (particulate/dissolved in kg/d) through a cross-section downstream from Hamburg Harbour (Nienstedten)

Tide No.	1	2	3	4	5	6	total
Water	1100	1220	680	1330	15	660	845
S.p.m.	42	65	- 36	8	- 112	- 67	- 16
Cr	430/150	750/160	- 600/100	- 30/180	- 1750/25	- 1180/90	- 390/120
Ni	170/1030	290/1160	- 180/660	20/1290	- 560/25	- 350/630	- 100/810
Cu	430/370	650/420	- 180/240	160/470	- 870/10	- 530/230	- 50/290
Zn	2410/2430	4120/2740	- 2990/1560	30/3050	- 8950/60	- 5890/1480	1820/1910
As	150/560	250/630	- 160/360	13/700	- 500/13	- 315/340	- 90/440
Br	320/16800	510/19000	- 240/10800	100/21100	- 730/390	- 380/10200	- 65/13200
Cd	5/6	10/7	- 10/4	- 2/8	- 30/0.2	- 21/4	- 8/5
Hg	20/7	35/8	- 30/4	- 2/9	- 90/0.2	- 60/4	- 20/6
Pb	380/n.d.	640/n.d.	- 420/n.d.	30/n.d.	- 1280/n.d.	- 820/n.d.	- 240/n.d.

As may be seen by the temporal behaviour of Ni transport, illustrated in Figure 6 as an example of the elements listed in Table 3, the transport of dissolved and particulate trace elements is mainly determined by the flow rates of water or se-

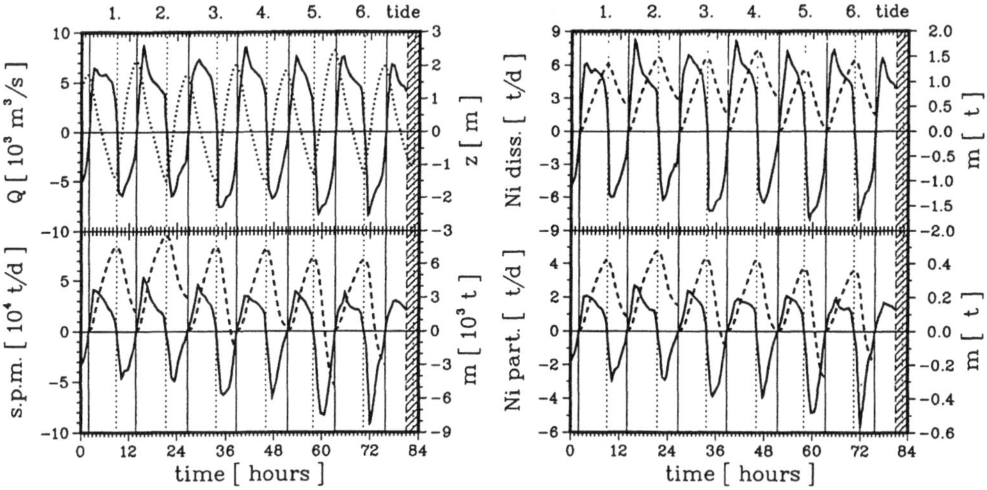

Figure 6: Multi-tidal flow rates at Nienstedten for water (upper left), s.p.m. (lower left), dissolved (upper right) and particulate nickel (lower right). Dotted curve: elevation (tide gauge Seemannshöft); dashed curve: accumulative tidal transport of s.p.m. and Ni

ston, respectively, in spite of the fact that the concentrations of elements show tidal variations of up to a factor of 3.5 (Table 1). Thus it is not surprising that the average transport is directed downstream for the dissolved, but upstream for the particulate phase of the trace elements. According to the specific partition ratio of elements, the output flow from the Hamburg harbour prevails for nickel, copper, arsenic, bromine, whereas the input flow predominates for chromium, cadmium, mercury, lead. A position of equilibrium is kept by zinc. The data give an excellent illustration of the necessity to analyse the filtrate and solid fraction of water samples separately in order to determine correct transport rates for heavy metals.

The most striking result of this experiment suggests that the Hamburg harbour acts as seston and heavy-metal sink from both upstream and downstream. Although local effects and a special wind and tidal situation (spring tide on Sept. 20) cannot be excluded to be the cause of this behaviour, the results correspond closely with the evidence for landward particle transport as found in mineral and isotopic investigations for a longitudinal section of the Elbe river [11].

REFERENCES

[1] Michaelis W (1989) The BILEX concept - research supporting public water quality surveillance. This volume, chapter III

[2] Michaelis W, Fanger H-U, Müller A (eds) (1988) Die Bilanzierungsexperimente 1984 und 1985 (BILEX'84 und BILEX'85) auf der oberen Tideelbe. GKSS 88/E/22

[3] Fanger H-U, Kunze B, Michaelis W, Müller A, Riethmüller R (1987) Suspended matter and heavy metal transport in the tidal Elbe river. In: Awaya Y, Kusuda T (eds) Specialised Conference on Coastal and Estuarine Pollution. Kyushu University, Fukuoka (Japan), p 302 - 309

[4] Kappenberg J, Fanger H-U, Männing V, Prange A (1989) Suspended matter and heavy metal transport in the lower Elbe river under different flow conditions. This volume, chapter IV

[5] Franz H, Kolb M (1989) Transport of rhodamine-B in the Elbe River following a short release upstream of the Port of Hamburg. This volume, chapter IV

[6] Fanger H-U, Kuhn H, Kappenberg J, Maixner U, Milferstädt D (1989) The hydrographic measuring system HYDRA. This volume, chapter V

[7] Neumann L J R (1985) AMO - Die automatische Meßstation Oortkaten. Die Küste 42, p. 151 - 161

[8] Danish Hydraulic Institute (1988) Transport Dispersion Model for Elbe River and Hamburg Harbour - Simulation of Rhodamine Experiment. Strom- und Hafenbau, Hamburg

[9] Prange A, Niedergesäss R, Schnier C (1989) Multielement determination of trace elements in estuarine waters by TXRF and INAA. This volume, chapter X

[10] Christiansen H, Strom- und Hafenbau Hamburg (1988), priv. comm.

[11] Knauth H-D, Schwedhelm E, Sturm R, Weiler K, Salomons W (1987) The importance of physical processes on contaminant behaviour in estuaries. In: Awaya Y, Kusuda T (eds) Specialised Conference on Coastal and Estuarine Pollution. Kyushu University, Fukuoka (Japan), p 236.

TIDAL ASYMMETRY AND FINE SUSPENDED SEDIMENT TRANSPORT IN THE EASTERN SCHELDT TIDAL BASIN (THE NETHERLANDS)

W.B.M. ten Brinke
Department of Physical Geography, University of Utrecht

Postal address: Ministry of Transport and Public Works, Rijkswaterstaat, Tidal Waters Division, P.O. Box 8039, 4330 EA Middelburg, The Netherlands

1. INTRODUCTION

The Eastern Scheldt is a tidal basin in the southwestern part of The Netherlands (Figure 1). In recent years the Eastern Scheldt has been subject to changes in the hydraulic conditions due to the building of a storm surge barrier in the western part, and two dams in the eastern and north-eastern part of the basin. By closing the Philipsdam in April 1987, and so creating a new fresh water lake, the Eastern Scheldt obtained its new shape.

The civil engineering works have influenced hydro-dynamics and thus the environmental conditions in the Eastern Scheldt. In order to investigate these changes in the environment much research is being carried out concerning water quality, biology, morphology and sediment transport. This paper deals with some investigations concerning fine suspended sediment transport. This research is important because fine particles partly consist of organic carbon, food for all sorts of organisms.

Fresh water input into the Eastern Scheldt is very small compared with the average tidal discharge. Thus gravitational circulation, a typical process in estuaries, is of no importance in the Eastern Scheldt. In fact, the Eastern Scheldt is not an estuary but a tidal basin in which the main process for fine suspended sediment transport is based on the tidal asymmetry.

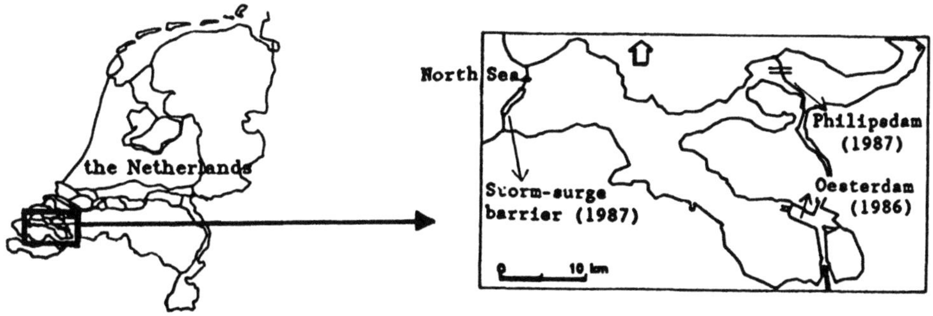

Figure 1. The location of the Eastern Scheldt tidal basin in The Netherlands.

2. TIDE-INDUCED RESIDUAL SEDIMENT TRANSPORT: A TRANSPORT MODEL

A theory explaining landward transport of suspended sediment in tidal basins with a very small input of fresh water (less than 1 % of the average tidal discharge) was developed by Postma (1961, 1967). This theory is based on the behaviour of sediment particles, the process of settling and resuspension. Postma assumed a sinusoidal shape of the tidal wave and a linear decrease of current velocity in the landward direction. In reality, however, a tidal wave will be asymmetrical because of deformation within the tidal basin (Allen et al., 1980). The direction and size of the residual transport are mainly determined by differences in duration of high (HWS) and low water slack (LWS). Using Postma's theory, Dronkers (1984, 1985, 1986) derived a one-dimensional transport model, formulating the tide-induced residual sediment transport as a product of the amount of sediment that is deposited at water slack (μ) and the relative displacement to which sediment particles are subjected (λ):

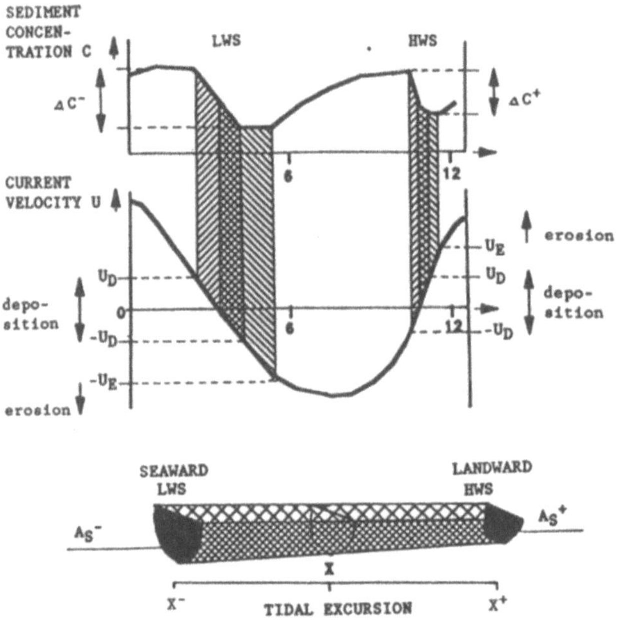

$A_s(x^{\pm},t^{\pm})$ = stream cross-section at water slack
$\Delta c^{\pm}(x)$ = sediment settled at water slack
U_e = critical velocity for eroding the sediment particles
$\Delta t_e^{\pm}(x^{\pm})$ = time interval in which $0 \le$ current velocity $\le U_e$
U_d = critical velocity for settling of the sediment particles
$\Delta t_d^{\pm}(x^{\pm})$ = time interval in which sedimentation occurs
HWS = high water slack
LWS = low water slack

Figure 2. Sketch defining the parameters used for calculating the tide-induced residual sediment transport (modified after: Dronkers, 1985).

$$\mu^{\pm}(x) \;=\; A_s(x^{\pm},t^{\pm}) * \Delta C^{\pm}(x)$$

$$\lambda^{\pm}(x) \;\approx\; \tfrac{1}{2}U_e * \Delta t_e^{\,\pm}(x^{\pm}) \qquad \text{(Figure 2)}$$

In the equations + means HWS and - LWS. The transport through a crosss-section at position X is calculated. Assuming that the tidal excursion is L, at HWS the sediment is deposited at location $X^+ = X + \tfrac{1}{2}L$ and at LWS at location $X^- = X - \tfrac{1}{2}L$. The difference between the landward transport $\mu^+ \lambda^+$ and the seaward transport $\mu^- \lambda^-$ is the transport during a tidal cycle through a cross-section at position X.

3. TIDAL ASYMMETRY IN THE EASTERN SCHELDT

Before dam building: The channels in most of the Eastern Scheldt are rather deep (up to 50 metres below mean sea level). In the period before dam building this morphology resulted in a tidal asymmetry in which the time derivative of the flow velocity is larger at HWS compared with LWS (Dronkers, 1986) (Figure 3A). A large time derivative of the flow velocity at water slack means a relative short period of time with low velocities. Hence it appears that for most of the Eastern Scheldt in its old shape a seaward transport of fine sediment was favoured. In the eastern part of the basin the morphology is dominated by large tidal flats, where the tidal asymmetry was reversed (Figure 3B) and a landward transport was favoured.

After dam building: Man's impact on the Eastern Scheldt system has changed hydro-dynamics quite substantially. Though current velocities are strongly reduced, the shape of the velocity curve in most of the Eastern Scheldt is still the same and so is tidal asymmetry (Figure 3C). This is not unexpected because tidal asymmetry is mainly the result of deformation within the tidal basin and thus depends on the morphology of the basin. Though the morphology of the basin will change in the next hundred years (Mulder et al., submitted), it has hardly changed at the moment.

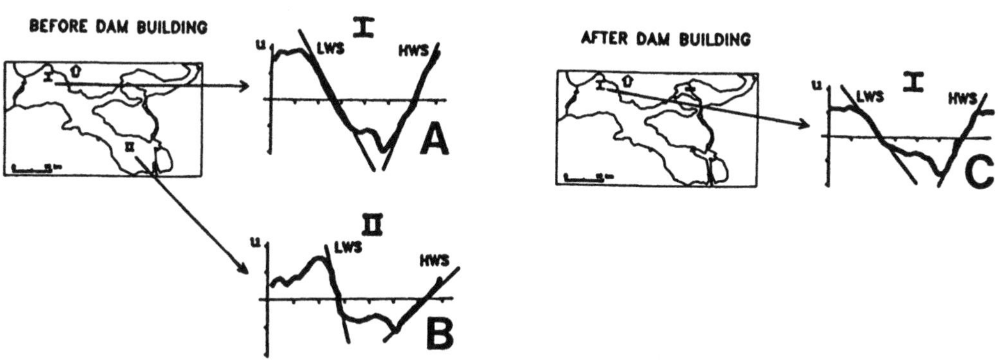

Figure 3. Asymmetry of the current velocity curve in the Eastern Scheldt before (3A and 3B) and after (3C) the construction of civil engineering works.

4. FIELD MEASUREMENTS: INPUT DATA MODEL

The input data necessary for calculating the tide-induced transport were derived from measurements with vessels during calm weather. During these measurements a certain water volume, marked by a float, was followed for 13 hours (a tidal cycle) and fine sediment concentrations and current velocities of this volume were recorded at regular time intervals.

A total of 13 measurements was carried out in 4 different channels of the Eastern Scheldt during 1985 - 1987. Six measurements were carried out in the period before the major changes in hydro-dynamics. The other measurements were carried out in the period between the first major changes and the completion of the engineering works.

5. FINE SEDIMENT TRANSPORT IN THE EASTERN SCHELDT: OUTPUT MODEL

The results from the measurements carried out in the period before the first major changes ("old" Eastern Scheldt) and between the first and the last major changes in hydro-dynamics are shown in Figure 4. For the Eastern Scheldt in its old shape the transport directions calculated from the measurements disagree with the expectations based on the tidal asymmetry. For the changing Eastern Scheldt a seaward transport was calculated for all 7 measurements that were carried out.

6. SEDIMENTATION-EROSION BALANCE: A COMPARISON

In order to evaluate the results of the measurements, transport of fine suspended sediment in the Eastern Scheldt was also calculated from a detailed study of sources (erosion) and sinks (sedimentation) in different parts of the tidal basin.

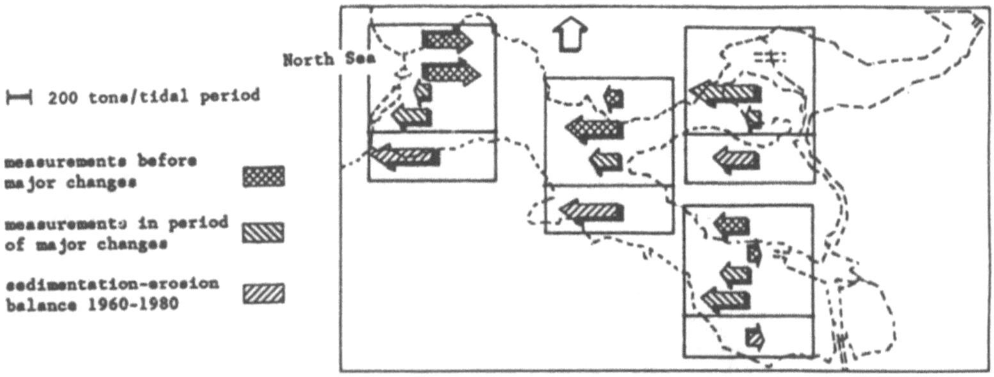

Figure 4. Fine sediment transport in the Eastern Scheldt calculated from measurements with vessels (section 5) and from a sedimentation-erosion balance (section 6).

The erosion of fine sediment in different parts of the basin was estimated from echo soundings over the period 1960 - 1980 and lithological information from borings throughout the Eastern Scheldt. Also data on the deposition of fine sediment in these parts during the 20-year period were collected.

The results of this balance study (Figure 4) agree with the expectations based on tidal asymmetry.

7. DISCUSSION AND CONCLUSIONS

Several complicating factors cause differences between results of measurements with vessels, and the interpretation of sedimentation and erosion in the basin. Of course, all measurements were carried out during calm weather whilst storm effects are considered in the sedimentation-erosion balance. Probably more important is the fact that the one-dimensional transport model is based on a number of assumptions. The most important are:

- no dispersion and diffusion processes
- no sidelong sediment exchange between the main channels and the shallow regions
- no net bottom erosion or bottom accretion on the trajectories of the floating measurements
- a unique relationship exists between the bottom shear stress and the cross-sectionally averaged current velocity.

These assumptions do not hold for most of the Eastern Scheldt. For example, seaward transport directions were calculated for the eastern part of the basin probably because erosion took place on the trajectories of the floating measurements.

Looking at tidal asymmetry, a seaward transport of fine sediment in most of the Eastern Scheldt in its new shape seems logical. Recent observations, however, indicate that sedimentation of fine particles has increased in different parts of the basin after completion of the engineering works. These observations disagree with a seaward transport of fine sediment in the main channels. In fact, the one-dimensional model cannot be used to calculate fine sediment transport in the Eastern Scheldt in its new shape. Because of strongly reduced current velocities, averaged current velocity is not an indication of bottom shear stress.

From the investigations the following can be concluded. In the period before the construction of the engineering works the large scale transport of fine sediment took place in a seaward direction for most of the Eastern Scheldt. A change in direction of net sediment transport as a result of the completion of the engineering works cannot be concluded from the floating measurements.

The eastern part of the basin functions as a sediment trap as described by Postma's theory.

REFERENCES

Allen GP, Salomon JC, Bassoullet P, Du Penhoat Y, De Grandpré C (1980) Effects of tides on mixing and suspended sediment transport in macrotidal estuaries. Sed Geol 26: 69-90

Dronkers J (1984) Import of fine marine sediment in tidal basins. Proceedings of the International Wadden Sea Symposium, Texel 1983. Neth Inst for Sea Res Publ Series 10: 83-105

Dronkers J (1985) Tide-induced residual transport of fine sediment. In: Van de Kreeke J (ed) Physics of Shallow Estuaries and Bays. Lecture Notes on Coastal and Estuarine Studies 16. Springer, Berlin - Heidelberg - New York - Tokyo, p 228

Dronkers J (1986) Tidal asymmetry and estuarine morphology. Neth J Sea Res 20(2/3): 117-131

Mulder JPM, Van der Spek AJF, Berben FML (submitted) Closure of tidal basins: geomorphological consequences and environmental impact. Env Geol Wat Sc

Postma H (1961) Transport and accumulation of suspended matter in the Dutch Wadden Sea. Neth J Sea Res 1: 148-190

Postma H (1967) Sediment transport and sedimentation in the marine environment. In: Lauff GH (ed) Estuaries Am Ass Adv Sci, Washington, p 158

FLUXES OF WATER, SOLIDS AND METALS THROUGH THE MERSEY ESTUARY (UK) AS MEASURED BY A CONTINUOUS DATA RECORDING STATION (MIDAS)

P C Head, P D Jones
North West Water Authority, Warrington, UK
and K Whitelaw
Water Research Centre, Medmenham UK

INTRODUCTION

With the increasing concern about the environmental consequences of waste discharges to the aquatic environment, attention is being focussed on how to determine the quantities of various potential pollutants discharged to the sea from land-based sources. Recognizing this, North West Water (NWWA) and the Water Research Centre (WRc) used the information gained from a series of tidal cycle surveys of the Mersey estuary [1] to develop a long-term data recording station (MIDAS) to attempt to measure the flux of water and solids in to and out of the estuary (see [2] and Figure 1).

Figure 1.

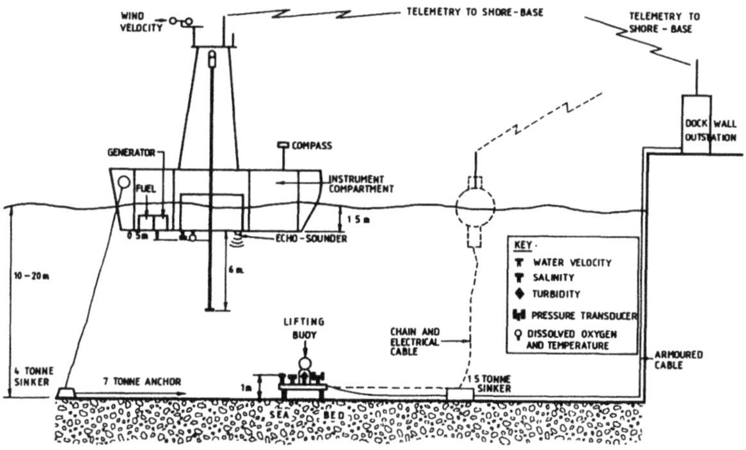

Mersey Inshore Data Acquisition Station - MIDAS

MIDAS was deployed, in 1986, to measure current speed and direction, temperature, salinity, and suspended solids every 15 minutes at three depths at a point near to the position of maximum flux, as determined from the tidal cycle surveys (see [2]). Because of various operational problems, data for all three depths were only obtained for an approximately 8 month period between November 1987 and July 1988.

FLUX CALCULATIONS

Relationship between MIDAS and cross-sectional data

To examine the relationship between measurements collected from MIDAS and cross-sectional information from the tidal cycle surveys, each was subjected to a simplified tidal analysis [3] which showed that the two data sets were comparable; depth mean estimates of the dominant M_2 current were 1.25 m s^{-1} for the MIDAS data and 1.15 m s^{-1}, for a similar position on the cross section, for the tidal cycle data. The tidal cycle data suggested that the MIDAS current data should provide a reasonable basis for estimating cross-sectional fluxes in the absence of persistent lateral circulation. An examination of the residual currents derived from the tidal analysis showed the predominant flow structure to be the classical estuarine regime of a seawards surface flow and a bottom landwards flow. The residual flows, of between 0.018 m s^{-1} landwards and 0.058 m s^{-1} seawards, for the four surveys may be compared to a typical fresh-water induced flow of about 0.003 m s^{-1} at this point in the estuary.

Measurements of suspended solids during the tidal cycle surveys showed that concentrations varied considerably with depth, position and tidal state. Overall the was a 4.5 fold increase with depth, and the depth average concentrations were about 3.5 times greater on the eastern side of the estuary. Half-tide solids fluxes calculated from the MIDAS data ranged from around 2 500 tonnes for a neap tide up to 205 000 tonnes for an extreme spring tide, which are in reasonable agreement with figures derived from the tidal cycle surveys and previous work [4]. Estimates of residual solids fluxes were found to be very sensitive to the assumed depth distribution of solids concentrations.

Water Fluxes

Water fluxes at the MIDAS position were calculated as the product of the depth-mean longitudinal velocity and water depth, for each of 15 minute values. This yielded maximum flood and ebb fluxes of around

20 m^3 s^{-1} per meter width for spring tides, and about half this figure for neap tides. By using a low-pass filter to remove the tidal influences a long-term non-tidal seawards flux of 1.15 m^3 s^{-1} was obtained. Landward non-tidal fluxes occurred only during periods of strong up-estuary winds. The non-tidal fluxes at the MIDAS position yield an average cross-sectional flux of around 1 500 m^3 s^{-1}, which although considerably larger than the probable fresh-water inflow, is consistent with previous data [5].

Solids Fluxes

As had been expected from previous work in the estuary, the concentration of suspended solids was closely related to tidal range. Peak concentrations ranged from undetectable (<50 mg l^{-1}) at both surface and mid depth during neap tides, to around 600 mg l^{-1} at the surface and 800 mg l^{-1} at mid depth during spring tides. A similar pattern was found for the near-bed data, with peak concentrations of around 100 mg l^{-1} during neap tides and usually more than 1 000 mg l^{-1} at spring tides.

Estimates of solids fluxes per meter width varied from around 300 kg s^{-1} at neap tides to about 10 000 kg s^{-1} at spring tides. These figures are equivalent to cross-sectional half-tide fluxes of between 6 000 and 130 000 tonnes for neap and spring tides respectively, and are comparable with the multi-boat exercises and previous work [6].

Metal fluxes

Estimates of metal fluxes were derived by multiplying the water and solids fluxes by typical metal concentrations, obtained from the tidal cycle surveys and previous investigations. The figures given in Table 1 show the extent to which instantaneous fluxes vary between neap and spring tides and the degree to which the various metals are partitioned into dissolved and particulate fractions. The mean fluxes were directed seawards, reflecting the net cross-sectional water and solids fluxes derived from the MIDAS data.

The dominant mode of transport for Hg, Pb, Cr, Cu and Zn was in association with solids, whereas for Cd dissolved forms are more important. In the case of Ni particulate and dissolved forms contributed almost equally to the transport.

Table 1:

Metal	Spring			Neap					
	g s^{-1}						kg d^{-1}		
	Diss.	Part.	Total	Diss.	Part.	Total	Diss.	Part.	Total
Hg	0.5	27	27.5	0.25	0.8	1.3	3	45	48
Cd	3.4	10	13.4	1.7	0.3	2.0	22	15	37
Cu	68	1000	1068	34	30	64	440	1600	2040
Zn	500	4100	4600	250	125	375	3240	6500	9740
Pb	44	1400	1444	22	40	62	285	2250	2535
Cr	13	1200	1213	6	35	41	80	1920	2000
Ni	72	300	372	36	10	42	470	480	950

Fluxes of dissolved (Diss.), particulate (Part.) and total metals

The fluxes calculated from the MIDAS data are, in general, about an order of magnitude greater than flux estimates obtained previously (See Table 2). The flow and concentration fluxes were obtained by multiplying the mean freshwater flow into the estuary by a mean total metal concentration obtained from the tidal cycle surveys, those derived by Campbell and Riley [6] are the product of the metal versus salinity gradients and river flows for particular surveys.

Table 2:

Metal	Mean estuarine flux			Mean input
	MIDAS data	Flow x conc.	Campbell & Riley [6]	Head [7]
Hg	48	3.3	NA	<4.8
Cd	37	2.3	1	<4.1
Cu	2040	130	70	245
Zn	9750	605	1410	848
Pb	2535	155	130	412
Cr	2000	140	NA	NA
Ni	950	60	140	156

Comparison of MIDAS fluxes with previous estimates - all figures are kg d^{-1}. NA - estimate not available.

The lack of agreement between these four sets of data is not unexpect -ed given the differences in the types of measurements made and the assumptions used in deriving the flux estimates. Given the over-estimate of the non-tidal water flux in relation to the size of the fresh-water inflows and the fact that the mean solids flux obtained is in the opposite sense to that required to satisfy the long-term infilling of the estuary [8], it would be surprising if the MIDAS data did not indicate considerably larger seawards fluxes than those suggested by the other methods.

The flux investigations in the Mersey well illustrate the problems involved in trying to determine the net flux of materials in to or out of an estuary, particularly one with a substantial tidal range. Net movements of dissolved or suspended materials are small in comparison with the volumes of water and particulates transported in to and out of the estuary by the tidal currents. The data from the tidal cycle surveys, collected from five depths at approximately 200 m intervals across the estuary, showed the extent of variability on representative tidal cycles. Although these data suggested that measurements at the MIDAS position would enable reasonable estimates of fluxes to be made for the cross section as a whole, they could be subject to significant errors associated with transverse circulations and non-uniform transports across the cross section.

As with previous measurements at this point in the estuary the MIDAS data show residual water movements to be seawards in the upper part of the water column and landward near the bed. However, with only three measurements in the vertical it is not possible to characterize the vertical distribution of velocity sufficiently well to be confident that an appropriate depth averaged value can be calculated. Given that characterizing the vertical distribution of solids is even more difficult the validity of the size and direction of the solids fluxes must be viewed with caution.

The MIDAS data give a very good indication of the variability of water flows, and solids and metal fluxes in the vicinity of the data station but extrapolating the information to give estimates for the cross section, even in this relatively regular section of the estuary, yields estimates which are relatively imprecise. It is probable that the half-tide rates of transport are a reasonable indication of the true situation but the precision with which the non-tidal signal can be established remains poor.

ACKNOWLEDGEMENT

This study was part funded by the Department of the Environment of England and Wales. We wish to thank all who assisted with the gathering and processing of the data.

REFERENCES

1 Whitelaw K, Cole JA, Head PC, Jones PD (1985) Tidal fluxes of metals through the Mersey estuary, 1982-1984. Wat Sci Technol 17:1363-1367

2 Jones PD, Head PC, Whitelaw K (1989) A data recording station to measure water and solids fluxes through the Mersey Narrows. In: McManus J (ed) Proc Symp Marine and Estuarine Methodologies Sept 1987. Olsen & Olsen, Copenhagen

3 Prandle D, Murray A, Johnson R (to be published) Analyses of flux measurements in the River Mersey. In: Proc Int Conf on Physics of Estuaries and Bays. 29 Nov-2 Dec 1988, Asilomar, California

4 Halliwell A R, O'Connor B A (1975) Water and sediment movement in the Mersey estuary and Liverpool Bay. In: Liverpool Bay; an assessment of present-day knowledge. Natural Environment Research Council Publs Series C, No. 14, London

5 Bowden K, Sharaf El Din SH (1966) Circulation, salinity and river discharge in the Mersey estuary. Geophys J R Ast Soc 10:383-399

6 Campbell JA, Riley JP (1984) The study of the distribution of selected trace metals in the waters and suspended particulate matter of the Mersey estuary. Final Contract Rep. to NWWA, Liverpool University

7 Head PC (1985) An assessment of the input of water and various dissolved materials to the estuaries and coastal waters of the Irish Sea from the area of the North West Water Authority. NWWA Rep No PL-SRD-85-1

8 O'Connor BA (1987) Short and long term changes in estuary capacity. J Geol Soc , London 144:187-195

SUSPENDED MATTER AND HEAVY METAL TRANSPORT IN THE LOWER ELBE RIVER UNDER DIFFERENT FLOW CONDITIONS

J. Kappenberg, H.-U. Fanger, V. Männing, A. Prange
Institute of Physics, GKSS Research Centre Geesthacht GmbH
P.O.Box 1160, D-2054 Geesthacht, Federal Republic of Germany

INTRODUCTION

The Elbe river in northern Germany is considered to be a major contributor to the pollution of the North Sea. Although at its mouth isolated measurements of pollutant concentrations do exist, quantitative data on the pollutant freight are scarce and notoriously unreliable, being based on the fresh water discharge at Neu-Darchau 200 km upstream.

The results of ealier investigations of pollutant and suspended particulate matter (s.p.m.) transport are available from the upstream [1,2] and downstream [3] ends of the Hamburg harbour area, which is located halfway down the river from Neu-Darchau. These measurements show considerable spatial and temporal variations along a reach of some 50 kilometers and over a period of one week. Thus it may not be expected that, on the basis of these measurements, realistic transport rates can be deduced for the river mouth 100 km downstream.

In 1986 a series of campaigns was initiated by GKSS, which should shed some light on the transport of s.p.m and pollutants at the mouth of the river near Brunsbüttel, and prepare the ground for a major BILEX-experiment [5] forthcoming in September '89 .

THE EXPERIMENTS IN 1986, 1987, 1988

The measurements were conducted on four cross-sections near Brunsbüttel using the measuring system HYDRA [4] on board the LUDWIG PRANDTL, and applying the moving boat technique in a similar way as during BILEX'86 [3]. Due to the greater width (ca. 2 km) of the river, the measuring frequency had to be reduced to hourly profiles. A further modification was the reduction of the balancing period to just over one tide per campaign, because the navigational hazards of slowly traversing a heavily shipped waterway at night (often accompanied by fog) could not be taken. Thus the measuring period did not include three slack water phases, so that the beginning and end of the tide could not be defined in the conventional way. Also the daily tidal inequality could not be accounted for in the measurements. Moreover, sensor failures and measuring gaps due to shipping ruined 4 out of 9 campaigns. Usable measurements were made in April and September '87 and in September and October '88. Their relation to the fresh water discharge can be seen from Figure 1 in [3]. The April '87 campaign coincided with extremely high water transport (2500 m^3/s): the fresh water discharge in September '87 corresponds to the long-term mean (700 m^3/s), while the measurements in '88 were performed under low flow conditions (400

The concentration of s.p.m. was deduced from optical attenuation data, calibrated by samples [4] and using an individual (linear) fit for each campaign. Coinciding with slack water and the fully developed ebb and flood current, these samples were taken in two depths and at three locations along the cross-section four times during one tidal cycle.

The samples were subsequently analyzed for their contents of trace elements, separately for the dissolved and the particulate fraction [6]. Only the results from the September 1987 campaign are presented in this paper.

RESULTS AND DISCUSSION

Table 1 gives the results of the campaigns at the cross-sections Brunsbüttel (Elbe km 695.5) and Elbe km 691.3, the site of a measuring platform anchored near the northern bank to be used for long-term monitoring of water quality and transport rates.

campaign	cross-section	Time interval	Neu-Darchau Q_0	Measuring site Water discharge	S.p.m. transport
B1	Brunsbüttel	870414, 7:30 - 19:50	2350	5020	70600
B2	Brunsbüttel	870916, 8:00 - 20:24	664	3230	49560
P1	Platform site	870929, 7:55 - 20:18	700	2690	34620
B3	Brunsbüttel	880921, 7:20 - 19:55	418	4900	174400
P2	Platform site	881012, 7:41 - 20:44	410	10	4150

Table 1 : Net transport rates of water in m^3/s and suspended matter in t/d

The water transport rates show poor correlation with the official fresh water discharges at Neu-Darchau. This is to be expected since vast amounts of water pass the cross-section during ebb and flood-tide, exceeding by far the quantities which flow down the Elbe at Neu-Darchau during 6 hours. The contribution of tributaries along the river from Neu-Darchau to Brunsbüttel may be neglected amounting to less than 100 m^3/s. The transport rates at the mouth of the Elbe river are clearly controlled by meteorological and tidal conditions. Secondly, travelling times between Neu-Darchau and Brunsbüttel and the large Hamburg harbour area acting as an unpredictable storage have to be taken into consideration.

The transport of suspended matter seems to reveal some similarity with the water transport rate, but is influenced by the position of the turbidity maximum as well as by possible dumping and dredging activities which are well known to take place in this area. From the overall transport rates no conclusions can be drawn on the spatial range of the transport. High s.p.m. concentrations (especially near the bottom) can be the product of local erosion or advective transport. In this respect something may be glimpsed from the time series of vertical profiles of s.p.m. concentration in Figure 1. Here, position 2 is located in the center of the cross-section, while positions 1 and 3 are about halfway from the center to the northern and southern bank, respectively. During the initial ebb-tide water with a high s.p.m. concentration is passing the cross-section. Towards the low water slack point, the s.p.m. is gradually removed from the water column and settles on the resulting in strong vertical gradients of the concentrations. After the maximum of the flood current we may see some of the s.p.m. again which passed the cross-section during ebb-tide. At the time of slack water, s.p.m. is again removed from the water column. The corresponding vertical profiles of salinity in the lower half of the figure confirm this interpretation. Water with high salinity is transported towards the sea at the beginning of the measurement. The salinity then remains fairly constant until the end of the flood when a different water volume passes through. The following ebb-tide reveals an interesting detail. While in the middle of the stream (position 2) a homogeneous cloud of s.p.m. seems to be carried downstream, vertical stirring processes seem to be responsible for a strong vertical gradient at positions 1 and 3. The time series from the other campaigns show a similar behaviour.

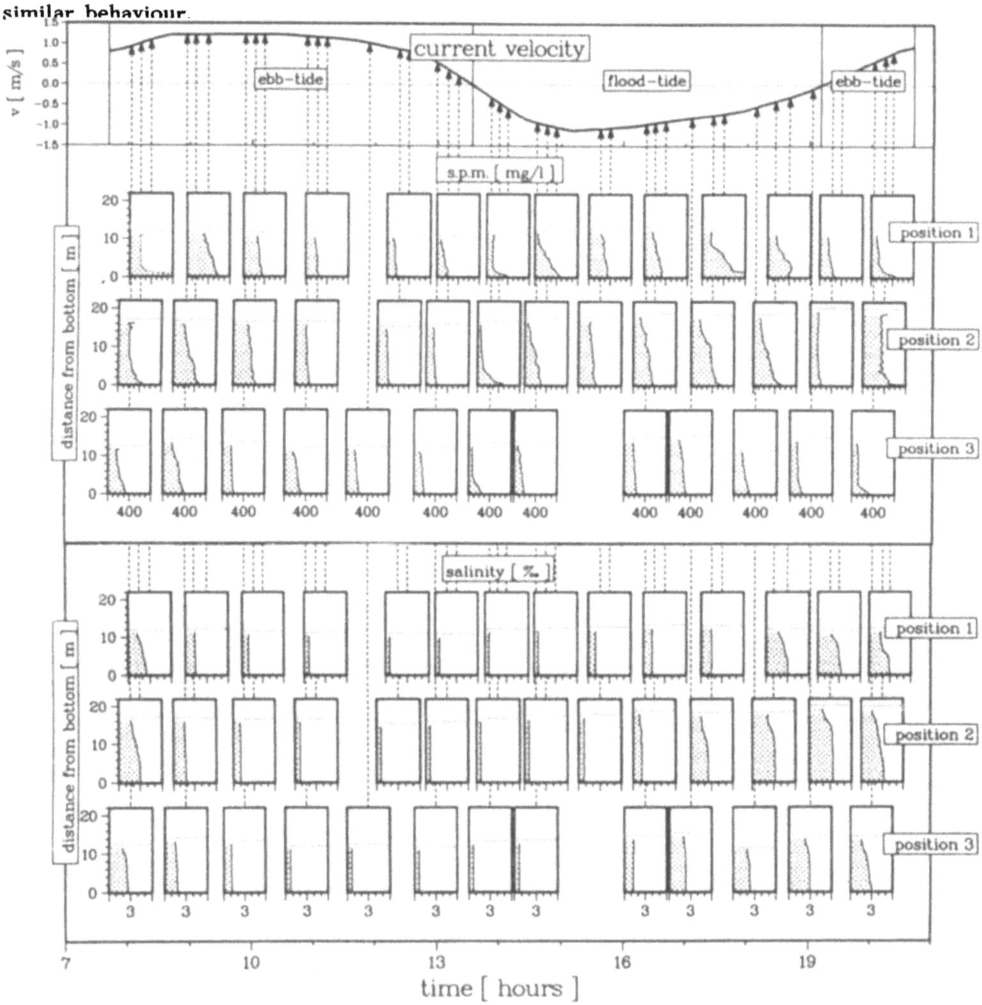

Figure 1: Time series of vertical profiles of s.p.m. and salinity at the platform site cross-section during campaign P1 on Sept 29, 1987.

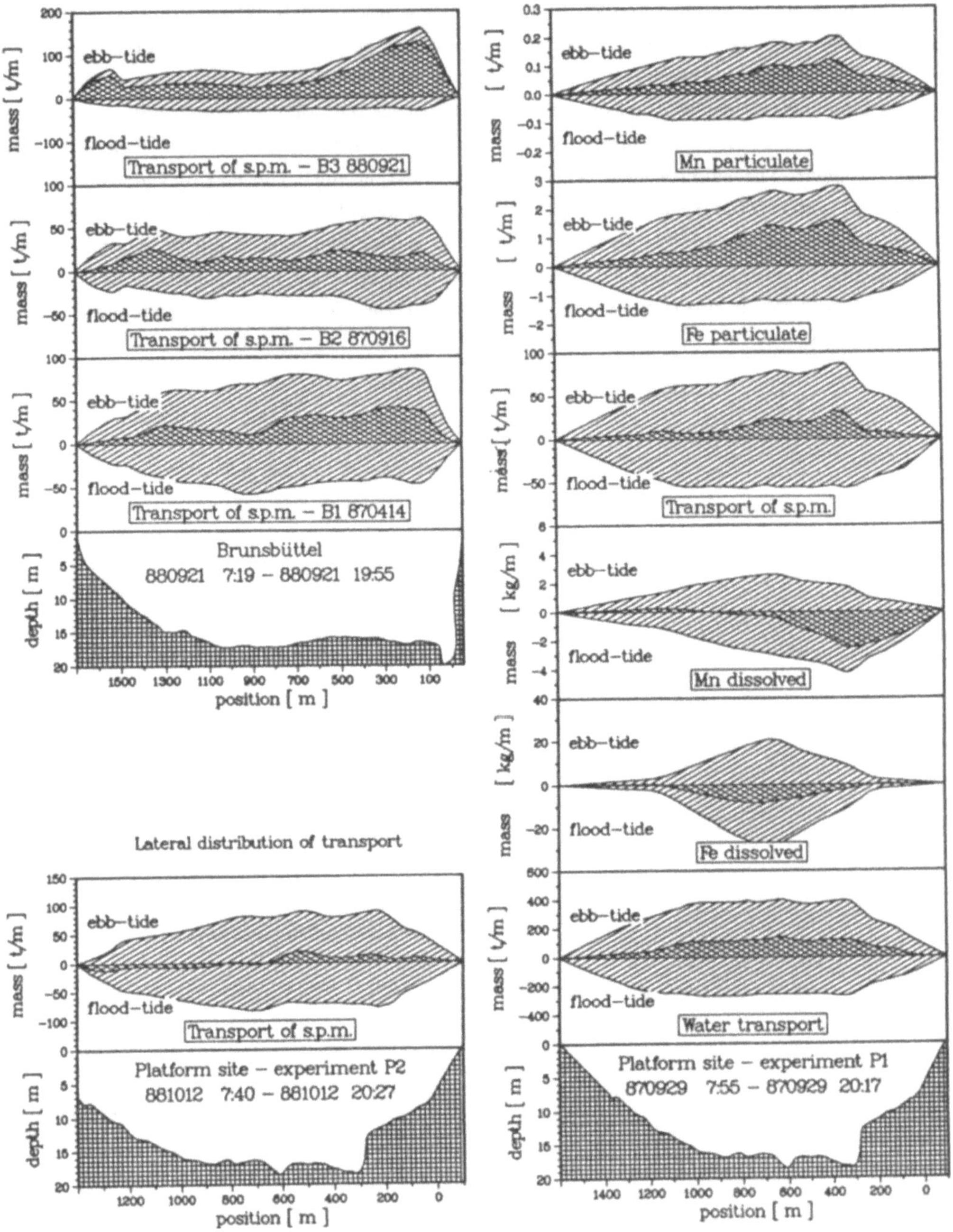

Figure 2: Lateral distribution of transport rates during the campaigns ; shaded area corresponds to integrated ebb and flood transport per meter of the cross-section. Crosshatched areas account for net transport.

Turning our attention to the lateral distribution of the s.p.m. transport integrated over ebb, flood, and the total tide as presented in Figure 2, a different distribution of the net transport for each campaign is found. The mass of s.p.m. passing a 1 m wide slice of the cross-section at a fixed lateral position during ebb or flood-tide is roughly correlated with the mean water depth. For the net transport no such general statement can be made. At the platform site in September '88 we observe upstream transport on the southern side while through the navigation channel at the northern side s.p.m. is transported towards the sea. In this case, a small shifting or stretching of the integration period will tilt the balance. At Brunsbüttel in September '87, the net transport is directed towards the sea but a maximum occurs at the shallow southern side due to the high s.p.m. concentrations. At the platform site 13 days later there is some similarity between the lateral distribution of net transport and depth. The B3 campaign at Brunsbüttel in September '88 reveals a peak of net transport near the southern bank but the lateral maximum of s.p.m. transport occurs at the northern side. Thus, there appears to be no general rule for the lateral distribution of the net transport of s.p.m., and measurements of s.p.m. transport rates on a platform near the bank are insufficient for the determination of the cross-sectional transport.

	B2 - Brunsbüttel Sept 16, 87		P1 - Platform site, Sept 29, 87	
Element	dissolved	particulate	dissolved	particulate
S	24800 t/d	214 t/d	13000 t/d	266 t/d
K	8950	681	3790	727
Ca	26770	1648	19150	1902
Ti		142		172
V	650 kg/d	5360 kg/d	422 kg/d	6990 kg/d
Cr		4937		5820
Mn	940	131 t/d	- 2080	165 t/d
Fe	1671	1872	- 8300	2264
Ni	1440	2925 kg/d	1330	3270 kg/d
Cu	1560	2842	1310	2840
Zn	1770	18236	1270	22300
Ga		620		740
As	1770	1620	760	2060
Rb	7580	5022	3600	5720
Sr	241 t/d	8900	165 t/d	10470
Mo	800 kg/d		570 kg/d	
Y		954		1250
Cd	< 5	77	< 5	85
Ba		16300		19970
Hg		125		173
Pb	60	3835	- 40	4390
U	670		420	

From the transport of trace elements as shown in Table 2 it can be seen that even within a fortnight's period very different conditions prevail. Even the peculiar upstream transport of Mn and Fe in the dissolved phase at the platform site was not observed at Brunsbüttel 13 days earlier and 4 km downstream. Also the ratios between the separate elements are by no means fixed. The transport rates for the particulate fraction during the two experiments are amazingly similar, taking into account that the s.p.m. transports differ by a factor of 1.4 . For most elements in the dissolved fraction there is a remarkable dissimilarity between the lateral distribution of trace element transport and water transport. Trace element concentrations are by no means homogeneous and they vary considerably with the state of the tide.

From these isolated campaigns no coherent picture of the transport processes in the lower estuary of the Elbe river emerges. There are several trends which can be observed during each campaign in slightly modified form, such as the settling and resuspension of s.p.m. near slack water. For the important distinction between vertical and advective transport of s.p.m. new measuring strategies have to be conceived which must involve measurements on several cross-sections not too far apart. Nevertheless the results of these campaigns show that transports of water and s.p.m. at the mouth of the river have little or nothing to do with transports measured far inland, and that pollution loads derived from a few samples are too inaccurate to calculate the total input into the sea. Furthermore, the high variability of the lateral distribution of the transport renders it doubtful that reliable estimates of the overall pollution load of the river might be gained from measurements taken from platforms moored near the banks. The only hope lies in gaining an understanding of the transport processes, may their nature be physical, chemical, or biological.

REFERENCES

[1] Fanger H-U, Kunze B, Michaelis W, Müller A, Riethmüller R (1987) Suspended matter and heavy metal transport in the tidal Elbe river. In: Awaya Y, Kusuda T (eds) Specialised Conference on Coastal and Estuarine Pollution. Kyushu University, Fukuoka (Japan), p 302

[2] Michaelis W, Fanger H-U, Müller A (eds) (1988) Die Bilanzierungsexperimente 1984 und 1985 (BILEX'84 und BILEX'85) auf der oberen Tideelbe. GKSS 88/E/22

[3] Fanger H-U, Kappenberg J, Männing V (1989) A study on the transport of dissolved and particulate matter through the Hamburg harbour. This volume, chapter IV

[4] Fanger H-U, Kappenberg J, Kuhn H, Maixner U, Milferstaedt D (1989) The hydrographic measuring system HYDRA. This volume, chapter V

[5] Michaelis W. (1989) The BILEX concept: research supporting public water quality surveillance in estuaries. This volume, chapter III

[6] Prange A, Niedergesäß R, Schnier C (1989) Multielement determination of trace elements in estuarine waters by TXRF and INAA, This volume, chapter X

SEASONAL VARIATIONS OF MINERALOGICAL COMPOSITION AND HEAVY METAL CONTAMINATION OF SUSPENDED MATTER IN THE RIVER ELBE ESTUARY

M. Vollmer, B. Hudec, H.-D. Knauth, E. Schwedhelm*
GKSS Research Center, Institut für Chemie, D-2054 Geesthacht, FRG

1. Introduction

1.1 Objectives

Previous investigations of various natural tracers and contaminants in the suspended matter and sediments of estuaries elucidated that sediments of marine origin are transported upstream the tidal river across the salt wedge [1,2]. Clay minerals, in particular, were used to trace the upstream transport of marine matter in the Elbe estuary [3,4]. Additional studies on water, suspended particulate matter (SPM) and sediments indicate that the distribution of particle bound pollutants in the Elbe estuary is predominantly influenced by the mixing of marine and fluvial sediments resulting in a seaward directed dilution. It could be shown that water discharge and, to a lesser extent, windstress situations typically affect the transfer of fine grained solids [5,6].
A quantitative approach must allow for the determination of net transport rates of dissolved and suspended substances through selected river sections as performed by GKSS Bilex experiments [7]. Thus far, variations in the clay content of SPM were obtained at two different discharge rate situations [3]. The objective of this study was to enlarge the data base by sampling and analysing SPM along the estuary upstream and downstream of Hamburg harbour.

1.2 Database

Four sites in the estuary were sampled from March 1986 to May 1987 upstream of Hamburg harbour (Schnackenburg, km 475; Bunthaus, km 610) and downstream of Hamburg harbour (Seemannshöft, km 682; Blankenese, km 635). From April 1988 onwards, one new sampling location further downstream was added; Grauer Ort, km 660, which is still in the limnic part of the estuary (Fig. 1).

The samples were collected over a period of one month in SPM traps set up by ARGE-Elbe water authority [8]. The smectite, illite, kaolinite and chlorite contents were determined by x-ray diffraction for the grain size fraction <2μm [1,2]. The heavy metals Cd, Pb, Cu, and Zn were analysed in the fraction <2μm by AAS. The ARGE-Elbe determined the heavy metal content of the same samples in the fraction <20μm and these data were used to complete the interpretation.

*present address: Labor für Geoanalytik, Richthofenstr. 29, D-3200 Hildesheim, FRG

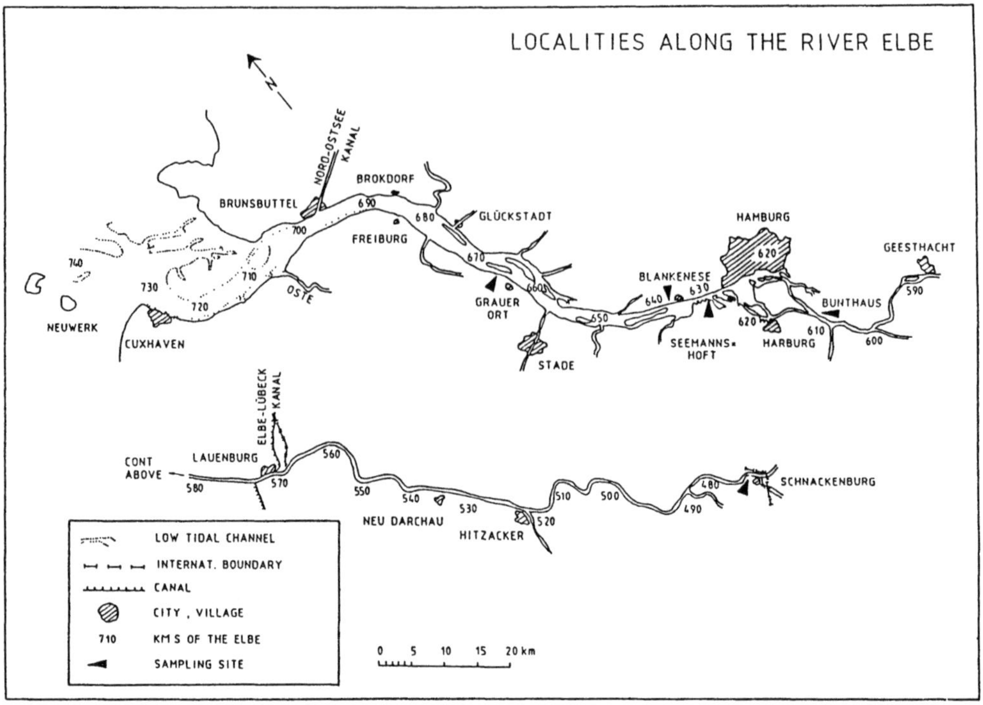

Figure 1: Sampling Locations along the River Elbe

2. Mineralogical Composition of the Suspended Matter

As shown by Schwedhelm [3], the amounts of illite and chlorite did not vary essentially along the river course. The values ranged between 37% to 48% and 9% to 13% respectively. But a decrease of kaolinite (30% to 13%) and an increase of smectite (9% to 36%) were observed. Thus the clay mineral composition of the sediments and SPM upstream of Hamburg was characterized by smectite/kaolinite-ratios < 1. In the lower Elbe section, downstream of Hamburg, the analyses of clay mineral contents showed higher smectite concentrations and simultaneously lower kaolinite values. Here the smectite/kaolinite-ratio was < 3. The transition zone between marine and fluvial sediments, as reflected by differences in natural tracer contents, is likely to be in the area of Hamburg harbour as already postulated from previous work [3].

Fig. 2 shows the smectite/kaolinite-ratio of the SPM at four different sampling sites for the period of March ´86 to June ´87. Apparently the clay mineral content at the two sampling sites Schnackenburg and Bunthaus did not essentially change: the smectite/kaolinite-ratio varied over the entire year in the range of 0.5 to 1.0. Results for the stations downstream of Hamburg, however, showed a different trend. At Seemannshöft, situated at the harbour outlet, and at Blankenese, 7 km further downstream, the ratio was as low as approximately one from March to the end of June, at a time of comparatively high freshwater discharges (Fig. 3). Simultaneously with a sharp decrease of the discharge, followed by a period of lower discharge rates, the smectite/kaolinite-ratio began to increase to values of 2 and 3.5 in November/December at Seemannshöft and Blankenese, respectively. Sediments and SPM with such high smectite/kaolinite-ratios were found to be typical of sediments of the outer estuary and the inshore area of the North Sea [3,9]. The smectite/kaolinite-ratio again decreased to values of between 1.7 and 1.4 in conjunction with the high water discharge up to 2300 m^3/s during the spring of 1987.

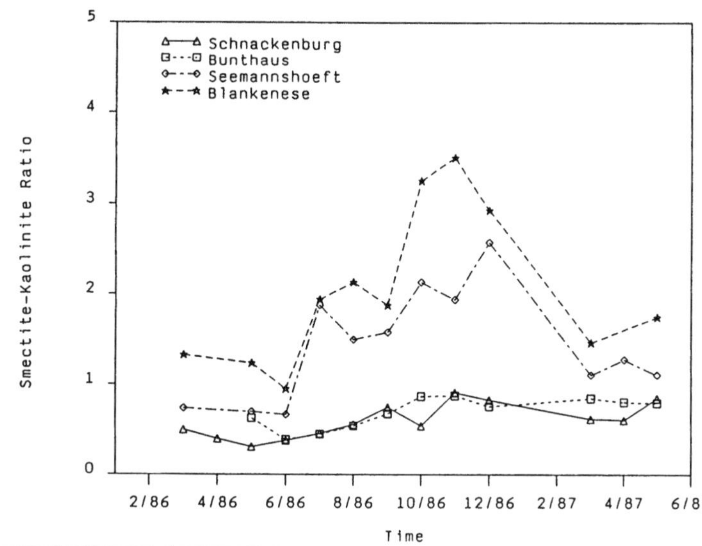

Figure 2: Smectite/Kaolinite-Ratio in the SPM (<2μm) at Schnackenburg, Bunthaus, Seemannshöft and Blankenese March '86 - May '87 (In Jan. '87 drift ice prohibited sampling)

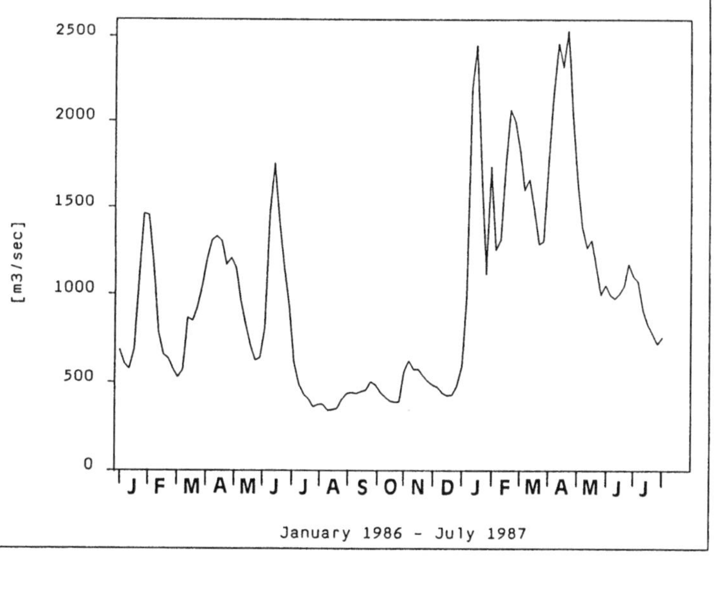

Figure 3: Discharge of the River Elbe (Water gauge Neu-Darchau) Jan. '86 - July '87

3. Heavy Metal Content of the Suspended Matter

Fig. 4 shows the cadmium and copper content of the SPM from Bunthaus and Blankenese. A comparison of the level of Cd and Cu in SPM reveals that, during the time of high freshwater discharge, i.e. up to the end of June, the differences in the particle-bound Cu and Cd concentrations between Bunthaus and Blankenese were comparatively low. Simultaneously with the sharp freshwater discharge decrease, the concentrations at the downstream position dropped leading to comparatively high differences in the seston bound heavy metal concentrations between both stations. Together with a new high freshwater discharge rate at the very beginning of 1987 the Cu and Cd concentrations at Blankenese increased again up to the values of the upstream station. These data confirm the results of a recent study [5].

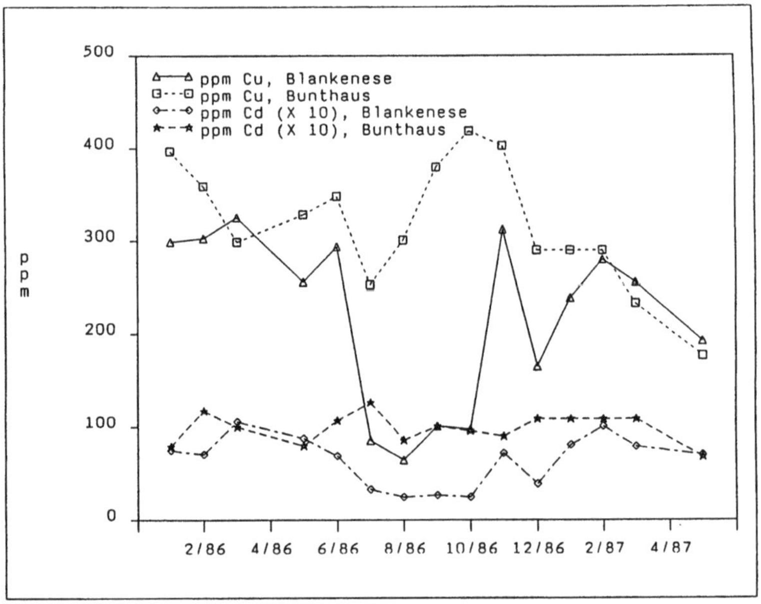

Figure 4: Content of Cadmium and Copper in SPM (<20µm) at Bunthaus and Blankenese Jan. ´86 - May ´87 (Data obtained from ARGE Elbe)

From April ´88 to January ´89 samples were taken at the station Grauer Ort (Fig. 5). A decrease in the heavy metal concentrations over the summer period ´88 with minimum values in Sept. ´88 was observed. This was simultanous with a relatively low water discharge of less than 500 m^3/s (June - end of November ´88). At the beginning of December ´88, a higher discharge (up to 1500 m^3/s) was measured. However, from October onwards the heavy metal content in the SPM had already increased notably. An unequivocal interpretation of this phenomenon can not be given as yet. A possible explanation for this phenomenon might be the remobilisation of biomass bound heavy metals in regions upstream of Grauer Ort at the end of October, followed by a subsequent reabsorption onto different SPM at the end of the annual growth period [10].

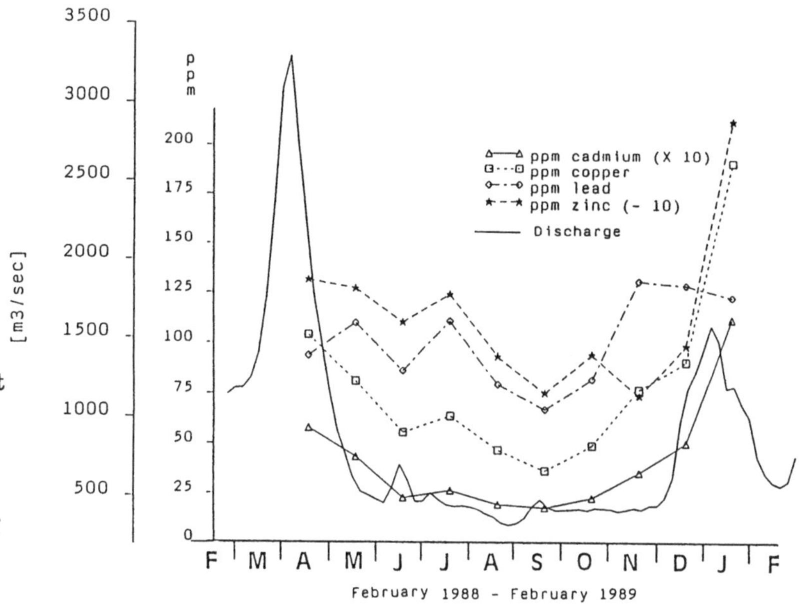

Figure 5:
Heavy Metal Content of SPM (<2µm) at Grauer Ort, April ´88 - Jan. ´89 and the discharge of the River Elbe

4. Conclusions

The clay mineral composition and the heavy metal concentration of the SPM show distinct dependencies on the freshwater discharge of the river Elbe. Upstream of Hamburg, the composition of the SPM did not vary with the freshwater discharge. Confirming the results of previous investigations, an immediate response of the heavy metal concentrations and of the smectite/kaolinite-ratio to the discharge rate at the stations downstream of Hamburg, was apparent. At high discharge rates, the differences between the concentrations upstream and downstream of Hamburg were found to be small, while at low discharge rates they varied notably. This can be interpreted as due to different sedimentation and mixing behaviour of suspended material at different conditions.

Our investigation showed two main trends:

- At low freshwater discharge the heavy metal content of SPM decreased relatively at stations downstream of Hamburg, simultanously with an increase of the smectite/kaolinite-ratio.

This can be interpreted by a comparatively high sedimentation rate of SPM in the Hamburg harbour area, that originated from regions upstream of Hamburg, combined with a dilution by material transported upstream.

- At high freshwater discharge the picture is completely different. Under these conditions small concentration gradients over the Hamburg area were found for seston bound heavy metal concentrations as well as for the smectite/kaolinite-ratio. This means that than highly polluted limnic material was carried through Hamburg harbour due to a higher energy input and thus introduced into the outer estuary.

Acknowledgements

Special thanks are due to the ARGE ELBE* water authority who supported us with useful information and provided us with the samples. A special mention should be made for Mr. K.Weiler who did the analyses of the heavy metals.

REFERENCES

[1] MEADE RH (1969)
Landward transport of bottom sediments in estuaries of the Atlantic coastal plain.
J Sed Pet, 39, pp 222 - 234

[2] EISMA D, BERNARD P, BOON JJ, GRIEKEN RE van, KALF J, MOOK WG (1985)
Loss of particulate organic matter in estuaries as exemplified by the Ems and Gironde estuaries. Mitt Geol Paläont Inst Univ Hamburg.
SCOPE/UNEP Sonderb H 58, pp 397 - 412

[3] SCHWEDHELM E, SALOMONS W, SCHOER J, KNAUTH H-D (1988)
Provenance of the sediments and the suspended matter of the Elbe Estuary.
GKSS Report 88/E/20, pp 1-76

[4] SALOMONS W, SCHWEDHELM E, SCHOER J, KNAUTH H-D (1987)
Natural tracers to determine the origin of sediments in estuaries.
JAWPRC/JSWPR Specialised Conference 'Coastal and Estuarine Pollution .
Fukuoka (Japan), pp 119-132

[5] KNAUTH H-D, SCHROEDER F, IRMER U (1986)
Investigations on the Heavy Metal Variability in the River Elbe Estuary Upstream and Downstream of Hamburg Harbour. 2nd Intern. Conference 'Environmental Contamination', Amsterdam, pp 186-188

[6] FÖRSTNER U, SCHOER J, KNAUTH H-D (1989)
Metal pollution in the tidal Elbe River.
The Science of the Total Environment (in publ.)

[7] MICHAELIS W, FANGER H-U, MÜLLER A (1988)
Die Bilanzierungsexperimente 1984 und 1985 auf der oberen Tideelbe.
GKSS Report 88/E/22, Geesthacht

[8] ARGE ELBE (1987, 1988)
Wassergütedaten der Elbe von Schnackenburg bis zur See, Zahlentafel 1986, 1987, Hamburg, 193p and 191p resp.

[9] IRION G, WUNDERLICH F, SCHWEDHELM E (1987)
Transport of clay minerals and anthropogenic compounds into the German Bight and the provenance of fine-grained sediments SE of Helgoland.
Journ Geol Soc, London, V, pp 236 - 240

[10] IRMER U, KNAUTH H-D, WEILER K (1988)
Influence of suspended particulate matter regime on the heavy metal pollution of the tidal river Elbe at Hamburg. Z Wasser-Abwasser-Forsch. 21, pp 236 - 240

SUSPENDED MATTER IN THE OUTER ELBE ESTUARY AND IN THE WADDEN SEA: DISTRIBUTION AND TRANSPORT

P. H. Koske
Institut für Angewandte Physik, Universität Kiel
2300 Kiel

1. INTRODUCTION

The transport of anthropogenic pollutants by rivers into tidally influenced estuaries and coastal waters has frequently been discussed but in spite of considerable research efforts not yet satisfactorily understood. Quantitative estimates of the freight of specific pollutants and their discharge into the adjacent seas are usually performed by using the average flow of the river and the analytically determined concentrations of the components of interest. As long as the dissolved substances can be considered as conservative constituents in solution this calculation is probably correct and the estimated inputs have a fairly high degree of reliability.

If in contrary dissolved components do not behave like conservative constituents but undergo various reactions, changing their chemical speciation and/or their physical state and phase, the input estimates as determined by the calculation of flow rate and concentration can be totally wrong.

This is particularly true when certain constituents are present in the water, partially dissolved in solution and partially adsorbed on suspended particulate matter, which usually can be found in estuarine waters. If a considerable fraction of the man made pollutants exists in the adsorbed state, the problem of river inputs into the sea is transformed to the question, how the suspended material in estuaries and coastal waters is transported by the tidal action.

The present paper deals with this question. It describes measurements and observations which were performed to establish some approximate figures for the amount of suspended material in the water column and for its main directions of transport during the tidal cycle, thus providing some of the background data needed for calculating the inventory of adsorbed pollutants in these estuarine and coastal waters.

2. MEASUREMENTS AND OBSERVATIONS

Since 1984 the Christian-Albrechts-Universität zu Kiel (CAU) and the GKSS Forschungszentrum Geesthacht GmbH (GKSS) have performed several joint field studies in the Elbe estuary and in the Wadden Sea off the west coast of Schleswig-Holstein. Within their collaboration agreement CAU used its research cutter LITTORINA, whereas GKSS was operating its research vessel LUDWIG PRANDTL.

Both ships used identical equipment for CTD- and turbitity-measurements consisting of an in situ system for the vertical profiling of temperature, salinity and light transmission as well as a shipboard system for continuous recording of the same parameter at the surface along the course line of the ship. For calibration purposes water samples were taken on both ships and suspended matter was determined gravimetrically.

With regard to the tidal movements of the water and the transport of suspended matter different methods were applied: drifting DECCA- buoys, suspended current meters at anchor stations and electromagnetic methods (GEK) from the moving research vessels.

The two-ship-campaigns usually were performed according to one of the following schemes:

1. Both ships were steaming on parallel courses at a distance of about 10 km (5.4 nautical miles) apart, starting in the outer Elbe estuary and continuing northwards along the west coast of Schleswig-Holstein. The more inshore vessel LUDWIG PRANDTL was following approximately the 10 m-bottom contour line which meant NNW-course about 10 - 15 km from the outer edges of the coast or the islands stretched out in front of it. The second vessel LITTORINA was heading the same course 10 km to the west of LUDWIG PRANDTL. On both ships the surface concentration of suspended material was recorded continuously along the course line by measuring the light transmission of the surface water. In addition both vessels performed hydrographic stations with vertical profiles of the above listed parameters and water sampling in selected depths according to the signals of the light transmission sensor. Thus a fairly synoptic survey of these coastal waters with regard to their content of suspended matter was achieved.

2. Both ships, about 10 nautical miles apart from each other (the LUDWIG PRANDTL more inshore, the LITTORINA further out) deployed drift-buoys and tracked them for more than 6 hours, thus producing tidal trajectories including the change from ebb-tide to flood and opposite. During this tracking procedure vertical profiles of the above listed parameters were recorded from the water surface to the bottom, in order to identify and describe quantitatively the surrounding water body, together with water sampling at selected depths according to the signals of the light transmission sensor.

Besides these two-ship-operations, RC LITTORINA carried out several research cruises of her own during the past few years in the outer Elbe estuary and in the Wadden Sea, performing similar observations while steaming along a pre-planned track with hydrographic stations, or drifting together with drift-buoys, or at anchor stations in some tidal channels between mud flats. So a rather comprehensive set of observations and data has been collected (for example several hundred hydrographic stations) which is used and partially evaluated in this paper.

3. RESULTS AND DISCUSSION

Typical profiles of temperature, salinity, and light transmission, which were recorded at two stations in the German Bight off the west coast of Schleswig-Holstein between the outer Elbe estuary and the island of Helgoland in August 1988, are shown in Figures 1 and 2. At both stations which are only about 2.5 nautical miles apart nearly identical conditions were recorded, though the time difference between the measurements was 4 days. The temperature was constant from the surface to the bottom. A similar homogeneity of the water column was observed in the traces for salinity, though there is a slight linear increase with depth at station 2 (Fig.1). These profiles of temperature and salinity indicate the strong mixing influence of the tides. In spite of the fresh water outflow of the river Elbe and the reduced salinities there is no layering of water bodies to be observed in these coastal waters.

Quite divergent vertical profiles are shown in the two figures for the light transmission, which indicates a completely different behaviour of the suspended material in spite of the well mixed water column. At both stations there was a surface layer of 5.5 m thickness with relatively high light transmission values of about 80 - 85 %. Below this rather transparent top layer there are several distinguishable layers with

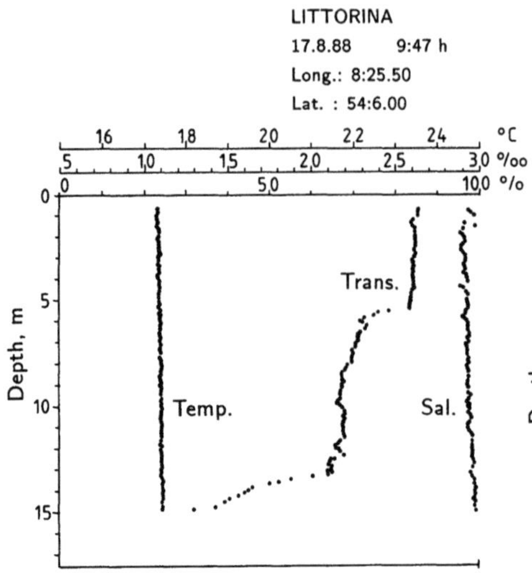

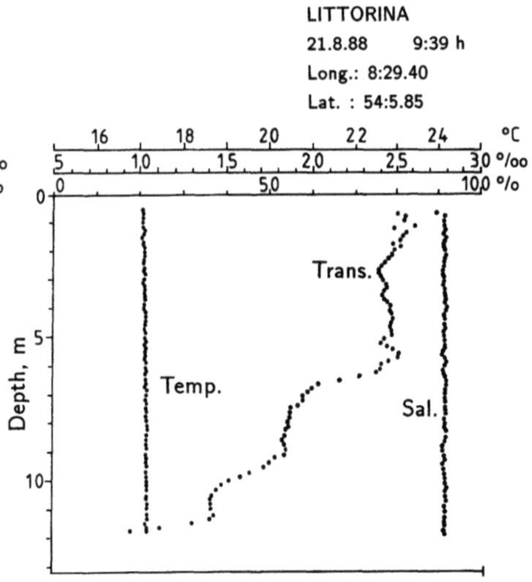

Figure 1: Profiles at station 2

Figure 2: Profiles at station 6

lower light transmission at both stations down to a 'near bottom' value of 30 % at station 2 and less than 20 % at station 6.

This decrease in transparency in the water column from top to bottom is typical for the coastal waters in the German Bight. Here the resuspension of particulate material at the sediment/water interface, brought about by the tidal currents, is effective. The gravimetric determined values for the concentration of suspended matter in the water, based on water samples collected simultaneously with the in situ records, were 20 mg/l respectively 28 mg/l in the surface layers of stations 2 and 6, and 48 mg/l at a water depth of 13.5 m at station 2 and 208 mg/l at 11 m at station 6.

As is evident from such differing numbers and from the continuous plots for the light transmission in Fig.1 and 2, a detailed knowledge of the vertical distribution of suspended matter in the water column is essential for an estimation of the total amount of material in suspension in these coastal waters.

Though there is a continuous periodic variation in the concentration of suspended matter in the water because of the change in velocity and direction of the tidal currents, a rough estimate for the situation of maximum flow is attempted here. For this purpose the two boxes in Fig.3

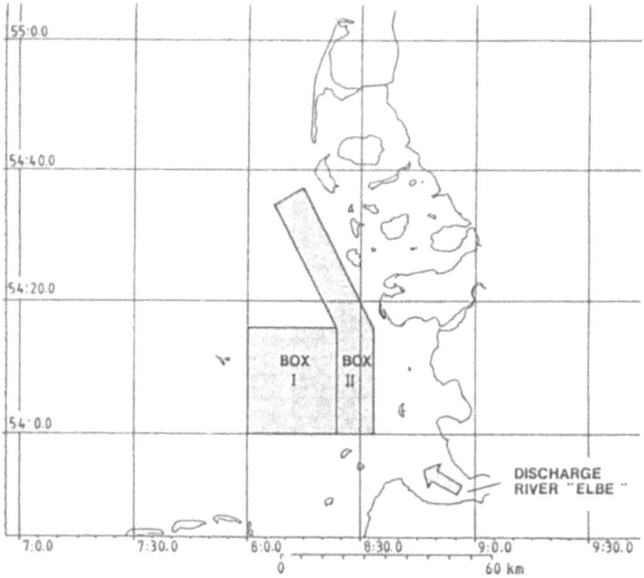

Figure 3: Boxes for the calculation of suspended matter

are used, for which sufficient station plots and surface recordings are available in the presently discussed data set. Box I represents a water volume of about 20 x 10^9 m^3 with a total content of suspended material of 200 000 - 300 000 t. Box II is smaller with regard to its water volume but considerably higher in its concentration of suspended matter. An estimated volume of 11 x 10^9 m^3 contains 400 000 - 450 000 t of suspended matter. With these estimated numbers the total content of both boxes adds up to between 600 000 and 750 000 t. This amount is comparable with the annual transport of particulate matter by the river Elbe which at the weir of Geesthacht is estimated to be 800 000 - 850 000 t.

Because of its sedimentation during slack water and its resuspension and transport during maximum tidal flow the trajectories of tidal currents are of primary importance for the fate of the suspended material in the water and the adsorbed pollutants. Therefore current measurements with drift-buoys were performed in order to collect some information about the nature of tidal trajectories in the German Bight and the occuring velocities. A typical result of such observations is presented in Figure 4.

At maximum ebb current the drift-buoy for the surface layer (sail depth 0.5 - 2.5 m) moved with nearly 2 knots (0.94 m/s) to WNW in agreement with the main outflow of the Elbe, whereas during "slack water"

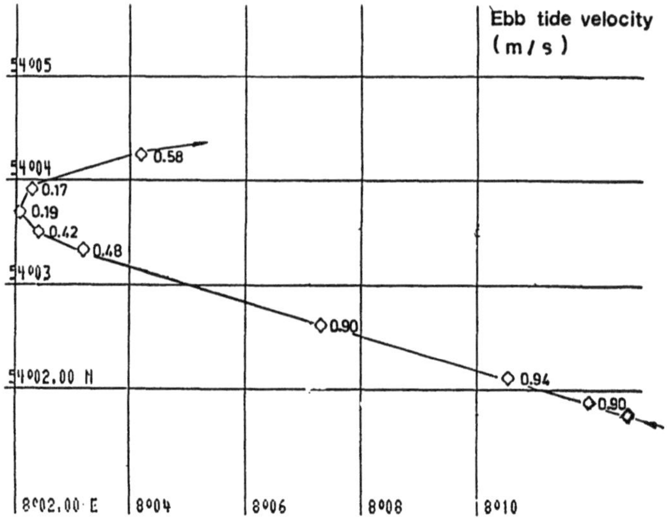

Figure 4: Tidal current trajectory in the German Bight during ebb tide

the current kept moving with the relatively strong velocities of 0.19 and 0.17 m/s. Similar current conditions were observed for the near bottom layer (sail depth 10 - 14 m; trajectory not depicted here). At low water the measured velocities were 0.23, 0.16, and 0.25 m/s, again as in Fig.4, combined with a change in direction of about 315 degrees. This situation is completely different during high water. Here the current trajectories are not elliptical but needle or pin like with a 360 degree change of direction and the velocities at slack water approach zero (0.04 m/s).

In concluding the following can be stated: A considerable amount of particulate matter is in suspension in the coastal waters of the German Bight at times of maximum tidal flow. This mass of suspended matter is of the same order of magnitude as the annual transport of the river Elbe across the weir at Geesthacht. If loads of anthropogenic pollutants in these waters are to be calculated, the adsorbed portion present on the suspended material might constitute an important percentage, absolutely as well as relatively. Because of the characteristic differences in the current situation between low water and high water, in combination with increasing sedimentation at reduced current velocities, a net inshore transport of suspended matter is to be expected together with a deposition of the material on the shallow mud flats in the Wadden Sea near the coast or in the Elbe estuary.

INVESTIGATIONS ON SUSPENDED MATTER TRANSPORT PROCESSES IN ESTUARIES AND COASTAL WATERS

Walter Puls, Manfred Lobmeyr, Agmar Müller, Matthias Schünemann (Inst. f. Physik)
Herbert Kühl (Institut für Anlagentechnik)
GKSS-Forschungszentrum, D–2054 Geesthacht

Abstract

Three findings from field surveys and numerical modelling were derived from extensive field investigations into the settling behaviour of mud flocs in the Elbe estuary. An apparatus for erosion experiments on natural mud cores and the appropriate methods were developed. First results exemplified a complex erosion behaviour. The simulation of suspended matter transport with a depth averaged 2-dimensional model gave results which agreed fairly well with measured data.

1. Introduction

The fate of contaminants in waters is closely related to the fate of suspended matter. Like dissolved matter, suspended matter is transported by advection and by turbulent diffusion. In addition, the fate of suspended matter is determined by settling and deposition, as well as by 'bed processes', e.g. consolidation, bioturbation and erosion.

In this paper, recent GKSS results on settling and erosion are presented. The importance of bioturbation in coastal areas is emphasized. A depth-averaged model computes the behaviour of suspended matter in a tidal section of the Elbe estuary.

2. Settling velocity

Between 1984 and 1988, the settling velocities of suspended matter (mud flocs) in the Elbe and Weser estuaries were measured during more than 50 campaigns. As a field instrument we used the so-called Owen tube (Owen 1976, Puls 1988). Between September 1985 and December 1987 long-term measurements (22 separate campaigns) were carried out at Grauerort, Elbe-km 660 (see accumulation of data points in Fig. 1).

The following findings are based on GKSS measurements in the Elbe estuary:

(a) Processing the data of each single campaign resulted in proportionalities between the median settling velocities 'w_s^{50}' and the suspended matter concentrations 'c' (e.g. Puls et al. 1988, Thorn 1981). These proportionalities were explained by: 'high suspended matter concentration means a high frequency of floc collisions and thus the formation of large (fast settling) flocs'. After recent data processing at GKSS our (new) explanation

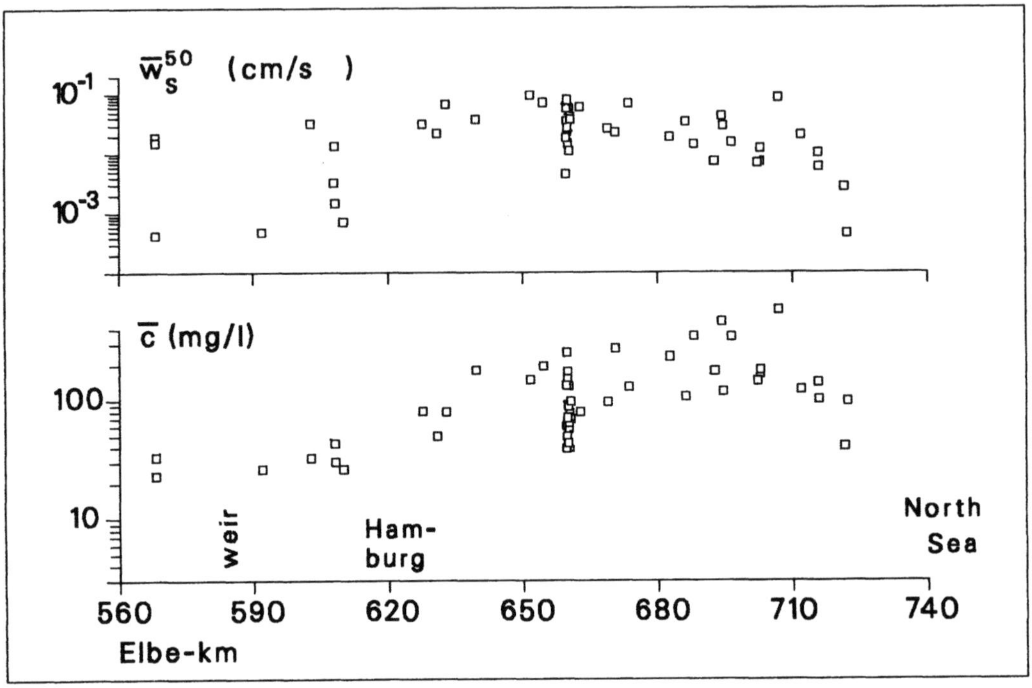

Fig. 1 Mean values of the suspended matter concentration $\overline{c}$ and the median settling velocity $\overline{w}_s^{50}$ in the tidal Elbe estuary

for $w_s^{50} \sim c^x$ $(0.5 < x < 3)$ in the Elbe estuary is found in the patchiness of suspended matter concentration. This patchiness is a result of spatial and temporal variations of large flocs. If by chance a greater amount of large flocs is sampled, both c and w_s^{50} are above average, thus causing $w_s^{50} \sim c$. The consequence: because numerical models do not resolve the patchiness, formulas of the form $w_s^{50} = \alpha c^\beta$ (as they were for instance recommended by Puls et al. (1988)) should not be used in numerical models of the Elbe estuary, but instead the averaged data from each campaign and for each station (and time).

(b) The sorting coefficient $(w_s^{75}/w_s^{25})^{0.5}$ for the mass distribution of settling velocities in the Elbe estuary is in the order of 3 to 30. In view of such large sorting coefficients, a characterization of the settling velocity distribution using one value (e.g. w_s^{50}) is not adequate. The w_s-distribution is only adequately represented by at least 4 or 5 fractions (see section 5).

(c) Fig. 1 shows logarithmic mean values of the suspended matter concentration $\overline{c}$ and the median settling velocity $\overline{w}_s^{50}$, plotted against the Elbe km (for every campaign a mean value was calculated, slack water data were left out of consideration). The standard deviations of $\overline{w}_s^{50}$ were between 0.5 and 1 magnitudes. Landward of Elbe-km 610 the ebb current dominated. A distinct flood current developed between km 610 and km 620. Due to tidal pumping (Uncles 1985, Dyer 1988) the concentration of suspended matter increased here (when the fluvial discharge was low). The Hamburg section can thus be regarded as the landward limit of the turbidity zone. The seaward limit is at about km 710. For the data between km 620 and km 710 the correlation between $\overline{c}$ and $\overline{w}_s^{50}$ was –0.16 (n = 46). The scatter of the settling velocities was due to (i) variations of the fluvial discharge (ii) biological influences (Greiser 1988) (iii) variable measurement sites with their hydrodynamic conditions and (iv) sampling conditions.

A great number of settling velocity measurements like those made at GKSS is justified only when special research interests exist. To cover the range of settling velocities, the measurements should be carried out (i) at extremely high and low fluvial discharges (ii) at extreme water temperatures (summer and winter) (iii) in late autumn (input of phytodetritus) (iv) at situations considered 'normal'.

3. Erosion

In modelling transport processes, it is necessary to have a knowledge of the quantitative, three-dimensional distribution of mud and sand as well as the characteristics of sedimentation and erosion. While the erosion behaviour of non-cohesive sediments is understood in principle (Shields 1936; Unsöld 1984), we investigate the properties of naturally formed and biologically active mud in the laboratory. The principal idea underlying the device for the erosion of cores by water in the laboratory can be outlined as follows.

Before the actual erosion experiment begins, the vertical density profile of the mud is measured with a gamma probe for a later correlation with the erosional characteristics.

The energy-input to induce erosion in the newly developed erosion device is provided by a calibrated propeller. The propeller, mounted 3 cm above the core, together with six evenly spaced vertical baffles, induces turbulences in the water above the core, just as turbulent currents do in nature.

For the calibration of the propeller, sand fractionated by closely spaced sieving was used in the following manner: from the 'Shields-diagram' (Shields 1936, Unsöld 1984) the critical shear stress for each of the sand-fractions was known; we increased the rounds per minute of the propeller until the first movements of sand-particles were perceived. Thus, corresponding to the diameter of the sand fraction in use, and its critical shear stress, a defined shear stress could be assigned to the number of rotations per minute of the propeller. Multiple observers minimized subjectivity and facilitated error estimation.

For the actual experiments, a core was mounted in a vice below the propeller. The measurements during the erosion experiments primarily resulted in the attenuation of light in a large vessel, containing a suspension of the material eroded so far.

The attenuation was then calibrated by sampling the suspension from the large vessel at roughly even time intervals. An influence of particle size distribution, particle-colour and colouring by dissolved agents on the development of attenuation over time was thus excluded. The data from the erosion experiments were recorded immediately and stored for further treatment in a computer.

The immediate results of the experiment were (i) the shear strength of the sediment at various times (and depths) and (ii) the development of the attenuation-curve in the vessel (Fig. 2), which was a function of the rate of erosion. These parameters were used for the numerical simulation of sediment erosion.

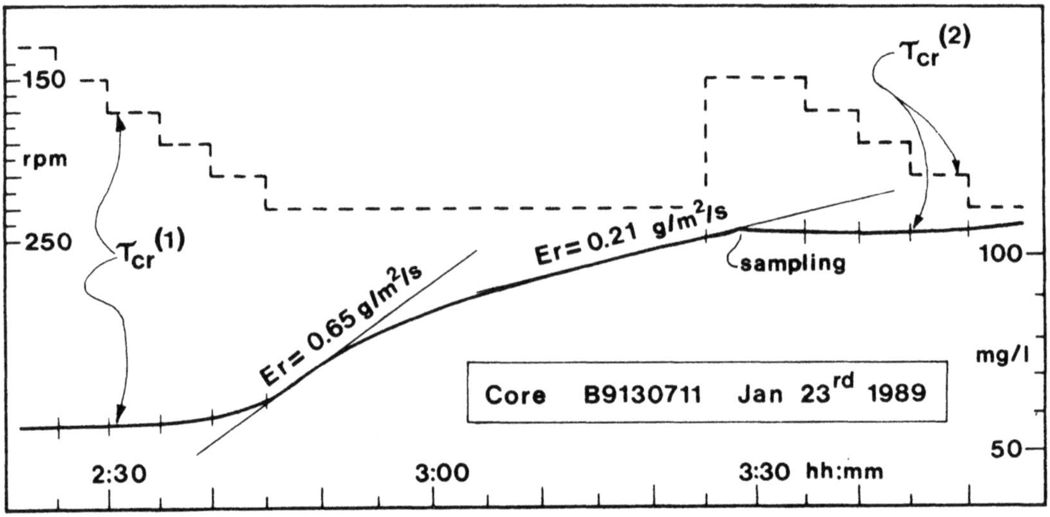

Fig. 2 Erosion run of core B9130711 from Elbe km 699, southern shore, Jan 23rd 1989 (excerpt). Revolutions per minute of the propeller (dashed line, scale on left vertical axis) and light attenuation converted to 'eroded sediment in suspension' [mg/l](solid line, scale on right vertical axis) are the principal data plotted against time (horizontal axis). Starting with 130 rpm at a suspension concentration of 52 mg/l, the propeller speed was increased at steps of 20 rpm in 5 minute intervals. At 150 rpm (equivalent to a bottom shear stress $\tau = 0.40\ N/m^2$), no bed erosion was discernible. At 170 rpm ($\tau = 0.49\ N/m^2$), erosion was noticeable. The critical bed shear stress $\tau_{cr}^{(1)}$ was thus between 0.40 and 0.49 N/m^2. The revolutions were further increased to 190, 210 and 230 rpm ($\tau = 0.79\ N/m^2$), where they remained unchanged for 40 minutes. During this interval the erosion rate (Er) decreased from 0.65 $g/m^2/s$ to 0.21 $g/m^2/s$ although the applied shear stress remained constant. Accordingly, the bed shear strength must have increased with time (and depth). The erosion rates are of the same magnitude as those given by Puls (1984). During the time interval of 40 minutes the concentration of the suspension approximately doubled to 106 mg/l. Next, the revolutions were reduced to 150 rpm (which is well below the level of the critical bed shear stress) and again increased in steps of 20 rpm's and 5 minute intervals to evaluate the new shear strength of the currently exposed bed surface. The bed shear strength $\tau_{cr}^{(2)}$ was now determined to be between 0.58 and 0.66 N/m^2.

The non-linear erosion-characteristics warrant further investigations on the changes in erodibility for a given shear stress ('self-armouring'), as well as on the influence of wave action, flooding at high tide and emergence at low tide and its resulting influence on the sediment by compaction, solidification, partial desiccation and biological activities.

4. Bioturbation

The importance of bioturbation (Kersten 1988) in the fate of suspended matter in the North Sea has been shown by Puls (1987). During summer, suspended matter settles in the Oyster Grounds (south of the Doggerbank). If bioturbation is not present, the newly deposited material stays at the bed surface. The winter storms will then be able to re-erode it and to transport it to the Norwegian trench. If, however, bioturbation is present in the Oyster Grounds, the deposited material is mixed into deeper bed layers, out of the reach of the winter storms. This means that contaminants adsorbed to suspended matter stay in the Oyster Grounds and thus will enter (via the benthic fauna) the food web.

5. Modelling of suspended matter transport

Suspended matter dynamics in the middle Elbe estuary (between km 644 and 654) are simulated by a time dependent, 2-dimensional depth averaged model (grid size 100 m); for the flow computations of this model see Krohn and Lobmeyr (1988). The results are compared with field data of the 'BILEX82' experiment (Michaelis 1989).
The advection - diffusion equation for suspended matter includes a sedimentation (Krone 1962) and a resuspension term (after Müller 1988):

$$sedimentation: \quad w_s \cdot c_b \cdot \left(1 - \frac{\tau}{\tau_{cr,d}}\right) \qquad\qquad resuspension: \quad f_1 \cdot D \cdot \frac{\tau^2}{b_1 + \tau^2}$$

τ is the bottom shear stress, c_b is the suspended matter concentration at the bottom, $\tau_{cr,d} = 0.1\ N/m^2$ was used as the critical shear stress below which deposition occurs; the sedimentation term is only valid for $\tau < \tau_{cr,d}$. In the resuspension term, D is the mass (per m^2) of non-consolidated solid matter on the bed and $f_1 = 6 \cdot 10^{-4}\ s^{-1}$ and $b_1 = 1.25 \cdot 10^{-3}\ N^2/m^4$ are constants.

The vertical distribution of suspended matter in the water column is the sum of four exponential profiles. These are necessary because the measured settling velocity mass distribution is represented by four fractions in the model, each defined by its settling velocity (cm/s) and its percentage amount of the total mass. These fractions are (i) 0.001/12%; (ii) 0.01/20%; (iii) 0.1/37%; (iv) 0.7/31%. The profiles react 'immediately' to temporal variations of the vertical turbulent mixing coefficient $A_z = 0.05 \cdot H \cdot u_*$ (Bowden 1967), where H is the water depth and u_* is the bed shear velocity. A_z cannot fall below 40 cm^2/s; this lower limit results from the processing of our settling velocity data.

No efforts were made to 'tune' the model – only measured data and formulas from literature were used. The exception is the factor f_1 in the resuspension term, which is three times higher in this model than as given by Müller (1988).

Fig. 3 shows the comparison of modelled and measured time series of the suspended matter concentration. The succession of sedimentation and resuspension was well evinced by the observed bottom concentrations. Compared to the measured values, the computed variation of the bottom concentration is underestimated. Close to the surface the computed concentrations were mostly higher than the measured ones, indicating that the chosen A_z was too high.

Fig. 4 shows isolines of the depth averaged concentration at low tide slack water. The isolines show, both, great longitudinal and lateral variations of the concentration. The lateral variations were mainly determined by the water depth: concentrations were low where the water depth (i.e. the 'height of fall' for suspended matter) was small. The longitudinal variations were determined by the current velocities, i.e. the concentration increases with increasing current velocity. An impression of the current pattern between

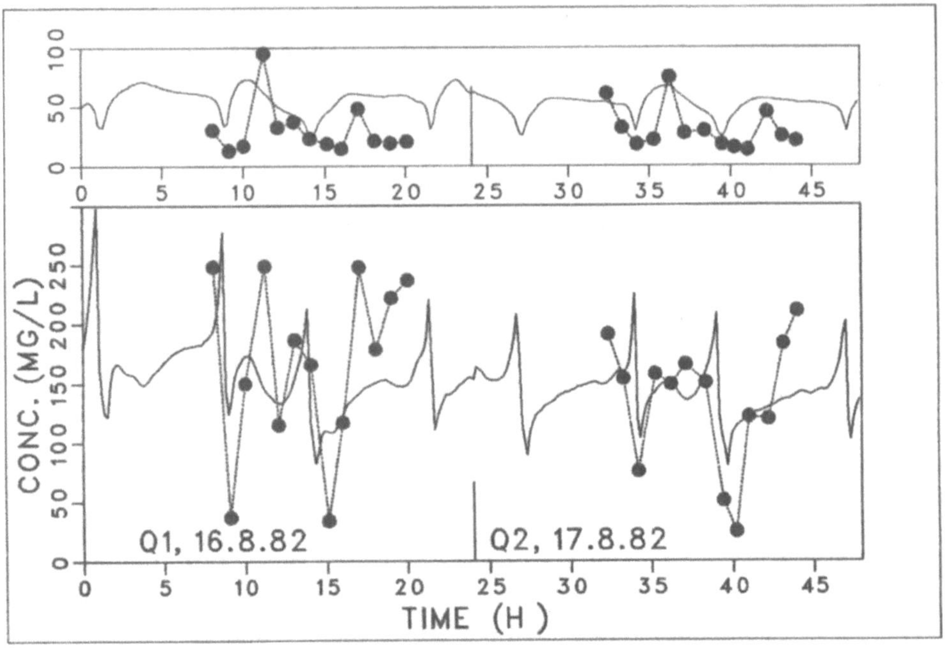

Fig. 3 Computed time series of sediment concentration at the middle of the cross sections Q1, date 18.8.82 and Q2, date 17.8.82 (solid lines), compared with measured values (dashed and marked lines). The positions of Q1 and Q2 are shown in Fig. 4. The upper plot shows the concentration 1.5 m below the water surface, the lower shows the concentration 1 m above the bottom (total water depth: 15 - 18 m)

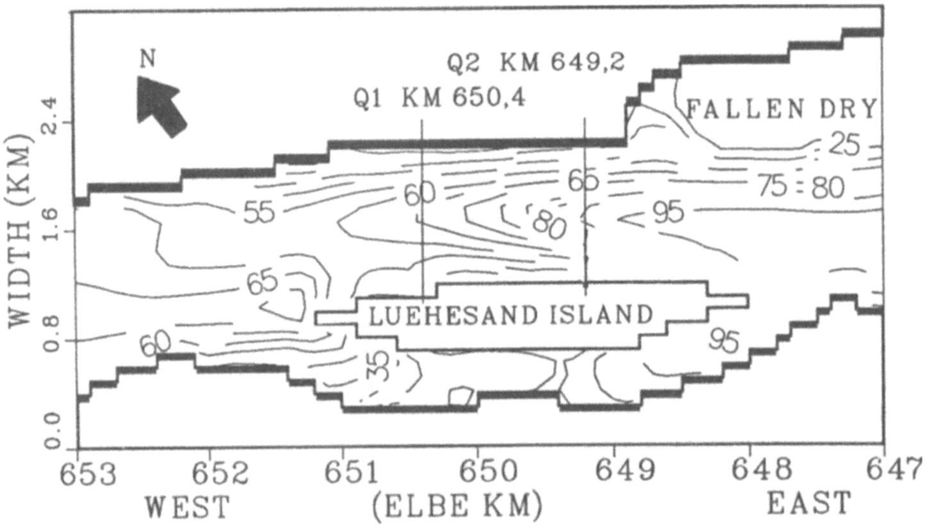

Fig. 4 Isolines of depth averaged sediment concentration (mg/l) at low tide slack water; time 8.50 a.m., date 16.8.1982. Flood current enters from the west, while ebb current still persists in the east. Minimum concentrations are 60 mg/l in the fairway and 30 mg/l in the river branch between Luehesand Island and the mainland

Elbe km 645 and 655 at low tide slack water (but at an earlier stage of the same tidal phase) may be gained from Fig. 4 (upper plot) of Krohn and Duwe (1989).

References

Bowden KF (1967) Stability effects on mixing in tidal currents. Phys Fluids Suppl 10: 278-280

Dyer KR (1988) Fine sediment particle transport in estuaries. In: Dronkers J and van Leussen W (eds) Physical Processes in Estuaries, Springer, Berlin Heidelberg New York, pp 295-310

Greiser N (1988) Zur Dynamik von Schwebstoffen und ihren biologischen Komponenten in der Elbe bei Hamburg. Dissertation im Fachbereich Biologie der Universität Hamburg, Hamburger Küstenforschung, Heft 45, 170 pp

Kersten M (1988) Geobiological effects on the mobility of contaminants in marine sediments. In: Salomons W, Bayne BL, Duursma EK, Förstner U (eds) Pollution of the North Sea, Springer, Berlin Heidelberg New York, pp 36-58

Krohn J, Duwe KC (1989) Mathematical modelling of hydrodynamics in the Elbe estuary. This volume, chapter II

Krohn J, Lobmeyr M (1988) Application of two- and three-dimensional high resolving models to a small section of the Elbe estuary – A comparative study. GKSS 88/E/52, 46 pp

Krone RB (1962) Flume studies of the transport of sediment in estuarial shoaling processes. Final Report.Hydraulic Engr Lab Sanitary Engr Res Lab, Univ of California, Berkeley 110 pp

Michaelis W (1989) The BILEX concept: research supporting public water quality surveillance in estuaries. This volume, chapter III

Müller A (1988) Das numerische Modell FLUSS. In: Michaelis W, Fanger H-U, Müller A (eds) Die Bilanzierungsexperimente 1984 und 1985 (BILEX 84 und BILEX 85) auf der oberen Tideelbe. GKSS 88/E/22: 24-28

Owen MW (1976) Determination of the settling velocities of cohesive muds. Hydraulics Research Report IT 61, Wallingford UK, 8 pp

Puls W (1984) Erosion characteristics of estuarine muds. Hydraulics Research Report IT 265, Wallingford UK, 38 pp

Puls W (1987) Simulation of suspended sediment dispersion in the North Sea. ICES Report CM 1987/c:37, 24 pp

Puls W, Kühl H, Heymann K (1988) Settling velocity of mud flocs: results of field measurements in the Elbe and the Weser Estuary. In: Dronkers J and van Leussen W (eds) Physical Processes in Estuaries, Springer, Berlin Heidelberg New York, pp 404-424

Shields A (1936) Anwendung der Ähnlichkeitsmechanik und der Turbulenzforschung auf die Geschiebebewegung. Mitt Preuss Versuchsanst für Wasserbau und Schiffbau 26: 1-42.

Thorn MFC (1981) Physical processes of siltation in tidal channels. Proc Hydraulic modelling applied to engineering problems, ICE, London pp 47-55

Uncles RJ, Elliott RCA, Weston SA (1985) Observed fluxes of water and suspended sediment in a partly mixed estuary. Estuarine Coastal Shelf Sci 20: 147-167

Unsöld G (1984) Der Transportbeginn feinstkörnigen rolligen Sohlmaterials in gleichförmigen turbulenten Strömungen: eine experimentelle Überprüfung und Erweiterung der Shields-Funktion. Reports Sonderforschungsbereich 95, Univ Kiel, 70, 144 pp

STATISTICAL PROPERTIES OF TURBULENCE IN ESTUARIES

K. Oduyemi
DUNDEE INSTITUTE OF TECHNOLOGY, DUNDEE, U.K.

INTRODUCTION

The understanding and prediction of the effects of engineering works on the flow of water and the transport of suspended solids in estuaries can be facilitated by using mathematical models. The solution of the transport equations for momentum or suspended solids is often strongly influenced by turbulent fluctuations which may be generated by a channel bed, by wind or internally. It is important to understand the properties of turbulence in estuaries if a full utilization is to be made of the potential of mathematical models to predict the effect of engineering works on estuarine environment.

In the past, the properties of turbulence have been studied extensively in wind tunnels and laboratory open channels [1,2,3,4]. These analyses have been extended to measurements made in the sea and the atmosphere [5,6]. However, few turbulence measurements have been made in estuaries [7,8,9,10]. This is probably due to technical and logistical problems (e.g. physical difficulties in deploying suitable transducers in rapid tidal flows). The complexities of estuarine flows (e.g. irregular channel cross-section, bends, bed structures, tidal oscillations, density gradients induced by heat, solutes and suspended sediments) made them to be potentially different from those found in laboratories and other geophysical flows. Thus, there is a well defined need to investigate the properties of turbulence in estuaries.

Some recent work has considered the statistical properties of momentum and solute fluxes in estuaries. The main purpose of this paper is to extend the previous work to include and elucidate some of the statistical properties of the suspended sediment fluxes.

DATA COLLECTION AND PROCESSING

Turbulence data were collected near to the centre of the channel in the upper reaches of the Tamar estuary, South West England (Fig. 1, location map) at both Calstock and 500 m south of Cotehele Quay. At high water, the channel at both locations is 100 m wide and fairly straight for about 250 m upstream and downstream of the measurement points. Commercially available transducers for velocity (55 mm diameter EMCM sensors, Colebrook Instrument Development Ltd.) and suspended solids (SDM 10 and S 1000 siltmeters, Partech Ltd.) were used to perform turbulence measurements. The turbulence sensors were mounted on a bed frame. The transducers for the velocity were fixed at approximately 0.50 m and 1.25 m above the bed, with the distance between the suspended solids and velocity transducers being 0.07 m. The depth of flow at the study locations varied between 1 m and 5 m, whilst the mean suspended solids concentration varied between 0.05 and 4 kg/m^3.

The analogue signals of the velocity components and suspended solids concentration in the Tamar were monitored by a telemetry system, which permits the measurements of 12 channels of turbulence signals and their transmission to a shore based microcomputer. The analogue signals underwent an analogue to digital (A - D) conversion, four channels were then multiplexed and the data transmitted on one of the three frequencies in the "458.5 - 458.8 MHz band". On shore, the signals were recorded on a Racal Store 4FM tape recorder, for subsequent analysis.

Calibrations for the EMCM transducers were on average 270 mV/(m/s) for both the longitudinal and vertical components of velocity. The noise level measured in still water did not exceed a root mean square value of 0.003 m/s for either component. Calibration for the silt meters varies with the set value of the full scale deflection, e.g. for the SDM 10 silt meter with a full scale deflection of 2 kg/m^3 the calibration is 0.5 V/(kg/m^3).

The criteria used in selecting record lengths in this study are similar to those used by Soulsby [12]. The lengths of record used are 410 sec. and 820 sec., in comparison to 600 sec. suggested by Soulsby [12]. Stationarity tests [13] applied after linear trend removal showed that stationarity at the 95% confidence limit was obtained for all runs.

DATA ANALYSIS AND DISCUSSION OF RESULTS

Examples of fluctuations of co-variance terms (Fig. 2) show that short intervals of more active than average activity exist. An important point to emerge from this study is that the contributions to the suspended solids fluxes are intermittent in a similar manner to the intermittent nature of the production of Reynolds stress. The intermittent nature (Fig. 2) suggests that the concept of "bursting", as previously observed in the laboratory may be applicable to estuarine conditions.

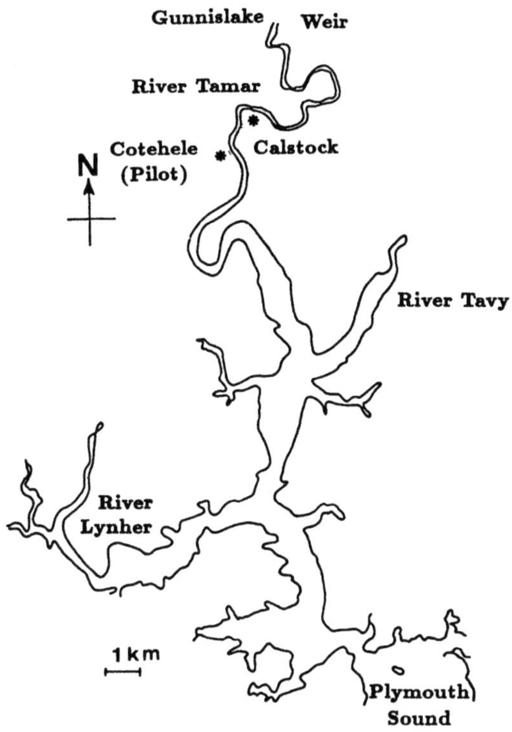

Figure 1: Sketch of the Tamar estuary showing station positions (*)

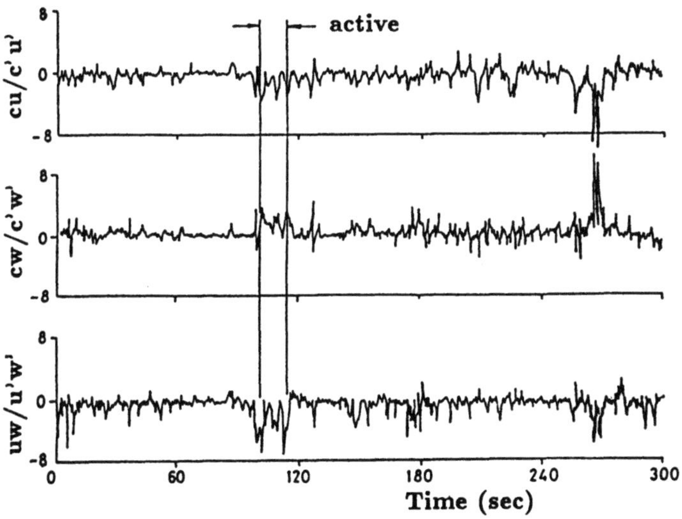

Figure 2: Examples of turbulent fluxes

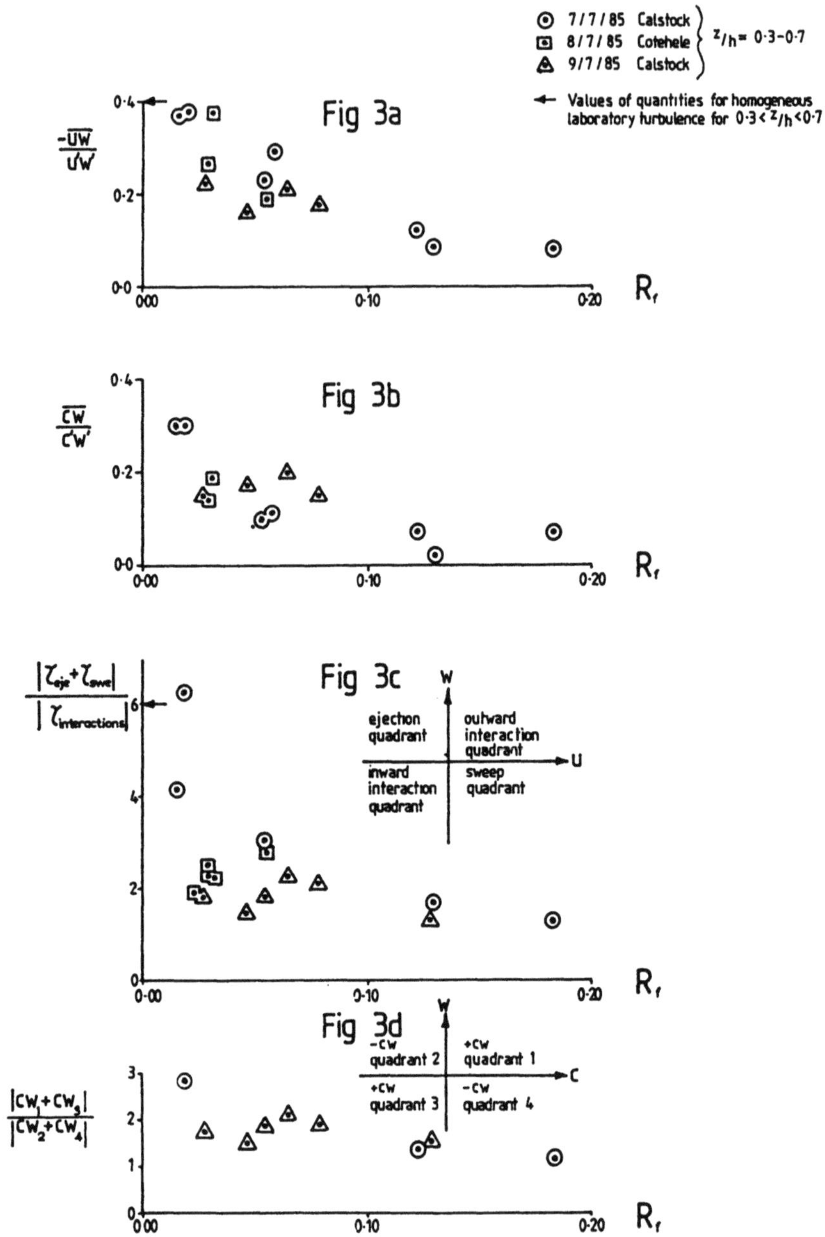

Figure 3: The effect of stability as a function of flux Richardson number on turbulent fluxes

The momentum and vertical particulate flux correlation coefficients are shown against the flux Richardson number, R_f [$= g\, \overline{\rho_c w} \,/\, \overline{uw}\, \frac{du}{dz}\, \rho$], in Figs. 3a and 3b. The two vertical transport terms reflect the mechanisms of a decrease in turbulent fluctuations and an increase in short period wave-like motions as R_f increases. The reductions in the vertical transport correlation coefficients are associated with an increase in velocity gradient and subsequent decreases in the eddy viscosity and particulate eddy diffusivity for a constant boundary shear stress [10].

The intermittent characteristics of the Reynolds stress, which is associated with the "bursting" mechanism observed in the laboratory [2] and in the sea [5,6] have been investigated by conditional sampling technique [10].

The average results for the magnitude of the contributions to the Reynolds stress for nearly homogeneous conditions (flux Richardson number, $R_f < 0.02$) and for $z/h \approx 0.3$ (Fig. 3c) show that the ejection and sweep events account for 120 - 130% of the local Reynolds stress. This leaves -[20 - 30%] of the local Reynolds stress to the negative contributors (inward and outward interaction events). Previously [2], values of approximately 120% and - 20% were respectively obtained for the positive and negative contributions. For large R_f values, the contributions in all the four quadrants increase the values greater than those for homogeneous conditions. These characteristics have been observed for salinity induced vertical density gradients in estuaries [14] and were thought to be due to short period wave-like motions which are typical of stratified conditions. This interpretation is consistent with the correlation coefficient reductions as R_f increases. A conditional sampling analysis of the suspended sediment vertical turbulent flux (Fig. 3d) also reveals a similar trend to that found in (Fig. 3c).

The attempt here in this report to quantify the effect of suspended solids induced stratification on mixing lengths respresents, to the author's knowledge, the first.

The momentum mixing lenghts were evaluated from:

$$- \overline{uw} = l_m^2 \left| \frac{\partial u}{\partial z} \right| \frac{\partial u}{\partial z}$$

where $\frac{\partial u}{\partial z}$ = velocity gradient in the vertical direction, using the Braystoke velocity meter data to determine $\partial u/\partial z$. The momentum mixing length for any vertical density gradient condition compared with the momentum mixing length for the homogeneous density conditions (i.e. $R_f = 0$) was found to decrease as the mixing efficiency decreases (i.e. as the stability parameter, R_f, increases (Fig. 4a). This suggests that the suspended solids induced vertical density gradient reduces the ability of turbulence to transfer momentum.

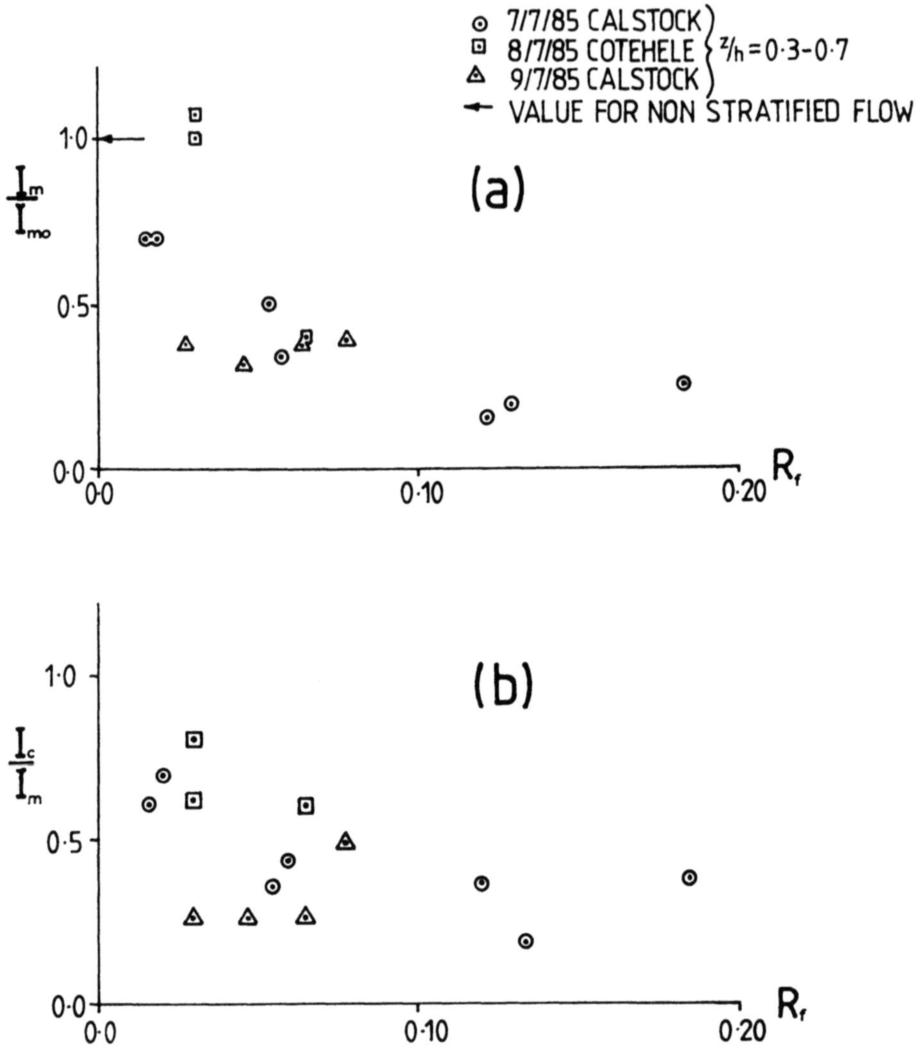

Figure 4: Mixing length in stratified flows

The ratio of the suspended solids mixing length to the momentum mixing length, which is equal to the ratio of the eddy diffusivity of suspended solids to the eddy viscosity was found to be lèss than 1 for the observed data. This suggests that the vertical transport of suspended solids is more suppressed than the vertical transport of momentum. It will be possible in future, after the present data have been reproduced for another site, to provide empirical formulae that will relate the eddy viscosity and eddy diffusivity of suspended solids to the gradient Richardson number. This is believed to be very useful for modellers.

CONCLUSIONS

The direct measurements of Reynolds stress and vertical suspended solids flux in partially mixed estuaries has led to a significant improvement in our knowledge of vertical turbulent transport of momentum and suspended solids. Vertical turbulent transport correlation coefficients for momentum ($-\overline{uw}/u'w'$) and suspended solids ($\overline{cw}/c'w'$) have been shown to decrease with increasing flux Richardson number (R_f). These reductions are due to the reduction in turbulent mixing as R_f increases. The suppression of vertical transport of momentum and suspended solids by suspended solids induced stratification have been reasonably quantified. It will be beneficial to investigate the effects of different fluvial and marine inputs on the quantification of the mixing lenghts.

ACKNOWLEDGEMENTS

The author greatfully acknowledges the help of the Tidal Water Research Group, University of Birmingham, PML, Plymouth and University of Bangor in the field.

REFERENCES

[1] LAUFER J (1951) NACA Report 1053, USA

[2] NAKAGAWA H and NEZU I (1977) J Fluid Mech 80: 99-128

[3] KOMORI S, UEDA H, OGINO F and MIZUSHINA T (1983) J Fluid Mech 30: 13-26

[4] NEZU I and RODI W (1986) J Hyd Eng 112: 335-355

[5] GORDON CM (1974) Nature 248: 392-394

[6] HEATHERSHAW AD (1979) Geophys JR Astr Soc 58: 395-430

[7] BOWDEN KF and HOWE MR (1963) J Fluid Mech 17: 271-284

[8] SOULSBY RL, SALKIELD AP, HAINE RA and WAINWRIGHT B (1985) Proc Euromech 192, Transport of Suspended Solids in Open Channels, Munich, 183-186

[9] WEST JR and SHIONO K (1986) Proc Int Symp on Physical Processes in Estuaries, The Netherlands, 9-12 Sept. 1986

[10] WEST JR and ODUYEMI KOK (1988) Accepted for publication by the American Society of Civil Engineers

[11] BRIERLEY RW, SHIONO K and WEST JR (1986) Proc Int Conf on Measuring Techniques of Hydraulics Phenomena in Offshore, Coastal and Inland Waters, London, England, 9-11 April, 359-366

[12] SOULSBY RL (1980) J Phys Ocean 10: 208-219

[13] BENDAT JS and PIERSOL AG (1971) Random data: analysis and measurement

[14] WEST JR and SHIONO K (1988) Estuarine, Coastal and Shelf Science 26: 51-66

RECENT DEVELOPMENTS IN LARGE-SCALE TRACING BY FLUORESCENT TRACERS

Jo Suijlen and Wim van Leussen
Tidal Waters Division, Rijkswaterstaat
P.O. Box 20907, 2500 EX The Hague, The Netherlands

ABSTRACT

Water quality management requires more insight into the transports of pollutants and nutrients over long periods. Fluorescent tracers have been shown to be an effective tool in determining these transports. However, determinations on large scales require extremely low detection limits and techniques for distinguishing the tracer from background fluorescence. Measuring techniques have been developed in The Netherlands based on on-line trace enrichment and on the separation of the concentrated fluorescent dyes using HPLC. With these techniques it is possible to detect the tracer (rhodamines) down to concentrations of about 10^{-12} kg m^{-3}, and to detect a number of tracer materials simultaneously. Some applications to surface waters are given.

1. INTRODUCTION

Water quality management of estuaries and coastal waters needs a thorough knowledge of the transports of pollutants, nutrients and other constituents. To make assessments of the distribution of dissolved substances in the water under various circumstances, it is essential to have sufficient insight into the water circulation and mixing processes.

Field experiments by tracers have made important contributions to the knowledge of large-scale water movements, as well as smaller-scale water movements resulting in dispersion of the dissolved and suspended substances. Tracer experiments are especially suitable for the calibration and verification of the turbulent diffusive models. Tracer experiments are also powerful tools for the simulation of proposed waste discharges and accidental releases.

Although the first experiments with fluorescent dyes were carried out before the beginning of this century, the applications increased enormously at the beginning of the sixties, when accurate fluorometers (Turner model 111) became available (Pritchard and Carpenter, 1960). Tracer experiments can be hindered by background fluorescence which results from dissolved or suspended matter. Large-scale experiments in such fluorescing surface waters need intolerable amounts of tracer material. Therefore, a long-term programme was started in the Netherlands in the seventies to develop techniques, with which accurate measurements can be made down to much lower concentrations.

Recently, measuring techniques became available based on on-line trace enrichment and on the separation of the concentrated dyes by HPLC (High Performance Liquid Chromatography). Through these methods it was possible to detect the tracer down to concentrations 1000 to 10.000 times lower than those obtained by current methods. These new measuring techniques also make it possible to determine a number of tracers simultaneously.

2. DEVELOPMENTS IN FLUORESCENT DYE TRACER TECHNIQUES

Much attention has to be given to the preparation of experiments in areas with varying background fluorescence, both in time and space. Measurement of the background fluorescence in Lake IJssel showed fluorescent background variations of between 50 to 300 ng l^{-1} rhodamine B (Suijlen, 1975), which means that for long-term simulations of the dispersal processes, large amounts of tracer would be needed. The reduction of this background by passing the water through an ultraviolet photochemical reactor, as proposed by Pritchard (1979), did not lead to sufficient improvements in this case. Therefore, a new fluorometric detection technique has been developed, based on the Synchronous Modulation Fluorometry (SMF) (Houpt, 1980). The SMF provides not only a great improvement in the detection limit in waters with a high fluorescent background but also for the maximum emission wavelength of the background fluorescence. Therefore, SMF allows a much greater degree of certainty concerning the identity of the signal than standard fluorometry.

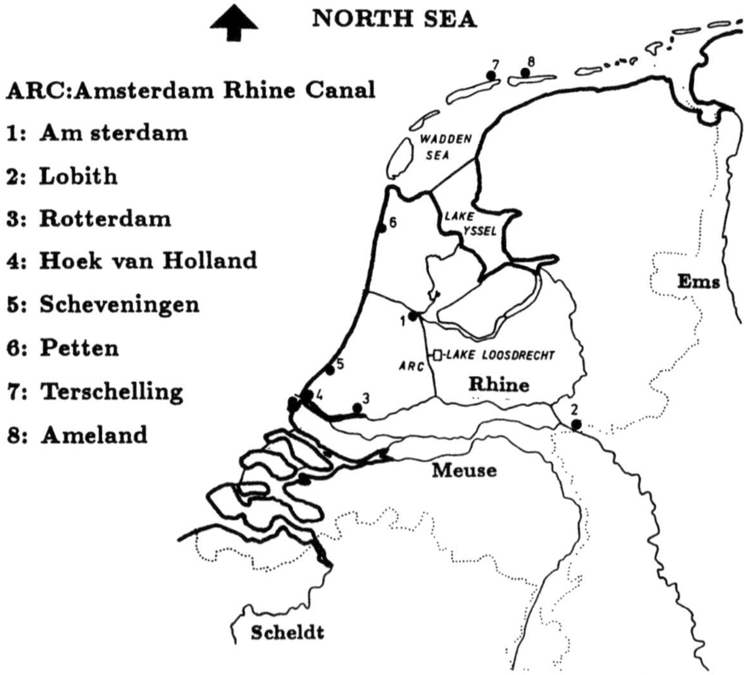

Figure 1: Location map of water systems in the Netherlands.

Through this SMF-method the detection limit of rhodamine B in clear water was improved to $2 \cdot 10^{-9}$ kg m^{-3}. Application of this method to a dispersion experiment in the Lake IJssel means a reduction of the background fluorescence to between 20 to 80 ng l^{-1}. This persisting background had a fluorescence maximum wavelength corresponding to that of rhodamine B. These improvements resulted in a reduction of the required amounts of trace material for a dispersion experiment in the Lake IJssel from 1000 to 100 kg rhodamine B.

The application of the SMF-method in the Dutch coastal zone showed a similar equivalent fluorescent background (10 to 20 ng l^{-1}). Further improvements were achieved by applying liquid chromatography, together with an efficient procedure for enrichment of the rhodamines B and WT. The basic step of the enrichment procedure was made by Laane et al. (1984). The implementation of the on-line enrichment and the separation of the rhodamine was done by Suijlen et al. (1987).

Liquid chromatography could be used in tracer studies because of the rapid development of this technique in the seventies, through which samples could be analysed in minutes instead of hours or even days. The procedure of the measuring system is outlined in Fig. 2. Firstly, the rhodamine in the sample is enriched in the solid phase extraction column (μBondapak C-18, 10 μm particles of chemically modified silica-gel). Secondly, the rhodamine is carried through the separation column (Novapak C-18, 4 μm particles) by the eluent or separation liquid (a mixture of methanol and water with chemical modifiers). The sample compounds migrate through the column at different rates, depending upon their attraction for the chemical packing inside the column.

The compounds are measured in the detector (fluorometer). The height of a peak in the chromatogram is proportional to the amount of the corresponding compound in the sample. An example from measurements in the North Sea is given in Fig. 3, showing peaks of sulphorhodamine B and rhodamine B. The third stage is cleaning

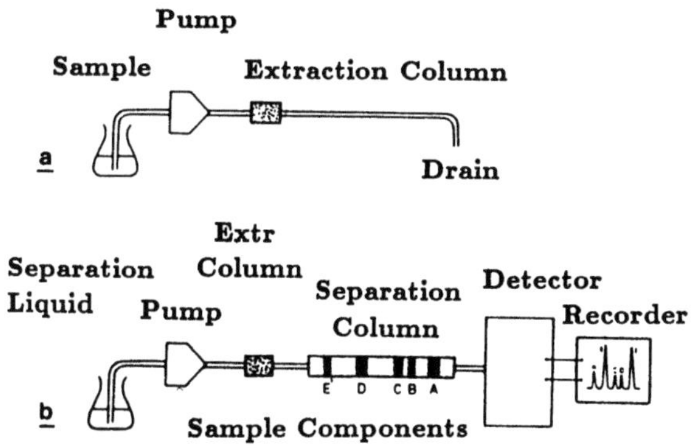

Figure 2: Principle scheme of the concentration determination by HPLC-techniques. a) enrichment phase b) phase of separation and detection of the various rhodamines.

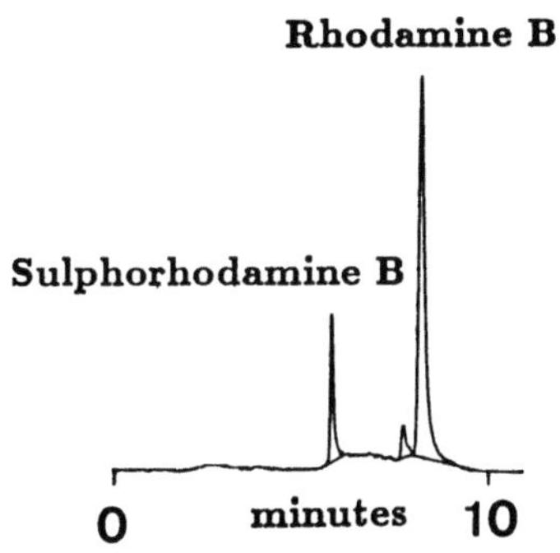

Figure 3: Chromatogram of a North Sea sample, showing peaks for sulphorhodamine B and rhodamine B.

up the solid phase extraction column. The mass detection limit is about $2 \cdot 10^{-12}$ g for rhodamine WT dissolved in pure sea water and $6 \cdot 10^{-12}$ g in lakes in which more rhodamines are dissolved. The analysis in the HPLC-system takes much more time than the classical fluorometry or the SMF-method. Experience shows 4 to 7 minutes at a required concentration detection limit of 50 pg l^{-1} (40 ml sample) to approximately 30 minutes for determining a concentration of 2 pg l^{-1} (1000 ml sample). Therefore, both the classical fluorometer technique or the SMF-method and the HPLC-method are being used in a dispersion experiment. In the first period after discharge the dye patch is followed in real-time by a fluorometer or the SMF-method, while at later stages, when the concentrations are much lower and may be masked by background fluorescence, the HPLC-system is applied.

3. SOME APPLICATIONS IN THE NETHERLANDS

3.1 Rivers and estuaries

On November 1, 1986 a fire in the Sandoz-industry in Basel resulted in a discharge of extinguish-water with a large amount of fluorescent material into the River Rhine, mainly sulphorhodamine B and rhodamine B.

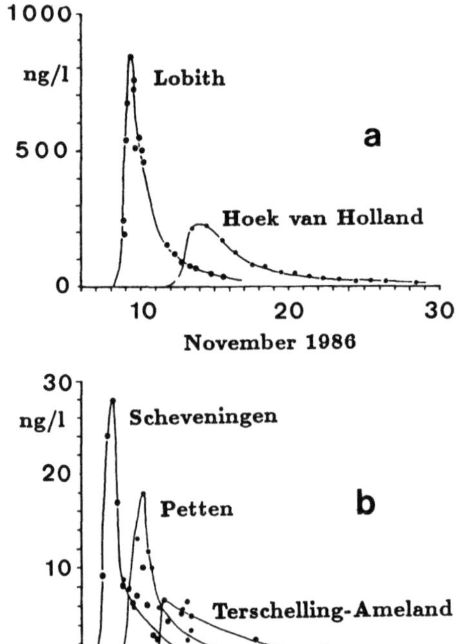

Figure 4: Passage of the sulphorhodamine B from Sandoz release on November 1, 1986 (Suijlen, 1987) a) in the Rhine at Lobith and at Hoek van Holland, b) along the coast (cf. Fig. 1 for station locations).

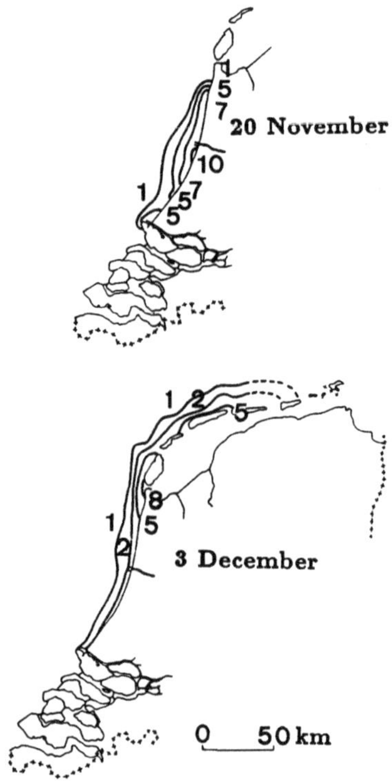

Figure 5: Distribution of sulphorhodamine B (ng l^{-1}) along the Dutch coast on November 20 and December 3, 1986 (Suijlen, 1987).

This material reached The Netherlands at Lobith (Fig. 1) on November 9, 1986. Here, the total fluorescence of sulphorhodamine B and rhodamine B was observed in real-time by a Turner 111 fluorometer.

The new HPLC-technique was used to accurately determine the concentrations of the two rhodamines. Amounts of rhodamines measured were 290 and 80 kg, respectively. In this case the sulphorhodamine B was a good representative for the other dissolved substances that were discharged during the accident at Basel. Some results of these measurements are presented in Fig. 4. The figures show a dilution by more than a factor of 100 along the trajectory from Lobith to Ameland.

The spreading of the polluted water along the coast on November 20 and December 3, 1986 is given in Figure 5. Integration of these concentration distributions gives a total amount of 175 kg sulphorhodamine B. The total sulphorhodamine B discharge through the Rhine Estuary at Hoek van Holland (Fig. 1) was estimated as being 220 kg. Because some parts in the sea area, as for example the Wadden Sea, were not incorporated in this analysis, it may be concluded that there is a fair agreement between these amounts. Rhodamine B could not be detected in the estuarine and coastal zone because its signal was lost in the natural background. This example thus illustrates the strength of the new tracing technique, with which the various rhodamines can be detected separately down to very low concentration levels.

3.2 River discharges into the North Sea

In studies on the pollution and eutrophication of the North Sea it is important to know the distribution of the various river waters in this area, because it gives a first impression of the various contributions of micro-pollutants, nutrients and other dissolved substances and can give direction to the process of looking for improvements of the water quality.

Most of the rivers, flowing into the North Sea, transport a composition of rhodamines,especially rhodamine B and sulphorhodamine B (Suijlen, to be published). For example, the river Rhine has a mean load of 10 kg rhodamine B and 1.5 kg sulphorhodamine B per day. When conducting tracer experiments in the North Sea one should be aware of these inputs. To get more insight into the appearance of some rhodamines in the Southern North Sea, exploratory surveys were conducted in January and June of 1985. These surveys, which included the Rhine, Scheldt, Ems, Weser, Elbe and Thames, showed that most of the rivers contain an analogue composition of fluorescent dyes, although in different ratios (Suijlen, to be published).

The observation of various photodecaying rhodamines and conservative substances, including salinity, prompts the interesting prospect of determining the percentage and age of dispersed river water (Suijlen, to be published). Some examples of these computations for the rivers Scheldt and Rhine are presented in Fig. 6, showing the application of the new tracer technique for the determination of the spreading of dissolved substances on large scales. This application of the HPLC-technique demonstrates again the advantage of being able to distinguish many fluorescing substances.

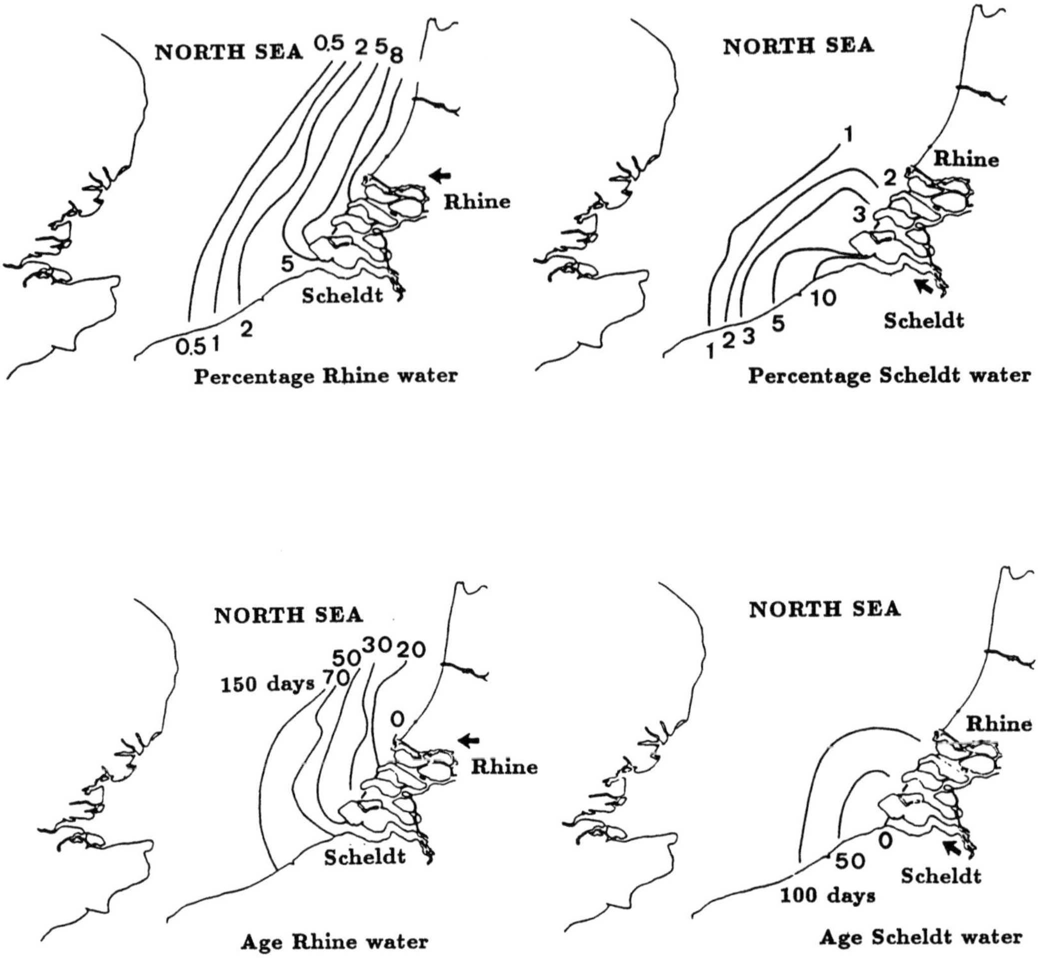

Figure 6: Age and percentage of river water from Rhine and Scheldt in June 1985 (Suijlen, to be published).

4. CONCLUDING REMARKS

Dye tracer experiments have been shown to be an effective tool in obtaining insights into water movements and dispersion processes which are needed in water quality studies. In these studies there is an increasing interest in following the released dye over a long period. Therefore, detection techniques are required, with which the fluorescent tracer can be measured down to very low levels. At these levels the measurement is greatly hindered by the natural background fluorescence, which appears to be varying both in space and time.

Because of these requirements, new measuring techniques were developed in The Netherlands during the last decade, with which accurate measurements can be made down to much lower concentrations. Through Synchronous Modulation Fluorometry (SMF) the detection limit of rhodamine B was lowered to $2 \cdot 10^{-9}$ kg m^{-3}. More recently, measuring techniques became available based on on-line trace enrichment and on the separation of the concentrated fluorescent dyes by HPLC. Through these methods it was possible to determine the tracer down to concentrations 1000 to 10.000 times lower than that achieved by current methods.

This HPLC-technique also makes it possible to determine a number of tracers simultaneously. The possibility of distinguishing more fluorescent tracers is of importance because several water systems seem to be polluted by some fluorescent substances. An example is the river Rhine, which carries remarkable amounts of rhodamine B, resulting in relatively high background fluorescence in Lake IJssel and in the Dutch coastal zone. In this case, a tracer with a low background concentration has to be applied.

Applications of these tracer techniques in rivers, estuaries and coastal waters in the Netherlands are given. They show the strength of these methods for the investigation of transports on large space and time scales.

Estuarine water quality management needs numerical models, with which predictions can be made of the transport of dissolved and suspended constituents. Tracers are useful for validating the results of these calculations. Besides data for testing the models, the tracer experiments also provide the essential information of Lagrangian currents and dispersive transports. Therefore, in coming years these new techniques, which have the possibility of detecting very low concentrations and of distinguishing various fluorescent materials, will play an essential role in the development of reliable models, because it is now possible to apply these techniques at much larger scales in both space and time.

REFERENCES

Houpt PM (1980) Synchronous Modulation Fluorometry. T.N.O. Innovatie 39: 8-9 (in Dutch with English Summary)

Laane RWPM, Manuels MW, Staal W (1984) A procedure for enriching and cleaning up rhodamine B and rhodamine WT in natural waters, using a Sep-pak C18 cartridge. Water Res 18, 2: 163-165

Pritchard DW, Carpenter JH (1960) Measurements of turbulent diffusion in estuarine and inshore waters. Bulletin of the International Association of Scientific Hydrology 20: 37-50

Pritchard DW (1979) Results of a preliminary study of the fluorescent background problem. Chesapeake Bay Institute, The Johns Hopkins University, Special Report 72, pp 36

Suijlen JM (1975) Turbulent diffusion in the lake IJssel at Medemblik, measured by rhodamine B tracer. Rijkswaterstaat, Report FA 7501 (in Dutch)

Suijlen JM (1987) Dispersion of the sulphorhodamine B release at Basel (Sandoz calamity) in the Dutch coastal zone and Lake IJssel in the period November 8, 1986 to February 23, 1987, Rijkswaterstaat, Tidal Waters Division, Report GWAO-87.012 (in Dutch)

Suijlen JM, Lemmen A (1987) A HPLC-based measuring method for fluorescent substances in surface waters. Rijkswaterstaat, Tidal Waters Division, Report GWAO-87.037 (in Dutch)

Suijlen JM (to be published) The dispersal and photochemical decay of river rhodamine in the Southern North Sea

TRANSPORT OF RHODAMINE-B FOLLOWING A SLUG RELEASE INTO THE RIVER ELBE ABOVE THE PORT OF HAMBURG

M. Kolb, Institute of Physics, GKSS Research Centre Geesthacht GmbH
P.O.Box 1160, D-2054 Geesthacht, FRG
H. Franz, German Hydrographic Institute (DHI),D-2000 Hamburg, FRG

1. INTRODUCTION

The port of Hamburg consists of numerous harbour basins and channels inside a ramification of the Elbe river and is exposed to tidal flow reversal (see Fig. 3 in [1]). The resulting abatement of the current velocity leads to sedimentation of suspended particulate matter in the harbour which necessitates expensive dredging activities. Experimental data on the length of time the river water stays in the harbour area may help to validate models. Therefore, as part of the BILEX '86 experiment [2], a substantial amount (120 kg) of an artificial tracer (Rhodamine B, for short "rhodamine" in the rest of the paper) was almost instantaneously (within 5 minutes) released into the river some 10 kilometers upstream from the bifurcation before the harbour.

2. MEASUREMENT STRATEGY

A survey of concentration fields in time and space which is necessary, for example, for the deduction of gradients and hence for the determination of dispersion coefficients requires a vast amount of fixed sampling points or moving measuring instruments. It was decided that one instrument (the in-situ sensor of SUAREZ [3]) should survey a downstream cross-section (Nienstedten) simultaneously with the hydrographic system [see 4] in order to measure the amount of tracer leaving the port area. Nienstedten is situated about 5 kilometers downstream from the junction of the Elbe's two branches.

A second moving fluorometer (the TURNER Design, operating aboard the REINHARD WOLTMANN vessel) was used to measure the dispersion and transport of the tracer throughout the port. The fluctuations in the tracer's concentration over time and space were expected to be large due to the complex channel network within the region of the harbour of Hamburg itself and the strong tidal currents present. Therefore, the (sampling/measuring) strategy had to be designed in such a way, that the tracer plume could be covered as often and as completely as possible.

The expected broad longitudinal extension of the plume did not permit a reasonable measurement of its entire structure across the stream. Moving between the left and right banks as often as possible was the only feasible strategy for approximately smoothing the fluctuations and structures over the cross-section. Due to the general traffic, even this strategy could not always be followed.

The fluorescence signal was recorded continuously on paper and digitized for later processing. Fig. 1 shows a typical, short plot. The ship's position was noted with the aid of landmarks and a watch. Positions between these points were obtained by linear interpolation (Fig. 1; dashed line). A change of crew at a fixed point in the harbour every 8 hours further restricted the continuity of measurements. Additionally, some points were selected throughout the harbour area where water was sampled in flasks.

3. MEASUREMENT TECHNIQUES

The selection of the fluorescent dye rhodamine was made as a result of the experience gained by the DHI during studies of open sea turbulence [5]. Accordingly, rhodamine sensors on hand from these studies were used for the present experiment. However, the water in the river Elbe is far more turbid than sea water. Moreover, the Elbe's turbidity varies strongly with topography and tidal current. It was therefore necessary to eliminate any interference caused by turbidity. To achieve this, certain measures were undertaken with the in-situ sensor SUAREZ [see 3], which are outlined in the following.

The in-situ sensor excites the fluorescence by a pulsed light beam of several flashes per second (filter 390 to 530 nm) and detects it synchronously with a sensor (filter 575 to 595 nm) orthogonally facing the exciting beam. According to [3] the small nominal separation (= 45 nm) of the two filters will only allow measurements of rhodamine content down to a level of 0.05 mg/m³ in fairly clear water (that is, water having an attenuation coefficient below 0.5 m^{-1}). However, attenuation coefficients c as high as 7 to 21.5 m^{-1} were encountered during BILEX '86. The light scattered from the exciting beam thus leads to a background signal comparable to the signal expected from rhodamine. (This experience started the development of a fundamentally new system, see [7].)

The background signal, U_s, proved to be linearily correlated with the signal, D, of the transmittance meters (see [4]) thus allowing for a straightforward correction. Fig. 2 presents the regression between D and U_s as measured in situ at Nienstedten when the rhodamine had passed. Likewise, Fig. 2 shows the instrument's sensitivity in measuring different amounts of rhodamine diluted in a basin of river water (suspended particle content about 20 g/m³ resulting in U_0 = 0.65 V) down to the low concentrations found at Nienstedten where a detection limit of about 0.1 mg/m³ was reached.

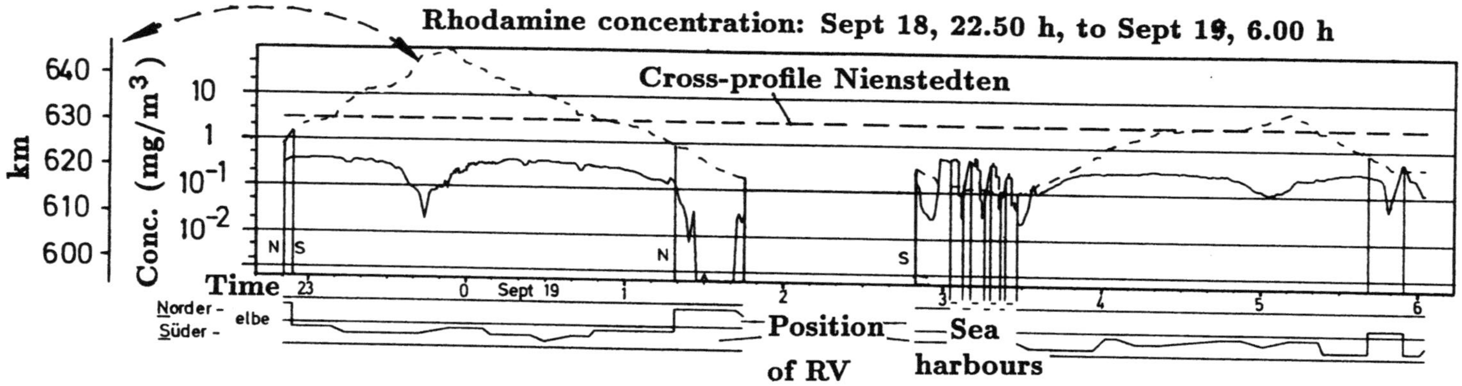

Figure 1: Rhodamine concentration recorded together with the ship's longitudinal (km = distance along navigable river) and lateral position (schematic, below)

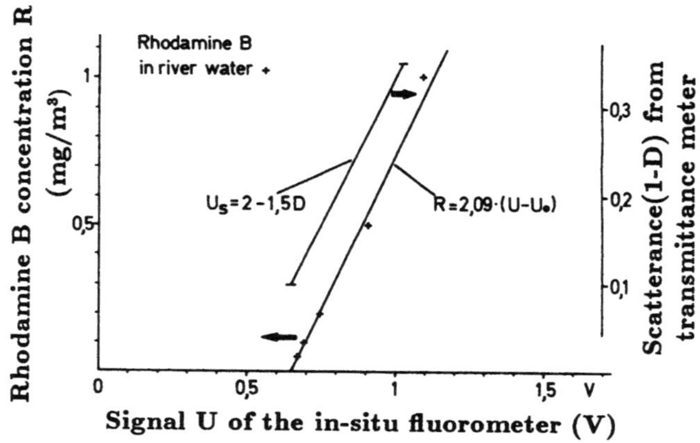

Figure 2: Signal of the SUAREZ sensor relative to known concentrations, R, of rhodamine under laboratory conditions (U, left scale) and to turbidity, respectively. Scatterance (1 - D) measured in-situ (U_S, right scale).

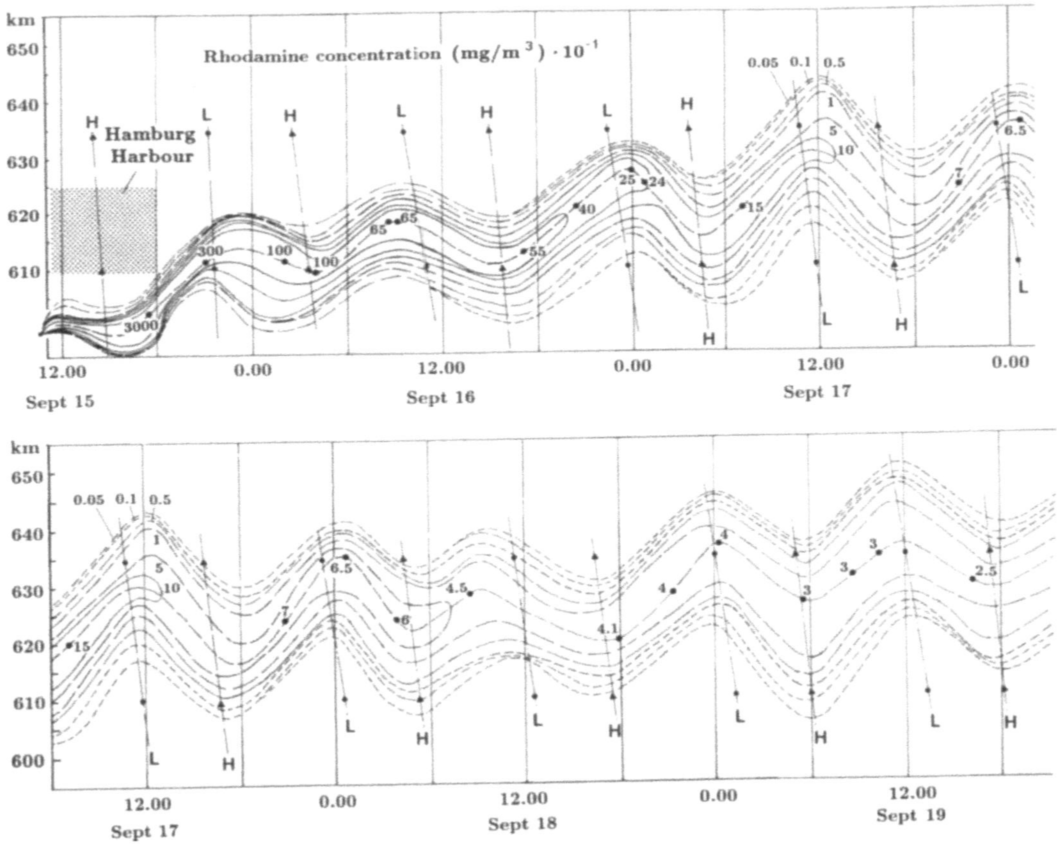

Figure 3: Isopleths of rhodamine concentration vs. time and longitudinal river coordinate (km) together with observed tidal crests (H) and troughs (L).

4. MEASURED RHODAMINE CONCENTRATIONS

4.1 The longitudinal distribution

The in-situ sensor was used to measure vertical profiles of the rhodamine concentration at different longitudinal and lateral positions of the river during the first 14 hours after the release because the rhodamine could not arrive earlier at Nienstedten. These early measurements confirmed "that vertical mixing is instantaneous compared to transverse mixing" (p. 113 [8]) and showed that the latter was far from complete when the rhodamine reached the bifurcation of the river some 10 kilometers below the point of release.

The recording of concentrations of rhodamine (Fig. 1) aboard the REINHARD WOLTMAN was smoothed by hand. Hence the measured space-time points of given concentrations were plotted in a space-time diagram and interpolated for each individual concentration separately. For this purpose the tide, as given by continuous records of the water level at 10 gauges within the region of the harbour, was also taken into account. Finally, the interpolated curves of the individual concentrations were drawn together on one graph and corrected at the points where their lines intersected. Fig. 3 displays the result for the Süderelbe and neighbouring regions of the Elbe. A corresponding diagram has also been constructed for the Norderelbe. Because the measuring track oscillated across the width of the river, it is justified taking these results as averages over the cross-section.

4.2 Continuous survey of a downriver cross-section

The main task assigned to the in-situ system was to partake in a four-day survey, taking hourly samples, over the cross-section at Nienstedten [2]. The vertical mixing of the rhodamine proved to be constant. Fig. 4 shows the transverse distribution of the rhodamine during the first two days. Thereafter, the lower rhodamine concentration values in the tail of the plume lead to increased scattering. They were, therefore, averaged across the river, taking into account the varying depth (Fig. 5).

The tidal movement and the dispersion of the rhodamine at the Nienstedten cross-section can be seen from Fig. 4. Furthermore, the initial asymmetry towards the southern bank shows that the transport through the shorter and deeper southern branch of the river is faster.

5. CONDENSED RESULTS

5.1 Balance of rhodamine transport

Absolute water velocity vectors were obtained for the Nienstedten cross-section simultaneously with the rhodamine concentration measurements [2]. This allows calculation of the momentary transport of rhodamine through the cross-section to be made. Fig. 6 shows the resulting amount of rhodamine transported down and upstream

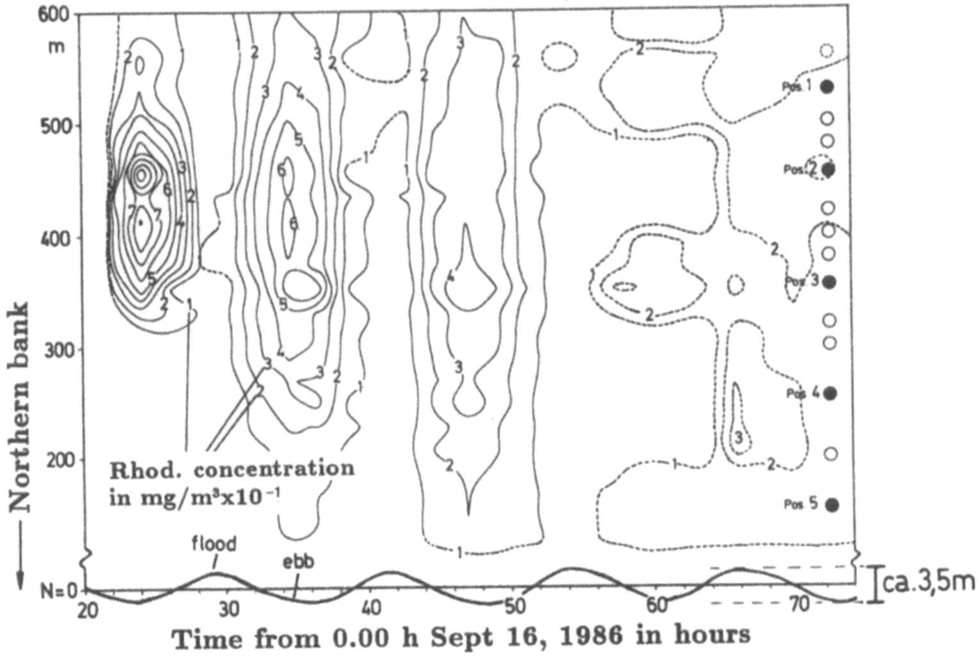

Figure 4: Time dependence of the rhodamine concentration for the Nienstedten cross-section together with tidal cycles (i.e. water level sketched around time axis)

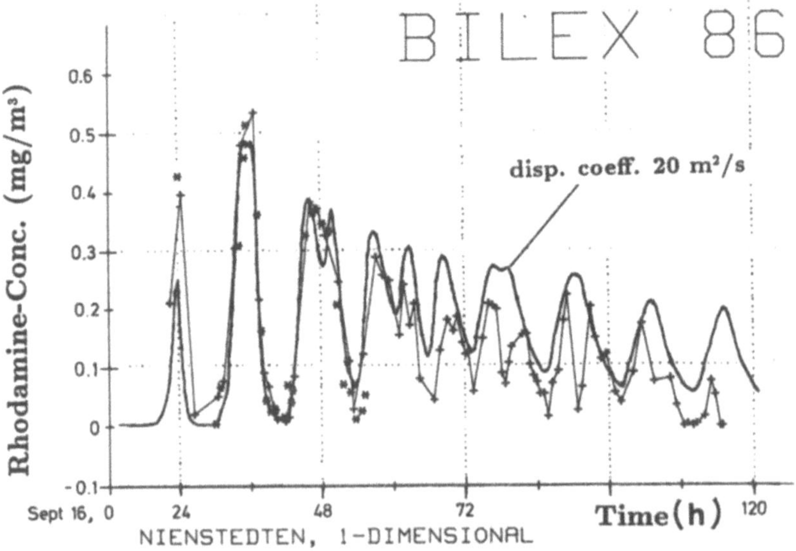

Figure 5: Time dependence of the cross-section averaged rhodamine concentration (+,*) at Nienstedten compared with simulation

during ebb and flood phases over the first seven tides after the arrival of the dye cloud. The errors reflect ± 10 % for the water discharge plus a relative error of ± 0.05 mg rhodamine/m³. The net outflow may be found by summing over one tidal cycle, i.e. adding ebb and flood transport. The accumulated net outflow indicates that more than half the rhodamine injected (i.e. 120 kg) had left the harbour after these 3 1/2 days, and that the rest seems to leave very slowly, if at all.

The balance of all the rhodamine recorded in the Süder- and Norderelbe becomes a function of time by multiplying the concentration distributed in time and space (Fig. 3) with the tide-dependent river cross-sections. Later, the coefficient of dispersion is determinable also as a function of time and space by means of a numerical treatment which has been developed. The evaluation of measurements carried out for individual harbours (e.g. see Fig. 1 about 3 o'clock) and connecting channels is expected to give direct results (e.g. turnover times for rinsing after contamination), and may also give some idea on transport by advection and dispersion processes.

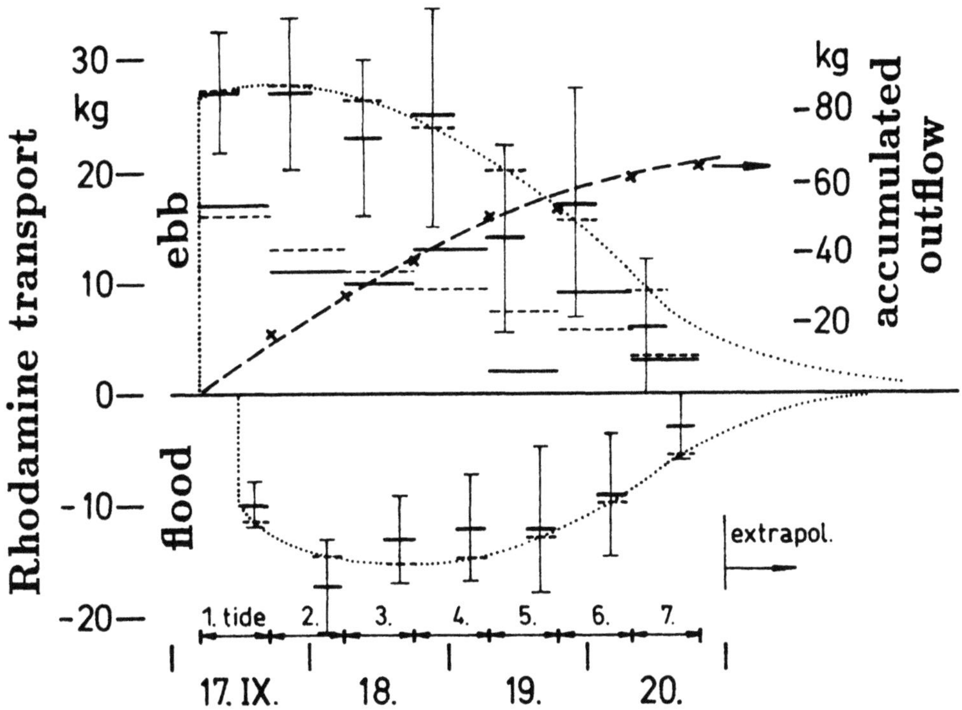

Figure 6: Left scale: Rhodamine transported through the Nienstedten cross-section as measured for ebb and flood phases of 7 tides (±); resulting net outflow (—)for every tide; and smoothed values (----).
Right scale: Accumulated outflow (x)

5.2 Comparison with a transport dispersion model

The Danish Hydraulic Institute operates a 1-D model routinely for the port authority and did particular runs for the rhodamine experiment with three different dispersion coefficients [9]. The hydrodynamic simulation was driven by the actual inflow, Q, (deduced from gauge measurements 95 km upstream from Nienstedten) and the actual water level variation 94 km downstream at the mouth of the Elbe. The model describes the fluctuating transport of the rhodamine plume through the Nienstedten cross-section quite satisfactorily (see Fig. 5). However, to achieve an absolute agreement with the measured concentrations the amount of dye injected had to be divided by three. The following arguments help to clarify this discrepancy, and to account for the 2/3 of the injected rhodamine "missing".

(1) The model neglects any loss of rhodamine which has settled in stagnant dead zones, has become attached to the sediments and banks of the river/channels/basins, or has decayed.

(2) The measurements made at Nienstedten [2] show that the net water outflow (i.e. freshwater discharge) is about twice as high as the Q used by the model. Doubling the water flow would halve the model's dye concentrations, i.e. only 1/3 of the injected amount would then be "missing".

REFERENCES

[1] Michaelis W (1989) The BILEX concept - research supporting public water quality surveillance in estuaries. This volume, chapter III

[2] Fanger H-U, Kappenberg J, Männing V (1989) A study on the transport of dissolved and particulate matter through the Hamburg Harbour. This volume, chapter IV

[3] Früngel F, Knütel W, and Suarez JF (1971) Impulslicht-Fluorometer in der ozeanologischen Meßtechnik. Meerestechn 2: 241 - 247

[4] Fanger H-U, Kuhn H, Maixner U, Milferstädt D (1989) The hydrographic measuring system HYDRA. This volume, chapter V

[5] Franz H, Klein H (1986) Some Results of a Diffusion Experiment at a River Plume Front. Dt hydrogr Z 39: 105 - 111

[6] Böttcher G, Kolb M, Puch B, Reimann H, Roscher O-M (1987) Messung von Rhodamin B in der Elbe bei Hamburg vom 15. bis 20 September 1986. GKSS 87/I/9

[7] Jahnke J, Michaelis W (1989) Laser fluorometer for the detection of rhodamine B in estuarine waters. This volume, chapter V

[8] Fischer HB, Imberger J, List EJ, Koh RCY, Brooks, NH (1979) Mixing in Inland and Coastal Waters. Academic Press, New York, pp 105-145

[9] Danish Hydraulic Institute (1988) Transport dispersion model for Elbe river and Hamburg harbour. Simulation of rhodamine experiment

FEASIBILITY OF TRACER STUDIES OF LARGE SCALE COHESIVE-SEDIMENT TRANSPORT

R. Spanhoff and J.M. Suijlen
Tidal Waters Division, Rijkswaterstaat
P.O. Box 20907, 2500 EX The Hague
The Netherlands

INTRODUCTION

The number and amounts of chemical substances entering the estuarine and marine systems influenced by human activities have increased dramatically since the beginning of the industrial revolution, up to levels at which negative effects are likely or proven. These substances originate from many sources such as deliberate legal or illegal industrial spills, from accidents and rather continuous urban wastes, from more diffuse sources such as agriculture and atmospheric desposition, etc.

Measures to reduce or prevent the inflow of chemicals into the aquatic environment all have their price, to be paid by industry, or by society on a national, local, professional or individual level. It will be clear that every party required to pay or ask this price has to be convinced in some sense of the anticipated effects of such measures preferably within a few years or a generation.

Only in a few cases are these effects so evident that they may lead to an immediate ban upon further releases. In most cases they are relatively small, when for instance the environment is already loaded up with the contaminant, or uncertain due to lack of knowledge about sources, transport and fate of the substance or about its biological effects. Our complex society, so dependent on chemicals as it is, will continue for decades to come to release directly or indirectly various potentially hazardous substances into the aquatic environment, and tools to monitor and predict their distribution and impact are needed more than ever.

Many classes of chemicals are adsorbed on silt present in the water. The transport of these sediments is dominated by water movements, but may deviate markedly from the transport of dissolved substances which follow those movements. This is due to effects related to the sediments such as sedimentation, consolidation and erosion, processes which at present are hard to model deterministically; more progress in that respect has been made in the modelling of the mere water movement.

As a consequence there exists a great need for field data on fine sediments, (i) to answer in a direct manner questions of the authorities on mainly the present situation, to provide (ii) knowledge about processes in sediment transport in the field and about their relative importance for modelling, as well as (iii) means of validation of water-sediment transport models under construction. These models then should be able to predict future situations, among others as a function of various possible scenarios in water management.

The aim of the present paper is to draw attention to the activities in The Netherlands concerning the development and use of cohesive-sediment tracer techniques and to stimulate thinking and discussions on this subject.

A FLUORESCENT TRACER FOR COHESIVE SEDIMENT

Fine sediments may travel large distances in estuarine and marine environments from their source to their final destination where they settle permanently and accumulate. The sources may be natural, erosion for instance, and anthropogenic, such as dumping or a change in the flow conditions by the construction of a harbour or the deepening of a channel, etc. To determine the impact of a specific source it is always necessary to distinguish its sediments from the background from other sources. Seldom does the studied sediment have a clear distinct fingerprint enabling this, which brings about the need for an artificial tracer.

Good tracers are not readily at hand, since radioactive tracers face resentment from the public. Neutron activation of specific elements to label the studied sediment has its practical limitations. A promising alternative is the use of fluorescing pigments as tracers, as first suggested and explored [1] by the Netherlands Organisation for Applied Scientific Research (TNO), and currently being developed under contract from the Rijkswaterstaat.

The tracer consists of discrete particles, in the micrometer range (see below), of an inert, virtually transparent resin containing 3 % fluorescing dyes; it is a base material to produce inks and coatings. Several types with differing colours can be purchased from Day-Glo Color Corp. An appropriate sediment tracer has to fulfil various requirements, such as (i) it has to follow the transport of the fine sediments, (ii) it must hardly change its properties in the timescales studied, (iii) it may bring no harm to the environment, etc. Here only recent developments and ideas can be discussed as well as a few aspects of the present tracer, for the remainder we refer to an earlier account [2].

DETECTION OF THE TRACER

Having injected a known amount of tracer material into the water system under study, for instance as a mixture of tracer and natural sediments, bottom samples are taken at various, possibly remote, distances from the source, after appropriate times have elapsed. They must be analysed with respect to their tracer content in order to obtain the desired information about the sediment transport and accumulation. Particles greater than, say, 63 µm are removed by sieving and a sufficiently dilute suspension of the bottom sample is pumped through the tracer detector. In a certain measuring time, N_{tr} of all the pigment particles passing the instrument are detected one by one. In addition a number N_{bg} of background counts is registered, due to instrumental effects or spurious light from, e.g., other fluorescing substances. Thus, $N_{tr} = N_t - N_{bg}$, with N_t being the total number of counts in the total measuring time. Poisson statistics apply to these numbers. In practice the statistical uncertainties are readily approximated by Gaussian distributions, leading to a standard deviation (67 % confidence interval) $\sigma(N_{tr})=\{\sigma^2(N_t) + \sigma^2(N_{bg})\}^{½}$ $= \{N_t + \sigma^2(N_{bg})\}^{½}$. So, when $N_{tr} \equiv N_t >> \sigma^2(N_{bg}) = O(N_{bg})$ we obtain $\sigma(N_{tr}) \equiv N_t$, leading to a relative error $\sigma(N_{tr})/N_{tr} = N_t^{-½}$. For $N_t = 100$ we find a 10 % statistical error, for instance. Reducing it with a factor 2 implies a fourfold measuring time at a given tracer concentration, etc., so 30 minutes is a reasonable practical limit. When the number of tracer counts is of the order of or much smaller than the background, good statistical accuracies are harder to obtain. In a field test of course, other variations play their role in the total uncertainties, such as inhomogeneities in the sampled areas.

Given in practice that the amount of tracer material to be used and the measuring time per sample are limited, a detection system has to meet several, often conflicting requirements, such as

(i) the detector must be efficient, counting as many passing tracer particles as possible

(ii) relatively high sediment concentrations must pass the detector, without masking the tracer, so as much as possible sample material is probed in the measuring time

(iii) background counting must be low, so the detector must among others be highly selective for the tracer.

DEVELOPMENTS IN TRACER DETECTION

A first-generation detection system based on a fluorescence microscope and a xenon lamp as light source was used in a field test conducted with labelled dredged material dumped in the Ems-Dollart estuary [2]. The study of the fate of the dumped silt was partially successful and the test provided several clues for improvements. As one of them, a second-generation detector was built, now with an air-cooled Ar-ion laser and new optics. The set-up is sketched in figure 1. The integrated system of laser, optics and electronics is so compact and robust that it can be used on board a vessel. Two selected, differently coloured pigments can be measured simultaneously. They are both efficiently excited by the laser, with relatively broad, partly overlapping, fluorescence spectra peaking in the orange and the green. Optical filters select for each a small wavelength band to pass exclusively to their respective photomultipliers. Thus only a fraction of the fluorescent

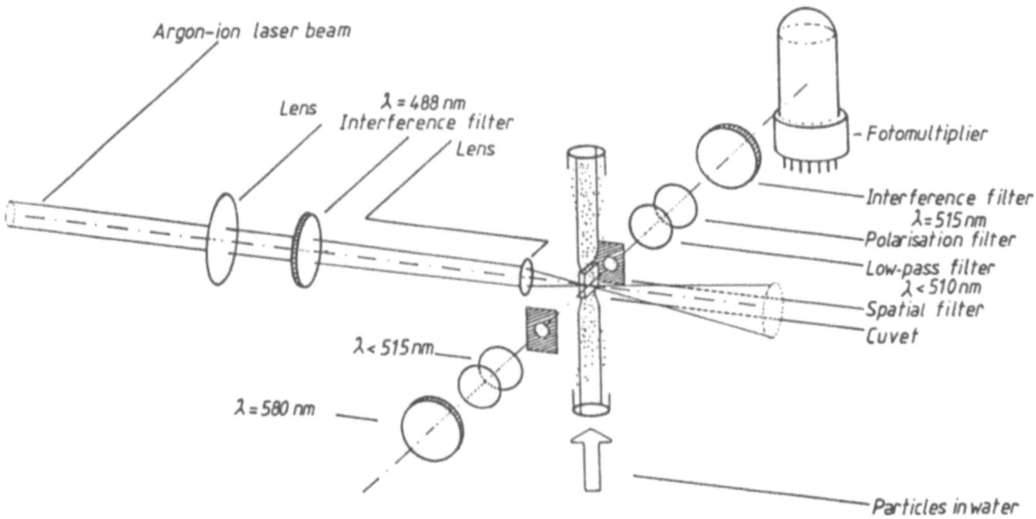

Figure 1: Dual-channel fluorescent particle detector

yield is used. Additional filters, including polarization filters, serve to block other light from reaching the photomultipliers, such as the abundantly present scattered laser light.

When a particle passes the laser beam, the corresponding photomultiplier gives a pulse. Its width and height are determined by the constant flow rate and by the particle's volume crossing the beam spot, respectively. Pulses exceeding an electronically set threshold are counted. Not all particles passing the detector meet this requirement. (i) They may be too small; a monotonically decreasing size distribution was measured, with circa 480 particles per nanogram and ranging virtually from 0.2 µm to 5 µm with a strong preference for the smaller sizes, and experiments with test suspensions indicate that only particles larger than circa 1.2 µm, i.e. 12 per cent, are detected. (ii) The particles are only partially excited or escape the laser completely, since the beam spot (order of tens of micrometers for a high light intensity) covers only a small fraction of the detection cell, which reduces the total efficiency further, to about 0,4 %. This figure is a factor of 2 better than for the previous system. A much more important improvement is that now mud suspensions up to 10 gram per liter can be measured, compared with 400 mg per liter, without noticeable masking of the tracer by the silt. So in the same measuring interval significantly more sample material can be probed (flow rate 0,3 liter per minute). The efficiency is fairly linear over at least four decades of pigment concentration, up to 10 µg per liter; at the highest concentrations a known dead-time correction for more particles seen as one may be applied.

The experiments up till now with this new set-up were done in the laboratory with mainly kaolinite, and they show a high sensitivity, defined with the above equations for a 30 minutes measuring time, namely 3 ng and 60 ng per liter for the orange and green pigments respectively, with hardly any cross-talk for the green particles in the orange channel but with a high one for the reverse.

Although almost ready, no field trials have been done yet, only some measurements with natural muds labelled in the laboratory with the tracers, since first an expected improved detection method that deliberately uses the observed cross-talk, is being investigated.

Relatively broad, mutually overlapping spectral filters in both detector channels, with wavelengths centered around the fluorescence maxima of the respective pigments, replace the earlier narrow-bandpass filters in the orange and the green. Both detectors will now give a pulse in case of a passing pigment particle, with a characteristic pulse height ratio for each pigment used. Selection and successive counting takes place by on-line processing with a microcomputer of both pulses and by comparing the height ratio as well as the pulse shapes with those of a standard pigment-particle pulse. More efficient use is made of the fluorescent light of the particles, so more particles can in principle be detected, since now smaller ones are included and/or the beam spot can be spread over a larger part of the detection cell. In addition, background due to spurious light and electronic noise may better be suppressed, since they give for instance deviating pulse shapes or trigger only one channel.

APPLICATION OF THE TRACER TECHNIQUE

The present tracer technique is notably developed for relatively large-scale experiments, such as into the impact on the Wadden Sea of the Rhine sediments or of the dredged material from the entrance of the Rotterdam Waterway dumped off the Dutch shore. Here the steady-dilution method [3] is appropriate, in which one adds regularly small fractions of the total tracer amount to the Rhine water and to the dredged material respectively. By doing so for one or a few years and by subsequently, on comparable timescales, sampling the accumulation areas a few times, the measured tracer distribution is representative for the long-term accumulation pattern of the silt from the studied source. Accidental effects due to specific environmental conditions at the time of an individual release are averaged out.

A conservative estimate, assuming that only 10 % of the tracer is deposited in the upper 10 cm of the bottom (accounting for bioturbation) of some 1000 km^2 of the Wadden Sea, suggests that of the order of 1000 kg material is needed to exceed a detection limit of 10 ng per liter suspension, see above (thinner layers of the samples will be analysed separately). With one or a few thousand kilograms to be released in the order of a year, the proposed experiments are well feasible. It should be noticed that tests have shown [2] that the tracer is inert and harmless, and that the possibly noxious fluorescent substances enclosed in it make up only 3%

of the materials weight. As a drawback of this inertness the aquatic system studied, which now is free of the tracer, gets loaded up with it, diminishing the sensitivity for later experiments. Therefore, the chances of a successful experiment have to be optimal, implying that first a program of preliminary tests [2], both under laboratory and small-scale field conditions, must be completed, such as comparing the tracer and silt transport under turbulent conditions and optimizing the reliability for almost real-time measurements and the sensitivity (smaller amounts of the tracer) of the detector.

Small-scale experiments that by their geography do not interfere with the large-scale ones, or that form a part of them, are easier to conduct and answer other existing questions, such as about a possible return of the dumped sludge to the Rotterdam Waterway.

In addition, specific processes in the sediment transport can be studied by sampling on smaller time and space scales after such individual releases; sampling the water phase with an ultra-centrifuge with respect to suspended sediments in transport is considered as well. Obtaining this process-related information may benefit from the use of more than one pigment colour, also in case of the large-scale experiments, for instance by labelling the dredged-material dumps under ebb and flood conditions differently, or by discriminating between summer and winter conditions or between low and high Rhine discharges, etc. The present detector permits no more than two colours to be distinguished, but it is conceivable that the data processing under investigation allows more.

ACKNOWLEDGEMENT

The authors gratefully acknowledge the collaboration with Dr. Houpt and co-workers of the Department of Electro-Optical Systems of the Netherlands Organisation for Applied Scientific Research (TNO), where the ideas concerning the use of the present tracers were born and implemented to the described detection systems.

REFERENCES

[1] Draaijer A, Tadema Wielandt R, Houpt PM (1984) Investigations on the applicability of fluorescent synthetic particles for the tracing of silt. Netherlands Organisation for Applied Scientific Research, Report R 84/152 (in Dutch)

[2] Louisse CJ, Akkerman RJ, Suijlen JM (1986) A fluorescent tracer for cohesive sediment. In: Pounsford (ed) Int Conf Measuring Techniques of Hydraulics Phenomena in offshore coastal and inland waters. 9 - 11 April 1986. BHRA, The Fluid Eng Centre Cranfield UK, p 367

[3] Crickmore MJ (1976) Tracer techniques for sediment studies; their use, interpretation and limitations. In: Diamond Jubilee Symp on Modelling Techniques in Hydraulic Eng, Vol 1. Central Power and Water Research Station, Paper A13

THE INFLUENCE OF A SALT WEDGE AND TIDAL FLOW DYNAMICS ON CONTAMINANT PATHWAYS IN THE FRASER RIVER ESTUARY, BRITISH COLUMBIA

J. H. CAREY
National Water Research Institute
Canada Centre for Inland Waters
Burlington, Ontario, Canada L7R 4A6

INTRODUCTION

The Fraser River Estuary is one of Canada's most productive and valuable wildlife habitats. The estuary supports the largest population of wintering wildfowl in Canada and is used by about half the birds migrating along the Pacific flyway. The river is renowned for its salmon runs; the five species of salmon that live in or pass through the estuary produce the largest natural salmon run in the world. The value of the commercial and recreational fishery supported by the Fraser is estimated at $90 million. At the same time, the estuary is home to more than 1.25 million people and is intensively used for heavy industry and resource-related enterprises. In an attempt to balance the natural environment with ever increasing industrial, commercial, recreational and residential pressures, an inter-governmental, inter-agency program called the Fraser River Estuary Management Program (FREMP) was established. One of the activities under FREMP is the development of a water quality plan for the estuary which involves the setting of site-specific water quality objectives and the establishment of a monitoring program. Although much work on contaminant identification and distribution has been completed, the development of site-specific objectives and the establishment of the monitoring program has been delayed, partially because of uncertainty with respect to the complex mixing processes in the estuary and their impact on contaminant dynamics.

The Fraser River drains the western slope of the Rocky Mountains in British Columbia, Canada, discharging via the Strait of Georgia into the Northeastern Pacific. The average discharge of the river is 3000 m^3/sec with peak discharge rates during freshet of about 10000 m^3/sec and minimum flows during the winter months of about 1000 m^3/sec. About 30 km inland, the river splits into two arms, with the Main Arm receiving about 75 % of the flow. The tidal range is large with a maximum of 4 metres. A typical mean-depth tidal velocity is 1.0 m/sec while the mean outflow of the river discharge is 0.4 m/sec (Geyer, 1985). This combination of substantial freshwater discharge and large amplitude tides results in the formation of a distinct salt wedge within the estuary. This paper examines the interaction of Fraser River mixing and flow dynamics on contaminant pathways and their impact on our ability to design a monitoring program.

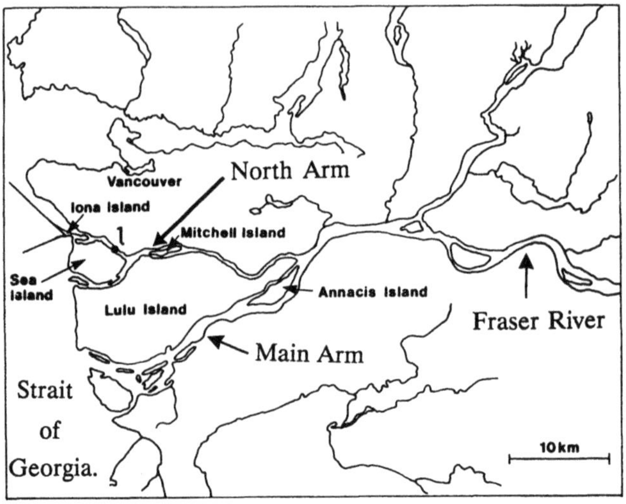

Figure 1. The Fraser River Estuary

CONTAMINANT SOURCES

Although sewage plant and industrial discharges and agricultural runoff all contribute contaminants to the estuary, the most important contaminants identified to date are chlorophenolics resulting from forest industry operations. Carey and Hart (1988) examined the identities and temporal variations of chlorophenolics in the estuary and concluded that pulp mills upstream from the estuary were important sources of a group of chlorophenolics dominated by chloroguaiacols, whereas distributions of 2,3,4,6-tetrachlorophenol (2346-TeCP) and pentachlorophenol (PCP) indicated specific sources of these compounds within the estuary. These latter chlorophenols are used in the lumber industry as fungicides to protect freshly cut lumber from sapstain and mould fungi. Concentrations of these compounds are highest along the North Arm (Carey et al., 1988) where many lumber mills are located. Although these mills have no effluent discharges, chlorophenols enter the river from spills in the application and drip areas and in stormwater runoff from treated lumber in uncovered storage areas. Krahn and Shrimpton (1988) examined the occurrence of chlorophenols in stormwater runoff at five sawmills and two lumber export terminals. Whenever there was measurable rainfall, chlorophenol concentration in the runoff exceeded 100 ug/l. These concentrations are of concern because sapstain treatment solutions are lethal to fish at chlorophenate concentrations of 100 ug/l and affect fish growth and reproduction at concentrations as low as 2 ug/l. In addition to being related to weather, the release of these compounds to the river can be affected by tidal cycle. The runoff from many of the sites is collected in storm drains which are equipped with tide gates to prevent salt water entry during flood tides. Thus it is possible for stormwater to collect behind the tide gates and be discharged in a single flush when the gate opens during ebb. The net effect is that inputs of 2346-TeCP and

PCP to the estuary can be very episodic. Figure 2. depicts one such event. Over a three day period, concentrations of the two chlorophenols were observed to vary by more than a factor of 20 at a site well away from point sources. The input occurred as a pulse about 6 hours wide. It is clear that any monitoring program that was incapable of detecting such episodic contaminant inputs would be seriously flawed.

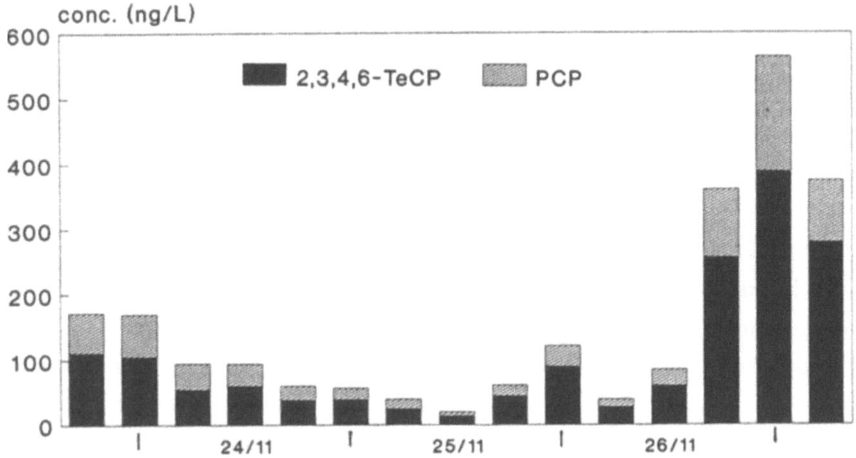

Figure 2. Concentrations of two chlorophenols in the North Arm of the Fraser River Estuary. Samples were 6 hour composites sampled hourly over a three day period in November, 1986.

CONTAMINANT SORPTION AND BIOACCUMULATION

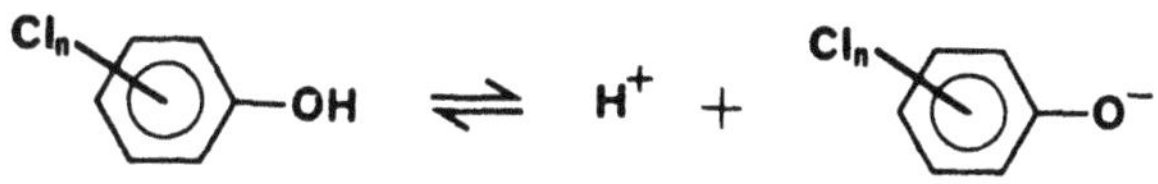

(1)

Chlorophenolics are different from many other organic environmental contaminants because they possess an ionizable proton and much of their environmental behaviour is controlled by the acid-base equilibrium shown in equation (1). At an average pH for Fraser River water of 7.76 the ratios of dissociated to undissociated forms are 725 and 290 for PCP and 2346TeCP respectively, ie. both chlorophenols exist predominantly in the dissociated form. The question of speciation is important because the dissociated forms of chlorophenolics are much more water soluble than the undissociated forms. For partitioning of a compound between two immiscible phases, most typically an aqueous and a non-aqueous phase, the position of the partitioning equilibrium is proportional to the solubility of the component in the two phases. Since aqueous solubility of the dissociated PCP is so much greater than that of the undissociated form, there is a large decrease on partition coefficient upon dissociation. In this way, acid-base equilibria have a significant effect on bioaccumulation and other pathways involving partitioning into lipophilic phases such as sorption into sediment.

Using suspended solids concentrations and fraction of organic carbon determined over two four day periods under high and low flow conditions (Carey & Hart, 1988), and the relationships of Karickhoff and coworkers (Karrickhoff, 1981; Karrickhoff et al. 1979), we calculated the ratios of the total analytical concentrations, or distribution ratios **D**, for PCP onto Fraser River suspended sediments. The distribution ratios were 12 and 40 for high and low flow conditions respectively. To check the accuracy of these predictions, we conducted adsorption experiments using suspended sediments collected with a sediment trap in the North Arm of the estuary under low flow conditions and found an average **D** of 727 (+/- 116) for PCP. The results underestimated actual sediment-water distribution ratios by more than an order of magnitude. A possible explanation for this difference is provided by the results reported by Schellenberg et al. (1984). These authors observed that the relationships used above were valid for low ionic strengths only and that sorption of chlorophenols onto sediments was dependent on ionic strength of the aqueous phase. These observations were further examined by Westall et al. (1985) who confirmed that in addition to sorption of the neutral species in the nonaqueous phase, sorption of ion-pairs into the nonaqueous phase also occurred. It follows from these results that important factor in determining the sorption of chlorophenols in estuaries is likely to be salinity distributions. Salinity may also play a role in determining extent of bioaccumulation of PCP and 2346TeCP in Fraser River fish. Carey et al. (1988) determined that bioconcentration factors for 2,3,4,6-TeCP and PCP in several species of fish from the Fraser River estuary were greater than those observed by other authors in field measurements involving a number of freshwater species but were comparable to those observed in a seawater bay. Since effects of contaminants are usually directly related to bioavailability, these observations indicate that salinity could be an important parameter in determining effects of chlorophenols in estuaries.

The hydrography of the salinity intrusion into the Fraser River estuary under various runoff conditions has been documented by Ages (1979). Salinity variations within the estuary occur on seasonal times scales with changes in runoff and on tidal time scales due to advection and mixing with the tidal flow. In freshet conditions, the salinity is confined to the outer 10 km of the estuary with the salinity residing at the mouth during most of the tidal cycle and a brief intrusion during flood. Under moderate runoff conditions, the salinity intrusion ranges from the mouth to 18 km inland depending on tidal amplitude and under low runoff conditions, the intrusion ranges from 15 to 30 km inland. In addition, it cannot be assumed that tidal flow merely translates but does not modify the salinity structure. Wright (1971) found evidence of greatly enhanced mixing during ebb flow for the salt wedge of the Mississippi River, indicating that tidal flow dynamics play a prominent role in exchange across the interface. Geyer (1985) reported similar processes occurring in the Fraser River estuary. He also observed that during low runoff conditions, an appreciable volume of mixed water resides between the head end of the salt wedge and the river mouth and conditions approached those of a partially mixed estuary. Under moderate flows, the estuary retains its stratfied structure although an appreciable volume of mixed water resides in the river. The accumulation of partially mixed water within the estuary depends on the extant of flushing that occurs during ebb.

We determined salinity distributions at a number of potential sites along the downstream portion of the North Arm under high, moderate and low runoff. Our observations indicate that apart from freshet, the North Arm exhibits characteristics consistent with partial mixing (Fig 3). In cases where there was appreciable discontinuity in the salinity profiles water was sampled for determination of chlorophenol concentrations. An inverse relationship between salinity and chlorophenol concentration was observed. This relationship suggests that there may be some possibilities for monitoring at these sites but much more information about mixing dynamics are needed. A pronounced asymmetry between ebb and flood would likely prevent the use of time averages in a monitoring program and the presence or absence of a salt wedge could give rise to uncertainty about the nature of the actual water sampled and its usefulness for contaminant monitoring.

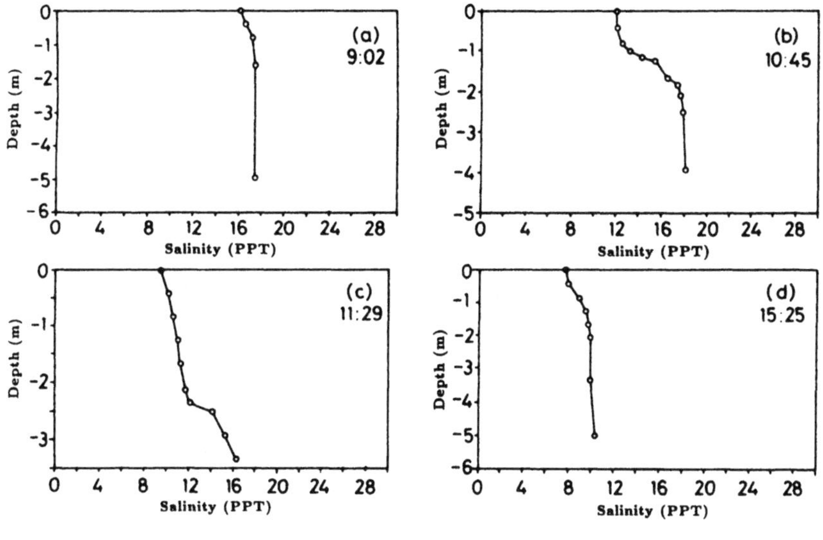

Figure 3. Salinity distributions in mid-channel at site **1** in Fig. 1 during tidal ebb on October 21, 1987.

CONTAMINANT TRANSPORT AND FATE

An important consideration in evaluating the pathways of contaminants in estuaries is the degree of 'conservativeness' of the compound. This can often be estimated by comparing the half-life for reaction with hydrodynamic residence time. Ages and Woollard (1988) have recently applied a one-dimensional model to simulate the path and residence time of a contaminant in the

Fraser River estuary. They estimated that the residence could vary between 6 and 30 hours depending on the time of year. For chlorophenols, chemical hydrolysis and oxidation are not important and since sorption is not important from a total mass viewpoint, the only reactions that need to be considered are biodegradation and photodegradation. In natural sediments, seawater and freshwater, 30 to 90 day half-lives for biodegradation of PCP have been observed. A number of studies have demonstrated the photodegradation of PCP. For a 1m deep system at 45° N in June, the NRCC (1982) estimated a photodegradation half-life of 4 days. In deeper systems, the half-lives are much longer. Given the short hydraulic residence times predicted in the Fraser River estuary, it is unlikely that much degradation can occur. In this system, chlorophenols likely are conservative and their main fate is to be transported in dissolved phase into the Strait of Georgia.

REFERENCES

Ages A (1979) The salinity intrusion in the Fraser River: salinity, temperature and current observations, 1976, 1977. Pacific Marine Science Report **79-14.** Institute of Ocean Sciences, Sidney, B.C.

Ages A, and A Woollard (1988) Tracking a pollutant in the lower Fraser River: A computer simulation. Water Pollution Research Journal of Canada **23**: 122-140

Carey J H, M E Fox and J H Hart (1988) Identity and distribution of chlorophenols in the North Arm of the Fraser River Estuary. Water Pollution Research Journal of Canada **23**:31-44

Carey J H, and J H Hart (1988) Sources of chlorophenolic compounds to the Fraser River Estuary. Water Pollution Research Journal of Canada **23**: 55-68

Geyer W R (1985) The time-dependent dynamics of a salt wedge. University of Washington, School of Oceanography, Special Report No 101. 199 pp

Karrickhoff S W, D S Brown and T A Scott (1979) Sorption of hydrophobic pollutants on natural sediments. Water Research **13**: 241-248

Karrickhoff S W (1981) Semi-empirical estimation of sorption of hydrophobic pollutants on natural sediments and soils. Chemosphere **10**: 833-846

Krahn P K, and J A Shrimpton (1988) Stormwater related chlorophenol releases from seven wood protection facilities in British Columbia. Water Pollution Research Journal of Canada **23**: 45-54

NRCC (1982) Chlorinated phenols: criteria for environmental quality. National Research Council of Canada. NRCC #18578. 191 pp

Schellenberg K, C Leuenberger and R P Schwarzenbach (1984) Sorption of chlorinated phenols by natural sediments and aquifer materials. Environ Sci Technol **18**: 652-657

Westall J C, C Leuenberger and R P Schwarzenbach (1985) Influence of pH and ionic strength on the aqueous-nonaqueous distribution of chlorinated phenols. Environ Sci Technol **19**: 193-199

Wright L D and J M Coleman (1971) Effluent expansion and interfacial mixing in the presence of a salt wedge, Mississippi River delta. J Geophys Res **36**: 8649-8661

CHAPTER V

ESTUARINE MEASUREMENT TECHNIQUES

THE HYDROGRAPHIC MEASURING SYSTEM H Y D R A

H.-U. Fanger, J. Kappenberg, H. Kuhn, U. Maixner, D. Milferstaedt
Institute of Physics, GKSS Research Centre Geesthacht GmbH
P.O. Box 1160, D-2054 Geesthacht, FR Germany

Initially planned as a ship-borne system for measuring the distribution and propagation of thermal plumes from riverside power plants, an efficient hydrographic and environmental research instrument (HYDRA) has been developed by GKSS over the past few years. The today's performance has been achieved in several step-by-step approaches fed by the experience of numerous major campaigns [1-2] in the tidal areas of the German rivers Elbe and Weser.

HYDRA is operated on board the GKSS research vessel LUDWIG PRANDTL which is a special flat-water going ship, 23 m in length and 1 m in draught. The measuring system consists of three main components: (i) an arrangement of sensors, assembled on a cantilever at the ship's bow, for horizontal profiling of current velocity, turbidity, temperature, conductivity, oxygen content and actual water depth; (ii) a vertical profiler with sensors of the same type bundled to a probe which is lowered from a starboard A-frame (the immersion depth being derived from an additional sensor for pressure); (iii) a high-precision positioning system using, in each case, two ad-hoc installed land-based RF stations.

Other auxiliary but indispensable devices are echo sounders, gyro and magnetic compasses as well as tilt meters. For the determination of suspended and dissolved matter concentration as well as for analysis of water samples for pollutant trace elements, horizontal samplers of the Niskin type are used in combination with the vertical profiler. Additional samples are taken with an OWEN tube, a technique yielding seston settling velocities [3].

A photograph of the vertical profiler is given in Figure 1. In the left foreground one may recognize the acoustic current meter for the three-component measurement of velocity. The underlying physical principle of this technique is the entrainment effect, which, for acoustic waves travelling in opposite directions within a flowing medium, results in phase differences dependent on the relative flow velocity. A fin (right-hand side) is necessary for the appropriate orientation and stabilization of the probe in the current. A swivel joint (with electric slip rings) serves as decoupler between the turning motions of the probe and the cable torque. Due to the rather compact structure of the probe, which combines a small drag area with a heavy weight by means of high-density ballast, the inclination of the probe does not exceed a value of 5 degrees at speeds of up to 5 knots. - The probe was developed jointly with an industrial partner, ME Meerestechnik und Elektronik, FRG.

One special feature of HYDRA is its ability to determine current speed and direction from a moving boat. This has been realized by a combination of the radio-positioning system, the fast high-resolution ultrasonic current meter, and the ship's gyro compass. Besides the due application of this technique with the cantilever sensor system for determining horizontal profiles, it has also proved to be indispensable for the necessary corrections in the evaluation of vertical profiles measured from a drifting vessel. Another significant feature of this system is the uninterrupted recording of seston concentration via light-attenuation measurements. This is possible through frequent calibrations regarding the natural variations in the optical properties of suspended particulate matter (s.p.m.). For calibration, the in-situ attenuation data measured by the vertical profiler have to be correlated with the s.p.m. values obtained by filtering (0.45 µm porewidth) of simultaneously taken water samples. In-situ and in-vitro light-attenuation values, measured with water of identical seston content, may differ considerably due to the destruction of flocs by the process of sampling; thus in-vitro calibration fails to produce correct seston concentration values via optical in-situ methods.

Depending on the number of sensors connected with the single-conductor cable telemetry, the sampling rate of hydrographic data is usually 8 samples per second

Figure 1: GKSS vertical profiler

for each parameter. Thus, enormous amounts of data have to be handled in the course of a multi-tidal campaign. Up to now, the main task of the on-board micro-processor system has been the acquisition and storage of data, while quick-look surveys had to be reduced to occasional controls. With the advent of fast multi-task real-time computers, this is now being changed so that part of the hitherto off-line evaluation and graphical presentation of the hydrographic situation will be carried out on-line.

A block diagram showing the existing as well as the modernized system is presented in Figure 2. The central on-board computer which, more or less, acted as a transfer link between the various heterogeneous signal sources and recording units, now achieves an active part in the new real-time data acquisition system. The survey computer handling the navigational and bathymetric task, is presently being replaced by a more flexible system with a higher capacity. Since it is one of the main tasks of the captain to find well defined cruise courses and positions repeatedly over a multi-tidal measuring campaign, navigational aids like path guidance and track plotter are most important.

From the field experiments of the last ten years, one may define four general areas of investigation to which HYDRA and the related analytical techniques may be applied:

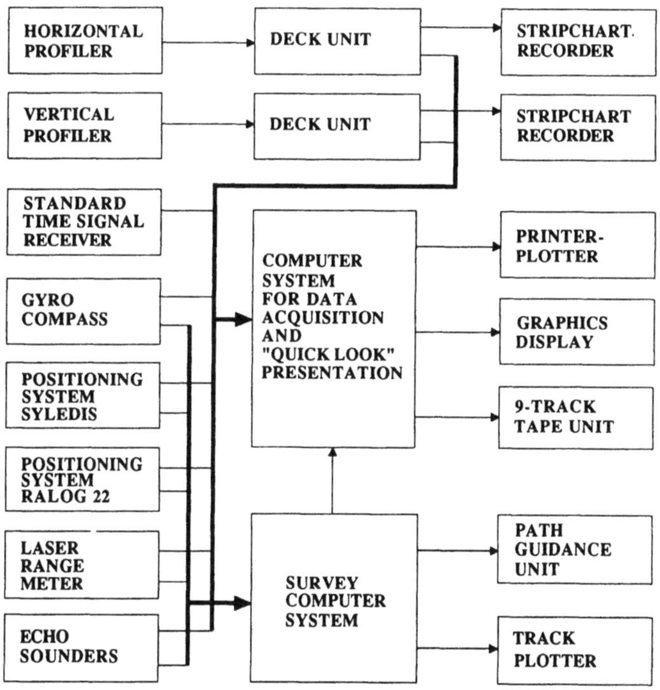

Figure 2: Block diagram of the HYDRA data acquisition system

(i) Lateral (cross-sectional) distributions of water, seston, nutrients, and pollutant transport in significant river sections, also in relation to existing or planned permanent measuring stations which yield only spot information.
(ii) Total discharge through strategic cross-sections (river, harbour, basins, confined marine areas).

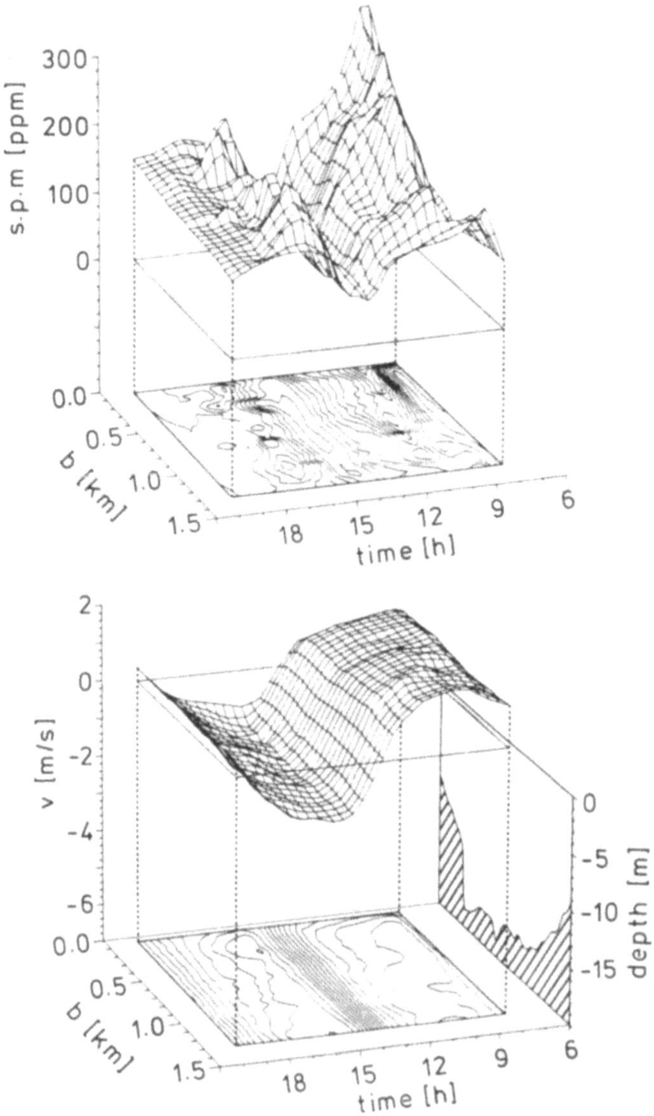

Figure 3: Sub-surface tidal current velocity and seston concentration across the Elbe river

(iii) Inflow-outflow balancing in the context of numerical modelling, calibration and verification of models (so-called BILEX concept).
(iv) Longitudinal distributions of hydrographic parameters in a stream, e.g. with respect to the site of a turbidity maximum or haline mixing zones.

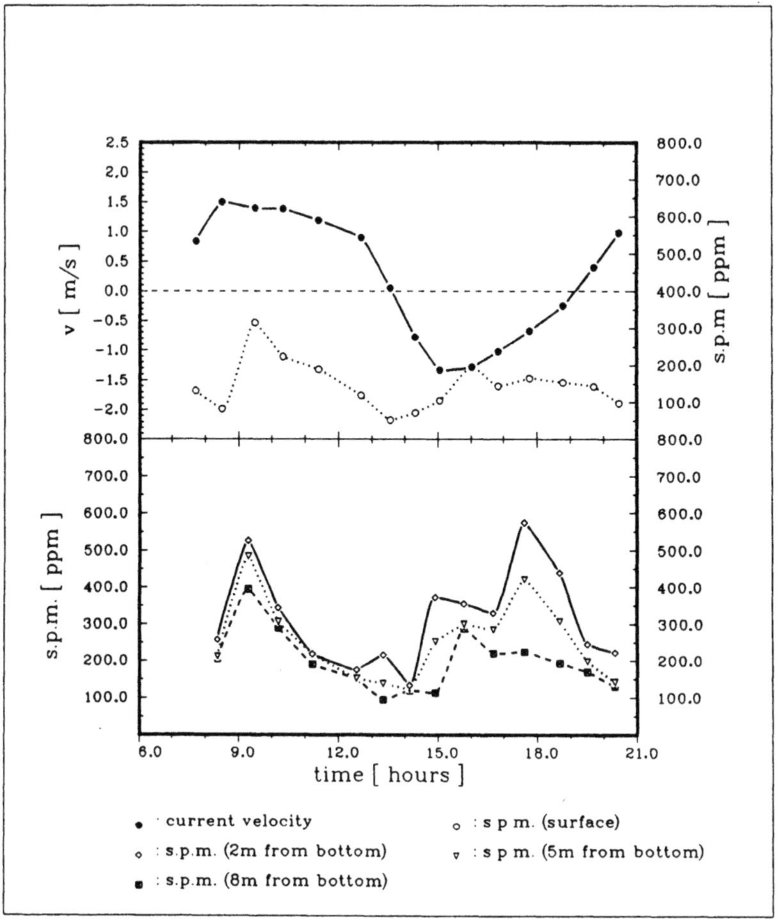

Figure 4: Tidal current velocity and seston concentration in various depths

As an illustration of its application, two results from recent GKSS campaigns are given. Figure 3 is a three-dimensional presentation of current velocity (lower part) and corresponding s.p.m. concentration (upper part) along a horizontal surface line crossing the Elbe river close to Brunsbüttel, displayed for the course of a tidal cycle. The two horizontal axes in Figure 3 are time in hours and distance in meters from the north banks; curves in the horizontal planes represent isolines for current and seston, respectively. The depth profile of the river bed along the cruise is indicated in the right vertical plane. The positive sign of current velocity has been defined for ebb tide. Whereas the current does not show any pronounced irregularities, the rather complex distribution of the seston content is evident. The s.p.m. maximum during fully developed ebb flow exceeds by far that at flood stream, and the lateral half-tidal distributions differ considerably. This investigation had been made to study the appropriateness of a platform station near the north banks with respect to the representativeness of measurements for the river cross-section.

For the platform position of the cross-section illustrated in Figure 3, the tidal curves of current velocity (uppermost curve) and the corresponding seston concentration at several depths (surface, 2 m, 5 m, and 8 m above ground) are shown in Figure 4. The expected gradient with depth is clearly revealed for each tidal phase. It is interesting to note that the decrease in seston concentration starts very early during the ebb current whereas a late near-bottom maximum is reached during the flood. The latter might be consistent with the assumption that resuspended material from the tidal-flat areas close to the river mouth arrives at the platform position with the end of the flood tide. However, the poor vertical mixing of seston is not well understood since no simultaneous stratification in salinity is observed; thus accidental local events cannot be excluded.

REFERENCES

[1] Michaelis W, Fanger H-U, Müller A (eds) (1988) Die Bilanzierungsexperimente 1984 und 1985 (BILEX'84 und BILEX'85) auf der oberen Tideelbe. GKSS 88/E/22.

[2] Fanger H-U, Kunze B, Michaelis W, Müller A, Riethmüller R (1987) Suspended matter and heavy metal transport in the tidal Elbe river. In: Awaya Y, Kusuda T (eds) Specialised Conference on Coastal and Estuarine Pollution, Kyushu University, Fukuoka (Japan), p 302.

[3] Puls W, Kühl H, Heymann K (1986) Settling velocity of mud flocs: results of field measurements in the Elbe and the Weser estuary. Int. Symp Physical Processes in Estuaries, 9-20 Sept 1986, Delft, The Netherlands.

CURRENT MEASUREMENT IN ESTUARIES BY ELECTRO-MAGNETIC METHODS AND DECCA-DRIFTERS

T. Knutz, P. Koske, J. Rathlev
Institut für Angewandte Physik, Universität Kiel
Olshausenstraße 40, 2300 Kiel

A profound knowledge of the current structure in tidally influenced estuaries is of fundamental importance for the development of monitoring concepts or for the modelling of the water dynamics in an estuary. This paper describes three different methods for recording tidal currents in estuaries and coastal waters:

1 - by current measurement from a moving ship using the earth's magnetic field (GEK - Geomagnetic Electrokinetograph)

2 - using an electro-magnetic current meter with permanent magnets (EM-current meter).

3 - with a DECCA based drift bouy.

The first two systems are based on Faraday's law of electro-magnetic induction. Seawater as a conductor moves through a magnetic field. This is either through the vertical component of the earth's magnetic field B_{ve} or through an artificially generated permanent magnetic field B_{pm}. Figure 1 shows the basic principle of an EM-current meter.

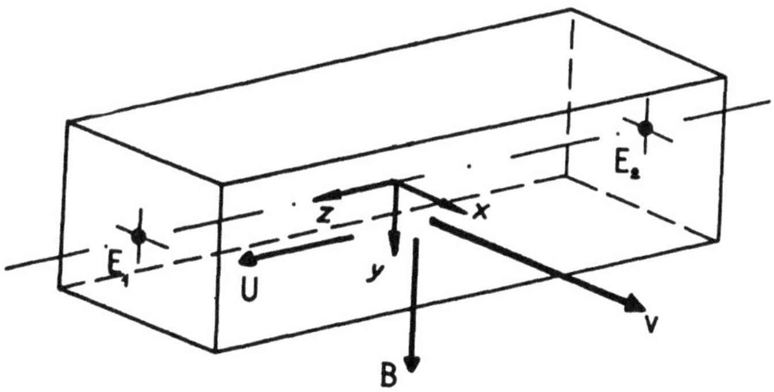

Figure 1: Principle of an Electro-Magnetic (EM) current meter

The magnetic field B (in y-direction) represents the fields B_{ev} and B_{pm}. An electric potential difference U (in z-direction), induced by the moving seawater (velocity v in x-direction), is recorded by the two electrodes E1 and E2. The GEK-method uses electrodes which are towed behind a moving ship (Figure 2). In this way current components perpendicular to the ship's heading are measured and recorded.

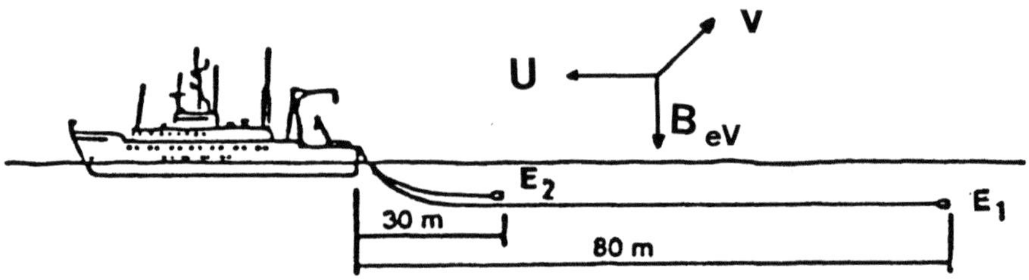

Figure 2: Sketch of the GEK-method used at the Kiel University

A new version of an in situ EM-current meter using a magnetic field with permanent magnets represents a mass-free current meter and is suitable for operation close to the bottom. The sensor shown in Figure 3 is the result of research and development activities, over the last few years, on the use of permanent magnets for EM-current measurements. The geometric dimensions have been optimized in accord with the size of commercially available standard permanent magnets so as to attain maximum sensitivity. As is demonstrated by the calibration diagram in Figure 4, the sensor signal is linearly dependent on the current velocity. It is therefore possible to carry out a two-point-calibration of the sensor.

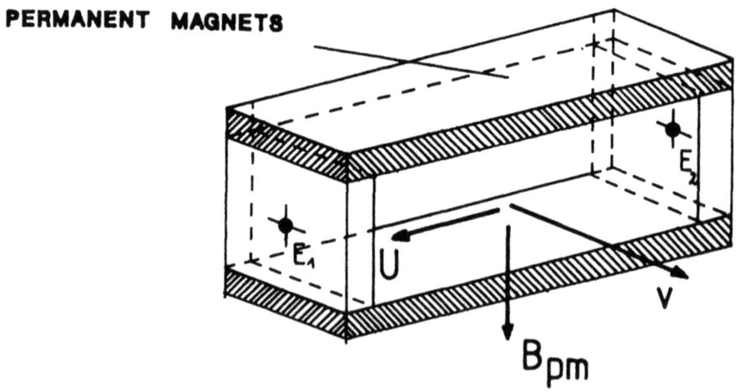

Figure 3: Permanent-magnet current meter

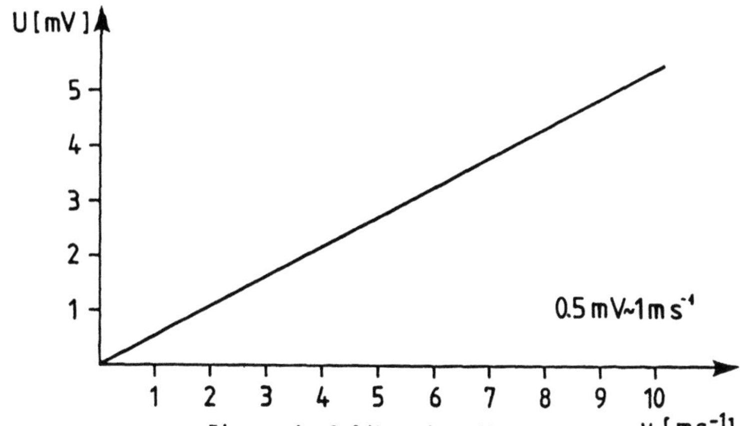

Figure 4: Calibration diagram

The first field measurements using these newly developed EM-current meters were carried out in May 1987 over a coastal area of 1000 m x 1000 m close to the shore of the island of Sylt in the North Sea. Six two-dimensional EM-current meters were placed directly on the sea floor for measurement of tidal currents and wave action and their influence on sand transport along the floor. The signals from these systems were transmitted by cable to a recording station on land.

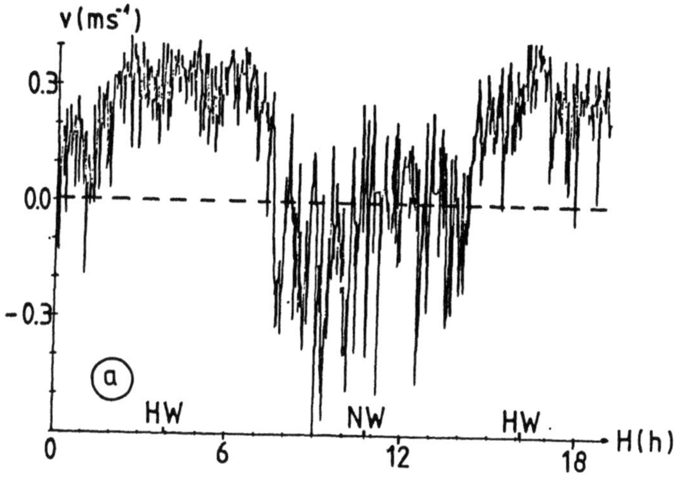

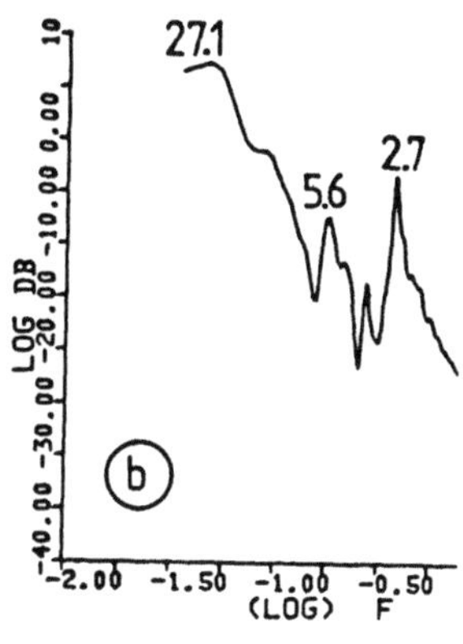

Figure 5a: EM-derived velocity (HW: high water, NW: low water, + northern direction, - southern direction).

b: FFT spectral series (numbers at the peaks give the periods in seconds).

Figure 5a shows an example of the EM-derived, digitally recorded velocities for a time period of 19 hours. In Figure 5b spectral series produced by Fast Fourier Transformation (FFT) from these recorded current data are depicted.

Some typical results from these field experiments for the Sylt location were a maximum velocity of 30-35 cm/s for the northern direction and a maximum velocity of 10-15 cm/s for the southern direction. The change of current direction did not coincide with the tide. At high tide the current turned 1-3 hours after maximum water level and at low water the current turned in phase.

All these results were obtained using signal averaging (1 minute with an average of 3 measurements per second). The data were in good agreement with trajectories recorded with the DECCA-drifter or radar-followed drifters. The DECCA-drifter operates on the basis of the navigational system as provided by DECCA-stations. It would also be possible to use navigational systems such as GPS or LORAN. The drifter gets its position once every minute by means of a built-in DECCA-navigator, stores this position in its memory and transmits (UHF) its momentary coordinates either to the research vessel or to a nearby receiving station on land (max. distance 5 sm). The drifter can be programmed for an operating time of up to 48 hours. After the buoy has been recovered the memory of the drifter is read by a PC (V24). This gives buoy positions with a high time resolution. A plot of such a drifter-derived trajectory for a time period of 16 hours close to the shore of Sylt is shown in Figure 6. The drifter moved predominantly parallel to the shore with the tides in north-south direction. This was in agreement with the records of the EM-current meters.

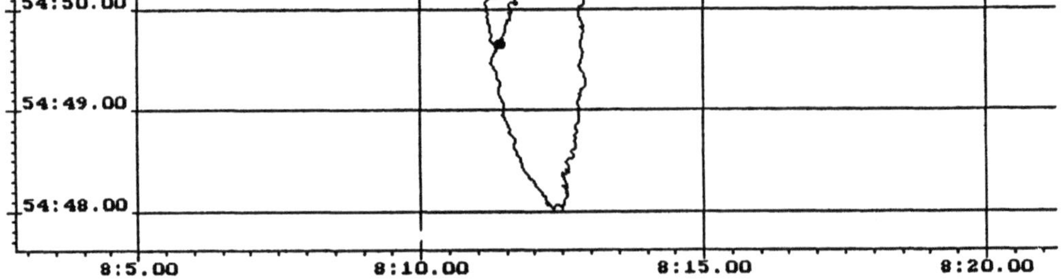

Figure 6: DECCA-drifter trajectory

ACOUSTIC DETECTION OF SUSPENDED MATTER IN COASTAL WATERS

B. Fahrentholz, U. Hentschke, P. Koske
Institut für Angewandte Physik, Universität Kiel
Olshausenstraße 40, 2300 Kiel

Estuarine and coastal waters in the German Bight are characterized by high concentrations of suspended matter. The main sources for this material are

- the runoff of the rivers Elbe and Weser, both of which discharge into the German Bight,
- the mud flats which are part of the so-called Wadden Sea, a unique type of coastal water and which are flushed periodically by tidal currents.

Besides the collection of water samples, the filtering and weighing, as standard procedures for the determination of suspended matter in water, there exists a considerable interest in improved techniques for a faster and more continuous method for the determination of the amount of material in suspension. One of the more successfull methods is based on the measurement of light attenuation in turbid seawater. This functions so long as a well defined relationship between suspended particulates and light intensity can be established. Although light attenuation measurements offer the option of continuous vertical profiling they still require hydrographic stations and cannot be used from a moving ship.
This contribution describes a method which is based on the backscattering of acoustic signals by suspended particulates in the water column and which can be implemented from a moving research vessel.
The system described in the present work consists of the components as shown in fig. 1:

- a 700 kHz transducer (angle of radiation +/- 1.3°) with the operating electronics
- an amplifier
- a rectifier
- the ADC and DAC units.

The transducer electronics (3) are triggered by a pulse from the 12-

bit-DAC (2) which is controlled by the computer (1). The transducer (4) sends an acoustic pulse with a duration of 0.1 ms which is reflected by the different targets. The back-scattered signals are amplified 375 times by a resonance amplifier (5), rectified by fast diodes (6) and digitized (7) by an ADC in steps of 13.6 μs. This represents a depth interval of about 2 cm. The digitized data are stored by the computer (1).

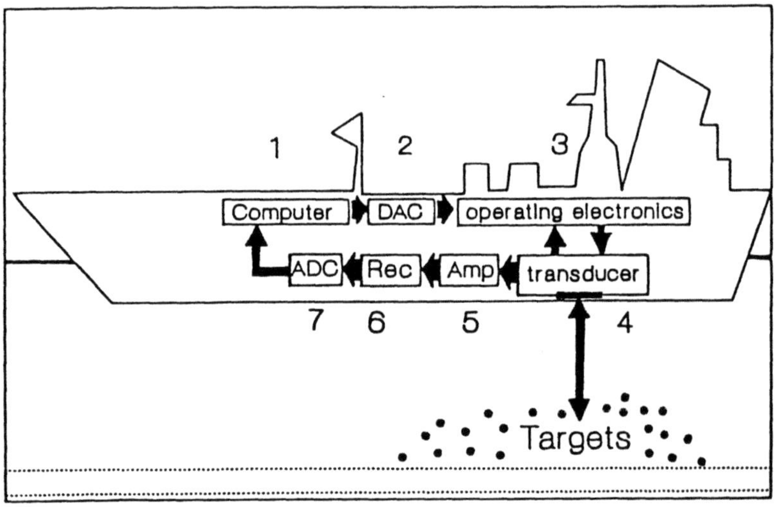

Fig. 1: Components and flow diagram of the acoustic system.

A data series for a seven hour time period which was taken at an anchor station of the vessel "LITTORINA" in December 1987 is presented. The station was located in the Elbe estuary at the Neufeld Reede, in an area with salinities between 0 to 10 and positioned within the turbidity maximum of the Elbe river.

A vertical profile was taken on a half hourly basis by a fast CTD-system which was in addition capable of measuring optical transmission. Simultaneously the back-scattered signals were recorded using the developed acoustic system and the current velocities were taken from the ship's log. In order to obtain a correlation between the optical transmission, the back-scattered acoustic signal and the total amounts of suspended matter present water samples were also collected and filtered. The results of these measurements are presented in fig. 2.

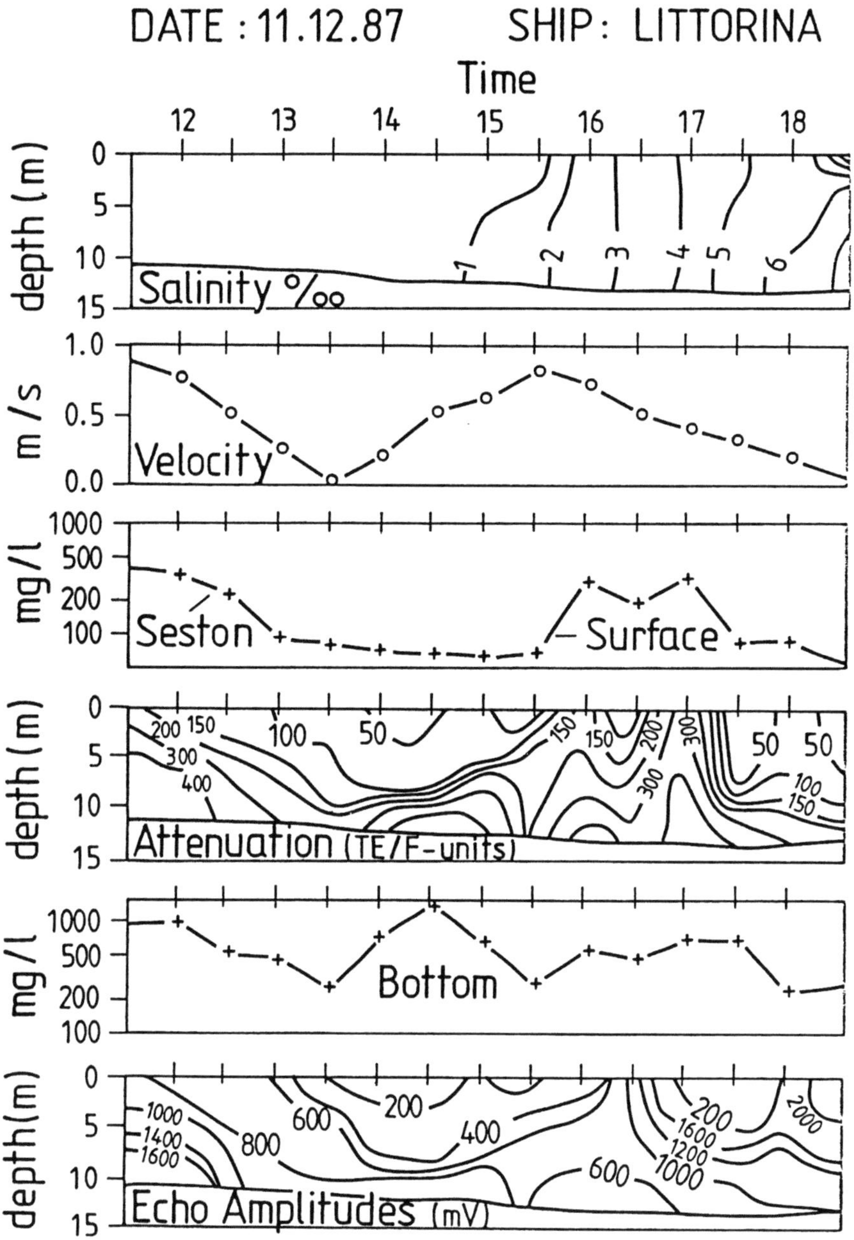

Fig. 2: Anchor station Neufeld at 52:53.30 N, 9:02.40 E ; low water 12:02 h, high water 17:23 h.

It can be seen from the upper two plots that the ebb-tide ended at approximately 13:30 h (flow velocity zero), and that for the remainder of the afternoon incoming tide showing a continuous increase in salinity from less than 1 to more than 6 and a maximum velocity of 0.8 m/s (1.6 knots) prevailed. The graph for seston content shows that close to the surface (plot 3) the seston amounts are closely correlated with the flow velocity. When the velocities decreased the seston concentration was reduced from about 400 mg/l to values of approximately 50 mg/l and less. After maximum flood velocity the values increased once again. The seston content in the water close to the bottom (plot 5) was considerably higher at all times, with values of between more than 1000 and 250 mg/l. It also was less sensitive to the varying water velocities.
The light attenuation profile (plot 4) as well as the acoustic back-scattering record (plot 6) were good reflections of this different behaviour. Both show high values at the beginning of the sampling with a good correlation between high ebb velocities and the high seston concentrations during this time. Both plots show minimum values at the surface from 13:30 h (zero velocity) to 15:00 h again. This is in accordance with the seston content at the surface. Shortly after the maximum flow velocity, between 16:00 h and 17:30 h, both plots again indicate high turbidity in the whole water column from top to bottom.

In conclusion it may be said that the acoustic detection system presented here provides an adequate remote sensing technique allowing observation of the horizontal and vertical distribution of suspended matter in estuarine waters at both anchor stations and from a moving ship. Future work will be concentrated on the use of several frequencies simultaneously in order to obtain further information of a more quantitative nature on the concentration of particulates in the water column.

LASER FLUOROMETER FOR THE DETECTION OF RHODAMINE-B IN ESTUARINE WATERS

J. Jahnke and W. Michaelis
Institute of Physics, GKSS Research Centre Geesthacht
D-2054 Geesthacht, FRG

1. INTRODUCTION

Fluorescent dyes are often used to study transport phenomena in surface waters. Several examples are described in chapter IV of this volume. Since rhodamine-B is soluble in water and since dilute solutions show only low toxicity (lethal dose LDLo = 500 mg/kg, orl-rat [1]), this substance is preferred for investigating the advective transport and the turbulent diffusion of conservative constituents in the water body. Several highly sensitive fluorescence sensors for in-situ applications are on the market. In general, they use a light source with a white spectrum and optical filters to select the exciting and the fluorescent radiation at the wavelengths λ_e and λ_f, respectively. In the case of rhodamine-B the maxima for excitation and emission are only 21 nm apart (554 nm, 575 nm). As a consequence, these sensors work quite efficiently in oceanography, but in estuarine waters the performance is rather unsatisfactory. In sea-water the extinction coefficients are usually < 0.5 m^{-1} so that minimum detection limits, defined as the threefold standard deviation of the background, of the order of $5 \cdot 10^{-11}$ g/ml or even lower are achieved. In estuaries, the high content of suspended particulate matter leads to extinction coefficients between 5 and 100 m^{-1}, the separation of fluorescence and Mie scattering is no longer complete and the sensitivity deteriorates to levels as high as 10^{-8} g/ml. By applying cumbersome scattering corrections with the aid of an in-situ light attenuation sensor [2] the detection limits may be improved to 2 or $3 \cdot 10^{-10}$ g/ml. However, the errors in the concentration measurements remain rather high. This situation stimulated experiments with the aim of optimizing the detection system for turbid water by employing different components and concepts. One promising opportunity is to apply monochromatic light from a laser to facilitate separation of the scattered and fluorescence signals.

2. EXPERIMENTAL

Fig. 1 shows a block diagram of the laser fluorometer which was used in laboratory experiments on artificial and natural water samples [3]. The light source was a tunable dye laser pumped by a cw argon ion laser. The dye was coumarin 6 which allows tuning between 520 and 575 nm. A mechanical chopper in combination with lock-in detection served for noise and daylight suppression. The incident beam was focussed by a lens (L1) into the measuring volume. The fluorescence radiation was observed with detection channel 1 at an angle of 90°. Combinations of interference filters were tested with regard to optimum transmission and blocking ratios. Channel 2 served for normalizing the signal to the laser intensity. The light attenuation in the sample was measured with channel 3. An avalanche photodiode (D1) and Si-PIN-diodes (D2, D3) were used as detectors. The d.c. signals from the lock-in amplifiers and from detector 3 were digitalized by an analog-to-digital converter with a sample rate of 2.5 kHz.

For studying spectral relations, the experimental setup was modified by inserting two monochromators, one of them in the primary beam in front of the beam-splitter, the other in channel 1 replacing the filter component.

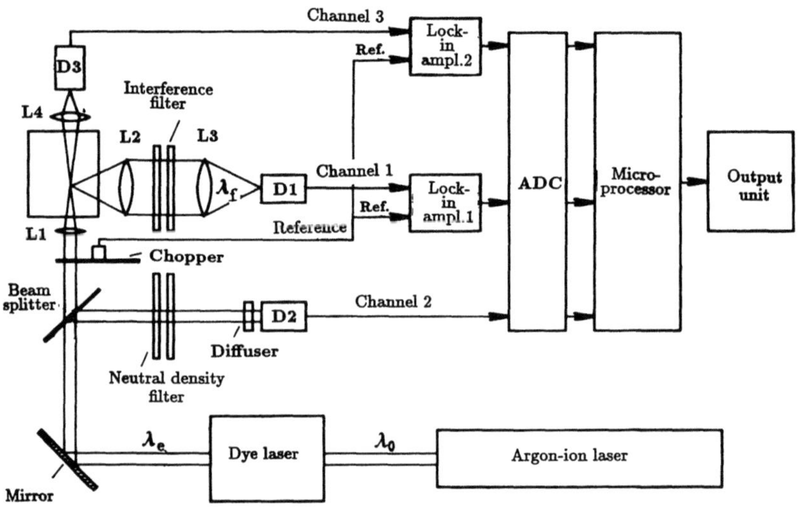

Figure 1: Block diagram of experimental setup

3. RESULTS

Extinction experiments on samples of filtrated and unfiltrated water from the Elbe Estuary as well as on artificial samples of rhodamine-B in demineralized water suggested the use of the wavelength at the maximum of the rhodamine absorption for excitation, i.e. λ_e = 554 nm. The response signal from natural water samples shows a strong decrease with increasing wavelength. The best intensity ratio for the signals from the tracer and unfiltrated water was observed at 577 nm using a combination of two interference filters (Ealing 35 - 3748; Corion P 10 - 577). Under these conditions the signal-to-background ratio was 2.44 for a rhodamine concentration of 10^{-10} g/ml.

The residual interfering signal from turbid water is no longer markedly due to Mie scattering by suspended particulates. This was demonstrated in a series of experiments with natural samples and (non-fluorescent) fine-grained $CaCO_3$ suspensions. In these studies detector 3 (Fig. 1) was arranged at an angle of 90° to the axis of the incident beam. The Mie signal was always below that of an equivalent rhodamine-B concentration of 10^{-12} g/ml and was thus negligible. The residual interference is probably caused by other fluorescent constituents in the river water, in particular yellow substance. This background restricts the sensitivity, if only one fluorescence detection channel is used.

For all samples taken upstream from Elbe-km 630 (cf. [4]) the minimum detection limits were 3 to 4 · 10^{-11} g/ml. The lower value was achieved by measuring in each case the background signal, the higher one resulted when using an average blank value. In the case of a sample from the turbidity zone (km 680), the detection limit increased to 6 · 10^{-10} g/ml due to the high content of fluorescent material. This feature is, of course, inherent to all single channel detection schemes.

Based on the experience gained in the laboratory studies a concept for an in-situ sensor is under development. By applying fiber optics the main optical and electronic components can be integrated on board the vessel. The probe itself can

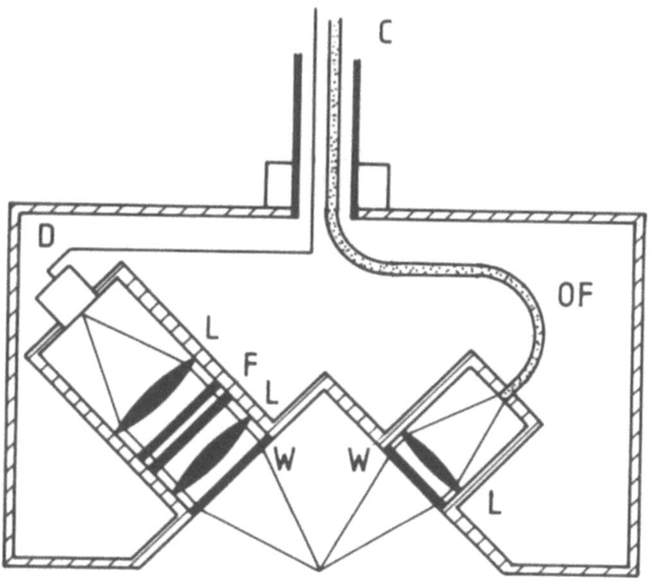

Figure 2: Sketch of an in-situ probe with single channel detection. OF = optical fiber, L = lens, W = window, F = filter, D = diode and C = cable

be constructed in a compact manner and only contains a few lenses, the interference filters and, depending on the version to be used, one or more semi-conductor diodes (Fig. 2) . Multispectral detection or excitation-detection schemes may be used either to provide automatic correction procedures for interfering fluorescent compounds or to detect several constituents simultaneously. A typical example is the continuous detection of chlorophyll and yellow substance.

4. CONCLUSIONS

The application of monochromatic light from a laser in combination with suitable interference filters can considerably improve the quantitative determination of rhodamine-B in turbid water. Interference from Mie scattering is almost completely eliminated. The detection limits approach those obtained with conventional systems in clear ocean water without cumbersome evaluation procedures. The sensitivity is inherently restricted by the presence of other fluorescent compounds. The use of multispectral detection or excitation-detection schemes may further enhance the performance.

REFERENCES

[1] US Department of Health and Human Services (ed.) (1979) Registry of Toxic Effects of Chemical Substances, Vol 1, Cincinnati, Ohio

[2] Kolb M, Franz H (1989) Transport of rhodamine-B in the Elbe River following a short release upstream from the Port of Hamburg. This volume, chapter IV

[3] Jahnke J (1989) Untersuchungen zur Fluoreszenz von Rhodamin B in Elbwasserproben. Diplomarbeit, Universität Hamburg (to be published)

[4] Michaelis W (1989) The BILEX concept: research supporting public water quality surveillance in estuaries. This volume, chapter III

QUASI-SYNOPTIC UNI-SENSOR IN-SITU 3D CURRENT MEASUREMENTS IN ESTUARIES

R. Spanhoff, T. van Heuvel, J.C. Borst
Tidal Waters Division, Rijkswaterstaat
P.O. Box 20907, 2500 EX The Hague, The Netherlands

H. Verbeek
Institute of Meteorology and Oceanography
University of Utrecht, The Netherlands

INTRODUCTION

Currents in estuaries are governed by various factors such as bottom topography, tidal movement and an often variable river run-off. As a result, water velocity patterns are inhomogeneous in time and space. Depending on the interplay between geometry and the degree of mixing of fresh river and salt ocean water, the current gradient is more or less pronounced over the length, width and/or depth of the estuary.

Virtually all questions in proper estuarine management require current information, in a momentary or in a statistical or time-averaged fashion. The current may be the direct parameter of interest, such as in ship-traffic guidance or in determining the required strength of constructions subject to currents, or it is a basic, often dominant factor in derived processes such as the generation of turbulence, mixing and dispersion that influence the fate and transport of suspended and dissolved substances in and through the estuary.

A synoptic picture of the current regime in a large part of an estuary requires measurements at many points, together with sampling at a sufficiently high rate with respect to tidal changes, i.e. at up to an hour. Apart from the HF-radar technique which is restricted merely to surface currents, such measurements are usually performed with a series of current meters at selected stations, either moored or deployed from a costly number of ships. The vertical current structure is then revealed by using several self-supporting meters on top of each other, respectively by a time-consuming lifting of the sensor over the vertical axis. However, these procedures, cumbersome as they are, cannot always be applied since they may interfere with other activities in the area, such as dense shipping traffic.

Specifically for such cases, the Rijkswaterstaat has developed a flexible-to-operate current-data gathering system. Its features and results are briefly described below.

THE EXTENDED ACCOUSTIC CURRENT-PROFILING SYSTEM

The system developed is based on the philosophy of using an easily manoeuverable boat with a direction- and depth-sensitive accoustic current sensor mounted under her (Figure 1), purchased from AMETEK, STRAZA Division. Depth discrimination is achieved by using a pulsed system and setting time windows on the back-scattered accoustic signals. Up to 32 consecutive layers are probed with each transmitted pulse. The water velocity at a selected depth is determined from the Doppler shifts of the sound back-scattered in that layer towards the small-aperture transmitting/ receiving element under the ship. The horizontal current vector is deduced from the combined velocity scalars measured by definition in each look direction of up to 4 sensors pointing downwards, at an angle to the vertical axis, in different directions. The underlying assumption is that the current is homogeneous in the area covered.

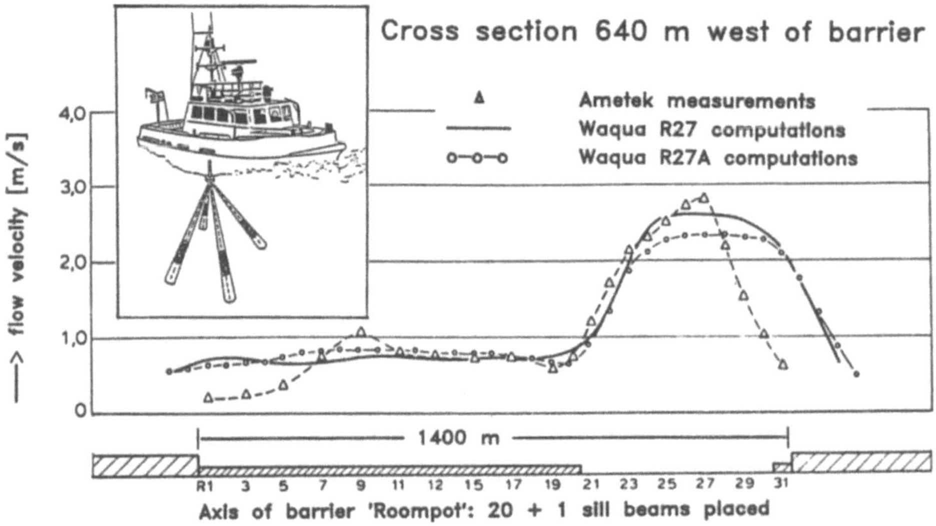

Figure 1: Maximum ebb currents in a 640-m downstream cross section parallel to the storm-surge barrier under construction in the widest gully of the Eastern Scheldt. The results of the extended AMETEK current-profiling system are shown together with some 2D WAQUA-model calculations for comparison. At the horizontal axis the barrier is sketched for reference; the numbers refer to the piers. Please note the asymmetry of the current pattern over the cross section due to the sills already in position. The insert sketches the 4 accoustic beams from the emitting element mounted under the survey vessel Molenvliet. One of the up to 32 measured water layers is indicated.

With in-house, mainly data-processing developments the system was extended for estuarine applications, originally to assist in the construction of the Eastern Scheldt storm-surge barrier (see below). Throughout, minimal standard AMETEK time windows are used, corresponding to 1.6-m thick water layers. So the total depth range of the 32 layers is of the order of 50 m, which is sufficient for the Dutch estuaries. In shallower waters, the strong sound reflection at the bottom prevents direct measurement of the lowest 1 or 2 layers; instead, the current therein is obtained by a downward extrapolation according to a logarithmic profile. The top layer to be measured is centered around 4.2 m below the water surface, because of the depth of the sensor itself and due to a dead time after the sound emission.

In the extended mode of operation developed a data-acquisition cycle is made every 0.6 s by a PDP 11/73 computer. The data are transferred to its hard disk. They consist of the Doppler-response time series of the 4 sensors for the whole measured water mass, the water depth from an independent echo sounder, the ship's heading and position readings from a gyroscope and an accurate TRIDENT positioning system, respectively. One full measurement, consisting of 100 cycles, takes 60 s. During this measurement the ship's position with respect to the earth reference frame is kept as constant as possible by manual navigation on the positioning device. In practice, this confines the drift to less than 20 m in 60 s, even in the strongest water currents.

After such a 1-minute measurement the boat is quickly transferred to the next station. Thanks to the high manoeuverability of the small MS Molenvliet and the navigational craftmanship, typically up to 15 to 20 stations can be measured in half an hour. Then the sequence starting with the first station can be repeated several times in 13 hours or so, in order to cover a tidal cycle. In this way a quasi-synoptic 3D current pattern for those 15 to 20 stations is obtained.

Data processing can be done off-line later, or semi on-line right after the measurement, taking circa 30 minutes for 20 minutes of data. It consists of a quality check of the accoustic data in which bad data, e.g. those including the strong bottom response, are rejected. The Doppler signals of the accepted pulses and water layers are processed to horizontal water current components that are corrected for the ship's velocity and rotation determined from the corresponding co-recorded data. Subsequently the corrected velocity results from up to 100 pulses are averaged and the full profile is drawn with respect to the water bottom.

The data are averaged over up to 100 sound pulses in order to reduce the instrumental statistical error to an acceptable 5 cm/s one standard deviation. The overall uncertainty is believed to be of this order thanks to the applied corrections. The absolute positioning accuracy of the boat is circa 2 m, with a relative error that determines her velocity one order smaller, which is negligible over a 60-s period. In situ comparisons of the present system with conventional current meters showed negligible differences not exceeding 7 %.

APPLICATION IN THE EASTERN SCHELDT

The construction of the Eastern Scheldt storm-surge barrier was the recent climax in the completion of the Delta Works that protect the SW part of The Netherlands against a repetition of the 1953 inundation. In three gullies, each up to 1.5 km wide, several piers were located whose mutual in-between vertical cross sections can be shut off in times of a surge, by lowering large steel gates on permanently installed underwater sills. The piers and sills significantly diminish the effective cross section for the water masses transported in and out of the estuary with the tide, thus increasing the local currents. This might lead to strong erosion of the sandy bottom and thus to a destabilization of the construction. Therefore, the bottom was protected first with a large-area synthetic mattress. The boundaries of the latter were of special concern during the construction. Experiments on a physical scale model, supported by 2D vertically integrated numerical model calculations (Klatter et al., 1987) that among others provided the boundary conditions for the scale model as well as general current information for the studied area, had predicted the local current strength for several scenarios in the sequence of the actual construction. These predictions had to be verified during the work in order to detect possible deviations and, if necessary, to switch to an alternative scenario, which in fact has happened.

Circa 15 such measurements above the mattress close to its borders were performed in the course of the work, each on two consecutive days in order to cover the maximum tidal currents in both directions downstream from the construction. In a line parallel to the latter, the Molenvliet measured about 15 stations within 25 minutes, in which the maximum currents were virtually constant. Notably with part of the sill beams placed, the predicted strong deviations from the original pattern were confirmed (Figure 2). Afterwards, the current data were used to adjust the above-mentioned 2D model. Later on, this was used at the closing of the eastern boundaries of the estuary to predict the current regime there, a delicate parameter (Bosselaar et al., 1987).

The present measuring technique was indispensable during the construction of the barrier, since conventional methods could not be used. Mooring of fixed current meters would damage the mattress or would interfere with the on-going activities that concentrated in a small area. Mechanical or electromagnetic current meters deployed from several unanchored ships would, apart from the practical problems, give poor results in the highly turbulent regions downstream from the construction.

APPLICATION TO THE ROTTERDAM HARBOURS ENTRANCE REGION

Current patterns are rather complex in the 25-m deep entrance area of the Rotterdam harbour region due to the latter's geometry and the relatively large Rhine discharge. A marked vertical salinity and current structure exists, with a salt wedge moving to and fro in the estuary with the tide. For nagivational purposes, series of current maps are used arranged in full hours with respect to the tidal phase and in separate series for the neap-spring tide cycle. The present currents deviate from the earlier maps due to recent man-made changes in the area such as the construction of the 'Maasvlakte'. Therefore, current measurements were necessary for new maps for e.g. the pilots who navigate among others the large 20-m deep tankers into the harbours. One is interested in the effective current forces on the ships, so in depth-integrated values. With the marked vertical structure obviously the measurements have to include several depths, one of the reasons for using the present system.

The area of interest was covered with 60 stations arranged in 8 sections, of generally 9 points in a row with a typical spacing of 600 to 1200 m. The measurements were done on 24 days selected with respect to the tidal difference in a 1-year period: each section was measured 3 times, at neap, mean and spring tide, respectively. One section could be done per day. Because of the larger spacing between the stations and the practical problems in this area, more time was needed per station than in the Eastern Scheldt case, namely 1 hour for a full section of 9 stations. Repeating this measurement 13 times, still a quasi-synoptic image of the current profiles over one tidal period was obtained. This image was extended to the whole area of interest by measuring the other sections on days with comparable conditions.

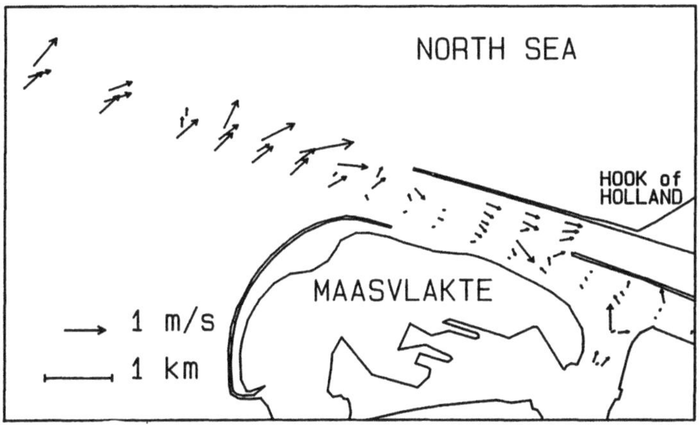

Figure 2. Example of a current map for navigational purposes obtained with the extended current-profiling system. Currents are averaged over the upper 20 m. Time is one hour before high water in Hook of Holland, in a mean-tide period.

Data analysis included an interpolation to fixed times with respect to the tidal phase for all stations, integration over the required depths, quality checks and presentation. As for the Eastern Scheldt, the present 3D data set will be used to validate and adjust numerical models that are being developed, in this case 3D models for inhomogeneous tidal waters.

The intense ship traffic in the region considered inhibits the use of moored meters and strongly hampers conventional measurements with many ships to cover a full tidal cycle, therefore the present technique was called for.

With the above applications in two distinct Dutch estuaries, the capabilities of the extended current profiler for 3D quasi-synoptic measurements under difficult conditions were clearly demonstrated.

REFERENCES

Klatter HE, Dijkzeul JMC, Hartsuiker G, Bijlsma L (1987) Flow computations nearby a storm surge barrier under construction with two-dimensional numerical models. In: Edge BL (ed) 20th Int Conf Coastal Eng, 9-14 Nov, 1986. American Soc Civil Eng, New York, p 1943

Bosselaar GJ, Thabet RAH, Roermund AGJM van, Bijlsma L (1987) Simulation of sand-fill building stages with numerical flow models. Ibidem, p 1001

CHEMICAL TRACERS - A NEW AND DESIRABLE TECHNIQUE ?

R. Spanhoff and J.M. Suijlen
Tidal Waters Division, Rijkswaterstaat
P.O. Box 20907, 2500 EX The Hague, The Netherlands

INTRODUCTION

Many of the man-made chemical compounds that enter the aquatic environment at present in more or less controlled quantities are hydrophobic and are, thus, largely adsorbed on the suspended sediments that transport them. These sediments may travel large distances before settling permanently in areas with specific hydraulic conditions, such as estuaries. The adsorbed pollutants accumulate to relatively high concentrations.

For proper estuarine and marine management one is concerned with the transport and fate, in the aquatic environment, of these possibly hazardous substances from various origins. A logical first step is to investigate and to model the transport of their carriers, i.e. the suspended sediments and the water. Knowing the physical-chemical behaviour of the substances of interest, with respect to the sediments in terms of ad- and desorption as a function of the dominant parameters such as water salinity and sediment composition (organic carbon content e.g.), one can incorporate the corresponding processes in the model and describe and predict the chemicals' fate. Obviously a lot of laboratory and field research is required to establish a relevant description of the physical-chemical processes and to parameterize and validate this pollutant-transport model. The uncertainties of the various modelled processes accumulate in the calculations. In this paper it is proposed to combine recent developments and findings in several fields to elaborate upon the possibilities of a more direct approach, introducing so-called chemical tracers.

CHEMICAL TRACERS

In our definition a chemical tracer is a substance that mimics the net transport of a specific pollutant or of a group of chemicals of interest, in having physical-chemical properties that closely resemble those of the latter. The tracer may be artificially released into the environment or it may happen to be already present 'naturally', accompanying the problem substances. The distribution pattern by the tracer in the areas to be studied then gives quantitative information about the paths and fate of the chemicals of concern, in as far as they can be considered to originate from a common source. A great advantage of artificially released tracers is that the impact of various sources can be studied separately. This is important in answering aquatic-management questions directly, as well as in assisting in the development and calibration of the above-mentioned pollutant-transport models.

A chemical tracer has to meet several requirements in order to be of practical advantage. Firstly, it must be much easier to detect than the corresponding problem substances. Secondly, it must resemble the latter closely in physical-chemical properties. Thirdly, when artificially released, it must have no, or negligible, toxic effects on the environment. Furthermore, the way it then enters the environment, such as adsorbed on sediments, should be representative for the local state of the problem substances; it should be, among others, virtually absent in the environment (low background) to allow a high effective sensitivity with limited tracer amounts to be used.

Of the many problem substances we address in the following, it is mainly PAH's and PCB's that at present draw much attention. We suggest looking for disperse dyes that meet the above requirements as a chemical tracer for those substances.

PHYSICAL-CHEMICAL AND TOXIC PROPERTIES

PAH's and PCB's are highly hydrophobic. They tend not only to adsorb onto sediments but they are also lipophilic, so they build up in the fatty tissues of fish and the like, and can thus be harmful to higher predatory species. The partition coefficient octanol/water P_{ow} is a measure of the lipophilicity, and a linear relation was found between the logarithms of P_{ow} and that of the bioconcentration factor B_f, respectively. Adsorption on suspended sediments can be described by the adsorption coefficient K_{oc} (normalized to the sediment's organic carbon content) and log K_{oc} is linearly proportional to log P_{ow}. Thus, a linear relation exists between log B_f and log K_{oc}. Therefore, any tracer with the same desired K_{oc} as the problem substance would as easily accumulate in fish. However, recent investigations have indicated that substances with sufficiently large molecules cannot pass through biological membranes so their B_f is dramatically reduced compared with the expectations based on P_{ow} and K_{oc} alone (Anliker et al., 1988). Therefore, it is possible to look for large-molecule disperse dyes which will not accumulate in living organisms when released into the environment, but will still have the same physical-chemical ad/desorption characteristics as the PAH's and PCB's to be studied.

When considering the toxic risks of a possible chemical-tracer experiment, one should compare the added tracer quantities with the amounts of similar compounds already present in the environment. For example, if one would perform an experiment in the Rhine-dominated Dutch coastal waters up to the Wadden Sea (Spanhoff and Suijlen, 1989), one could think of a continuous release during 2 years of 100 kg dyes in total. In a risk analysis these 100 kg must be compared with the 20 tons/y 'natural' PAH's in the Rhine and the 400 tons/y non-volatile halogenated hydrocarbons released in the last 10 years or so. Because of the persistency of most of these substances, the total rather than the annual releases must be considered. With the pessimistic assumption that the added tracer has the same toxicity as the problem substances and is persistent as well, an extra risk in the order of 10^{-5} results. This estimate is definitely several orders of magnitude too pessimistic, (i) because of the above low permeability of biological membranes for the tracer, and (ii) since a low genotoxic tracer material will be selected. In summary, the dyes to be used will thus provide negligible extra toxic risks. One should realize that not knowing the fate of the PAH's and PCB's in the aquatic environment presents a risk in itself that greatly surpasses the extra risk of the experiment. These considerations should be the basis of a discussion about the desirability of actually performing the experiments suggested here.

MODERN DETECTION OF DISPERSE DYES

Of the many dyestuffs that in principle can be used to mimic the transport of PAH's and PCB's, some certainly will display a highly fluorescent behaviour, since circa 5 % of all organic dyes do so. The progress made in modern laser-induced fluorescence detection has led to mass-detection limits of 10^{-20} g dyes, albeit in ideal laboratory circumstances, so our present dye-detection sensitivity (Suijlen and van Leussen, 1989) can be improved further, almost certainly by several orders of magnitude. As a conservative estimate based on our experience with field studies on rhodamine, we expect an extrapolated 10-fg detection limit to be feasible. Solid-phase extraction and separation with HPLC should then be combined with the use of microbore columns and laser excitation, techniques that are regularly applied nowadays (Trkula and Keller, 1985; Demas and Keller, 1985). The tracers adsorbed on the suspended sediments will be extracted according to the standard techniques for PAH's and PCB's (Duinker et al., 1983), but we propose to detect them with the microbore HPLC instead of with the common gas chromatography, thus making optimal use of the much greater selectivity and sensitivity specific for fluorescence detection.

To study the tracer's dissolved phase, 100 l sea-water samples can be used. Suspended sediments can, at present, be extracted from the water phase rather efficiently, thanks to modern high-capacity ultracentrifuges that can treat about 500 l water in half an hour, a realistic time for one sample. The resulting sediment amount should be sufficient for the present purposes. Samples taken from the upper layer of the bottom are much easier to measure because of the more abundant amounts of material.

APPLICATION PERSPECTIVES

Several applications of the above chemical-tracer techniques are conceivable. Only one is mentioned here, namely a possible combination with the physical tracer of our accompanying paper (Spanhoff and Suijlen, 1989) by adding chemical tracers as well. The physical tracer reveals the pure sediment transport, while a cocktail of about 5 different chemical tracers will account for the ad/desorption processes of the various PAH's and PCB's of interest. For a low bioconcentration, these chemical tracers should be carefully selected from those having a log P_{ow} smaller than 5 and larger than 8 (Anliker et al., 1988). On the other hand log P_{ow} has to be larger than 3 in order to have sufficient adsorption onto the sediments. As a result of the relation between log P_{ow} and log K_{oc} a wide range is thus covered in the physical-chemical processes, and their influence on the time and space scales of interest is probed independently once the sediment transport is known. By sampling the water phase along the transport routes, and the sediments suspended in the water column and on the bottom including the accumulative areas, the required complete quantitative picture is obtained. With the above sensitivities, 10 kg/y during 2 years of each chemical tracer would suffice. We stress the great advantage, from an aquatic-management and modelling point of view, of investigating the impact of a distinct polluting source such as the Rhine with tracers, because it is nearly impossible to monitor the actual pollutants themselves; e.g. significantly contributing diffuse sources such as atmospheric deposition into the North Sea may readily obviate the latter.

CONCLUSION

Recent developments, such as (i) sensitive detection techniques, (ii) efficient sediment-sampling methods, and findings like (iii) the discovery of a low bioconcentration of hydrophobic large molecules, make it worthwhile to investigate the possible use of so-called chemical tracers, notably of appropriate dyestuffs to mimic the transport of difficult-to-monitor PAH's and PCB's. Many technical and practical details have to be addressed, and a consensus about the application of the tracers has to be reached. However, the benefits seem to outweigh these efforts, especially when compared with other means of obtaining this ecologically important information.

REFERENCES

Anliker R, Moser P, Poppinger D (1988) Bioaccumulation of dyestuffs and organic pigments in fish. Relationships to hydrophobicity and steric factors. Chemosphere 17: 1631-1644

Demas JN, Keller RA (1985) Enhancement of luminescence and Raman spectroscopy by phase-resolved background suppressioon. Anal Chem 57: 538-545

Duinker JC, Hillebrand MTJ, Boon JP (1983) Organochlorines in benthic invertebrates and sediments from the Dutch Wadden Sea; identification of individual PCB components. Neth J Sea Res 17: 19-38

Spanhoff R, Suijlen JM (1989) Feasibility of tracer studies of large-scale cohesive-sediment transport. This volume

Suijlen JM, Leussen W van (1989) Recent developments in large-scale tracing by fluorescent tracers. This volume

Trkula M, Keller RA (1985) Reduction of Raman background in laser-induced fluorescence by second harmonic detection. Anal Chem 57: 1663-1669

MEASUREMENT OF SUSPENDED LOADS IN STREAMS BY MEANS OF HYDROCYCLONES

E. Renger
Landesamt für Wasserhaushalt und Küsten, Schleswig-Holstein
Saarbrückenstraße 38, D-2300 Kiel, FRG

INTRODUCTION

A method for continuously recording suspended sediment concentrations with high accuracy has been developed and tested over the past few years (Renger, 1984, 1986). It is proposed to apply this method in cases where a knowledge of non-steady particle transport is needed.

MEASUREMENT PRINCIPLE

At the sampling location in the stream water with suspended particulate matter is continuously sucked in by means of a pump and fed through a pipe into the SEDIWA (Sedimentwaage) automatic balance (Fig. 1). Here the sediment-water mixture enters the hydrocyclone tangentially under pressure. As a result of the high centrifugal

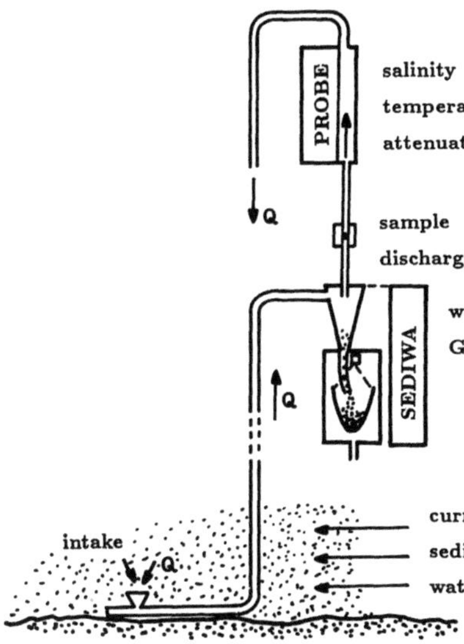

Figure 1: Main components of the SEDIWA-measuring system (hydraulic part)

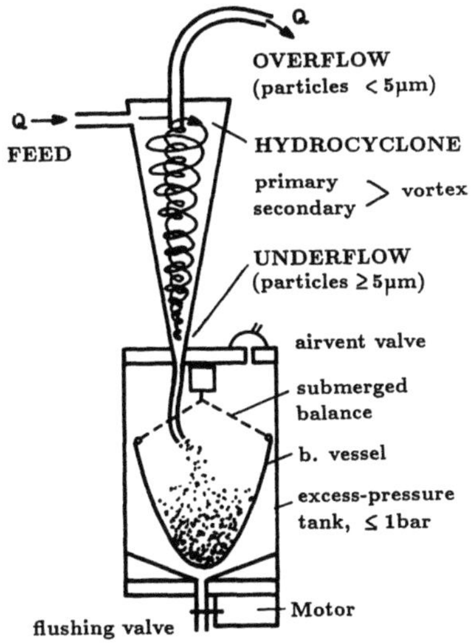

Figure 2: Hydrocyclone flow diagram and sediment balance (wet-weighing)

force, particles coarser than the cut-off diameter (d_T = 5 µm, in this case) concentrate within a primary vortex adjacent to the wall of the hydrocyclone and they move downwards on a spiral trajectory to the exit at the bottom by gravity (Fig. 2).

The extracted sediment flux is fed to a submerged balance vessel in a pressure - stabilized water tank and continuously weighed (wet-weighing). The mean concentration of suspended solids $\bar{c}(\Delta t)$ in mg/l within the time interval Δt is calculated in this direct measuring method from the weight increase $\Delta G(\Delta t)$ and the properly adjusted flow-rate Q (Fig. 3).

Particles finer than the cut-off diameter d_T migrate into a secondary upward-moving vortex, along the axis of the hydrocyclone, and leave with the water via the overflow. The quantity of this 'residual turbidity' is optically determined by means of an attenuation probe. The accuracy of this indirect measuring method is fairly high because the size distribution contains only very fine particles.

Some important specifications of the SEDIWA-system may be outlined as follows:

1. The flow-rate Q is controlled behind the overflow by means of an orifice.
2. The measuring system is completed by an electroacustic current meter and sensors for water temperature and electrical conductivity (salinity).

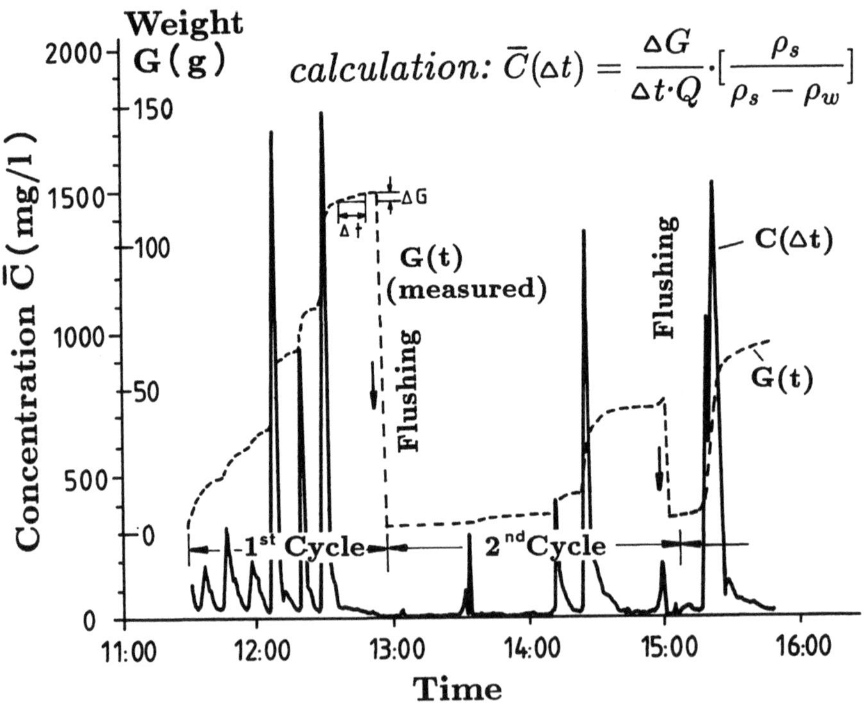

Figure 3: Simultaneous records of suspended sediment concentrations and sediment weight of the SEDIWA automatic balance and the calculation formula

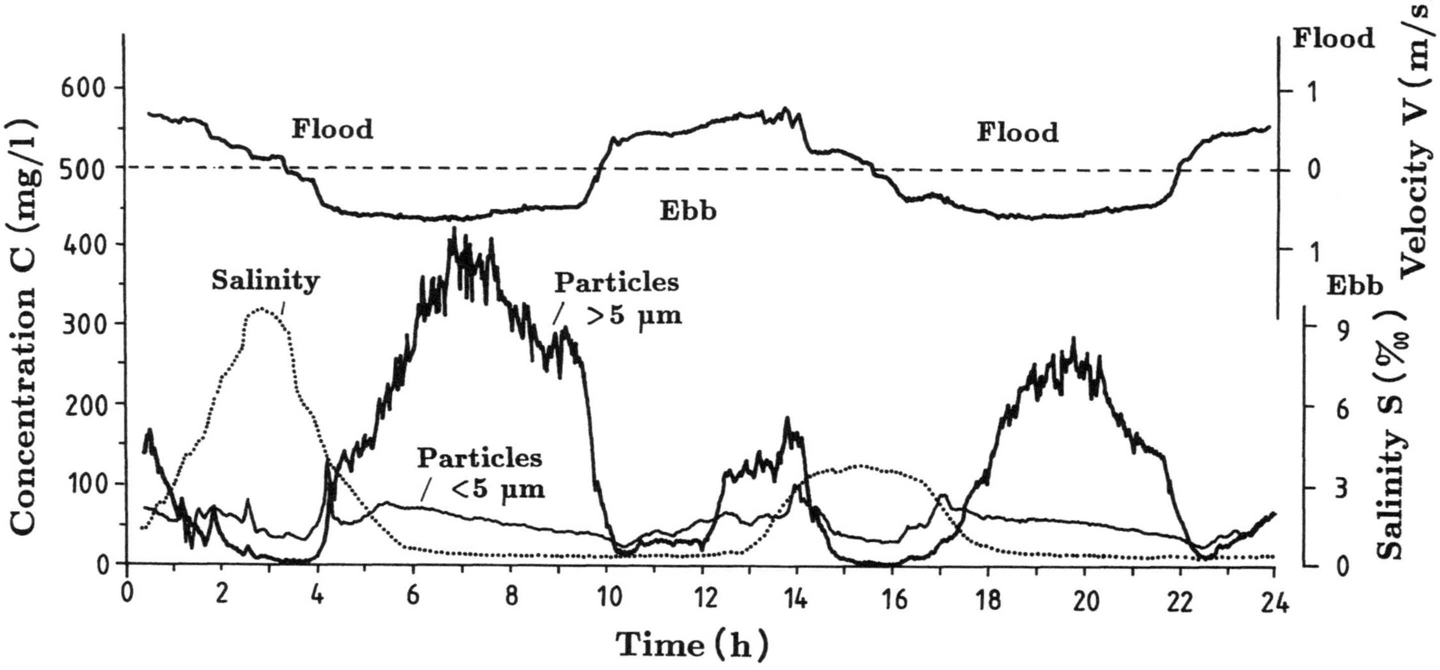

Figure 4: Simultaneous records of suspended sediment concentrations (coarser and finer 5 µm), salinity, and current velocity in the Eider estuary (Schleswig-Holstein)

3. An acquisition system stores the data of the connected probes and prepares them for further analysis. The sampling rate is up to three data sets per second. These data sets are averaged over selectable intervals between three seconds and five minutes, converted into physical units and presented for quick-look on a display. Via an analog output the data sets are then printed out. Furthermore, raw data are stored on cassette for later analysis (Fig. 4).
4. Emptying of the balance vessel is achieved by simply opening the flushing valve located at the base of the measuring chamber (excess-pressure tank, cf. Figs. 1 and 2). This operation is necessary when the material collected attains a weight of 200 g at the most or after a defined time period has elapsed (Fig. 3). By choking the overflow, the excess pressure of about 1 bar in the water tank falls off drastically. This leads to a strong flushing jet at the hydrocylone underflow which empties the full balance vessel within 2 to 5 minutes. Data acquisition is only possible 1 minute after the flushing operation, which is finished when the flushing valve is closed again.
5. Extraction and feed equipment for the hydrocyclone and the SEDIWA-system is realized by rotary pumps in combination with water-jet pumps. Abrasion is nearly negligible and the service life will be rather favourable.

ACHNOWLEDGEMENTS

This work was supported by the Kuratorium für Forschung im Küsteningenieurwesen (KFKI) and the Bundesminister für Forschung und Technologie (BMFT), which is greatly acknowledged. ME Meerestechnik-Elektronik GmbH contributed by optimizing the technical layout of the prototype.

REFERENCES

Renger E (1984) Development of a Sediment Transport Measuring System. Proc 19th Int Conf on Coastal Engineering (ICCE)

Renger E, Bednarczyk K (1986) Schiffahrtserzeugte Schwebstofftransporte im Neßmersieler Außentief. Die Küste 44: 89-132

CHAPTER VI

SEDIMENT-WATER INTERFACE

FLUID-MUD MOVEMENT AND BED FORMATION PROCESS WITH DENSITY CURRENT IN ESTUARIES

Tohru Futawatari and Tetsuya Kusuda
Faculty of Engineering, Kyushu University
Hakozaki, Fukuoka, 812, Japan

ABSTRACT

The movement of fluid mud, which consists of fine cohesive clay on a bottom slope, was investigated by the use of a flume under various conditions of initial concentration of suspension and gradient of the bottom slope. The velocity and density of the fluid mud layer on the slope were measured and the growing and decaying processes of the layer were analytically explained. Based on the experimental results, the mechanism and flow-rates of the fluid mud density currents with or without erosion or deposition were theoretically studied, and the horizontal flux of fluid mud was shown to be estimated by the use of a method developed in this study. Field data suggests that the fluid-mud movement with density currents exists.

1. INTRODUCTION

Fluid mud on the bottom of an estuary plays an important role in horizontal sediment transport as well as in bed formation, which may move downward as a result of density currents. When fluid mud flows downward gaining more potential and kinetic energy, due to both descent flow and supply of suspended solids from overlying water, than it loses, it increases in speed and may erode bed-mud. On the other hand, when it flows losing more energy, due to both viscosity and the settling of suspended solids contained in it, than it gains, it decreases in speed and thickness. The movement of fluid mud depends on the settling flux of suspended solids from the overlying water, the gradient of the bottom, the erosion of bed mud, the thickness of fluid mud which flows downward, and the flow velocity. Fluid-mud behaviour is of importance when discussing the transport of both sediments in water and soluble matters at the sediment-water interface.

Pantin [1] and Parker [2] mathematically explained turbidity currents with erosion and deposition where the only available material is medium to coarse silts and sands. However, in the sea, where fine cohesive sediments exist in large quantities, the sediments in suspension are deposited on a bottom slope with hindered settling, where they form bed mud or fluid mud. As well as this, fluid mud flows into deeper regions before accumulating relative to the topographical gradients present. The movement of fluid mud as a density current causes siltation of access channels. It is necessary to discuss the density current of fluid mud with sediment entrainment and bed formation on a bottom slope.

Experiments using a flume with a bed of variable gradient under various conditions of suspended solids concentration were conducted to study the following: 1) to examine the behaviour of fluid-mud density currents; 2) to determine the condition for the steady state between growth and decay of the density currents; and 3) to obtain bed formation rates and characteristics of the bed such as density profile and consolidation rate.

Moreover, the experimental and theoretical results are compared with the field data observed in a port, and the existence of fluid-mud movement as a density current is discussed.

2. EXPERIMENTAL MATERIAL, APPARATUS, AND PROCEDURE

The material used in this study was obtained at the Kumamoto Port in the Ariake Sea, Middle Kyushu, Japan. The Ariake Sea has the largest tidal range in Japan (at most 5 m), and so the extensive tidal flat area spreads over a distance of several kilometers at low water. The particle size distribution of the test material was measured using the hydrometer method. About 48 % of the material was in the clay range and 47 % in the silt range. The specific gravity was 2.65 and the ignition loss was 12.2 %.

Prior to the experiments, and in order to study the settling properties of the material, hindered settling tests of suspensions using several concentrations of suspended solids were performed in salt water of specific gravity 1.025. A result of the settling tests is shown in Fig. 1. When the concentration of suspension was less than 20 kg/m³, the interface was not distinctly observed throughout the settling period, and as a result the settling curve could not be obtained in this range. The overall settling velocity for each concentration was derived from the settling curve, and the settling flux was calculated as shown in Fig. 2. The set-

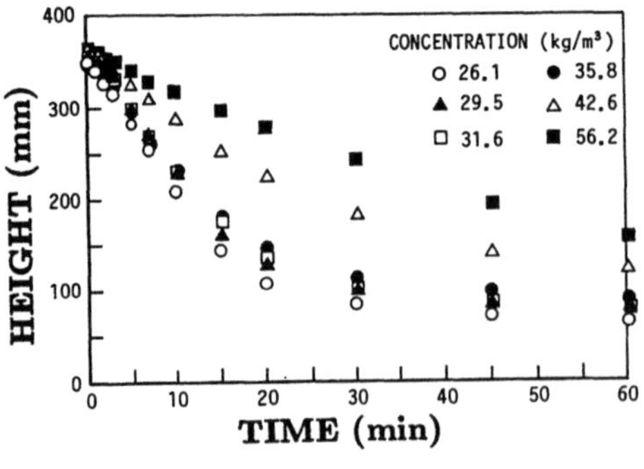

Figure 1: Settling curve

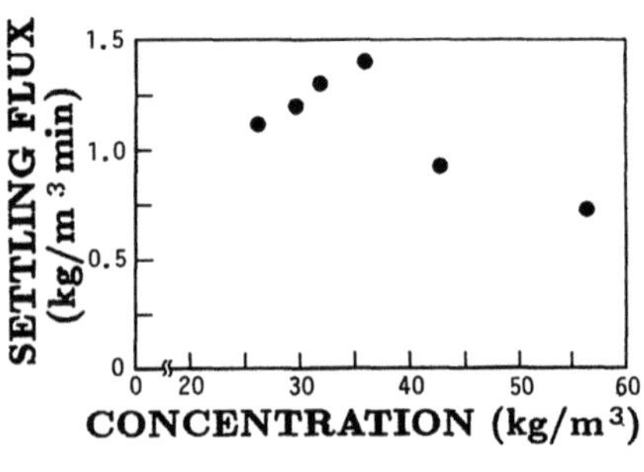

Figure 2: Settling flux

tling flux shows the maximum value at a concentration of 36 kg/m³. The reasons for this result are that the concentration was low even if the settling velocity was large on the left-hand side of the maximum value, and that the settling velocity in high concentrations becomes small because of hindering. Therefore, in the experiments on the density current of fluid mud using this material, initial concentrations of suspended solids, lower than the maximum, were adopted.

Figure 3 shows a schematic diagram of the flume used in this study. This flume is 2 m high, 1 m wide and 0.2 m deep. A movable inclined bed was installed in the flume, and it was possible to arbitrarily change the gradient of the bed. During experiments, density currents of fluid mud flow over the inclined bed, and fall into the pit at the end of the bed. Stainless steel sampling pipes of 2 mm diameter, were installed 0.7 m and 1.2 m above the bottom of the flume, and eleven pipes, at a distance of 3.5 mm, were also installed over the inclined bed, which is about 0.2 m above the bottom of the flume. These pipes were vertically arranged 0.17 m, 0.37 m, 0.57 m, and 0.77 m away from the upper edge of the inclined bed.

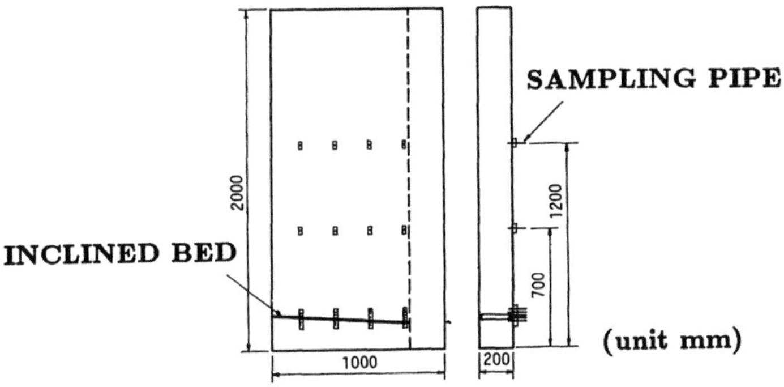

Figure 3: Experimental flume

A suspension was prepared for a certain concentration of suspended solids using salt water and was poured into the flume to a depth of 1.7 m. As soon as the gradient of the inclined bed was set, the experiment was started. Suspensions were then sampled through the pipes, and the concentration of suspended solids was measured with a turbidity meter and filtering membrane. At the same time, a picture of the density current was taken from one side with a video camera. The thicknesses of the fluid-mud layer and bed-mud layer, and the vertical profile of velocity in the fluid-mud layer were measured using video-tape records.

3. EXPERIMENTAL RESULTS AND DISCUSSION

In the density current experiments, gradients of the inclined bed and initial concentrations of suspension were changed as shown in Table 1.

The settling of suspended solids led to the movement of fluid mud as a density current. Whether a bed-mud layer was formed or not depended on the gradient and

Table 1: Conditions of density current experiment.

Gradient of inclined bed (%)	Initial concentration of suspension (kg/m³)	Water depth (m)
0.43	17.1	1.72
0.97	22.5	1.67
0.97	9.9	1.65
0.97	3.2	1.73
1.83	24.4	1.71
1.83	4.3	1.68

gradient and the initial concentration. In this study, the fluid-mud and the bed-mud layer were defined as the layer which flows parallel to the inclined bed and the stationary layer, respectively. The density current gradually increased towards the end of the inclined bed, but an end-effect existed in this region. In such instances, the measurements of the thicknesses of the layers and of the velocity of fluid mud were performed in a position without any end-effect. The thicknesses of the layers changed with time. Fig. 4 shows an example when the gradient was 1.83 % and the initial concentration was 24.4 kg/m³. In this case, the thickness of fluid mud rapidly increased within several minutes from the start of the experiment. When the bed-mud layer was formed after 25 minutes, the fluid-mud surface rose in the same way. The thickness of the fluid-mud layer, however, did not change. In other cases, when the slope was small, the formation of a bed-mud layer occurred within a minute. Rates of formation are shown in Fig. 5. In addition, the thickness of the fluid-mud layer in a steady state, when the formation of bed mud did not appear, was similar to that found in another case with the same gradient.

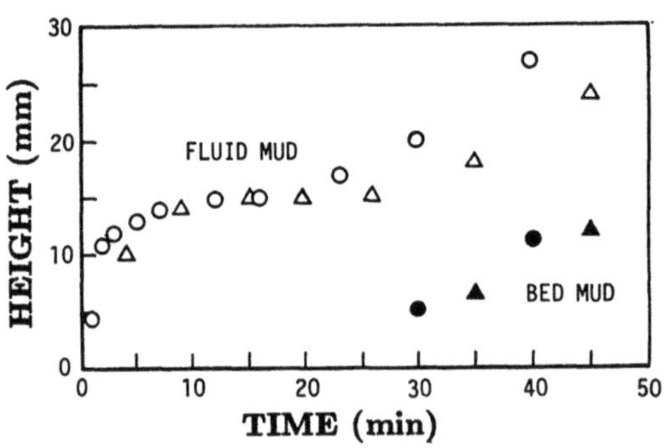

Figure 4: Thicknesses of fluid-mud and bed-mud layer.

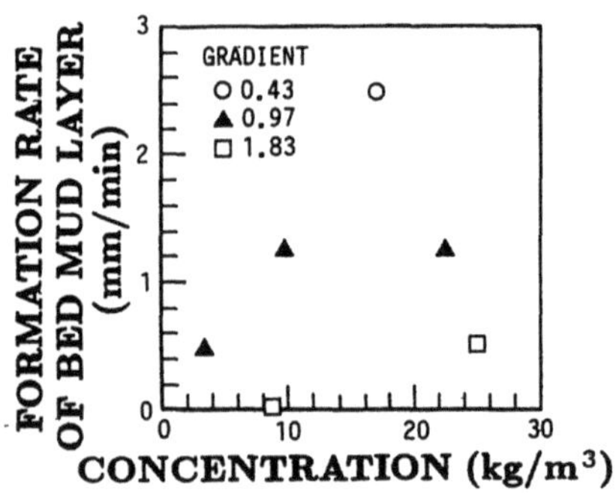

Figure 5: Formation rate of bed-mud layer.

Velocity profiles in the fluid-mud layer are depicted in Fig. 6. The maximum velocity at a given time decreased as time passed, and it then shifted upward as a whole because of the bed-mud layer formation. In the fluid-mud layer, a layered flow was noted for the movement of suspended sediment particles. Figure 7 shows the concentration distribution of suspensions. The concentration in the region near the bed increased as time passed, but a spatial difference was not easily observed in the fluid-mud layer. In other cases, the velocity and concentration profiles showed the same tendencies.

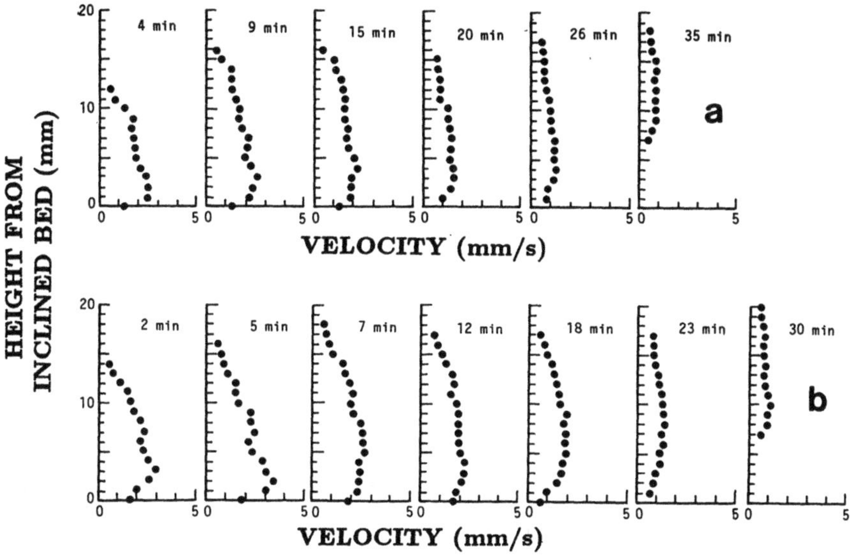

Figure 6: Velocity profiles in fluid-mud layer. Distance from upper end of inclined bed: a) 0.2 m; b) 0.4 m

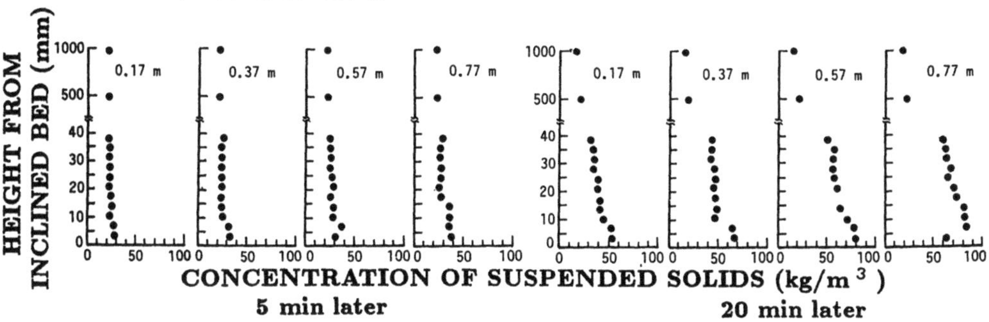

Figure 7: Concentration distribution of suspension.

4. MODELLING OF DENSITY CURRENT

Based on the experimental results, the modelling of the density current of fluid mud with sediment entrainment was carried out in order to determine the condition for a steady state between growth and decay of the density current. On the assumption that the flow is regarded as being layered, and that it has a layer-averaged velocity and concentration, the fundamental equations of motion and mass conservation are expressed as follows:

$$\frac{\partial(\rho_m hU)}{\partial t} + \frac{\partial(\rho_m hU^2)}{\partial x} = \rho' g\theta h - \frac{g}{2}\frac{\partial}{\partial x}(\rho' h) - D - (P+E)U \tag{1}$$

$$\frac{\partial(\rho' h)}{\partial t} + \frac{\partial(\rho' hU)}{\partial x} = (P+E)\,\frac{\rho_s - \rho_1}{\rho_s} \tag{2}$$

where x is the axis parallel to the bed, positive in the direction of flow; h is the local depth of the flow; ρ_m is the absolute density of the suspension and equal to $\rho_l + \rho_s(1-\epsilon)$, ρ_l is the density of the liquid, ρ_s is the density of sediment, and $(1-\epsilon)$ is a volumetric concentration of the suspension; ρ' is the effective density of the suspension and is equal to $(1-\epsilon)(\rho_s-\rho_l) = \rho_m - \rho_l$; U is the local velocity of the flow along the x-axis; θ is the angle of the slope (assumed to be small); D is the frictional force at the surfaces of the flow; P is the rate of deposition or erosion of sediment; and E is the rate of sediment entrainment from the overlying water layer, i.e. the settling flux of the suspension from this layer.

The rates and the frictional forces are represented as follows:

$$P = \alpha(U/U_{eo}-1) = \alpha'(U-U_{eo}) \qquad \text{for } U_{eo} \le U \tag{3}$$

$$= -\beta(1-U/U_{do}) = -\beta'(U_{do}-U) \qquad \text{for } 0 < U \le U_{do}$$

$$= 0 \qquad \text{for } U_{do} < U < U_{eo}$$

$$E = wC \tag{4}$$

$$D = \gamma\rho_m \nu U/h \tag{5}$$

where U_{eo} is the critical velocity for the erosion of sediments; U_{do} is the critical velocity for deposition; w is the settling velocity of sediments; C is the concentration of suspension; ν is the coefficient of the kinematic viscosity of the fluid mud; and α, β, γ are constants.

In this model, a steady state is assumed, with time-differentials equal to zero, and the concentration in the flow does not change along the x-axis. The thickness of the flow, h, depends on entrainment, deposition and erosion. The problem of how the velocity and thickness of the flow change is then investigated. Eqs. (1) and (2) become:

$$\frac{dhU^2}{dx} = \frac{\rho'}{\rho_m} g\theta h - \gamma\nu\frac{U}{h} = f_1 \tag{6}$$

$$\frac{dhU}{dx} = \frac{P+E}{(1-\epsilon)\rho_s} = f_2 \tag{7}$$

The first term in the middle of Eq. (6) is related to the supply of energy by the gravity force, and the second is the demand of energy by the friction force of the flow. In consequence, if f_1 is positive, the flow grows, and if negative, the flow decays. When the flow shows a steady state, f_1 is equal to zero, i.e., h is proportional to $\sqrt{U}$. In Eq. (7), f_2 describes the amount of sediment as the sum of entrainment and deposition/erosion. Combining Eqs. (6) and (7), we thus obtain:

$$\frac{dhU^2}{dhU} = U + hU\frac{dU}{dhU} = \frac{f_1}{f_2} \tag{8}$$

Finally, we obtain:

$$\frac{d\ln h}{d\ln U} = \frac{2Uf_2 - f_1}{Uf_2 - f_1} \tag{9}$$

The relationship between the thickness and the velocity of the flow is represented in Fig. 8. In this figure, the broken line indicates a steady state of the flow, and the flow eventually becomes asymptotic to this line.

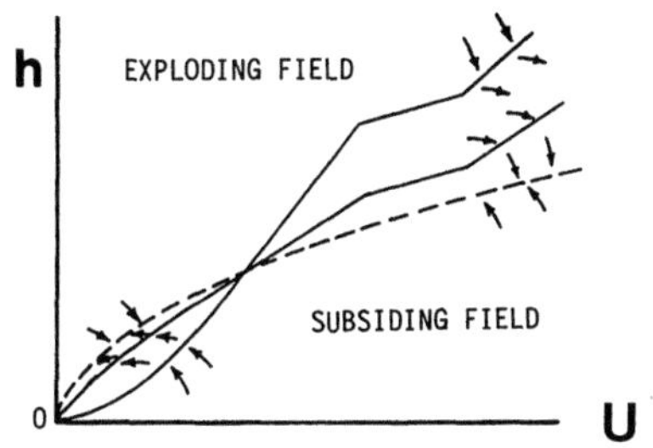

Figure 8: Relationship between thickness and velocity

5. COMPARISON WITH FIELD DATA

Field surveys of siltation were performed in the Kumamoto Port [3]. Based on these results, the existence of density currents with fluid mud is examined.

The field study involves a comparison of the degree of siltation in dredged trenches at three locations in the Port. The three trenches have the same area and profile. Each dredged trench has a rectangular area of 50 m x 70 m, the depth is 2 m and the side gradient is 1/5. Hence, the bottom area is 30 m x 50 m. Trench No. 1 is at 4 m water depth and No. 2 and 3 are at a depth of 2 m. No. 3 is enclosed by a submerged dike which is 1 m above the sea bottom and is 3 m wide. During the survey period, currents, waves, tides, and concentrations were also observed.

Figure 9 displays the siltation at the centers of the trenches; arrows indicate the time of rough seas. Severe siltations occurred in trenches No. 1 and 2, i.e. about 1.5 m in a year. In contrast, siltation in trench No. 3 was very small. At the lowest measuring point, the concentration of suspended solids at the bottom layer reached 1 kg/m³ during rough seas. This is not sufficient to supply the recorded amount of accumulated sediments to the trenches. Therefore, it can be said that siltation in the trenches could have occurred not only by sedimentation of suspended matter but also by the flow of fluid mud on the sea bed. The small amount of siltation in trench No. 3 is explained by obstruction of the fluid-mud flow by the submerged dike. The flow of fluid mud, however, was not observed directly, and the slope of the sea bed is very gentle in comparison with the experiments carried out in this study. It is therefore necessary to study further the fluid-mud movement with density current in both field and laboratory experiments.

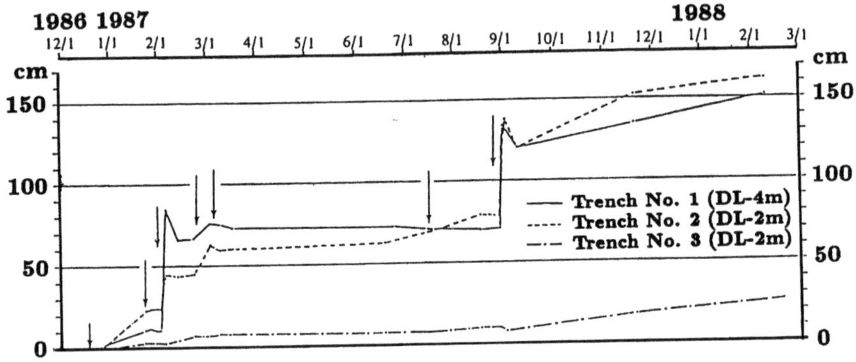

Fig. 9: Siltation at the centres of the trenches

6. CONCLUSIONS

The results of the experiments carried out on density currents of fluid mud indicated that the flow on a slope was a layered one. Sediment entrainment from suspension increased the thickness of the fluid-mud layer. The condition for a steady state of the flow was developed based on a theoretical investigation. The probability that fluid-mud movement as a density current could exist in the sea was also suggested by field data.

This study was performed with the aid of the Kajima Foundation Research Grant and supported by the 4th District Port Construction Bureau, Japan.

REFERENCES

[1] Pantin HM (1979) Interaction between velocity and effective density in turbidity flow: phase-plane analysis, with criteria for autosuspension. Mar Geol 31: 59-99

[2] Parker G (1982) Conditions for the ignition of catastrophically erosive turbidity currents. Mar Geol 46: 307-327

[3] Japan International Cooperation Agency (1988) A case study on siltation countermeasures in Kumamoto Port, Japan

BELL- JAR EXPERIMENTS AND SUSPENSION MEASUREMENTS FOR THE QUANTIFICATION OF OXYGEN CONSUMPTION AND NITROGEN CONVERSION IN SEDIMENTS OF THE ELBE ESTUARY

F. Schroeder, D. Klages, H.-D. Knauth
GKSS Research Centre, Institute of Chemistry
Max-Planck Str.1, D-2054 Geesthacht, FRG

Introduction

In the Elbe easily degradable organic matter and nutrients originating from sewage effluent, terrestrial runoff, or phytoplankton production are a major source of oxygen consumption. Mineralisation processes which take place primarily in sediments are an important factor required for the quantification of the nitrogen budget; e.g. in coastal areas benthic recycling can account for between 30-80% of the nitrogen required for phytoplankton production [1]. The coupling of nitrification and denitrification processes in sediments provide an ideal "shunt" of regenerated NH_4^+ away from those assimilative pathways which support photosynthesis. Between 43-55% of the total input of nitrogen can be removed by denitrification processes in sediments and transformed to N_2 [2]. Investigations in the Elbe showed high dentrification rates in sediments [3]; this is, inter alia, due to large nitrate concentrations, prevalent in Elbe waters.

The objective of our investigations was to carry out research enabling a partial quantification of the major benthic processes which affect the nitrogen budget of the Elbe estuary.

Methods

Oxygen and nitrate and fluxes across the sediment/water interface were directly measured using incubation methods. A diagenetic two-layer model [4,5,6] was applied in the calculation of concentration profiles with depth in porewaters, using measured reaction rates and assumed exchange coefficients as input parameters.

Study sites - The study was conducted in four typical shallow water areas in the River Elbe estuary. The most seaward stations *Neufelder Watt* and *Böschrücken* are about 25 km and the station *Schwarztonnensand* is 60 km from the estuary mouth. The most landward station, *Mühlenberger Loch,* is situated immediately downstream of Hamburg harbour, approximately 90 km from the mouth. Only the *Neufelder Watt* and *Böschrücken* sites are in the oligohaline reaches of the estuary, with salinities varying between 1 ‰ and 5 ‰. At the other stations the salinity is about 1-2 ‰. *Mühlenberger Loch* is a typical area of sedimentation with deposition rates of 5 -10 $cm \cdot yr^{-1}$ [7].

Flux measurements (in situ bell jar experiments)- In situ measurements of fluxes across the sediment/water interface were carried out with a hydraulically operated bell-jar system, designed especially for application in estuaries [8]. The bell-jar, a plastic hemisphere with a diameter of 70 cm, encloses about 80 l water above 0.3 m^2 of sediment. In order to prevent settling of suspended matter and the build up of concentration gradients inside the bell-jar, an air driven membrane pump circulates the water.The water is pumped in a second closed circuit from the bell-jar to a measuring loop on board a pontoon where oxygen and nutrient analyses may be carried out. A closed gas loop with a volume of about 200 ml is coupled to the measuring loop by means of a porous disk, preventing oversaturation of formed gases (N_2, N_2O) in the waterphase. The oxygen concentration in the measuring loop and therefore in the bell-jar can be maintained at a well defined level between 0.01-10 $mg \cdot l^{-1}$ O_2 by electronically regulated addition of oxygen into the gas phase.

Flux measurements (laboratory)- Incubations of undisturbed sediment cores were carried out in the laboratory for comparison with bell-jar experiments. The cores were collected with cylindrical acrylic tubes (10 cm diameter) either by hand in the shallow water regions or with a specially designed, hydraulically operated sediment corer [9] during cruises with R/V "Ludwig Prandtl" (tube with 20 cm diameter). Only cores with undisturbed stratification were used. The cores were kept at about 4°C until incubation (within 6 h after collection); special care was taken to preserve the sediment structure during sampling and transport. In the laboratory the tubes were sealed with an air tight lid, fitted with stirrer, O_2 and pH probes, as well as a sampling port [10]. Incubations of the sediment cores were carried out in the dark at *in situ* temperatures.

Rate measurements - Reaction rates were measured by incubating stirred suspensions of 10 g wet sediment in 500 ml autoclaved Elbe water. For the measurement of nitrification rates the suspension was aerated and for nitrate reduction rate measurements the suspension was held under anaerobic conditions with nitrogen [10]. Material for the incubations was acquired by cutting the collected sediment cores into oxic and anoxic slices, without disturbing the chemocline.

Sediment composition - Water contents were determined by drying sediment samples to constant weight at 110°C. Grain size analyses (200 µm and 63 µm) were carried out by wet-sieving. Organic carbon content was measured by combustion with oxygen and detection of the formed CO_2 with an IR analyser, model 5000 (ASTRO).

Chemical analyses - NO_3^- was determined using ionchromatography (model 2000i DIONEX) and NO_2^- determined colorimetrically [11]. NH_4^+ was measured with FIA using a gas-sensitive NH_3-electrode for detection [12]

Results

Sediment characteristics - The sediments from *Mühlenberger Loch* consisted mainly of soft mud (52-75 % silts and clays) with organic carbon values of 4-5 %, and water contents of 55-71 %. The sediments from the other locations were dominated by fine sands and silts (17-55% silts and clays) with 1.5-3 % of organic carbon and water contents of 37-50 %.

Rate measurements - Rate measurements were carried out in spring and autumn with sediments from the *Mühlenberger Loch* and *Böschrücken* sites. In all cases the nitrification and nitrate reduction reactions under *in situ* conditions were in the zero order range of Michaelis-Menten kinetics. The rate constants for *Mühlenberger Loch* sediments were relatively high, sometimes exceeding 17 $mmol \cdot d^{-1} \cdot g^{-1}$(dry weight) at 19 °C for nitrification and 20 $mmol \cdot d^{-1} \cdot g^{-1}$ for nitrate reduction.The rate constants for *Böschrücken* sediments were significantly smaller, ranging between 3-5 $mmol \cdot d^{-1} \cdot g^{-1}$ (18°C) for nitrification and 5-8 $mmol \cdot d^{-1} \cdot g^{-1}$ for nitrate reduction.

Flux measurements - Flux measurements (*in situ* and laboratory) were carried out for all sites. A typical diagram representing the concentrations inside the bell-jar over time is depicted in fig.1.

In the first 50 hours of the experiment oxygen was consumed with a flux of 1.5 g O_2 $m^{-2} \cdot d^{-1}$. The gradual decrease of NO_3^- during this period shows that a small

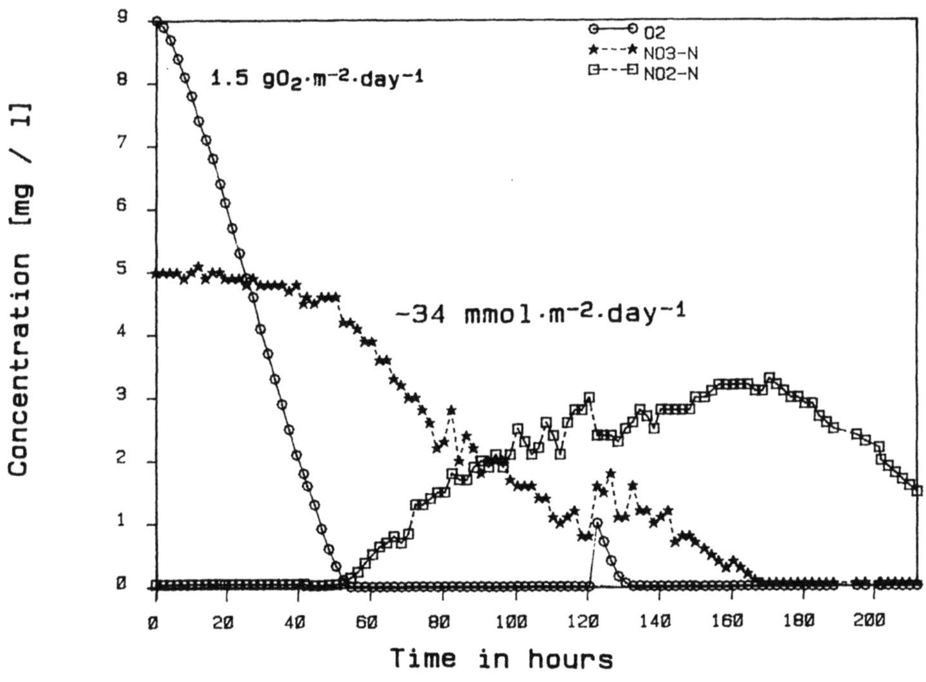

Fig.1. Concentration-time diagram for a bell-jar experiment -*Schwarztonnensand*, October 1986, [water temperature: 18°C] -

flux of nitrate from the water into the sediment occurred. This flux increased immediately after all oxygen had been consumed and the NO_3^- present was used as electron-acceptor. During the anoxic phase NO_2^- was formed initially and following the consumption of NO_3^-, it was used up. The resulting fluxes for NO_3^- were -3 mmol $N \cdot m^{-2} \cdot d^{-1}$ during the oxic and -34 mmol $N \cdot m^{-2} \cdot d^{-1}$ during the anoxic period. The NO_2^--fluxes due to NO_2^--formation and consumption were +25 mmol $N \cdot m^{-2} \cdot d^{-1}$ and -29 mmol $N \cdot m^{-2} \cdot d^{-1}$ respectively.

Bell-jar and laboratory flux measurements for the same site gave similiar results with non-systematic deviations of less than 30% between the two methods. These deviations are to be expected as the areas of the incubated sediments differed by a factor of 38 for the two methods.

The results from bell-jar experiments and laboratory flux measurements at the different sites are depicted in fig.2.

The O_2-fluxes varied between -109 mmol $O_2 \cdot m^{-2} \cdot d^{-1}$ for the muddy *Mühlenberg* area and -7 mmol $O_2 \cdot m^{-2} \cdot d^{-1}$ for an partially silty sample from *Neufelder Watt*. The NO_3^- fluxes during the oxic phase were between 0 and -9 mmol $N \cdot m^{-2} \cdot d^{-1}$. During the anoxic phase, in which no nitrification could occur, the measured fluxes were between -14 and -40 mmol $N \cdot m^{-2} \cdot d^{-1}$ (except in the case of sample BR 5/88 with -59 mmol $N \cdot m^{-2} \cdot d^{-1}$).

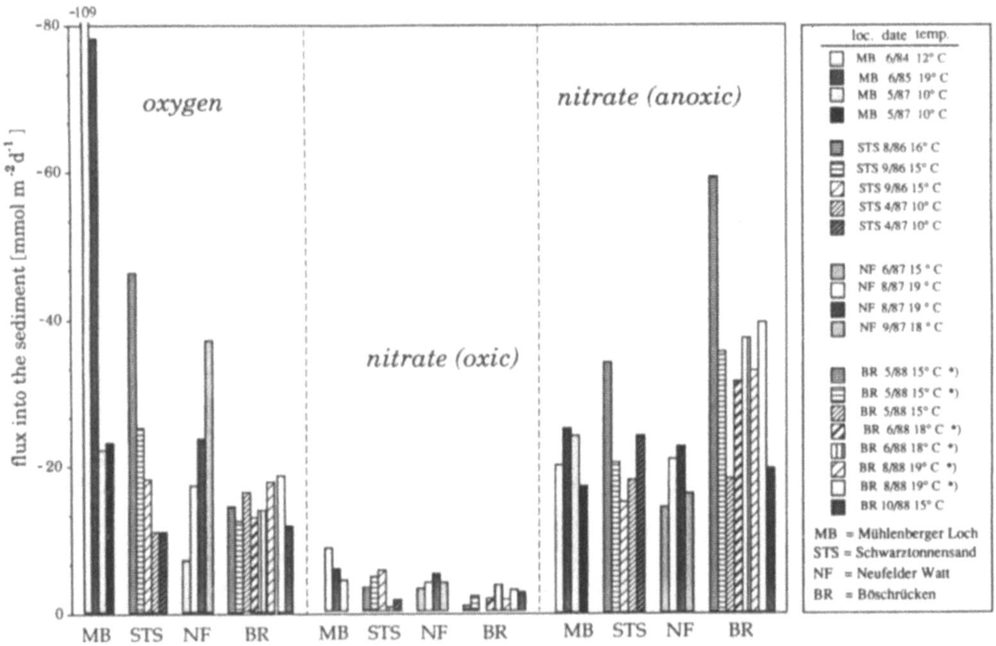

Fig.2. : Fluxes of O_2 and NO_3^- for different sites in the Elbe estuary. The NO_3^- fluxes were measured during the oxic and anoxic phases of bell-jar and incubation experiments. Experiments in which microbial population shifts were suspected, due to a long measuring period, are denoted by *).

Discussion

Oxygen consumption results exhibited high variations. These were between both the samples from different areas and between individual samples from one area taken on different dates. No correlation with temperature was found. The variation between replicate samples were small, but differences occurred between samples taken on different dates.This may be attributed to navigational inaccuracies, sediment heterogeneity and to temporal changes in sediment composition, particularly after storm events.

The nitrate fluxes measured under anoxic conditions were high in comparison with data from other estuaries [13]. The variation between the fluxes as measured for *Mühlenberg, Schwarztonnensand* and *Neufelder Watt* sites was only 36%, (with the exception of the first STS sample which in addition exhibited unusual high oxygen consumption). In some cases *Böschrücken* sediments, despite similiar sediment compositions, had higher fluxes than *Neufeld* sediments. In these cases the laboratory incubations took longer than 100 hours, so that there may have been microbial population shifts and bacterial growth on the walls of the reaction vessels.This assumption has been substantiated by the fact that the reaction velocities increased over time during these experiments.

The range of nitrate fluxes, measured under *in situ* conditions during the oxic period, was comparable with data found for other estuaries [14]. The measured fluxes were compared with computed fluxes from model calculations using the measured reaction rates and exchange coefficients from the literature [5,6]. The agreement between measured and calculated fluxes was satisfactory for most samples from the *Mühlenberg* and *Schwarztonnensand* area. Problems, however, occurred with some samples from the *Böschrücken* area where the fluxes measured under anoxic conditions were much higher than those resulting from model calculations. The reason for this could be that in the model two separate layers were assumed. Thus, the close coupling of sediment nitrification and nitrate reduction could not be accounted for [15]. The very effective coupling can be explained either by additional nitrate reduction processes in anoxic microniches in the upper sediment layer [16] or by additional nitrification processes in regions of enhanced oxygen penetration, e.g. macrofaunal burrows, within the reduced, lower sediment layers [17].

A preliminary estimation of the total nitrate consumption of Elbe sediments enables an assessment of the sediments to the total nitrogen budget of the Elbe estuary. The area of shallow water sediments (including tidal flats) between Hamburg and the estuary mouth is about 300 km^2. Assuming that 30% of this area has a nitrate consumption of -15 mmol $N{\cdot}m^{-2}{\cdot}d^{-1}$ (a very cautious assumption), a total nitrate reduction of 31 t $N{\cdot}d^{-1}$ results. This value may be compared with a total nitrogen load of 200-300 t $N{\cdot}d^{-1}$ which is transported by the water phase in summer [18].

Thus it is obvious that the nitrogen budget of the Elbe estuary is strongly influenced by nitrogen transformation processes occurring in sediments.

References

[1] Blackburn T H, Henriksen K (1983) Nitrogen cycling in different types of sediments from Danish waters. Limnol Oceanogr 28: 477-493

[2] Seitzinger P S (1988) Denitrification in freshwater and coastal marine ecosystems: Ecological and geochemical significance. Limnol Oceanogr 33: 702-724

[3] Wolter K, Knauth H-D, Kock H-H, Schroeder F (1985) Nitrifikation und Nitratatmung im Wasser und Sediment der Unterelbe. Vom Wasser Bd.65: 63-80

[4] Vanderborght J P, Billen G (1975) Vertical distribution of nitrate concentration in interstitial water of marine sediments with nitrification and denitrification. Limnol Oceanogr 20: 953-1061

[5] Vanderborght J P, Wollast R, Billen G (1977) Kinetic models of diagenesis in disturbed sediments. 2. Nitrogen diagenesis. Limnol Oceanogr 22: 794-803

[6] Billen G (1978) A budget of nitrogen recycling in North sea sediments off the Belgian coast. Est Coast Mar Sci 7: 127-146

[7] Petersen W, Knauth H D, Pepelnik R, Bendler I (1989) Vertical distribution of Chernobyl isotopes and their correlation to heavy metals and organic carbon in different sediment cores of the Elbe estuary. The Science of the Total Environment (in press).

[8] Schroeder F, Milde M P, Knauth H D (1989) A new bell-jar system for flux measurement studies in estuaries under defined conditions. (In preparation)

[9] Schroeder F, Milde M P, Petersen W (1989) A hydraulically operated sediment corer for undisturbed sampling. (In preparation)

[10] Klages D (1989) Nutrient fluxes and kinetics of nitrate transformations by nitrification and denitrification in sediments of the river Elbe estuary. Reports of the EEC commission (in press)

[11] Bendschneider K, Robinson RJ (1952) A new spectrophotometric method for the determination of nitrite in seawater. J Mar Res 11: 87-96

[12] Schroeder F, Kock H H (1985) Untersuchungen tidebedingter Konzentrationsschwankungen von Nitrat und Ammonium in der Unterelbe bei Lühesand. GKSS Externe Berichte, GKSS 86/E/8: 1-27

[13] Koike I, Sørensen J (1985)Nitrate reduction and denitrification in marine sediments. In: Blackburn T H, Sørensen J (eds) Nitrogen cycling in coastal marine environments. John Wiley, New York, p 251

[14] Henriksen K, Kemp W M (1985) Nitrification in estuaries and coastal marine sediments. In: Blackburn T H, Sørensen J. Nitrogen cycling in coastal marine environments. John Wiley, New York,:p 206

[15] Jenkins M C, Kemp W M (1984) The coupling of nitrification and denitrification in two marine sediments. Limnol Oceanogr 29: 609-619

[16] Jørgensen B B (1977) Bacterial sulfate reduction within reduced microniches of oxidized marine sediments. Mar Biol 41: 7-17

[17] Aller C (1985) Benthic fauna and biogeochemical processes in marine sediments: The role of burrow structures. In: Blackburn T H, Sørensen J (eds) Nitrogen cycling in coastal marine environments. John Wiley, New York, p 301

[18] Arbeitsgemeinschaft für die Reinhaltung der Elbe (Hrsg.): Wassergütedaten der Elbe 1986. Hamburg 1987

PHOSPHATE SORPTION EQUILIBRIUM AT SEDIMENT-WATER INTERFACES - CONSEQUENCES FOR THE REHABILITATION OF WATERS

Günter Schlungbaum and Günther Nausch,
Sektion Biologie, Wilhelm-Pieck-Universität Rostock
Freiligrathstraße 7/8, DDR-2500 Rostock 1

1. THE BALTIC COASTAL WATERS OF THE GDR

The 340 km coastline of the GDR includes a large system of internal coastal waters (Fig. 1). With a total area of some 1600 km², these boddens and haffs contain over 6 x 10^9 m³ of water. In view of their mean water depth of 3.5 to 4 m, they must be considered as being shallow.

Their estuarine character is indicated by increasing salinity from west to east and is a result of their morphology, fresh water inflow and their connection to the open Baltic (Schlungbaum, 1989).

The quality of their water is characterized by eutrophic processes caused by natural factors and anthropogenous influences. Primary production is usually intense owing to the relatively large and continuous nitrogen and phosphorous inputs.

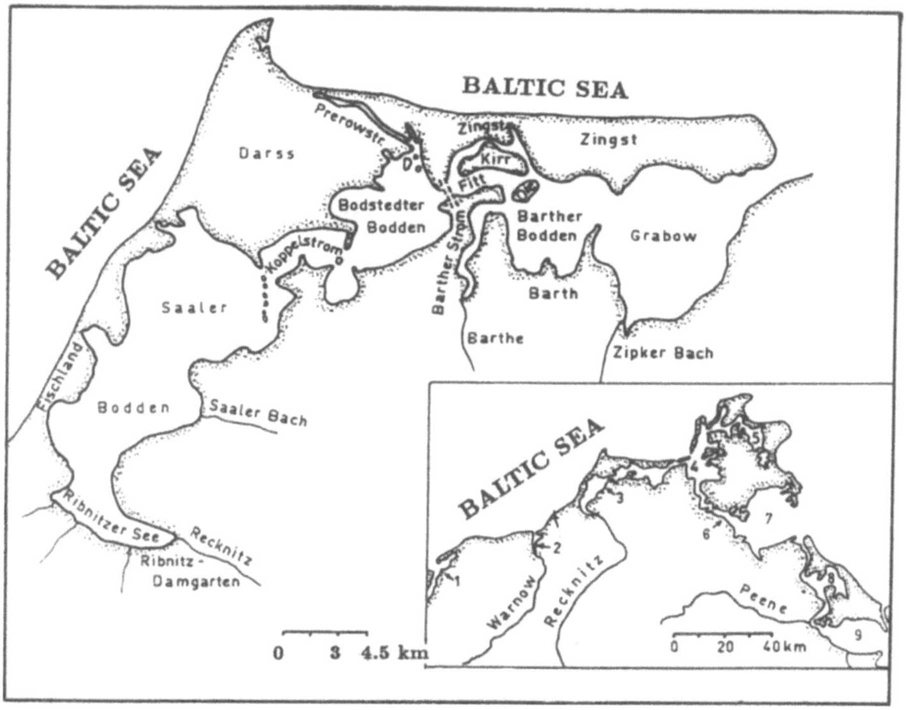

The natural food web in the bodden chain south of the Darss-Zingst Peninsula is shown in Fig. 2 (see also Fig. 1, surface area = 197 km^2, water volume = $397x10^6$ m^3, mean depth = 2 m). Disrupted food webs in which the biomass resulting from primary production is not fully utilized are typical of highly eutrophic waters. The surplus biomass is a source of faster sedimentation. As it settles, nutrient-rich organic matter accumulates at the bottom of the water, where it is constantly shifted by currents and wind induced turbulence in shallow waters and is subsequently deposited in hydrologically calmer areas or in depressions.

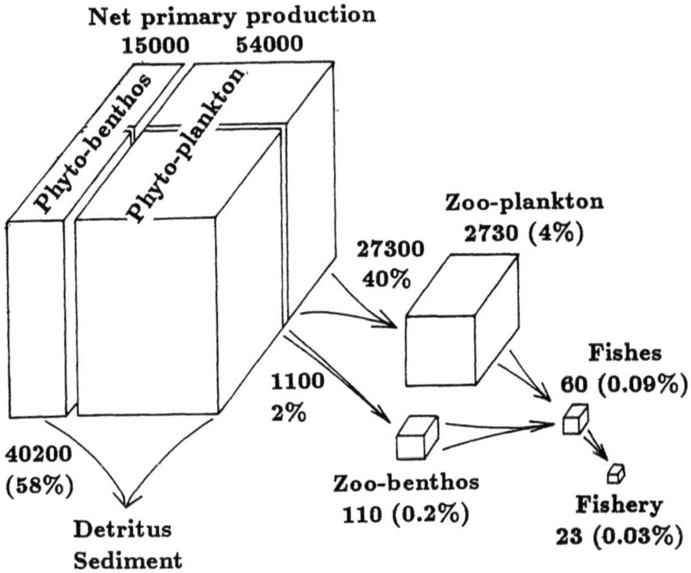

Fig. 2: Nutrient web balance in the boddens south of the Darss-Zingst Peninsula (figures represent produced biomass in $kg \cdot ha^{-1} \cdot a^{-1}$ after Schnese, 1978)

Both sediment and water quality depend on interactions between the sediment and water. In particular, the eutrophic status of a water is strongly influenced by the equilibrium between phosphate sorption and desorption (Schlungbaum, 1982).

The higher sedimentation rates generally act as constraints on ways in which surface waters are used (e.g. fisheries, recreation, transport), and their rehabilitation to safeguard their water quality and utilization involves the investment of considerable economic and technical resources.

2. FACTORS INFLUENCING EUTHROPHICATION IN SHALLOW COASTAL WATERS

The effects of anthropogenous factors vary considerably due to the natural differences between waters in terms of water depth, size and catchment area, freshwater inflow and water exchange conditions. Fig. 3 shows the magnitude of biological production in the bodden chain south of the Darss-Zingst Peninsula as affected by its catchment area.

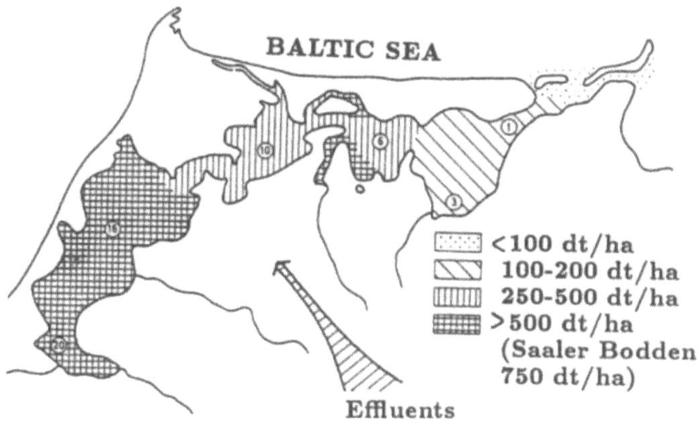

Fig. 3: Biological production in the bodden chain south of the Darss-Zingst Peninsula as influenced by its catchment area (after Schnese, 1978)

Nitrogen and phosphorus are generally needed in a (atomic) ratio 16 : 1 for primary production. Nutrients are not production-limiting factors in these coastal waters. In other words, the ecologically soundest way to manipulate the water quality is through nutrient input. It has been found that the easiest way to do this is by controlling the phosphate input. Nitrogenous compounds enter the waters with land runoff, but also through precipitation and as a result of bacterial fixation of atmospheric nitrogen.

3. THE PHOSPHATE STATUS OF SHALLOW COASTAL WATERS

The sum of the nutrient inputs into a water is reflected in its eutrophic status. Owing to intensive interactions between sediment and water, dissolved orthophosphate concentrations in the water body remain virtually constant throughout the year in shallow waters due to the phosphate sorption equilibrium (Schlungbaum, 1982). The equilibrium concentrations found in the GDR's bodden waters vary between 0.3 and 0.8 $\mu mol \cdot l^{-1}$. In other words, the measured orthophosphate concentration is not a measure of water status. The phosphate sorption capacity of the sediment (Schlungbaum, 1982; Schlungbaum and Nausch, 1988) controls the concentration by adjusting the water-to-sediment flux if the concentration in the water rises and the sediment-to-water flux if it decreases due, say, to increased biological production. It can thus be concluded that the high productivities of coastal waters are regulated by the high turnover rates in the cycle of matter and the equilibrium concentrations.

4. MEASURES FOR CONTROLLING PHOSPHATE AVAILABILITY IN COASTAL WATERS

In internal coastal waters, intrinsic inputs due to phosphate liberation and mineralization can exceed extrinsic inputs by a factor of 2 to 4 owing to the huge amounts of phosphate in the sediment. In other words, even the most efficient sewage treatment methods will not improve water quality for many years. Steps to reduce external loads must therefore be linked to measures to reduce the intrinsic load. Since the fixation of phosphate in the sediment is only partly reversible, the easiest and most definite way to do this is to remove the sediment. In the case of large waters, dredging of the whole bottom is not feasible for both economic and technical reasons. This also applies to the boddens.

Methods involving covering the sediment, nutrient fixation by precipitation or oxidative treatment of sediment by injecting calcium nitrate (RIPLEX methods, Ripl, 1978) yield only short term improvements in shallow waters. As shown in Fig. 4 for the Saaler Bodden, the distribution of nutrient-rich mud depends on the morphometry of the water and hydrological factors. This must be taken into account when planning rehabilitation work. We shall use the Saaler Bodden as an example to discuss a multifactorial rehabilitation programme for reducing intrinsic loads.

- Dredging

This expensive method will be needed, and therefore recommended, only in certain parts of the bodden, including Ribnitzer See (southern part of Saaler Bodden, oligohaline) and the bay at Wustrow. The organic sediment layer is up to 4 m thick

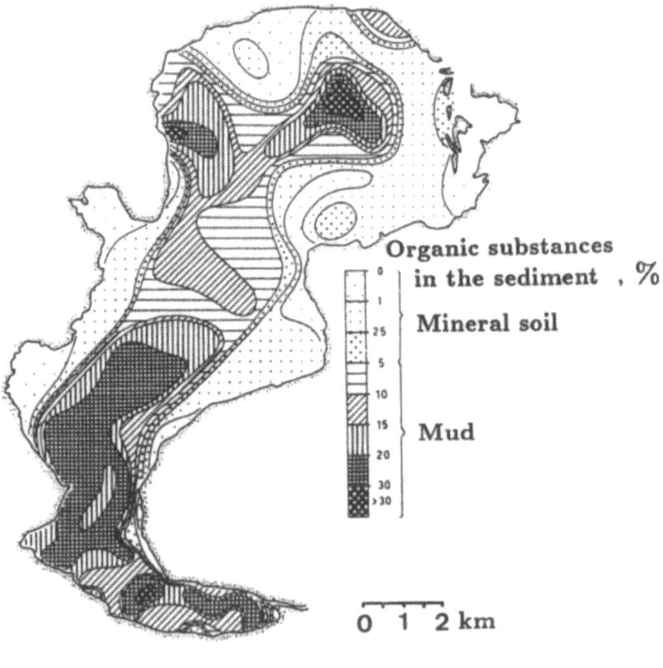

Fig. 4: Distribution of organic matter in the surface sediment of Saaler Bodden

in some parts of Ribnitzer See. The depth to which it should be removed must be decided on the basis of studies of the phosphate availability in the lower parts of the layer.

- Sediment traps

Sediment traps consist of depressions dredged in the bottom in front of river mouths. The depth should be at least one or two meters below the surrounding bottom. A trap of this kind was constructed in front of the mouth of the River Recknitz in the Saaler Bodden in 1987, and another will be dredged in 1989 where the River Körkwitz flows into the bodden. Further east in the bodden chain, a sedimentation trap has been constructed in front of the mouth of the River Barthe to protect the Barther Bodden.

The sediment trap principle is based on the assumption that most of the particulate nutrient input flows immediately above the bottom. Checks have shown that they are effective. They need regular attention to ensure that they do not function as active sludge pits, from which phosphate is released rapidly (Baader and Schlungbaum, 1982). The Sanieromat, an automatic suction dredger developed in the GDR, is ideal for this purpose.

- Sedimentation trenches

A sedimentation trench with a width of about 30 m will be laid from the mouth of the River Körkwitz to the opposite side of the bodden in the southern part of Saaler Bodden during 1989/90. The purpose of this is to protect subsequently dredged areas of the Ribnitzer See from renewed sedimentation due to the particle flow from the remainder of the bodden. Like the sediment traps, the sedimentation trench will also require periodic attention.

- Dredging of navigation channels

The removal of nutrient-rich organic sediment to permit navigation naturally also reduces the overall organic load and forms part of the multifactorial rehabilitation programme.

These measures foreseen for the Saaler Bodden are summarized in Fig. 5. Analogous programmes based on scientific studies have also been proposed for the other areas of the bodden chain south of the Darss-Zingst Peninsula (Schlungbaum, 1979; Nausch, 1981). The different measures discussed here as a means of reducing the intrinsic load all form part of a national programme that is being implemented jointly by water management authorities, industry, agriculture, and scientific establishments. They are being coordinated and checked by a group that has been established within the Society for Nature and Environment in the GDR's League of Culture to protect and ensure the continued utilization of the boddens and their surroundings.

While this work to rehabilitate the boddens and their surroundings is being implemented as a joint project, plans are under development to ensure that the different interests in the bodden waters (recreation, fisheries, shipping, nature conservation, etc.) are satisfied. The whole programme is based on long-term targets

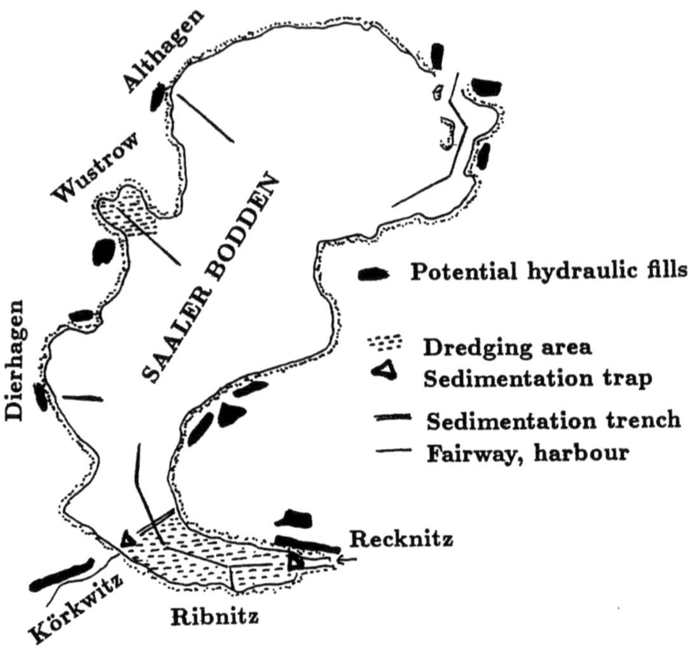

Fig. 5: Multifactorial rehabilitation programme for the Saaler Bodden involving dredging work

and is also part of a larger programme for the protection of the Baltic Sea. The use of the nutrient-rich mud as an agricultural fertilizer after desalination through temporary outdoor storage will contribute to the economics of the project.

REFERENCES

Baader G, Schlungbaum B (1982) Untersuchungen zur Phosphatfreisetzung und zur Sauerstoffzehrung an der Sediment/Wasser-Kontaktzone in flachen Küstengewässern. Acta hydrochim hydrobiol 10: 153-166

Correns M (1976) Charakteristische morphometrische Daten der Bodden- und Haffgewässer der DDR. Vermessungstechnik 24: 459-461

Nausch G (1981) Die Sedimente der Darss-Zingster Boddengewässer - Zustandsanalyse und Stellung im Phosphatkreislauf. Dissertation (A), Wilhelm-Pieck-Universität Rostock

Ripl W (1976) Biochemical oxidation of polluted sediment with nitrate. A new lake restoration method. Ambio 5: 132-135

Schlungbaum G (1979) Untersuchungen über die Sedimentqualität der Darss-Zingster Boddenkette unter besonderer Berücksichtigung der Stoffaustauschprozesse zwischen Sediment und Wasser. Dissertation (B), Wilhelm-Pieck-Universität Rostock

Schlungbaum G (1982) Phosphatsorptionsgleichgewichte zwischen Sediment und Wasser in flachen eutrophenen Küstengewässern. Acta hydrochim hydrobiol 10: 135-152

Schlungbaum G (1989) Steuerung beschaffenheitsbestimmender Parameter in inneren Seegewässern - dargestellt am Beispiel der Darss-Zingster Boddenkette. Wasserwirtschaft-Wassertechnik 39: 26-28

Schlungbaum G, Nausch G (1988) Nutrient turnover at the sediment-water inface in shallow eutrophic coastal waters. Kieler Meeresforschungen, Suppl., im Druck

Schnese W (1978) Produktionsbiologische Grundlagen für die Einbürgerung von Planktonfressern in der Darss-Zingster Boddenkette. III Wiss Konf Physiologie und Biologie von Nutzfischen, Rostock, Sept 1978, Suppl 6, pp 184-190

DENITRIFICATION: ELIMINATION OF AMMONIA AND NITRATE NITROGEN FROM TIDAL SEDIMENTS

Thomas Höpner, Hossein Ebrahimi, Holger Glaus, Annette Hemming, Bernd Krecke,
Ilona Kutsche-Schmietenknoop, Petra Lindenlaub and Elith Wittrock
Institut für Chemie und Biologie des Meeres (ICBM)
Universität Oldenburg, Postfach 2503, 2900 Oldenburg, Federal Republic of Germany

ABSTRACT

Laboratory experiments on denitrification in sediments have been performed. In denitrifying sediment cultures nitrite was formed temporarily. In comparable, but aerobic cultures nitrite was the main product of ammonia nitrification and a nitrite dependent nitrous oxide formation was observed.

1. INTRODUCTION

Denitrification is the only biochemical process which removes nitrogen from aquatic systems. The reaction site is in the narrow suboxic sediment layer in which oxidizable organic carbon (electron donor) and nitrate (electron acceptor) coexist. Nitrous oxide and gaseous nitrogen are the products. Nitrate enters from the aerobic layer or is formed by nitrification. We studied denitrification conditions and rates and nitrification in sediment cultures. The aim was to simulate and to modify the conditions present in the biochemically active sediment layers and to translate the depth profiles of reaction conditions into changes of conditions with time. In our studies, nitrogen-intermediates were considered and various carbon compounds were tested for their function as electron donors. The conversion of ammonia via nitrite to nitrous oxide, which has been described in soils by Seiler and Conrad (1981), could be verified in aerobic sediment set-ups.

2. EXPERIMENTAL

We used a standardized and homogenized sediment in seawater (sediment/water 5:3 w/w, salinity 30 ‰) as a biochemical reaction medium containing uncharacterizec mixed cultures. The sediment had a natural organic carbon and nitrogen content of about 1 and 0.1 %, respectively, and the water phase contained the about 28 mmol/ sulphate typical for sea water. The content of the various sulphur species in the sediment was higher. The set-ups were inoculated from pre-cultures obtained fror mud flat mesocosms which we have maintained for almost four years in a greenhouse like experimental laboratory. Reactions proceeded at room temperature in close serum bottles, open shaken vessels or Warburg flasks. The reaction rates were followed by (i) gas pressure (manometrically or by gas expansion before openin closed reaction vessels), (ii) enzymatic or gas-chromatographic analysis of carbon substrates, (iii) colorimetric analysis of inorganic electron acceptors or products. Denitrification was also studied in sediment-free suspension cultures and with bacterial isolates the identification of which is in progress.

3. ELECTRON DONOR: GLUCOSE

One mmol of glucose was completely degraded anaerobically within three days both with and without 2.5 mmoles nitrate. In the presence of nitrate, N_2-evolution attained its maximum rate on the 5th day and reached 1.25 mmoles after 20 days. Nitrite appeared on the first day, attained a maximum of 1.3 mmoles on the 4th day and could not be found after the 6th day. Without glucose, nitrate decreased by 0.4 mmoles within 7 days and thereafter remained constant, i.e. 0.35 mmoles of N_2 were formed within about 20 days. The maximum of only 3.5 µmoles nitrite was observed between the 2nd and the 8th day.

In the presence of glucose and nitrate, sulphide traces could be detected after 8 days; a significant formation began on the 15th day. Up to the 80th day 1.7 mmoles were formed. Traces of methane could be detected from the 30th day onwards and 0.1 mmoles were formed within 80 days. When nitrate was omitted, sulphide appeared on the 1st day and 2.8 mmoles were formed up to the 80th day. Methane appeared after the 20th day and reached about 1 mmole after 80 days.

Part of these results is summarized in Fig. 1. The conclusions are: A) glucose is converted into a relatively stable intermediate which serves as the direct electron donor. B) Intermediate nitrite is formed faster than it is consumed, and thus accumulates. C) Sulphide formation starts only after nitrate consumption. D) Methane formation coincides with sulphide formation.

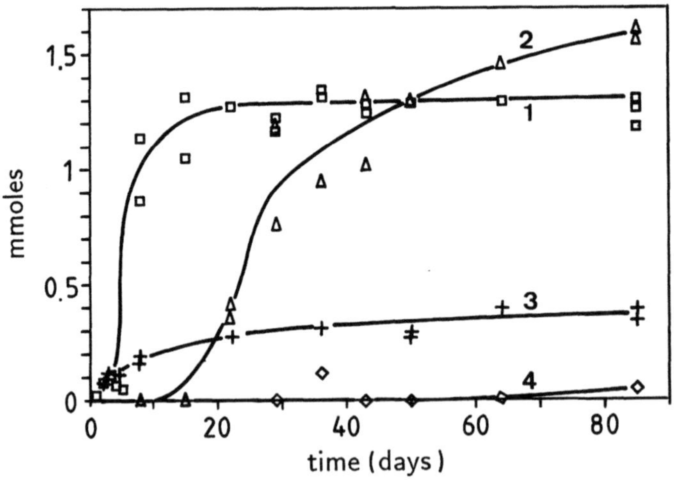

Figure 1: Reactions in 50 g sediment containing 2.5 mmoles nitrate. 1: N_2 in the presence of 1 mmole glucose. 2: Sulphide in the presence of 1 mmole glucose. 3: N_2 without glucose. 4: Methane in the presence of 1 mmole glucose

4. ELECTRON DONORS: METHANE, HYDROCARBONS AND OXYGENATED HYDROCARBON DERIVATIVES

There is no convincing evidence that methane and hydrocarbons can be degraded by nitrate of sulphate in the absence of oxygen. The aerobic and denitrifying degradation rates of cetyl alcohol and palmitic acid, however, are similar (Höpner et al., 1987). Within 13 days 0.4 mmoles methane were degraded aerobically after a lag phase of about 8 days (Fig. 2). We recorded an appearance of about 1.5 mmoles CO_2 in the gas space. After poisoning the reaction with mercuric chloride no methane decrease and no CO_2 evolution was evident.

We conclude that hydrocarbon decomposition is oxygen dependent because the initial mono-oxygenase reaction cannot be replaced by oxygen independent reactions. This also holds for methane. We would not like to extend this statement to aromatics with the same sense of unambiguouity because here our results are less clear cut.

5. ISOLATION AND EXAMINATION OF A STRAIN OF MARINE DENITRIFYING BACTERIA

Attempts to separate the bacterial denitifying activity from the sediment system and to isolate pure bacterial strains revealed two properties. The first was that the bacteria were dependent on the type of surfaces (the original sediment, glowed sand, glass beads) offered, at least after transfer to synthethic media. The second was that glucose was no longer a suitable carbon substrate and had to be replaced by acetate or lactate. The usual strain formed smooth convex slightly brownish colonies on beef extract agar, showed very slightly curved rods of 2 to 4 x 0.5 µm, was motile, gram-negative and did not form spores. In a saturated nitrate containing medium it produced N_2 while N_2O could not be found.

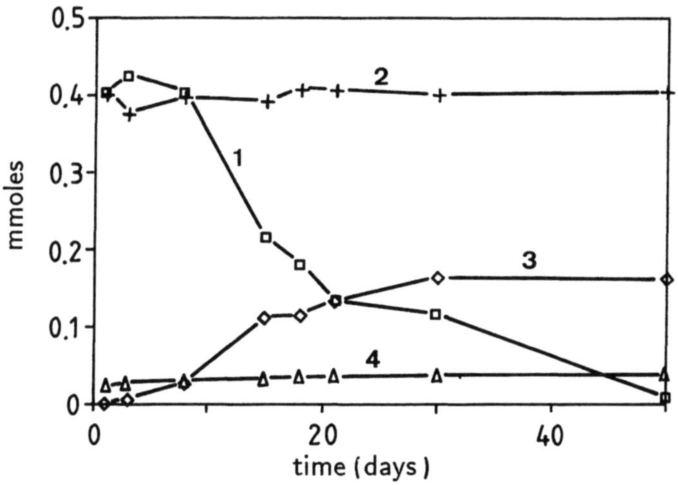

Figure 2: Aerobic methane conversion to CO_2 in 50 g sediment. 1: methane. 2: same as 1, but with $HgCl_2$. 3: CO_2. 4: same as 3, but with $HgCl_2$

If acetate and nitrate were added in the molar ratio of 2.5 : 4 (following the acetate dependent denitrification stoichiometry) to the growth medium, N_2 and nitrite were formed in the molar ratio of 1 : 2 during logarithmic growth. In the stationary phase nitrite decreased while N_2O appeared (Fig. 3). This was more significant when only half of the nitrate was used. If the acetate contents were lowered to half of the stoichiometric amount, the molar N_2/nitrite ratio also reached a value of about 1 : 2 and did not alter after acetate exhaustion.

The conclusion is that the free-nitrogen-intermediate-nitrite attaines considerable concentrations. This observation, gained from the mixed cultures, was confirmed by the pure suspension culture.

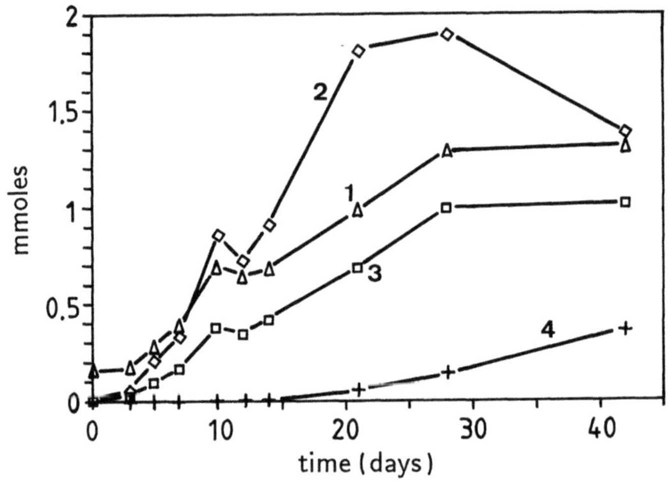

Figure 3: Nitrate conversion in the presence of acetate during growth of a pure denitrifying strain. 1: protein in mg/10 ml. 2: nitrate. 3: N_2. 4: N_2O

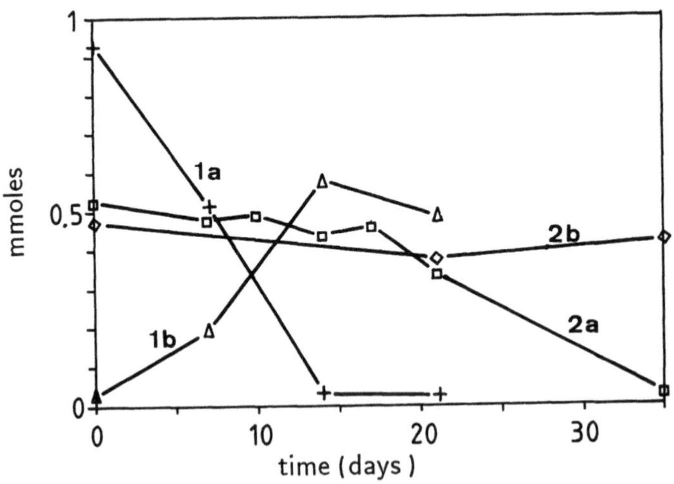

Figure 4: Aerobic conversion of ammonia (curves 1) and nitrite (curves 2) in 30 g sediment. 1a: decrease of ammonia. 1b: formation of nitrite from ammonia.

6. NITRIFICATION

Our experience has been that it was not possible to convert ammonia to nitrate in an intensively shaken aerobic sediment mixture. We found complete disappearance of ammonia up to the 14th day (Fig. 4) starting with 1 mmole ammonia. There was only a small ammonia loss if the mixture had been sterilized. Almost 0.6 mmoles of nitrite appeared within the same period of time, and no nitrate was found. When 0.6 mmoles nitrite were added to the mixture, instead of the ammonia, a complete decay was observed within 35 days. In this case the nitrite loss from a sterilized set-up was also small. While N_2O evolution from ammonia was slow and seemed to show a long lag phase, N_2O evolution from nitrite started immediately (Fig. 5). Note that the amounts formed are far below any stoichiometric expectation.

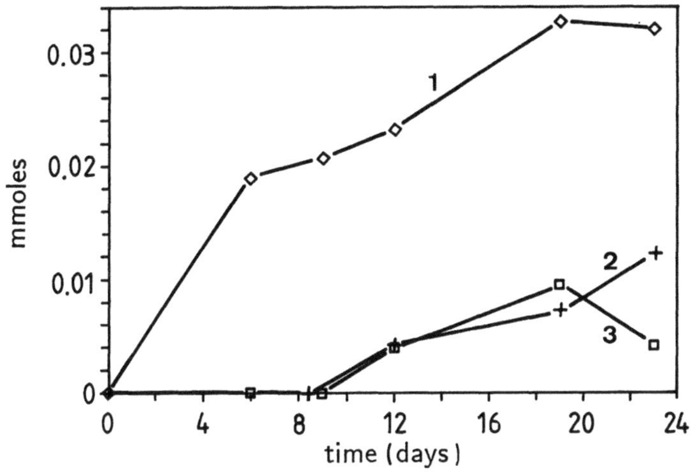

Figure 5: Aerobic N_2O evolution in 20 g sediment. 1: N_2O from 0.5 mmoles nitrite. 2: N_2O from 1 mmole ammonia. 3: N_2O without ammonia or nitrite (possibly from endogenous nitrogen)

The aerobic, shaken sediment set-ups did not show any visible clod formation, and there was no hint of the occurrence of suboxic niches. Nevertheless, nitrate could not be observed, and nitrite behaved as expected for both suboxic or anoxic conditions. We must conclude that in tidal sediments ammonia-nitrogen can be converted into gaseous nitrogen. In our shaken mixture it can be assumed that the oxygen support was better than in situ in a non-mixed sediment; even so the reaction does not run to the most oxidized end product, but veers away from the expected direction and results in the most stable, but less oxidized product.

7. DISCUSSION

When enrichment cultures were inoculated from our tidal flat mesocosm, the catalysis of various nitrogen conversion reactions by microbiological activity was shown. When mixed cultures were used, the natural organic carbon content of our sediment samples was mineralized both aerobically and anaerobically. Glucose could be used as a model of the carbon substrate. With glucose, denitrification was observed, even a long time after complete disappearance of the glucose. After the isolation of pure strains, glucose was no longer a suited substrate and was replaced by acetate or lactate. It may be assumed that glucose was converted into products similar to acetate or lactate and that these served as electron donors for denitrification. Sulphate reduction started only after nitrate consumption, but in the presence of small nitrite concentrations. Methane formation started after a sulphate decrease from 28 to, at least, approximately 10 or less moles/l.

We found that nitrite was a free intermediate of both denitrification and nitrification. Thus we propose that nitrite is the preferable intermediate between nitrification and denitrification, over nitrous oxide (as assumed by Seiler and Conrad, 1981). Under our conditions of vigorous aerobic shaking, the conversion of nitrite into nitrate was not observed; instead of this nitrite was converted to N_2O. The conclusion is that not only nitrate but also ammonia can be converted into gaseous nitrogen under oxic conditions and that nitrogen elimination from the sediment is not restricted to nitrate present in or immediately near to anoxic zones.

Thermodynamic aspects seem suited when explaining the role of nitrite as an intermediate of nitrification and denitrification. The free energy of nitrite oxidation to nitrate by oxygen is relatively small, i.e. - 64 kJ/mole. In the case of a reductant of the reducing power of hydrogen, nitrite conversion to N_2O has a much higher free energy, i.e. - 453 kJ/mole. The subsequent reduction of N_2O to N_2 (removing N_2O from the nitrite reduction equilibrium) is also favoured: - 341 kJ/mole. This means that the conversion path nitrite to N_2O to N_2 is preferable over the path nitrite to nitrate in the case of ambiguous conditions ("suboxic"). Because the ammonia oxidation by oxygen to nitrite has a free energy of - 275 kJ/mole, the ammonia conversion to nitrite is more likely under ambiguous conditions than the conversion of nitrite to nitrate, so that a nitrite accumulation is not unexpected. The energy values quoted were adapted from Thauer et al. (1977).

REFERENCES

Seiler W, Conrad R (1981) Mikrobielle Bildung von N_2O (Distickstoffoxid) aus Mineraldüngern - ein Umweltproblem ? Forum Mikrobiologie 6/81: 322-328

Thauer R K. Jungermann K, Decker K (1977) Energy Conservation in Chemotrophic Anaerobic Bacteria. Bacteriological Reviews 41: 100-180

Höpner Th, Dalyan U, Kanje K, Kant U, Kiesewetter K, Koopmann B, Michaelsen M, Kutsche-Schmietenknoop I, Weinert R (1987) Untersuchungen mit dem Ziel des Vorschlags biologischer Maßnahmen bei Ölverschmutzungen der Wattoberfläche. Abschlußbericht zum BMFT-Vorhaben 01 ZV 85070 an die DFVLR. Oldenburg, 157 S

THE BED SHEAR STRESS OF AN ANNULAR SEA-BED FLUME

Jerome P.-Y. Maa
Asst. Prof., Virginia Institute of Marine Science/School of Marine Science
College of William and Mary, Gloucester Point, VA 23062, USA.

INTRODUCTION

An accurate estimation of erosion and deposition rates is very important in the study of fine, cohesive sediment transport. In situ sea-bed flume tests would be the most reliable way to obtain these rates. Young and Southard (1978) used a sea-bed flume to study the incipient motion of sandy sediments. Recently, a more sophisticated sea-bed flume, the SEADUCT, has been deployed at the HEBBLE site (Nowell et al. 1985) to measure the erosion rate.

At the Virginia Institute of Marine Science (VIMS), we initiated the annular sea-bed flume to study both erosion and deposition rates. This paper presents the results of the first phase study: to understand the hydrodynamic characteristics of the proposed flume prior to fabrication. The objectives are to find the optimal dimensions for this flume and to examine the possible performance.

THE ANNULAR SEA-BED FLUME

The proposed annular sea-bed flume will function similarly to the annular flume used at Case Western Reserve University (Fukuda 1978), and The Carousel installed at Hydraulics Research Limited in England. However, the design will be completely different to allow the flume to be operated on the floor of shallow water environments. Figure 1 shows the conceptual drawing. Two steel cylindrical shells with diameters 2.1 m and 2.5 m, form the inner and outer walls of the flume. This flume has no bottom and will be lowered from a boat to penetrate into the sea-bed. An annular ring on top of the flume will rotate to induce a primary flow in the angular direction and a secondary circulation as well. The bed shear stresses can be calculated based on the flow velocities. The ring will be driven by a variable speed DC motor via a gear train. An on-board DC motor controller will be used to control the rotation speeds.

The choice of an annular flume is based on the following reasons: (1)an annular flume does not have a beginning or an end in the flume, and it does not have a pump to break the sediment flocs; therefore the aggregation process will not be interrupted; (2)because of the secondary flow, the suspended sediment will be fairly uniformly distributed in the entire cross section and no stratification problem will occur in the flume. The sediment sensors, which are used to obtain the total suspended sediment mass, do not need to be installed very close to the water-sediment interface. Only one or two sensors can be used to accurately obtain the total suspended sediment mass; (3)this study is concentrated on the processes that take place

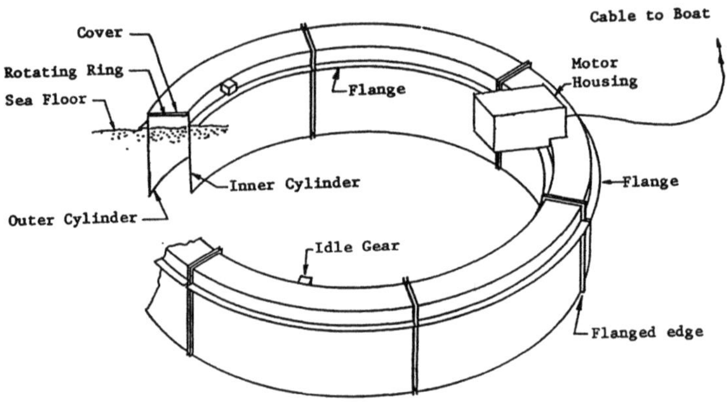

Fig. 1. Conceptual Drawing of the Sea-Bed Flume. The lifting device and the major bearing plate are not shown.

at the water-mud interface. We are interested in observing how the suspended sediment changes to the bed material, and vice versa. The settling, dispersion, and transport processes are not the goal.

The secondary flow also induces radial bed shear stress and makes the interpretation of data obtained more complicated. However, the significance of this radial component compared with the tangential bed shear stress is not well documented. Thus, one of the goals of this study is to address this potential problem.

GOVERNING EQUATIONS

The simplified Reynolds equations (Hinze 1959) with the Boussinesq's eddy viscosity model in cylindrical coordinates are given as follows. Tangential gradient terms and small terms have been omitted for this axial symmetrical flow.

$$\frac{\partial U_r}{\partial t} - \frac{U_\phi^2}{r} = - \frac{\partial p}{\rho \partial r} + \nu\left(\frac{\partial^2 U_r}{\partial r^2} + \frac{\partial U_r}{r \partial r} + \frac{\partial^2 U_r}{\partial z^2} - \frac{U_r}{r^2} \right) \quad (1)$$

$$\frac{\partial U_\phi}{\partial t} = - U_r \frac{\partial U_\phi}{\partial r} - U_z \frac{\partial U_\phi}{\partial z} + \frac{U_\phi U_r}{r} + \frac{\partial}{\partial r}\left(\epsilon_r \frac{\partial U_\phi}{\partial r}\right) + \frac{\partial}{\partial z}\left(\epsilon_z \frac{\partial U_\phi}{\partial z}\right) - \frac{2\epsilon_r}{r}\frac{\partial U_\phi}{\partial r} \quad (2)$$

$$\frac{\partial U_z}{\partial t} = - g - \frac{\partial p}{\rho \partial z} + \nu\left(\frac{\partial^2 U_z}{\partial r^2} + \frac{\partial U_z}{r \partial r} + \frac{\partial^2 U_z}{\partial z^2} \right) \quad (3)$$

The continuity equation is

$$\frac{\partial U_r}{\partial r} + \frac{U_r}{r} + \frac{\partial U_z}{\partial z} = 0 \quad (4)$$

where U_r, U_ϕ, and U_z are local mean velocities in the radial, tangential, and vertical direction, respectively; r, ϕ, and z are coordinates in radial, tangential, and vertical directions, respectively, ν is the kinematic viscosity of the fluid, g is gravity, ρ is fluid density, and ϵ_z and ϵ_r are eddy viscosity coefficient in the vertical and radial direction, respectively. Turbulence in the secondary flow was neglected because of the relatively insignificance. The boundary conditions are:

$U_\phi = r\Omega$ at $z = h$ ....(5)

$U_\phi = 0$ at $z = 0$, $r = d_1/2$, and $r = d_2/2$ ....(6)

$U_z = U_r = 0$ at $z = 0$, $z = h$, $r = d_1/2$, and $r = d_2/2$ ....(7)

where Ω is the rotation speed of the annular ring, $z = 0$ stands at the channel bed, and $z = h$ stands at the channel top, d_1 and d_2 are diameters of the inner and outer walls, respectively.

Inasmuch as the flow is axially symmetric, Eqs. 1 to 4 can be solved in a two dimensional grid system that fits the rectangular cross section of the flume. Mixing length theory was used to describe the eddy viscosity profiles both in the vertical and radial directions. An implicit numerical model was developed to solve these equations. Details can be found elsewhere (Maa, in review).

RESULTS AND DISCUSSION

The model simulated tangential velocity profiles and average bed shear stresses have been compared with those measured by Fukuda (1978) and agreed well (Maa, in review). Series calculations have been made to compare the flow fields and the bed shear stresses for various flume dimensions. The final selected flume dimensions (based on the requirements of a reasonable uniform bed shear stress, a stable flow field, and good maneuverability) are $d_1 = 2.1$ m, $d_2 = 2.5$ m, and $h = 0.1$ m.

Figure 2a shows the contour plots of the tangential velocity in the entire cross section of the flume at $\Omega = 7$ rpm. It reveals that despite the linearly increased tangential velocity at the top boundary (Eq. 5), the tangential velocities near the bottom are reasonably constant except near the two corners. The small tangential velocity near the two bottom corners is not favorable. Unfortunately, it is not avoidable for any kind of flume.

Figure 2b shows the vector plots of secondary flow. Notice that the maximum secondary flow velocity is about 4.5 cm/s near the top. Near the bottom the maximum radial velocity is about 2 cm/s. They are about 15% of the near-by tangential velocities. These secondary flow velocities is much larger than the settling velocity of cohesive sediments (e.g., kaolinite suspensions have a settling velocity about 0.1 -0.3 mm/sec). As a consequence, the suspended sediment concentration would be nearly uniform in the entire flume. The settling or dispersion process can not be studied by using this flume because of the relatively strong secondary flow.

Figure 3a shows the profiles of tangential, radial, and total bed shear stress. It indicates that the radial component is not significant, only about 15% of the tangential component). The dotted line in this figure shows the average bed shear stress. It represents the entire bed shear stress reasonably well except near the two side boundaries. The area affected by the side boundaries is about 15% of the total bed area and the influences of this may be neglected.

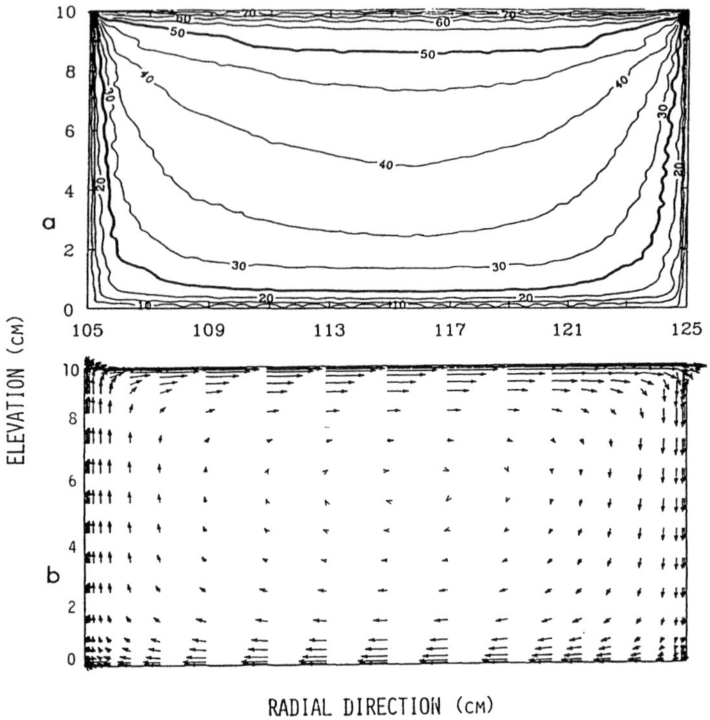

Fig. 2. Flow Field in the Proposed Sea-Bed Annular Flume with Ring Speed = 7 rpm. (a)Contours of Tangential Velocity in cm/s; (b) Vector Plots of Secondary Circulation. Maximum secondary velocity is 4.5 cm/s: ——— near top.

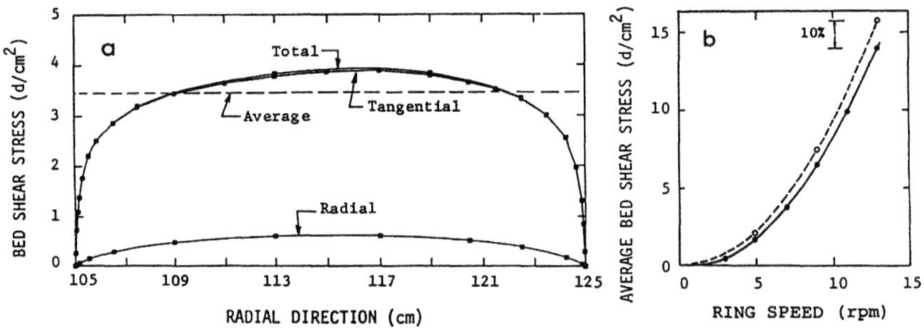

Fig. 3. Bed Shear Stress in the Proposed flume. (a) Profiles for Ring Speed = 7 rpm; (b)Average Bed Shear Stresses versus Ring speeds.

Partheniades (1986) stated that bed shear stress is the most important parameter that controls the deposition and erosion of cohesive sediments. The secondary circulation, which contributes little to the total bed shear stress, would hardly affect these two processes although it circulates the sediment particles and flocs. The secondary circulation, however, does not prevent sediment particles and flocs to contact with the bed. Thus by using this kind of flume, we are still able to study the processes that happen at the water-mud interface. In other words, the deposition process in this flume is not for describing the settling of sediment particles and flocs. It describes the process that changing suspended sediment to bed sediment. For an erosion process, this flume is used to observe the changing of the bed material to suspended sediment.

The solid line in Fig. 3b shows the calculated average bed shear stress versus possible operational ring speed for the designed channel depth, 10 cm. To answer an important question, "If the channel depth is not constant along the flume because of any reason, could we still obtain a reasonably uniform bed shear stress in the entire flume ?" we need a three dimensional model. Here we can only answer this question in a crude manner because of our model limitations. For example, we have the following case: the channel depth is 10 cm at one side and 8 cm at the other side. We may run the model for these two depths. The dashed line in Fig. 3b shows the average bed shear stress for the 8 cm channel depth. It indicates that a 10% difference could result because of the change of channel depth. This crude estimation is encouraging because of the small difference.

CONCLUSIONS

A numerical study of the flow field and bed shear stress of an annular sea-bed flume has been made. The tangential flow, the secondary circulation, and the associated bed shear stresses for the proposed sea-bed flume are solved for various operating conditions. Numerical experiments to find the optimal dimensions of the flume and simulate possible situations that might be encountered in fields reveal that reasonably uniform bed shear stress in the entire flume can be achieved. Because of the secondary circulation, the dispersion and settling of sediment can not be studied by using this flume. However, this flume would be indeed an excellent tool to observe the erosion and deposition processes that occurred at the water-mud interface.

ACKNOWLEDGMENTS

Sincere appreciation goes to Drs. R. J. Byrne and L. D. Wright for their continuous support and encouragement. This paper is a contribution of the Virginia Institute of Marine Science, No. 1506.

REFERENCES

Fukuda MK (1978) The Entrainment of Cohesive Sediments in Fresh Water, Ph.D. Dissertation, Case Western Reserve Univ., Cleveland, Ohio.

Hinze JO (1959) Turbulence, An Introduction to its Mechanism and Theory. McGraw-Hill.

Maa JPY (in review) Modeling The Hydrodynamic Characteristics of an Annular Sea-Bed Flume. Submitted to J of Hydraulic Engng., ASCE

Nowell ARM, McCave IN, Hollister CD (1985) Contributions of HEBBLE to Understanding Marine Sedimentation. Marine Geology 66:397-409

Partheniades E (1986) A Fundamental Framework for Cohesive Sediment Dynamics. in: Mehta AJ (ed) Estuarine Cohesive Sediment Dynamics, Lecture Notes on Coastal and Estuarine Studies, No. 14, Springer-Verlag, New York.

Young RN, Southard JB (1978) Erosion of Fine-Grained Marine Sediment: Sea-Floor and Laboratory Experiments. Geological Society of American Bulletin 89:663-672

PARTICULATE CONTAMINANT TRANSPORT IN RIVERS
- MODELLING OF NEAR BED EXCHANGE PROCESSES -

B. Westrich, J.M. Ham, Y. Kocher
Institut für Wasserbau, Universität Stuttgart
Pfaffenwaldring 61, 7000 Stuttgart 80, Federal Republic of Germany

SEDIMENTS AND WATER QUALITY

In coastal and inland rivers with high water pollution the transport, sedimentation and resuspension of fine sediments becomes important for the water quality and its management. Deposition and resuspension of polluted sediments are the most important sink and source terms in a contaminant budget. Maintenance dredging and flood erosion may cause remobilization of harzardous sediment-bound substances. Pollutant transport models are needed for short- and long-term prediction of water quality. This investigation concentrates on the hydromechanical aspects of the sedimentation, the particle exchange and mixing within the upper bed layer depending on flow, sediment and bed conditions.

TURBULENCE ENERGY AND DEPOSITION CRITERIA

Experiments with non-cohesive sediment (20 µm < d < 200 µm) have shown that there is a critical suspended sediment concentration $c_{crit,s}$ corresponding to a critical bed shear $\tau_{crit,s}$ at which sedimentation starts. The critical condition can be described by a parameter which represents the ratio of turbulent energy production per unit time to the rate of work to be done to keep the particles in suspension. This ratio can be expressed as

$$c_{crit,s} = k \frac{\tau_0 U}{(\rho_s - \rho)\, g\, h\, v_s} \qquad (1)$$

with flow depth h, settling velocity v_s, bottom shear τ_0 and mean flow velocity U. As the energy containing turbulent elements are strongly related to the large-scale boundary roughness, the bed forms have a significant influence on the deposition criteria. The magnitude of k and hence, $c'_{crit,s}$ is smallest for a flat bed, and $c''_{crit,s}$ is highest for large bed deformations such as dunes (Fig. 1). If the turbulence is produced by wall shear there is no special effect on the characteristics of turbulence. However, if bed dunes are established the free shear in the wake of the dunes enhances the turbulent mixing and hence the sediment concentration. In alluvial channels with unlimited sediment supply the equilibrium concentration c_{eq} is established which is the maximum possible concentration associated with an intensive sediment exchange at the bed where sedimentation and erosion are balanced.

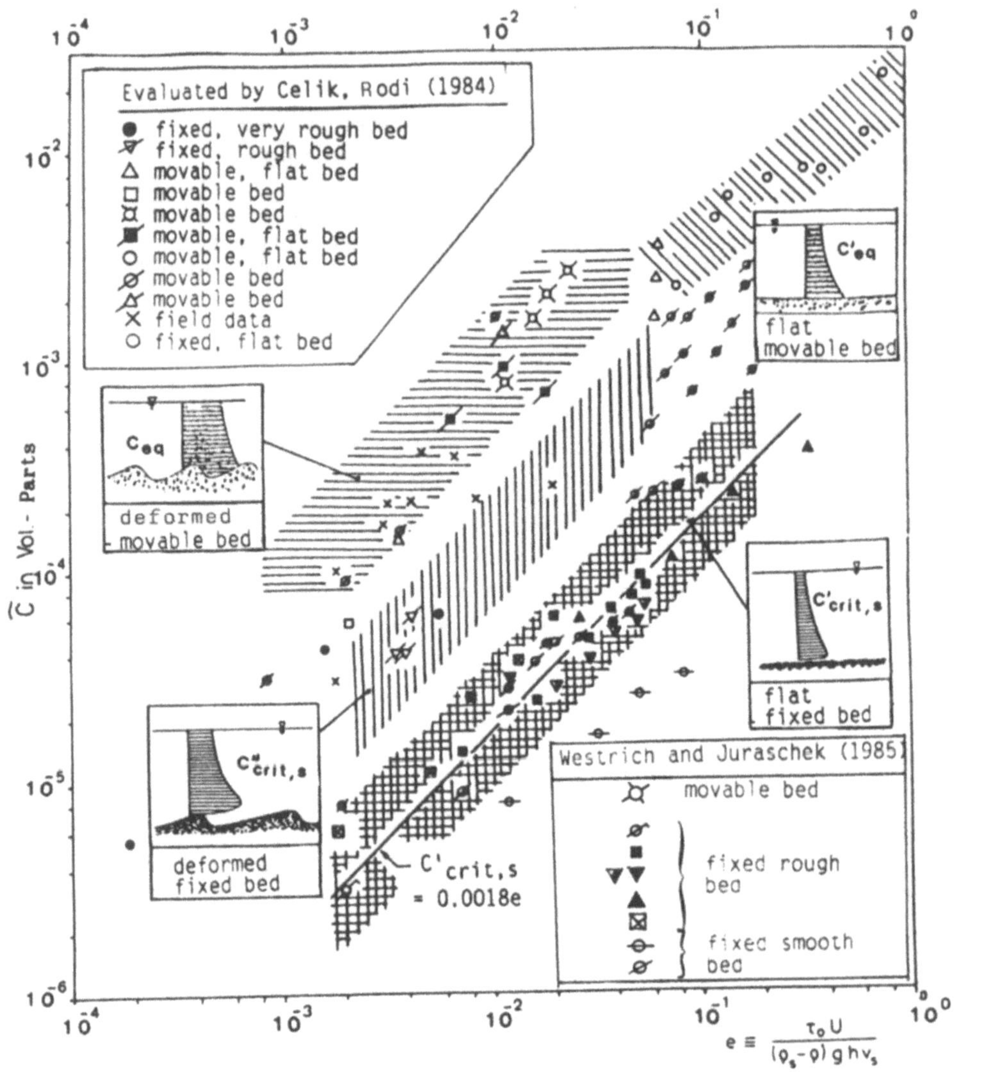

Figure 1: Criteria for the beginning of sedimentation. For symbols see Appendix

SEDIMENT INTERACTION WITH THE RIVER BED

If the bottom shear exceeds the critical erosion, shear sediment mixing in the upper bed layer takes place. The length and time scale of the sediment exchange (mixing depth m_{eq}; exchange time t_e) can be estimated by making use of the kinematics of bed forms (Fig. 2). The actual depth of particle penetration depends on the mobility of the dunes and the time of contact t*. The maximum possible penetration after a long enough time of contact is limited by the dune height. In inland rivers the maginitude of t* is immediately related to the length of the contaminant cloud and the flow velocity. However, in tidal rivers the residual flow velocity is significant for the time of contact. In inland rivers there are extended backwater flow sections without erosion and long consolidation time, whereas in tidal rivers sedimentation and resuspension occur in a short regular cycle causing less consolidation but more intensive sediment mixing at the bed.

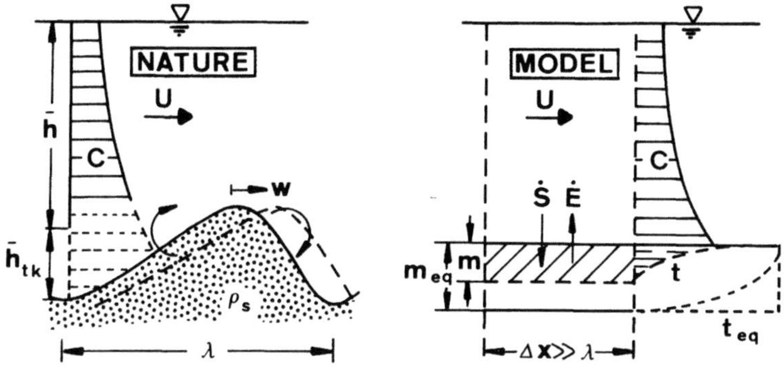

mean residence time of all transported particles within Δx: $\bar{t}$		estimation of exchange time of sediments in the movable bed layer: t_{eq}
$\frac{\Delta x}{u}$	immobile bed	very short exchange time, exchange only in a very thin top layer
$\frac{\Delta x}{w} \cdot \frac{1 + \frac{\rho_s}{c} \cdot \frac{h_{tk}}{h}}{1 + \frac{\rho_s}{c} \cdot \frac{h_{tk}}{h} \cdot \frac{w}{u}}$	suspended and bed load	short to long exchange time, depending on vertical sediment flux and dune migration
$\frac{\Delta x}{w}$	only bed load	very long exchange time, depending only on dune migration

partial mixing: $t^* < t_{eq}$; $m < m_{eq}$

total mixing: $t^* \geq t_{eq}$; $m = m_{eq}$

$\tau > \tau_{crit,E}$; $\tau < \tau_{crit,s}$

	flat bed	ripples	dunes
mixing depth scale m_{eq}	≈ (1-2)d (millimeter)	≈ (10-100)d (decimeter)	≈ (0.2-0.3)h (meter)
exchange or mixing time scale t_{eq}	≈ 0 (minutes)	$0 < t_{eq} < \frac{\lambda}{w}$ (hours)	$\approx \frac{\lambda}{w}$ (days)

Figure 2: Length and time scale for sediment mixing at the bed. For symbols see Appendix

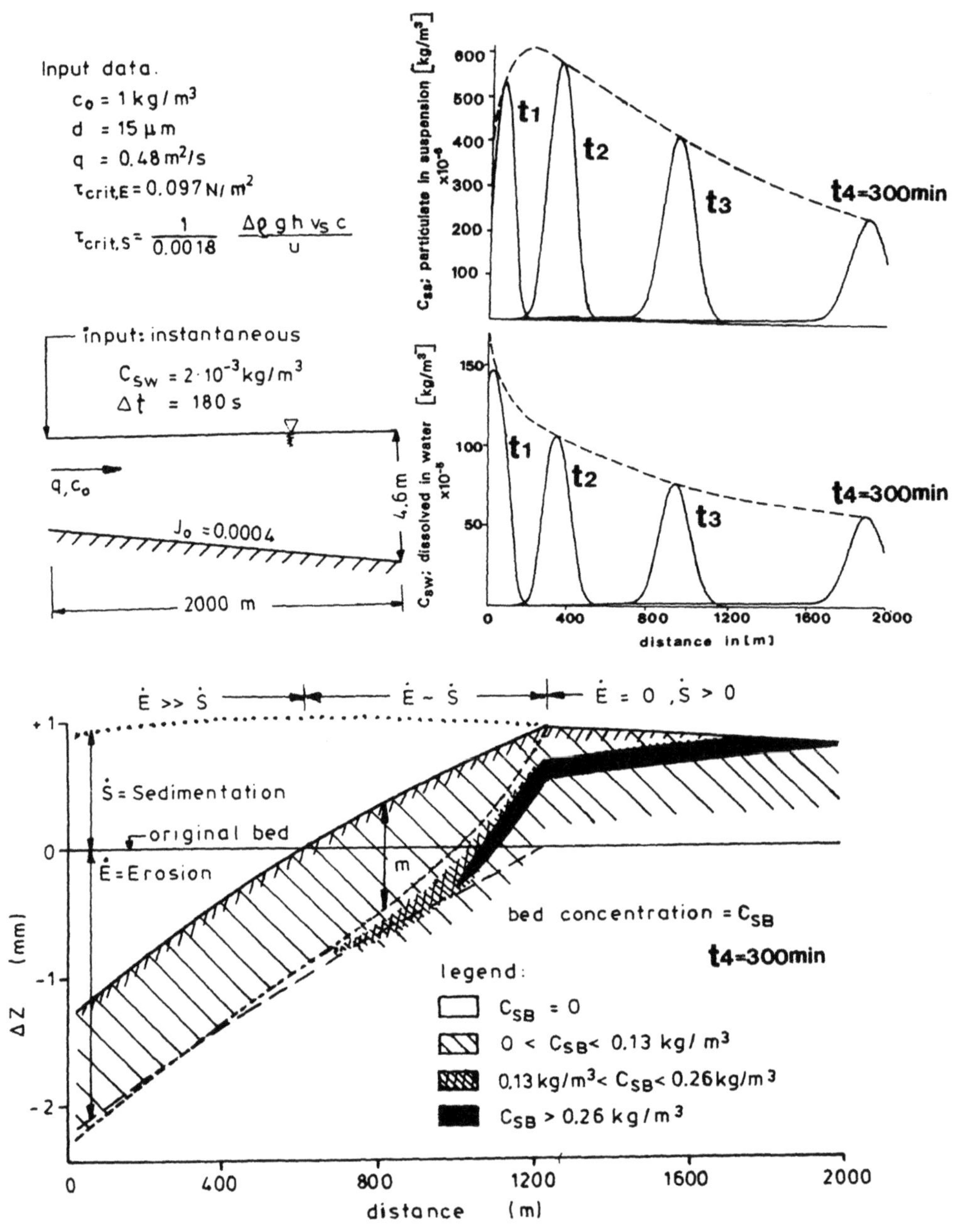

Figure 3: Unsteady contaminant transport in a backwater flow section. For symbols see Appendix

CONTAMINANT TRANSPORT MODEL

The unsteady transport of a dissolved and particulate contaminant cloud in a river is simulated in a one-dimensional transport model which accounts not only for the exchange process at the river bed but also for the lateral diffusive transport to near bank groyne cells with a large sediment trapping efficiency. To illustrate the different boundary effects on the fluvial transport some numerical results are shown for the transport through a backwater flow section in a river with an instantaneous input of dissolved adsorptive contaminants (Fig. 3). The results show the longitudinal profile of the concentration of absorbed and dissolved contaminants in the water column as well as the particulate contaminants in the top bed layer.

APPENDIX

List of symbols

C	particulate concentration	kg/m³
C_{eq}	equilibrium concentration	kg/m³
$C_{crit,s}$	critical sedimentation concentration	kg/m³
C_{sb}	pollutant concentration in the upper bed layer	kg/m³
C_{ss}	pollutant concentration absorbed on particles	kg/m³
C_{sw}	pollutant concentration dissoloved in water	kg/m³
d	grain diameter	µm
$\dot{E}$	erosion rate	kg/m²s
g	acceleration due to gravity	m/s²
h	flow depth	m
$\bar{h}$	mean water depth	m
$\bar{h}_{tk}$	mean dune height	m
J_o	bed slope	-
k	empirical constant	-
K_s	sediment roughness	mm
m	exchange depth at time t	m
m_{eq}	exchange depth in the equilibrium stage	m
$\dot{S}$	sedimentation rate	kg/m²s
t	time	s
$\bar{t}$	mean residence time of all moving particles	s
t_i	time elapsed from the beginning of pollutant spill	min
t_{eq}	time required for equilibrium mixing	s
t^*	time available for polluted sediments to be mixed with the bed	s
U	velocity of a uniform flow	m/s
v_s	settling velocity of sediment particles	m/s
w	migration velocity of bed dunes	m/s

λ	characteristic length of bed dunes	m
ρ	water density	kg/m^3
ρ_s'	sediment bulk density	kg/m^3
ρ_s	particulate density	kg/m^3
τ_0	bottom shear	kg/ms^2
$\tau_{crit,e}$	critical erosion bed shear	kg/m^2s^2
τ	critical sedimentation bed shear	kg/ms^2

REFERENCES

Celik I and Rodi W (1984) A deposition-entrainment model for suspended sediment transport. SFB 210 /T/ 6, Universität Karlsruhe

Salomons W, Förstner U (1984) Metals in the Hydrocycle. Springer, Berlin

Westrich B, Juraschek M (1985) Flow transport capacity for suspended sediments. Intern Congr Hydr Research, IAHR, Melbourne

Westrich B (1988) Fluvialer Feststofftransport - Auswirkungen auf die Morphologie und Bedeutung für die Gewässergüte. Oldenbourg Verlag, München

CHAPTER VII

MONITORING

LOWER WESER MONITORING AND MODELLING

A. Müller
Institute of Physics, GKSS Research Centre Geesthacht, D-2054 Geesthacht, FRG
M. Grodd
Brake Water Authority, D-2880 Brake, FRG
H.-P. Weigel
Bremen Water Authority, D-2800 Bremen, FRG

ABSTRACT

Based on a brief introduction to the Lower Weser River (UW) hydrography, the UW monitoring systems and the UW numerical model are outlined. The results of both are discussed with a view to the water quality management of this river.

1. LOWER WESER HYDROGRAPHY

The tide-influenced section of the Weser River between the weir at Bremen-Hemelingen and the Weser mouth into the North Sea downstream of Bremerhaven, 85 km in length, is called the Lower Weser River (Fig. 1), [1] to [8]. The UW drainage area is 7900 km², about 17.2 % of the entire Weser River catchment area. The main UW tributaries Ochtum, Lesum, Hunte, Geeste are tide-influenced and drain an area of about 6300 km². The long-term averages of annual mean, lowest and highest freshwater discharges at the weir are 326 m³/s, 120 m³/s, 1181 m³/s, respectively. The mean freshwater input from the tributaries is about 60 m³/sec. The mean ebb and flood passages at Bremerhaven are about 6500 m³/s. The freshwater limit of the UW under normal flow conditions is near Nordenham, about 28 km upstream of the river mouth.

The UW is the navigable waterway to the Bremen harbour. Therefore, over the last 100 years it has been continuously developed by water engineering to fullfill the requirements of navigation. Today the UW has a channel-like character and downstream from Bremen is artificially kept to a depth of 9 m below minimum water level in the navigable area. In general the bottom sediment is sandy with the exceptions of the turbidity zone near Nordenham and some very shallow or harbour areas where mud can occur. The UW tides are semidiurnal (M_2 + S_2) and asymetrical. The mean tide parameters for Bremerhaven and Bremen are given in Tab. 1. The water engineering alteration made to correct the UW have had a drastic effect at Bremen. The tidal currents in the UW, with peak values of about 1 m/s caused by the high tidal range, produce strong turbulence; the UW is therefore generally well mixed.

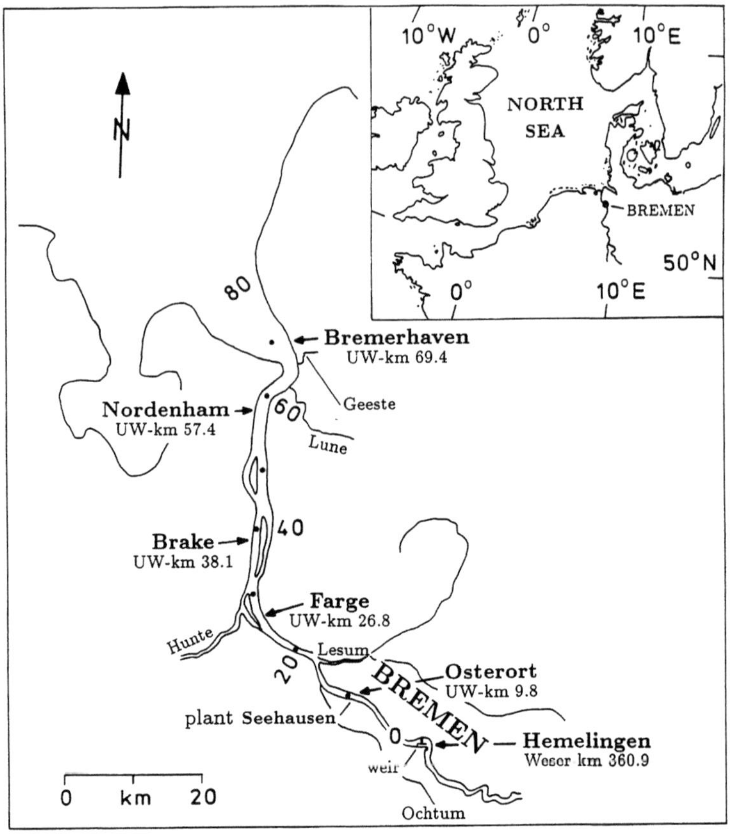

Figure 1: The Lower Weser River (← positions of the bank-side monitoring stations; 40: Lower Weser km 40, abbr. UW-km 40)

Typical water temperature values for the UW are 18 °C in summer and 5 °C in winter. The high salinity values of the freshwater discharge at the weir, which range between 0.6 to 2.0 ‰ depending on the volume of freshwater discharge, are a result of the potash mining and treatment processes in the catchment area of the

Table 1: Tide parameters before the water engineering UW correction and present day values (data before 1887 from [9])

	Bremerhaven		Bremen	
	before 1887	now	before 1887	now
mean tidal range [m]	3.5	3.6	0.2	4.0
mean ebb period [h:min]	6:28	6:30	9:35	7:30
mean flood period [h:min]	5:57	5:55	2:50	4:55

		before 1887	now
mean run time of the tide wave from Bremerhaven to Bremen [h:min]	H.W.	3:52	1:50
	L.W.	6:59	2:54

Weser headwaters; salinity values before mining were about 0,1 ‰. Salinity values at Bremerhaven are typically 4 ‰ at the end of the ebb stream and up to 20 ‰ at the end of the flood stream. The UW is a turbid water; suspended matter concentrations downstream of the weir are about 30 to 40 mg/l and in the Bremerhaven area, depending on the position of the turbidity zone and the tidal phase, recorded values range between 30 to 1500 mg/l. The penetration of sun light is between 1 to 1.5 m below the water surface.

The UW receives urban and industrial sewage and waste heat from power plants. Tab. 2 shows that the maximum input from the most important waste heat producers can reach nearly 5000 MW.

Table 2: Important waste heat discharges into the UW (data from the newest approval notices or [1], [3]; KW: power plant)

UW-km	discharger of waste heat	maximum intake of cooling water [m³/s]	maximum warm up span [°C]
7.9	KW Hafen	17.2	7.9
11.2	Klöckner	15.0	10.0
11.2	KW Mittelsbüren	17.1	8.3
25.9	KW Farge	12.2	8.5
51.4	KW Unterweser	61.0	10.0

2. LOWER WESER MONITORING

Movements of water and water constituents in tidal rivers are characterized by pronounced spatial heterogeneities and temporal variabilities, which are caused and directed by the tidal wave at the river mouth and by the run-off of headwater, therefore it is no simple monitoring strategy for tidal rivers. In practice, tidal river monitoring is determined by the questions to be adequately answered, the circumstances in the river section to be monitored and last but not least by the possibilities of available monitoring equipment and the expenditure that can be reasonably taken. There are three different types of monitoring for the UW

- discharge of effluent monitoring
- monitoring of water quality in the river
- special field investigations.

Discharge of effluent monitoring

Monitoring of effluents is done by the companies operating the outlet structures and, independently of them, by the official controlling agencies. The purpose of this monitoring is to obtain information on the type and amount of water constituents discharged with the effluent and to prove that the limits specified in the approval notices are kept. Results of effluent monitoring are also used to assess the impact of isolated effluent discharges on the river water quality.

Depending on the type of effluent, monitoring is done by determining the discharged sewage volume, periodically taking sewage samples for laboratory analysis and/or if possible, by on-line analyses. In Tab. 3 typical monitoring results are given for the most important isolated sources of nutrient discharges.

Table 3: Important BOD_5 and nutrient discharges into the UW (typical values, evaluated from the monitored discharge of effluent data; KA: sewage treatment plant)

UW-km	discharger of effluent	waste water m^3/s	BOD_5 mg/l	NH_4-N mg/l	NO_2-N mg/l	NO_3-N mg/l	org. bound N mg/l	O-PO_4-P mg/l	P_{tot} mg/l
				nutrient contents					
8.8	KA Seehausen	1.7	7.7	38.8	0.14	8.4	5.2	10.1	12.9
11.9	KA Delmenhorst	0.19	7.9	37.8	0.31	3.1	8.0	1.1	1.4
17.0	KA Lemwerder	0.012	7.9	16.8	0.62	20.7	8.0	9.4	12.0
21.5	Bremer Wollkämmerei	0.031	4.5	1.4	0.09	15.4	31.6	1.0	1.3
25.1	KA Farge	0.23	12.4	31.3	0.62	8.4	10.5	6.0	7.8
42.4	Fettraffinirie Brake	0.11	34.0	-	-	-	-	-	-
43.5	KA Brake	0.035	3.0	3.7	0.04	3.0	8.0	2.1	2.7
59.9	KA Nordenham	0.040	16.5	30.8	0.03	0.67	8.0	2.3	2.9
63.5	KA Bremerhaven	0.53	9.4	57.4	0.21	0.75	7.5	0.4	0.6

Monitoring of water quality in the river

The systematic monitoring of water quality, adhering to standardized guidlines, has been carried out on the UW since 1979 by the appropriate water authorities. The objectives of this monitoring are to document the actual water quality situation in the tidal river, to detect long-term changes (trends) in water quality and to obtain better information about the sources, distribution and sinks of water constituents as well as to find out environmental changes. Data obtained from this monitoring are the basis for all environmental impact assessments on the tidal river as well as for the water management plans and decisions concerning tidal river utilization and engineering works.

Fig. 1 displays the positions of the monitoring station Hemelingen above the weir and the five bankside monitoring stations on the UW. In addition, a meteorological station is located at Osterort. Fig. 2 schematizes the function of such a monitoring station. The parameters conductivity, oxygen content, temperature, pH-value are recorded on-line. The parameters BOD_5 and mercury content are obtained from water samples which are taken every 14 days and analysed as soon as possible after sampling. From samples collected and mixed over two weeks the following parameters are determined: contents of potassium, sodium, calcium, magnesium, sulphate, chloride, hydrogen-carbonate, total phosphorus, ortho-phosphate-phosphorus, ammonium-nitrogen, nitrate-nitrogen, COD, DOC, nickel, zinc, copper, chromium, lead, cadmium, iron, manganese and arsenic. In Tab. 4 mean annual values of the headwater nutrient and heavy metal load, evaluated from the data of the monitoring station at

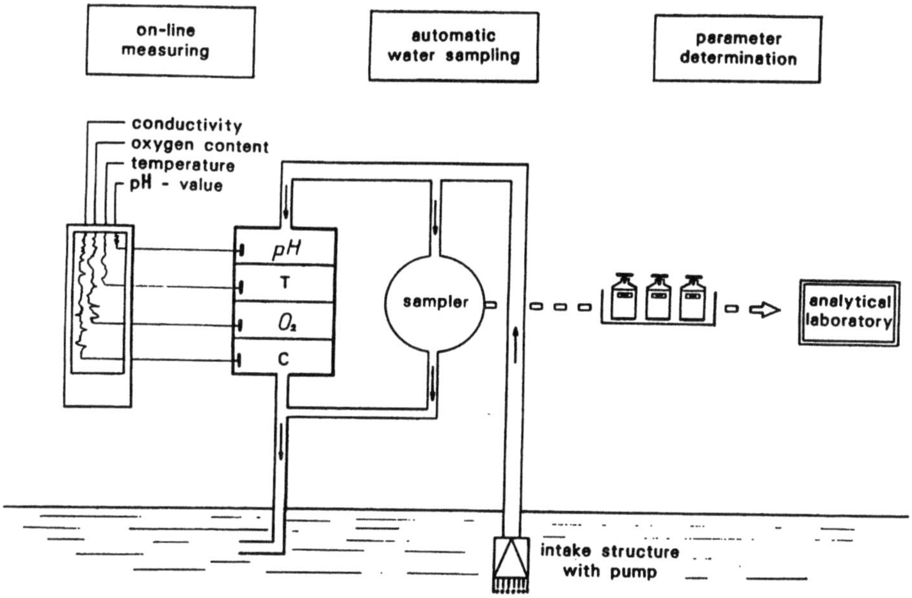

Figure 2: Scheme of a fixed monitoring station (re-drawn from [1])

Hemelingen are given. Fig. 3 illustrates the oxygen situation in the UW; time series of the oxygen content at the monitoring stations Hemelingen and Farge, together with trend curves (linear regression for the time intervals 1979 to 1985 and 1985 to 1987, respectively) are plotted. The Bremen municipal sewage treatment plant Seehausen, located between these two monitoring stations (Fig. 1), was equipped in 1985 with a biological stage. The resulting improvement in the oxygen situation at Farge, caused by the reduction of the easily degradable BOD input from Seehausen, seems to be significant.

The UW monitoring at bankside stations is completed by a weekly mid-river monitoring of the same parameters from a boat at a fully developed ebb current. Water sampling and on-line measuring is done at a depth of 1 m below the surface. Distances between two successive water sampling stations are about 2 km. Examples of the boat monitoring results are given in Fig. 4.

Table 4: Headwater nutrient, BOD_5 and heavy metal load (mean annual values)

year	head water [m³/s]	load [g/s] BOD_5	NH_4-N	NO_3-N	O-PO_4-P	P_{tot}	As	Cd	Cr	Cu	Fe	Mn	Ni	Pb	Zn	Hg
1987	481	1500	170	2710	90	200	0.6	0.2	1.6	3.4	770	86	2.7	2.9	19	0.07
1986	354	1000	120	2190	90	230	0.5	0.2	2.1	3.7	720	71	2.4	2.8	17	0.05
1985	275	900	100	1760	90	190	0.3	0.2	1.6	1.8	450	64	2.3	1.9	15	0.04
1984	365	1200	120	2110	110	240	0.6	0.4	3.8	3.5	870	290	3.6	3.5	22	0.04
1983	316	800	170	1760	90	250	0.5	0.2	4.2	2.1	450	84	2.4	2.2	17	0.04

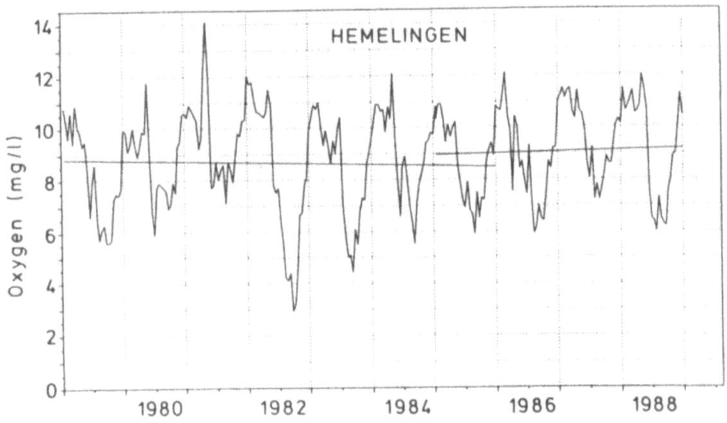

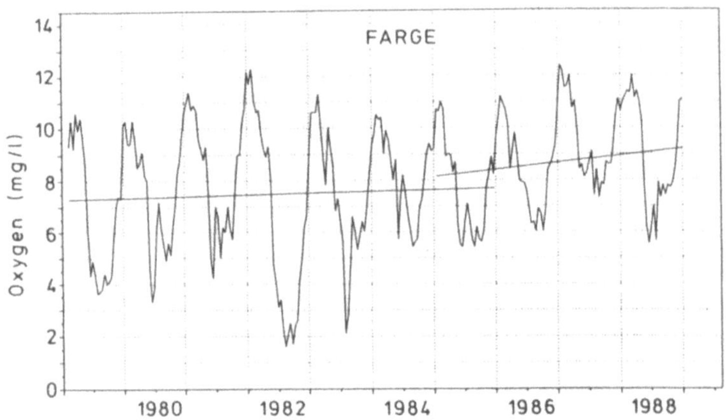

Figure 3: Time series and trend curves of oxygen content at the monitoring stations Hemelingen and Farge

Special field investigations

Many problems concerning the present status of water quality and pollutant transport in the UW, or estimates of the discharge of these pollutants into the North Sea, or predictions of the ecological trends in the UW estuarine area are, to-date, rather vague and indefinite. This is mainly due to the complex flow conditions in the UW, combined with physical, chemical and biological processes which effect both dissolved and suspended water constituents and the sediment. Consequently, there is a need for special field investigations of these topics. Such investigations are increasingly being carried out by the appropriate water authorities in cooperation with scientific institutions or by scientific institutions supported by the water authorities. A special concept for such investigations as developed by GKSS is outlined in [10]. The hydrographic measuring system belonging to this concept is presented in [11]. Some examples from the 1987 UW field experiment, jointly undertaken by the UW water authorities, the University of Bremen and GKSS Research Centre in May/June, are given in [12].

3. LOWER WESER MODELLING

In connection with tidal river monitoring, numerical modelling of transport processes in tidal rivers is becoming increasingly important for hindcasting and interpreting tidal river monitoring results, for environmental impact assessments of tidal river utilization and for the planning of engineering works.

To simulate transport processes in tidal rivers and estuaries, numerical models with different degrees of complexity (1-D, 2-D, 3-D models) have been developed at the GKSS Research Centre. All models are time-dependent and derived, basically, from the conservation equations for mass, momentum, energy, and water constituents (e.g. dissolved and suspended matter, trace elements, etc.). Some terms in the model equations have as yet to be parameterized by empirical relationships which are based on field observations and special measurements. Therefore, numerical modelling is carried out in close connection with the experimental investigations at the GKSS Research Centre, in order to firstly, verify the model assumptions, secondly, to adjust the model parameters and thirdly to get realistic boundary values. Although it is most desirable to simulate phenomena in tidal rivers by considering the time-dependence and all three spatial dimensions, 3-D and 2-D models are, as yet, not used in UW modelling for practical purposes, but are however used in special investigations, [13] to [16]. This is mainly due to several difficulties: (i) the lack of sufficient input data from field measurements, (ii) the problem of proper calibration, (iii) the necessity of additional parameterization, and (iv) the limitations in computer capacity and/or computer speed.

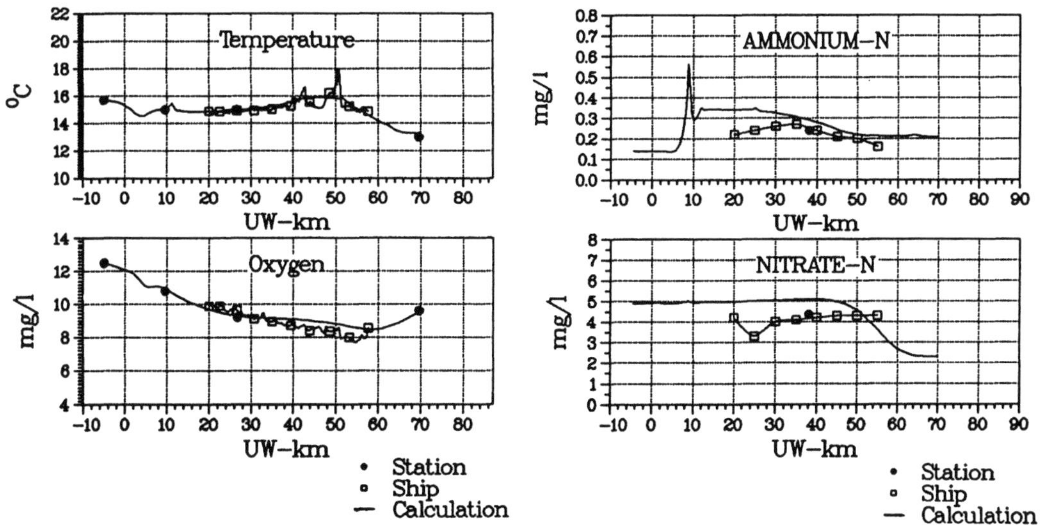

Figure 4: UW hindcast simulation compared with measured data of 25.5.1987 about 2:30 to 3:30 p.m. (ebb stream)

For the UW modelling at the GKSS and at the Bremen Water Authority a time-dependent cross-section averaged model based on the GKSS model FLUSS [17] to [21] is used. It computes time series at specified UW-km and UW longitudinal profiles at given times for the following parameters

- water level, water flow, salt content (concentration of a conservative parameter), temperature, suspended particulate matter, trace elements, dissolved oxygen, carbon-BOD, ammonium-nitrogen, nitrite-nitrogen, nitrate-nitrogen, organic bound nitrogen and chlorophyll a.

The model simulates the changes in the water quality situation and the transport processes in the UW between the weir at Bremen-Hemelingen and Bremerhaven by time-dependent cross-section averaged transport equations, using the values from these monitoring stations as boundary conditions. The physical, chemical and biological processes occur in the water body, on the water surface and on the river bed (e.g. heat balance on the water surface, sedimentation and resuspension of suspended matter, oxygenation through the water surface, oxygen depletion in the water body and on the river bed, nitrification etc.) are parameterized as realistically as possible given by the present state of our knowledge. Discharges of effluents are described by time functions. The space interval resolution is between 125 to 250 m. Typical computing times on the IBM 3090 are about 20 minutes per day simulation. The UW model is used for both analytical and prognostic studies, and especially to test the influence of isolated discharges of effluents on the water quality of the UW. Fig. 4 shows a hindcast simulation of the UW longitudinal profiles for certain parameters at a fully developed ebb stream compared with bankside and ship measurements under normal headwater conditions (315 m^3/s). Results from modelling and measurements agree well. In Fig. 5 a result of a warm-up study for

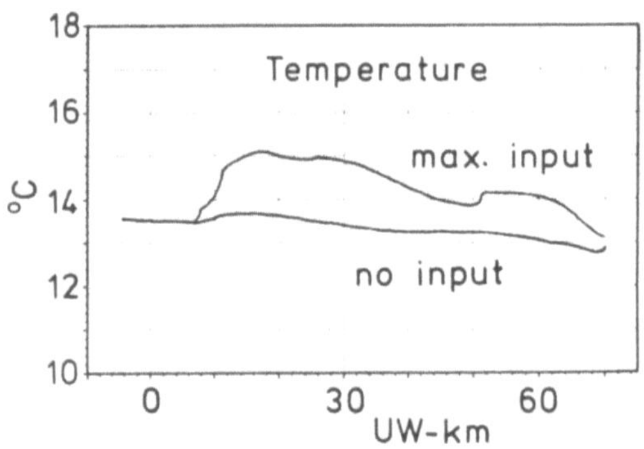

Figure 5: Calculated UW temperature profiles for a fully developed ebb stream with and without waste heat input under normal head water and meteorogical conditions (May 1987)

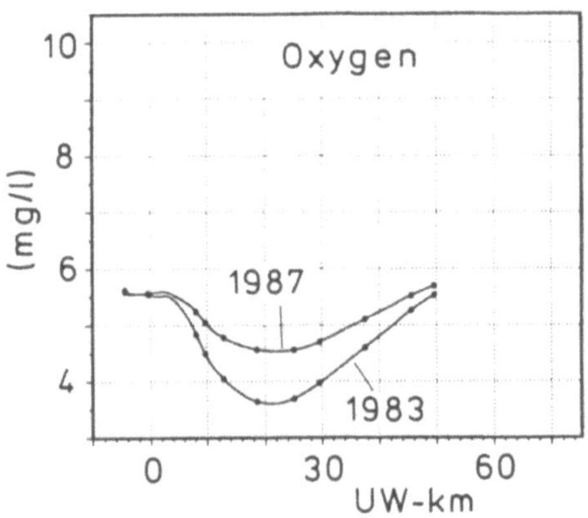

Figure 6: Calculated UW oxygen profiles for a critical summer oxygen situation before and after improvement of the Seehausen sewage plant with a biological stage

the UW is given. The calculated warm-up span is clearly below the maximum allowable warm-up span of 3 °C upstream and 2 °C downstream of the fresh water limit. In Fig. 6 a critical summer oxygen situation in the UW is plotted, where the computation is done under identical UW conditions but the different waste water discharge values from the Seehausen sewage plant before (typical BOD_5 input 1983: 553 g/s) and after (typical BOD_5 input 1987: 13 g/s) the installation of the biological stage, are used as input values for the model. The calculated improvement of approximately 1 mg/l oxygen agrees with the results shown in Fig. 3.

The monitoring and modelling of the UW are an effective complement to one another as tools for the UW water quality management. Therefore, it is planned to boost the cooperation between GKSS and the UW water authorities on this subject.

4. CONCLUDING REMARKS

The UW is the life-line of a large economic and housing region. Several millions of people live in its drainage area and use this river in manifold ways. It is the shipping channel to the Bremen harbour. Many factories, power stations and waste water treatment plants utilize this tidal river as a receiving body (Tab. 2 and 3). All this contributes to the water quality situation presently monitored in the UW. However, the predominant part of detrimental water constituents in the UW originates from upstream the weir (Tab. 4); for these loads of harmful substances the UW acts mainly as a transmission zone from the Weser headwaters and their catchment areas to the North Sea. Therefore, all countries, which share in the catchment area of the Weser River and its headstreams, are jointly responsible for the water quality situation in the UW and also in the North Sea.

REFERENCES

[1] ARGE Weser (Arbeitsgemeinschaft der Länder zur Reinhaltung der Weser) (1982) Weserlastplan 1982, Bremen

[2] Niedersächsisches Landesamt für Wasserwirtschaft (1987) Deutsches Gewässerkundliches Jahrbuch Weser- und Emsgebiet, Ablfußjahr 1985, Hildesheim

[3] Arbeitsgemeinschaft zur Reinhaltung der Weser (1974) Wärmelastplan Weser 1974, Bremen

[4] ARGE Weser (Arbeitsgemeinschaft der Länder zur Reinhaltung der Weser) (1979 to 1987) Zahlentafeln der physikalisch-chemischen Untersuchungen, Hildesheim

[5] Freie Hansestadt Bremen (1983). Meßprogramm Weser in Bremen (MEWEB), Bremen

[6] GKSS-Forschungszentrum Geesthacht GmbH (1980) Gewässeranalytische Untersuchungen der Unterweser im Herbst 1979, Geesthacht. GKSS 80/E/27

[7] Lüneburg H, Schaumann K, Wellershaus S (1975) Physiographie des Weser-Ästuars (Deutsche Bucht), Veröffentl. Inst. Meeresforsch. Bremerhaven 15, 195-226

[8] Wöbken K, Kunz H (1980) Beitrag zu Wassergütefragen der Unterweser, Wasser und Boden, 8 (372-377)

[9] Walther F (1954) Wasserstände und Gezeiten der Unterweser, Die Weser 28 (12), 133-136

[10] Michaelis W (1989) The BILEX concept: research supporting public water quality surveillance in estuaries. This volume, chapter III

[11] Fanger H-U, Kuhn H, Maixner U, Milferstädt D (1989) The hydrographic measuring system HYDRA. This volume, chapter V

[12] Grabemann I, Kühle H, Kunze B, Müller A, Neumann LJR (1989) Studies on the distribution of oxygen and nutrients in the Weser Estuary. This volume, chapter VIII

[13] Häuser J, Eppel D, Müller A, Nehlsen A, Tanzer F (1981) A thermal impact assessment model with measured field data applied to the tidal river Weser in: S. Sengupta, S. Lee (Eds.): Proceedings of the third Waste Head Management and Utilization Conference. Hemisphere Publishing corporation; Washington, D.C., USA

[14] Nehlsen A, Michaelis W, Müller A (1982) Tidal river flow calculations with measured velocities on the open boundaries. Proc. 4th Int. Conf. on Finite Elements in Water Resources, Hannover, 6-29

[15] Markofsky M, Lang G, and Schubert R (1986) Numerische Simulation des Schwebstofftransportes auf der Basis der Meßkampagne MASEX '83. Die Küste, 44:171-189

[16] Boehlich MJ (1982) Ein zweidimensionales Modell zur Berechnung der langfristigen Vermischungsvorgänge in schmalen Ästuaren, Diplomarbeit Universität Hamburg.

[17] Fiedler H, Müller A, Nolte D (1981) FLUSS - Ein eindimensionales Modell des Wärme- und Stofftransports in Flüssen. GKSS report 81/E/12, 1-52

[18] Fanger H-U, Michaelis W, Müller A (1985) Dynamic Processes in the Heavy Metal Load of Estuaries in: T.P. Lekkas: Proceedings of the International Conference on Heavy Metals in the Environment; Athens

[19] Fanger H-U, Kuhn H, Michaelis W, Müller A, Riethmüller R (1986) Investigation of Material Transport and Load in Tidal Rivers. In; Estuarine and Coastal Pollution: Detection, Research and Control. (Editors: D.S. Moulder, P. Williamson). Wat. Sci. Techn., Vol. 18, No. 4/5, pp. 101-110

[20] Michaelis W, Knauth H-D (Editors) (1985) Das Bilanzierungsexperiment 1982 (BILEX '82) auf der Unterelbe. GKSS report 85/E/3, 1-212

[21] Michaelis W, Fanger H-U, Müller A (eds) (1988) Die Bilanzierungsexperimente 1984 und 1985 (BILEX '84 und BILEX '85) auf der oberen Tideelbe. GKSS 88/E/22

[22] Fanger H-U, Kunze B, Michaelis W, Müller A, Riethmüller R (1987) Suspended matter and heavy metal transport in the tidal Elbe river. In: Awaya Y, Kusuda T (eds.) Specialised Conference on Coastal and Estuarine Pollution. Kyushu University, Fukuoka (Japan), p 302

MONITORING OF RADIOACTIVITY IN THE ELBE ESTUARY

Rolf-Dieter Wilken and Roland Diehl

GKSS-Research Center, Institute for Chemistry, D-2054 Geesthacht

INTRODUCTION

In Germany the monitoring of nuclear installations has been regulated by law since 1979 [1], but there is about 30 years of experience in monitoring environmental radioactivity [2]. The analytical standard of the monitoring program is guaranteed by two independent programs for each nuclear installation, which differ only slightly. Intercomparison tests are obligatory for laboratories involved in the programs.

In German rivers the water, the sediment and the fish are controlled for radioactive emissions from nuclear installations. In the Elbe, the γ-emitting nuclides are monitored as are the only β-emitting nuclides such as ^{90}Sr and the α-emitters such as $^{238,239/240}$Pu.

SAMPLING LOCATIONS AND METHODS

There are 4 nuclear power plants and one research institute, with two small nuclear reactors, along the river Elbe in the Federal Republic of Germany.

The locations are shown in figure 1.

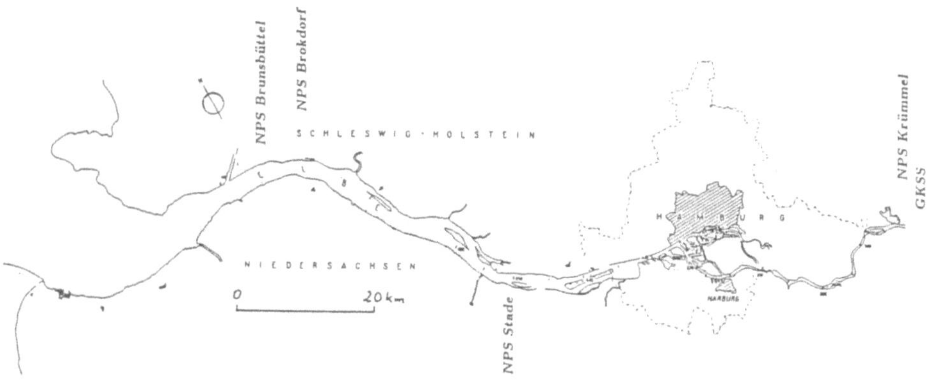

Fig. 1: Nuclear power stations (NPS) at the Elbe river

The NPS Krümmel, a boiling water reactor (BWR), is located in the non tidal influenced part of the river about 3 km upstream from the weir at Geesthacht (Elbe-km 580); the GKSS Research Center (Elbe-km 578) in the neighbourhood has two small research reactors (5 and 15 MW_{therm}). The other NPS are in the tidal area of the Elbe: that is downstream from Hamburg at Stade (Elbe-km 654) (PWR), then at Brokdorf (Elbe-km 683) (PWR) and the last station at Brunsbüttel (Elbe-km 693) (BWR). Brokdorf and Brunsbüttel are situated in the mixing zone of the estuary.

The monitoring is organized in accordance to an official program which starts three years before a nuclear installation begins to work, so that the pre-contamination of the site can be recognized.

To evaluate the influence of the nuclear installation, samples are taken before and after the facility. The nuclides monitored are, besides the γ-emitters such as ^{137}Cs, the β– and α–emitters ^{90}Sr, and, at some places, ^{238}Pu and $^{239/240}Pu$. ^{3}H-concentrations are also monitored but are in general below the detection limit of 5 Bq/l (without enrichment).

The method for determining ^{90}Sr is the nitric acid method, described elsewhere [3]. Usually 10 l of water or 0.1 kg of sediment are taken for analysis.

The determination of $^{238,239/240}Pu$ starts with 40 l of river water or 0.15 kg of sediment. ^{242}Pu is added as tracer for determining the yield of the separation. Plutonium is extracted by Tri-n-octylphoshinoxide (TOPO) in cyclohexane, purified by coprecipitation with LaF_3, separated by an anion exchange resin, electroplated on a high grade steel plate and measured in an α-spectrometer (surface barrier counter).

^{137}Cs is determined by γ-spectroscopy without enrichment. The measurements are performed by the GKSS Radiation Protection Group (SST).

Water samples are obtained usually automatically using a special device, which pumps 200 ml river water for a few minutes every two hours via a bypass into storage tanks. This is done upstream and downstream from the NPS; other water samples are taken from the inlet and outlet of the nuclear installation. From these 24 h samples, monthly mixed samples are prepared, without separating suspended particulate matter and water.

Other water samples are taken monthly in the river at certain times, especially when the storage tank in a NPS is emptied. Then 10 l are taken, concentrated and analysed.

Sediment samples and fish samples are taken twice a year at different places from a boat in the river upstream and downstream from each nuclear facility at certain places.

A description of the Elbe is here omitted. It is described in other contributions.

RESULTS AND DISCUSSION

^{90}Sr and ^{137}Cs - concentrations

Whereas the concentrations of ^{90}Sr in the whole estuary show no difference (2 - 40 mBq/l) [4,5] the ^{137}Cs - concentrations differ by a factor of 10 (inner estuary: 1 - 5 mBq/l, outer estuary 10 - 50 mBq/l) [4,5]. An explanation is that emissions from the reprocessing plants in Windscale / Sellafield, GB, and La Hague, France, into the North Sea reach the Elbe estuary [6].

The results of measurements made over several years are shown in an overview for ^{90}Sr and ^{137}Cs in figure 2.

The peak of ^{90}Sr and ^{137}Cs in May and June 1986 can clearly be seen. The concentration of ^{90}Sr in the river water before the Chernobyl emissions is due to the nuclear bomb tests; emissions of the NPS are negligible in comparison. The concentrations decreased significantly sharper than would be expected from the physical halflife of 28.1 years. From the data an ecological halflife of about 5 years for ^{90}Sr could be estimated.

The different peaks in that period could be explained by the small samples for the analysis and the statistical error of the results.

In June the concentration increased to about 60 mBq/l, that is by a factor of four.

The decrease of ^{90}Sr after the accident is twofold: there is a sharp decrease up to the fourth quarter of 1986 followed by a slow decrease thereafter. This also occurs with an ecological halflife of about 5 years, the same as after the nuclear tests. At the end of 1988 the values of end of 1985 are reached.

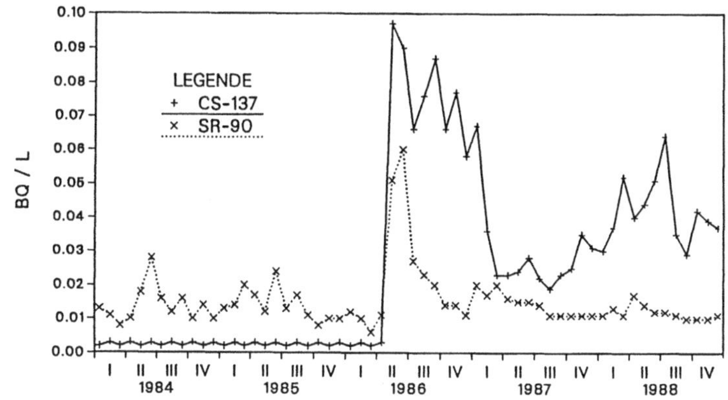

Fig. 2: ^{90}Sr and ^{137}Cs in Brokdorf from 1984 to 1988

The ^{137}Cs concentrations before the Chernobyl accident were often below the detection limits of 0.003 Bq/l.

During the emissions from Chernobyl the concentrations increased by a factor of 100 - 500. In the period after June '86 and in 1988 other maxima could be explained by the lower water discharge of the river and the water from the North Sea reaching the sampling location [11] or by input of sediments with higher contamination.

There is not such a decrease of the concentration after that as was the case for the ^{90}Sr. On the contrary: there is a constant or even increasing concentration during 1988.

The ^{90}Sr and ^{137}Cs - concentrations in the water show a characteristic trend during the Chernobyl emissions due to their different behaviour in a river. Figure 2 gives an example with temporal different peaks for ^{137}Cs compared to ^{90}Sr .

The different behaviour of the two nuclides could be explained by their different behaviour in the river. Whereas ^{137}Cs is bound mainly to the suspended particles, the ^{90}Sr is mainly in solution [8,9].

From this fact two consequences could be drawn:

1. the sampling of representative water samples for determination of ^{137}Cs contents depends mainly on the representative spm content in the samples, which is not an easy problem, because spm is distributed differently in a river with regard to time and location.
2. the transport of ^{137}Cs occurs mainly with the spm, which is different to the transport of the water phase. So ^{137}Cs is concentrated also in the sediment and is there deposited and also resuspended.

The concentrations of ^{90}Sr and ^{137}Cs in sediments are shown in figure 3. The concentrations of ^{90}Sr in the sediments from 1984 to 1988 lie between < 0.1 and 1.1 Bq/kg and show no significant trend. Especially the Chernobyl accident did not increase the concentration significantly.

The ^{137}Cs concentration in sediments on the other hand shows an increase of about a factor of 10 due to the accident at the end of 1986 and the beginning of 1987. The peak is very much broader than the peak for ^{137}Cs in the water phase. The second peak in 1988 corresponds to the ^{137}Cs peak in the water (fig. 2) and shows the spm and sediment interaction.

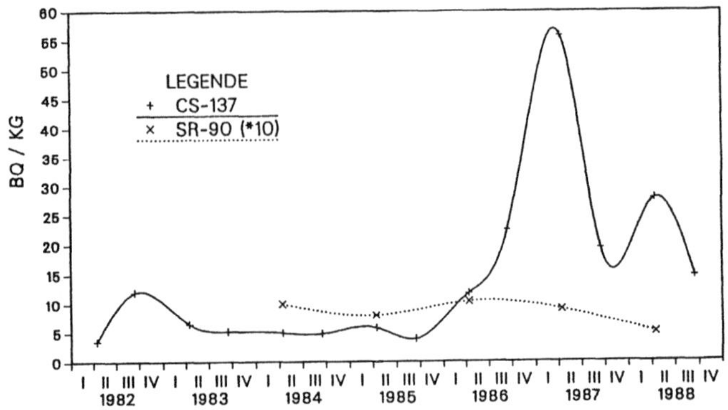

Fig. 3: ^{90}Sr and ^{137}Cs in sediments of the Elbe

The concentration of ^{90}Sr and ^{137}Cs in fish is shown in table 2. The values correspond to the meat only. Whole fish has a factor of 2 - 3 higher contents of ^{90}Sr . The reason for this is the uptake of ^{90}Sr in the bones of the fish, whereas ^{137}Cs is not bound. The values obtained depend therefore mainly on the amount of bones in the sample, and the error given in table 2 depends more or less on this factor. The error in the determination is a factor of 3 better. In meat the ^{90}Sr concentration shows no trend. The ^{137}Cs concentration increases after the accident by a factor of about 10 and decreases only slowly, which corresponds to the ^{137}Cs concentration in water and the biological halflife of ^{137}Cs in the fish.

Tab. 2: ^{90}Sr and ^{137}Cs in fish (meat) [Bq / kg (fresh)]

	Sr-90	Cs-137	
1982	0.027 ± 0.01*	0.91	± 0.1
1983	0.030 ± 0.01*	0.81	± 0.1
1984	0.062 ± 0.03*	0.36	± 0.3
1985	0.026 ± 0.02*	0.26	± 0.2
1986	0.034 ± 0.01*	3.2	± 1.5
1987	0.032 ± 0.01*	4.9	± 1.2
1988	0.038 ± 0.01*	3.1	± 1.4

* the error is due to different fish measured; it depends e.g. on the remaining parts of bones. The error in the determination is a factor of 3 lower.

Pu-concentrations

In fish the $^{238,239/240}Pu$ - concentrations were lower than the detection limit (< 0.1 mBq/kg for ^{238}Pu and also for $^{239/240}Pu$).

Tab. 3: ^{238}Pu and $^{239+240}Pu$ in water and sediments*					
in water [mBq/l]			in river sediments [mBq/kg]		
	^{238}Pu	$^{239/240}Pu$		^{238}Pu	$^{239/240}Pu$
1985	0.04	0.16	Apr. '86	30	164
1986	0.017	0.047	Oct. '86	37	159
1987	0.05	0.03	Apr. '87	39	181
1988	0.007	0.035	Jan. '88	22	167

* The standard deviation of the measurements are up to 100 % due to the values being near the detection limit and the different samples analysed.

CONCLUSION

The monitoring program for radioactivity in German rivers is very effective and, especially during the Chernobyl accident, could record very clearly

- the amount of contamination
- the chronological course and
- the different behaviour of the nuclides in an estuary

However, the results also show difficulties in taking representative water samples with regard to the spm, where ^{137}Cs is mainly found (which is, of course, of no importance for the NPS).

It is clear, that the ^{137}Cs -concentrations differ by a factor of 4, whereas the ^{90}Sr -concentrations differ only by a factor of 2 during the Chernobyl event. The reason for this is, that Strontium is more or less transported in the water phase [8,9], whereas the Cesium is bound to clay particles and is therefore transported in the spm phase [8,9]. Our own values from water samples (and not from the automatic sampling devices) and other results [6,10,11] show, that the ^{137}Cs concentrations from water samples taken by the automatic samplers are a factor of 2 to low. This is due to the unrepresentative sampling of spm in the water.

The different behaviour of the nuclides ^{90}Sr and ^{137}Cs is also seen in

- the concentration in fish:
  - ^{90}Sr is bound in the bones mainly, whereas
  - ^{137}Cs is found in the whole fish
- the concentration in sediments:
  - ^{90}Sr has no significant increase in the sediments because it is transported in solution, whereas
  - ^{137}Cs is concentrated in the sediments (with a temporal delay) in comparison to the water phase.

REFERENCES

[1] Anonymus (1979) Richtlinie zur Emissions- und Immissionsüberwachung kerntechnischer Anlagen. GMBl. Nr. 32, pp 665 - 683, Bonn 26.11.1979

[2] Der Bundesminister des Innern (ed.) (1986) 30 Jahre Überwachung der Umweltradioaktivität in der Bundesrepublik Deutschland, Bonn 1986

[3] Wilken RD, Diehl R (1987) Strontium-90 in Environmental Samples from Northern Germany before and after the Chernobyl Accident. Radiochim. Acta. 41:157-162.

[4] Feldt W, Kellermann HJ, Kanisch G, Lauer R, Melzer M, Nagel G and Brinkmeier G (1981) Berichte aus dem Isotopenlaboratorium der Bundesforschungsanstalt für Fischerei, Hamburg Radioökologische Studien an der Elbe. Teil I: Vorbelastung, Konzentrationsfaktoren.

[5] Arbeitsgemeinschaft für die Reinhaltung der Elbe (ed.) Radioaktivitätswerte der Elbe von Schnackenburg bis zur See 1978-1983.

[6] Wilken RD, Diehl R (1989) Strontium-90 und Cäsium-137 in der Elbe vor und nach dem Tschernobyl-Unfall. Vom Wasser 72: 65-81.

[7] Kautsky H (1987) Investigations on the distribution of ^{137}Cs , ^{134}Cs and ^{90}Sr and the water mass transport times in the northern North Atlantic and the North Sea. Dtsch. Hydrogr. Z. 40: 49-69.

[8] Hübel K, Laschka D and Herrmann H (1982) Anreicherung von Radionukliden in Sedimenten bei Kernenergieanlagen. In: Radioökologie und Strahlenschutz. Verlag E. Schmidt, Berlin, p. 157-161.

[9] Mundschenk H (1985) Zur Sorption von Cäsium an Schwebstoff bzw. Sediment des Rheins am Beispiel der Nuklide Cs-133, Cs-134 und Cs-137. Teil 1, Naturmessungen. D. Gewässerk. Mitt. 27: 12-20.

[10] Schoer J (1988) Zur Abschätzung von Transportzeiten in der Elbe unter Verwendung von Tschernobyl-fall-out als künstlichem Tracer. D. Gewässerk. Mitt. 32: 154-159.

[11] Schoer J (1988) Investigation of transport-processes along the Elbe river using Chernobyl-Radionuclides as tracers. Env. Tech. Lett. 9: 317-324.

REGULATIONS AND MONITORING FOR THE DISCHARGE OF COOLING WATER AND RADIOACTIVE SUBSTANCES FROM A NUCLEAR POWER PLANT INTO THE WESER ESTUARY

Hans Kunz
Forschungsstelle Küste des Niedersächsischen Landesamtes für Wasserwirtschaft
An der Mühle 5, D-2982 Norderney, Federal Republic of Germany

ABSTRACT

The operators of nuclear power plants apply for offical permission to discharge cooling water and radioactive liquid wastes produced in the plant into receiving waters. The complicated transporting and mixing processes in estuarine waters require sophisticated measuring equipment to obtain data necessary for control, monitoring and securing evidence. In the Lower Weser nuclear power plant (KKU) automatic measuring equipment is installed in all water streams that discharge cooling water and likewise in those through which radioactive wastes could be released. The regulations given by the water authority extend from the receiving water to these measuring stations. Receiving water is the Lower Weser Estuary in which a comprehensive automatic water quality measuring system has been installed. This paper describes how the discharge from the nuclear power plant of the pressurized water reactor type into a tidal river has been regulated, controlled and monitored. Requirements placed on the KKU measurement system, the technical configuration, experience and results are outlined.

1. LEGAL ASPECTS, DATA ON THE KKU AND THE LOWER WESER RIVER

Responsibility for deciding on the application to construct and operate a nuclear power plant rests with the Nuclear Inspectorate - regional authority acting on behalf of the federal authority [1]. Discharging liquid waste from the power plant requires approval from the water authority of the region concerned [2]. The authority must assume that the permits guarantee that, even in the event of malfunctions, no risks emanate from the plant [1, 3]. Any release of radioactive wastes from the plant must be detected and stopped automatically if possible. The discharge of cooling water must be automatically reduced, when the limits fixed for the cooling or the receiving water are reached.

The nuclear power plant KKU was constructed on the left bank of the Lower Weser (Fig. 1). It is equipped with a pressurized water reactor and has a net output of 1230 MWe. The average flow at the location of the KKU is 5000 m³/s, the average tide range is 3.50 m, the average inflows from the tide-free Weser river and the tributaries are given as 300 m³/s. Salt water from the North Sea regularly penetrates as far as this point. Fig. 2 shows the location of the KKU. Cooling water intake (E) is directly north of the mouth of the tributary "Schweiburg". All liquid wastes from the KKU flow into the seal pit (A) from where they pass through a siphon below the "Schweiburg" to the outlet structure (C). Approval was not given to discharge wastes into the "Schweiburg" because this contains sedimentation areas and because a 200 km² agricultural area is supplied with water from the Beckumer Sieltief ((1) in Fig. 2).

2. WATER FLOWS IN THE NUCLEAR POWER PLANT

The primary water heated by the reactor moves in a closed primary loop ((D) in Fig. 3). The turbine set is located in the secondary loop (B) which is closed as well. Heat transfer from D to B takes place in the heat exchanger. It is removed from the secondary loop in the condenser by transferring it to the main circulating water (H). From that the heat is released into the receiving water. Heat must also be removed from the system components of the primary loop, this being the purpose of the closed nuclear component cooling loop (C) and the secondary circuit (N).

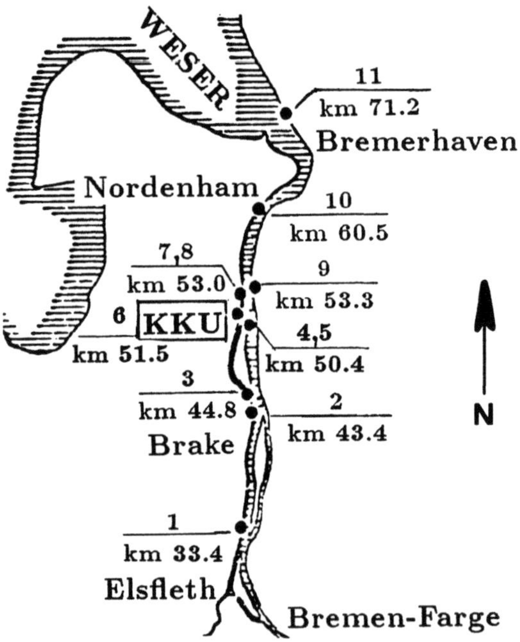

Figure 1: Location of the water quality measuring stations allocated to the KKU (1 to 11)

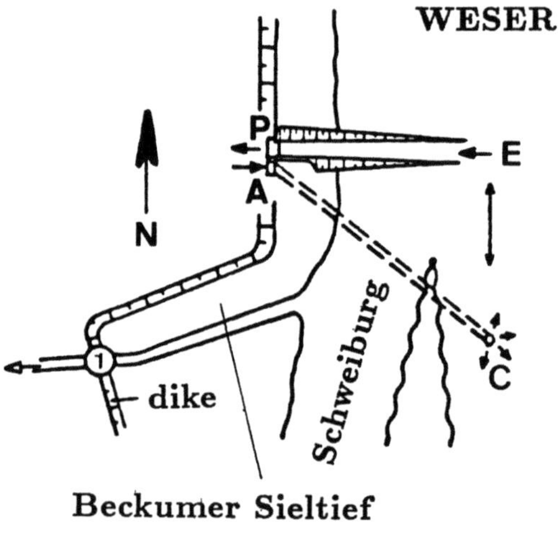

Figure 2: Structures in the nuclear power plant KKU for water intake from the Weser river and discharging cooling water and liquid wastes

The water in the primary loop is highly radioactive. The radioactivity is limited by decontamination measures, with radioactive liquids (d) being produced. The secondary loop is generally declared as being not radioactive. However, it may contain radioactive substances due to the radiation from the primary loop and if leaks are present. Then radioactive liquids (b) will be in the steam generator blowdown. If leaks exist in the heat exchanger between (D) and (C), in the nuclear component cooling loop radioactivity may be present which requires reprocessing (c). Leakages (e) from the components located in the monitoring area flow together in pump sumps and pass just like the contaminated liquids b, c, d to the decontamination unit (A). Radioactive liquid wastes are produced at this point in the form of residual liquids for which the nuclear plant operator then applies for a discharge permit. For this purpose he collects the liquid wastes in a transfer tank (Ü). Small quantities of secondary water escape from the secondary loop (B) in the Operating Monitoring Area and pass into pump sumps from where they are pumped out of the plant together with other liquids produced in the turbine house (3). If leaks exist in the condenser, main circulating water (H) will pass into the secondary loop due to the pressure condition existing. If the nuclear component cooling loop (C) exhibits radioactive substances and if the heat exchanger is leaking into the secondary cooling water (N), radioactive liquids will pass into the secondary cooling water via (4) due to the overpressure in (C). The liquid waste streams (1), (3) and (4) merge into the main circulating water and are then discharged into the receiving water (2).

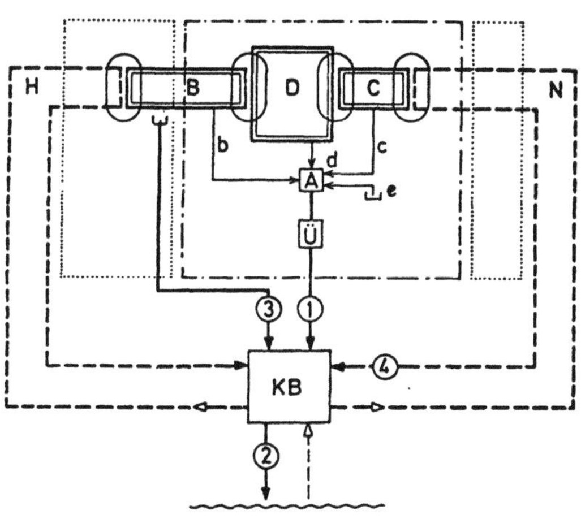

Figure 3: Simplified illustration of the water loops in a nuclear power plant with pressurized water reactor and the liquid wastes discharged from the plant

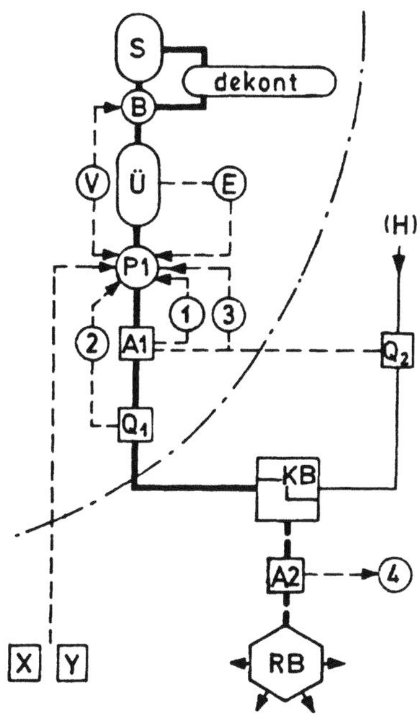

Figure 4: Sketch of location and logical linkage of the automatic measuring instruments for controlling and monitoring the discharge of radioactive wastes

3. CONTROL SYSTEMS IN THE KKU FOR OPERATIONAL LIQUID WASTE AND COOLING WATER

Application was made to discharge radioactive liquid wastes from the controlled area of the plant ((1) in Fig. 3). Such an application was not required for waste water streams (3), (4) because it was possible to limit the release of radioactive substances from the secondary loop and the nuclear component cooling loop by means of regulations of the Nuclear Inspectorate in accordance with [3]. Nevertheless, automatic measuring instruments (3), (4) have been installed in the header for the turbine house waste and the secondary cooling water streams to provide total measurement. They act as interfaces between the responsibility of the nuclear inspectorate and the water authority: These measuring stations enable the water authority to monitor unauthorized release of radioactive substances into the receiving water, while the monitoring activities of the nuclear inspectorate are based on measuring stations in components of the nuclear plant (B, C). These stations first of all detect the risks of unauthorized discharge and initiate operational countermeasures [8, 10]. The water control systems described for the KKU exceed the minimum requirements demanded in [4, 5]. They were based on the view derived from the principles that the activity load released from a nuclear power plant must be as low as possible.

The KKU was granted permission (1976) to discharge 950 Ci/a tritium and additionally 2 Ci/a dissolved radioactive substances from the controlled area into the Lower Weser. The discharge from the KKU controlled area ((1) in Fig. 3) is precisely controlled. Fig. 4 shows the automatic measuring stations between the transfer tank (Ü) and the intake structure (RB) as well as their logical linkage (broken line) to each other and to the transfer pump (P1) for discharge of the radioactive wastes. All the radioactive liquids produced in the KKU controlled area (b, c, d, e in Fig. 3) pass into the collecting tanks (S) located in the controlled area. After processing in the decontamination unit, they flow into the transfer tank (Ü) from where they can only be pumped (P1) into the transfer line to the seal pit (KB). The discharge time is limited as a function of the tide to the period from the flood flow reversal (K_f) to one and a half hours following. An automatic flowmeter (X) registers when the tide turns. Limiting the discharge time ensures that the radioactive wastes released are carried towards the North Sea as fast as possible [8]. The water level in the Weser must not rise more than 20 cm above the mean high water level during discharge. This control parameter is recorded by a tide level gauge (Y). Before discharge commences, confirmation must be provided that the activity concentration of the liquid waste in the transfer tank (Ü) is no greater than 5×10^{-4} Ci/m³ (^{137}Cs equivalent). While the liquid waste is being pumped out, an automatic interlock (V) ensures that valve (B) interrupts the flow to the transfer tank. An activity measuring station (A1) and a flow measuring station (Q1) are allocated to the pump (P1) for the discharge from the transfer tank.

(A1) is logically linked to flow measuring points in the main circulating water (Q2). Discharge from the transfer tank is automatically interrupted if the different limits are reached [8, 10]. After the radioactive wastes have mixed with the cooling water, they pass the activity measuring station (A2). This station is assigned a limit value of 1×10^{-7} Ci/m³ as an alarm value (4).

Comprehensive analysis programs are assigned to the system for automatically monitoring the discharge from the transfer tank. Balancing of the discharge loads is based exclusively on laboratory analysis and is nuclide-specific; non-detectable nuclides are balanced with the proof limit. As an example: in 1988 2.7×10^{-3} Ci detected nuclides, respectively 3.1×10^{-2} Ci balanced with the proof limit (limit is 2.0 Ci), and 314 Ci Tritium (limit 950 Ci) were discharged.

The thermal load of the cooling water and the receiving water has to be limited. The limit value for the heat-up of the receiving water (max ΔT) should relate to the (natural) temperature before thermal preloading (T_{w0}). If the receiving water has already been subjected to thermal loading from other heat discharges (T_{w1}), a variable temperature increase $\Delta T^* = (T_{w0} + \max \Delta T) - T_{w1}$ must be met instead of the constant temperature rise max ΔT. The limits for the KKU are: river water max T_{w2} = 26 °C, max ΔT^* = 1.7 K; cooling water max T_k = 30 °C, max ΔT_k= 10 K.

For the KKU these parameters have been combined in regulations. They automatically regulate the electrical power output which is possible within the given limits [8]. Since the regulations work well, no cooling towers had to be erected yet.

The requirements for operating the KKU specify that the temperature values ΔT^* and T_{w2} be continuously determined for different tide phases. This requires an automatic water temperature measurement system. A programme for securing evidence had been integrated. To enable the measuring stations to fulfill this task, they measure in addition to the temperature the oxygen content, the conductivity and to a limited extent the pH value [7, 9, 11].

The plant operator must control his plant so as to remain within the specified limits. There are difficulties in determining the applicable temperature of the receiving water after it has mixed with the cooling water (T_{w2}) and particularly the heat-up (ΔT^*) in a tidal river. In an estuary the heat builds up in line with the number of passages (i) over a multitude of successive times (Fig. 5). The result is a hill-shaped heat-up (Fig. 6) whose maximum is positioned at flood slack water (K_f) in the area of the discharge (E). The maximum is displaced with the ebb flow by the ebb distance Se(E). The reference temperature T_{w1} in an estuary can not be measured directly upstream of the discharge point. For the KKU the distance has to be almost 20 km. The heat-up can only be determined from the difference of T_{w2} and T_{w1}. In the case of KKU, T_{w1} and T_{w2} are obtained by means of several measuring stations which are positioned upstream and downstream of the discharge. The values are linked together as a function of the tide phase. KKU measuring stations 4, 7, 9 (Fig. 1) allow T_{W2} to be determined. To calcualate the reference temperature of the Weser in the KKU area (T_{W1}) requires incorporating measuring stations 1 and 11. Station 1 is subjected to a greater influence of thermal discharges, particularly in the Bremen area, compared to station 11. Calculating back to the KKU location on the basis of measuring station 11 is complicated by the influence of the sea water and adjacent tidal flats. Measuring stations 2, 3, 6, 10 were designed exclusively to secure evidence. Not all of them have been operated on a long-term basis.

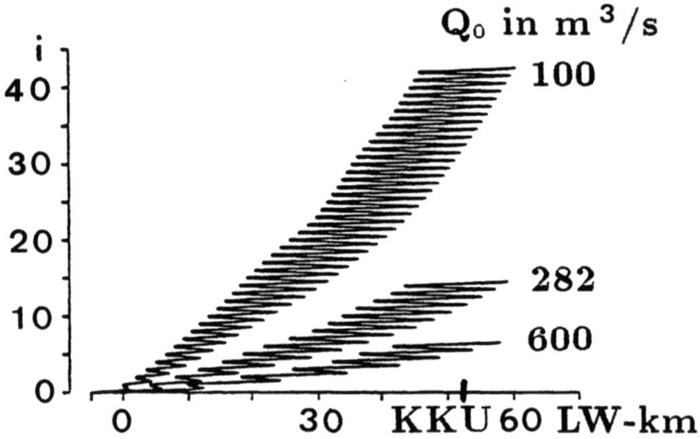

Figure 5: Movement of a "water package" from Bremen through the Weser estuary. Timescale: i x 12 h 25'; i = number of passages as a function of location, influenced by tide and run off (Q_0)

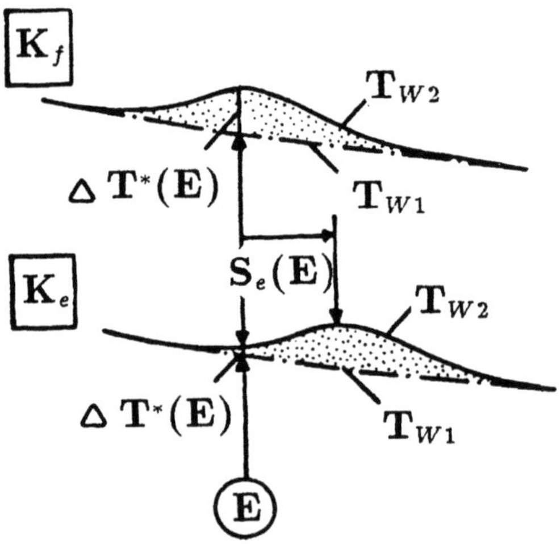

Figure 6: Principle sketch for heat-up (ΔT^*) of a tidal river due to discharge of cooling water (E) related to tidal phase between flood and ebb slack water K_f, K_e

The measuring stations respond to special addresses from the central station and are activated for data transmission. The computer polling cycle is normally 6 minutes [7, 11].

4. AUTOMATIC MONITORING BY AUTHORITIES

Since the KKU has been installed, the demands on authorities for monitoring nuclear power stations have become more and more extensive. The intention is to make use of remote monitoring to supervise the plant operation and the resultant environmental pollution and risk. Automatic remote monitoring systems have been installed in the KKU with respect to plant operation as well as the discharge of radioactivity (air, water). The water authority confines itself to assuring that the operator observes the regulations. To ensure automatic information transfer when limits are exceeded, without relying on the plant operator, and to convey up-to-date knowledge regarding the discharge of heat and radioactive substances, the KKU provides automatic data transmission to the authorities.

REFERENCES

[1] Gesetz über die friedliche Verwendung der Kernenergie und den Schutz gegen ihre Gefahren (Atomgesetz)

[2] Gesetz zur Ordnung des Wasserhaushaltes (Wasserhaushaltsgesetz)

[3] Verordnung über den Schutz vor Schäden durch ionisierende Strahlen (Strahlenschutzverordnung)

[4] Richtlinien für das Einleiten von Abwasser aus Kernkraftwerken mit Leichtwasserreaktoren in die Gewässer, Länderarbeitsgemeinschaft Wasser, November 1976

[5] Sicherheitstechnische Regel-Messungen flüssiger radioaktiver Stoffe zur Überwachung der radioaktiven Ableitungen (KTA - 1504), Juni 1978

[6] Grundlagen für die Beurteilung der Wärmebelastung von Gewässern, Teil 1 Binnengewässer, 2. Auflage, 1977, Druckschrift der Länderarbeitsgemeinschaft Wasser

[7] Grodd M, Krause C, Kunz H (1977, 1979, 1983) Berichte zum KKU-Meßprogramm Wasser, H. 1-3, Mitteilungen des Wasserwirtschaftsamtes Brake

[8] Kunz H (1979) Regelungen für die Abgabe radioaktiver Stoffe aus einem Leichtwasser-Kernkraftwerk vom Druckwassertyp in einen Tidefluß, Mitteilung des Franzius-Institutes, Universität Hannover, 49, pp 112-145

[9] Kunz H (1979) Wasserrechtliche Regelungen für die Einleitung von Kühlwässern in einen Tidefluß unter besonderer Berücksichtigung des Einsatzes von automatischen Meßsystemen, Mitteilung des Franzius-Institutes, Universität Hannover, 49, pp 197-243

[10] Kunz H (1981) Automatic control of the discharge of radioactive substances from light water nuclear power plants of the pressurised water reactor type into flowing streams, Wat Sci Tech 13: 561-566

[11] Kunz H (1981) Unterweser automatic water quality measurement system, Wat Sci Tech 13: 657-662

CHAPTER VIII

NUTRIENTS AND OXYGEN BUDGET

CORRELATION OF THE SEASONAL AND ANNUAL VARIATION OF PHYTOPLANKTON BIOMASS IN DUTCH COASTAL WATERS OF THE NORTH SEA WITH RHINE RIVER DISCHARGE

W.W.C. Gieskes and B.E.M. Schaub
University of Groningen, Dept. Marine Biology,
Box 14, 9750 AA Haren, The Netherlands

1. INTRODUCTION

Water quality parameters of the Dutch rivers and of Dutch coastal waters of the North Sea have been monitored regularly for many years by the Dutch government agency "Rijkswaterstaat". The results have been published every 3 months in a series of reports, from which we have compiled the concentrations of nutrients in the Dutch part of the river Rhine and of chlorophyll in Dutch coastal waters measured between 1972 and 1986. The purpose was to find trends in eutrophication and to analyse how the pelagic ecosystem of the North Sea, and especially the basis of this system, the phytoplankton, responds to the expected annual variability in eutrophication.

One often hears these days about pollution of the marine environment, and more often than not an increase, rather than a decrease, is reported. Recently, Beukema (1986) presented evidence of a continuing trend of increasing contamination of the Wadden Sea area, presumably through the Rhine. This is one of the most important rivers of Europe in terms of size of watershed and of human population living in the catchment area. Consequently, pollution loading is heavy. In Fig. 1 we present a map of the area. The reason why much of the eastern part of the North Sea is affected by the Rhine is that the residual current system in the North Sea carries the freshwater of the river plume northeastward, at the same time keeping the polluted water close to the continent (Fig. 1).

The concentration of nutrients in the Dutch part of the Rhine has been measured for a long time - of course with different methods in the old days, but on the basis of theoretical considerations one is allowed to compare the old data with the new ones. It is obvious from these data that in the fifties and early sixties the concentration of nutrients was still low, very low even compared to the situation in the seventies. However, in the eighties the situation changed again, with a tendency for concentrations to decrease (Fig. 2). This is rather surprising since in recent reports, both in scientific journals and in the press, and especially in "green" periodicals, one reads about "ever increasing plant nutrient loads" discharged into the North Sea, and an "ever increasing algal biomass" as a result,

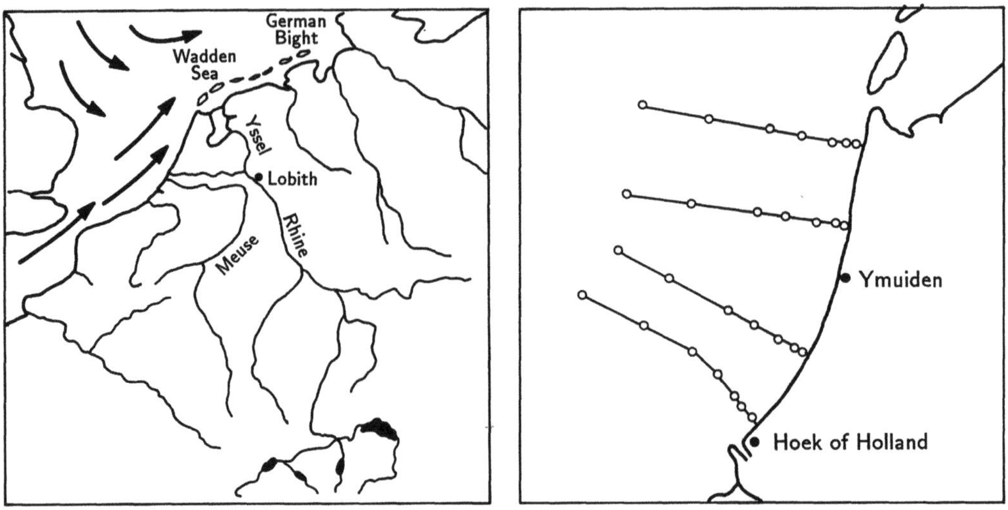

Figure 1: Map of the area, with sampling stations in the North Sea

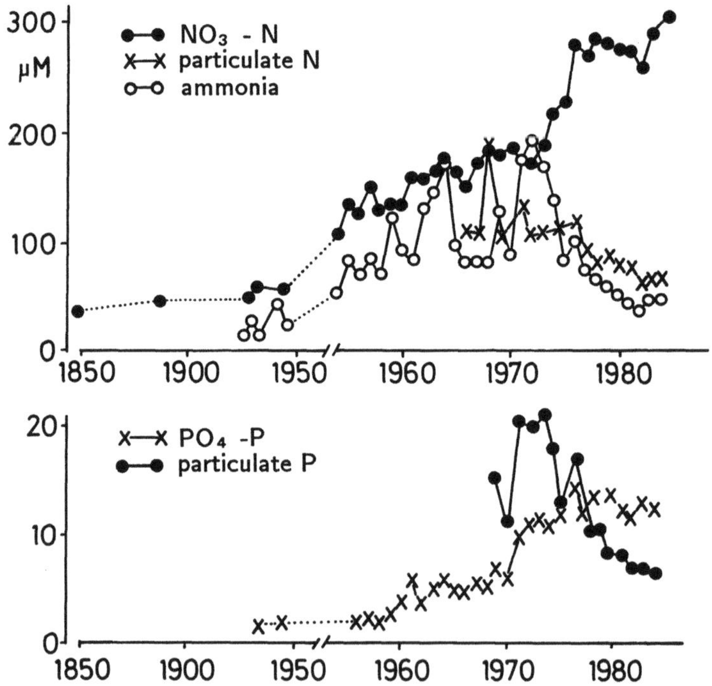

Figure 2: Nutrient concentrations in the river Rhine: before 1945 at Rhenen, after 1955 at Lobith. After Nelissen and Stefels, 1988: Fig. 27

while the measurements of the last decade do not seem to support this argument. Phytoplankton biomass trends recorded in Dutch coastal waters have not, until recently, been very clear, partly because the natural long-term variability cannot be clearly separated from variations caused by man's activities (Gieskes and Kraay, 1977; Owens et al., 1989).

2. RECENT LONG-TERM TRENDS IN EUTROPHICATION IN THE RHINE

The nutrient concentrations in the Rhine have, in fact, decreased rather than increased during the last 15 years, or at worst have remained stable. In Fig. 3 we show the seasonal fluctuations; the concentrations are means of values measured at all stations in the Dutch part of the Rhine between Lobith and the North Sea. The causes and consequences of these trends cannot all be treated here, but we will briefly discuss a few interesting ones. The ammonia concentration shows the most dramatic decrease. Admiraal et al. (1989) suggest that the decrease since the mid seventies is related to the decrease, in the seventies, in the discharge of toxic substances to the river water (Anon., 1986). This has allowed greater microbial activity and, therewith, higher rates of nitrification. Thus, ammonia and nitrite concentrations decreased due to this microbial oxidation of reduced nitrogen compounds. The nitrate concentration increased accordingly (Fig. 3).

With regard to silicate: no trends are obvious here (Fig. 3), basically because silicate is not related to man's activities, although when diatom growth is stimulated by eutrophication silicate may become more depleted than in the natural situation when light is not the limiting factor. It is obvious (see Fig. 3) that in the turbid Rhine, silicate concentrations decrease each summer to a low level, although silicate is not completely exhausted. Recent studies by E.de Ruyter van Steveninck, of the Dutch National Institute of Public Health and Environmental Protection (RIVM), indicate that, in the Dutch part of the Rhine, diatoms are major contributors to phytoplankton throughout most of the year. HPLC pigment analyses by G van Dijken (1989) support this observation. This diatom abundance in the Rhine is the reason for the neat seasonal pattern in the Rijkswaterstaat series (Fig. 4). By using silicate/carbon ratios recently published by Conley et al. (1989) it is possible to calculate diatom primary production from the Si depletion shown in Figure 4. This production estimate amounts to 180 g C $\cdot$ m^{-2} $\cdot$ $year^{-1}$ - a quite considerable organic matter loading from this autochthonous source alone.

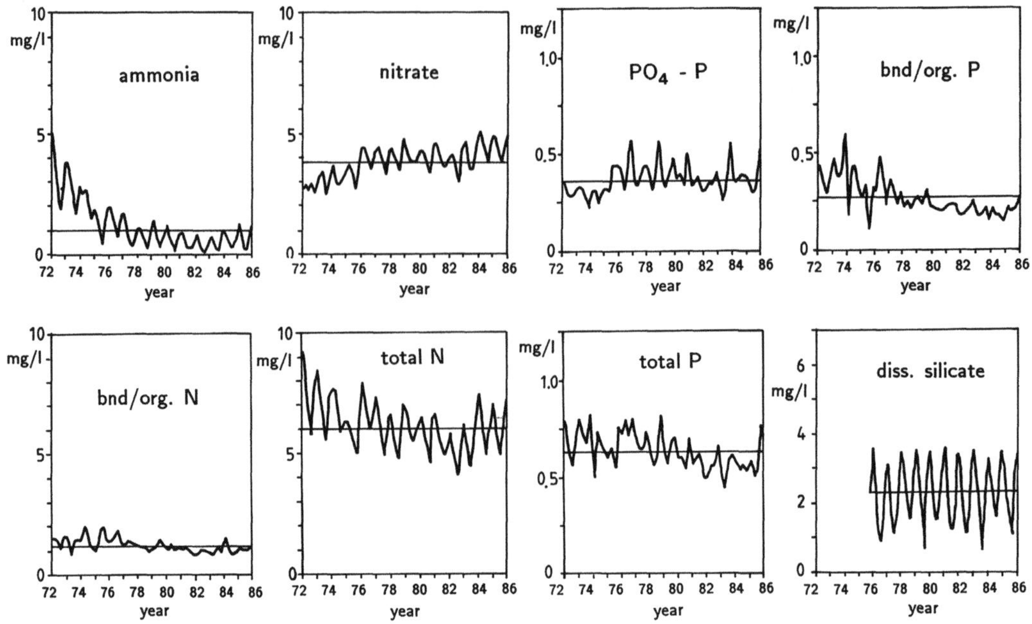

Figure 3: Seasonal and annual fluctuation of nutrient concentrations. Mean values for all stations along the Rhine in the Netherlands (Rijkswaterstaat Monitoring Programme)

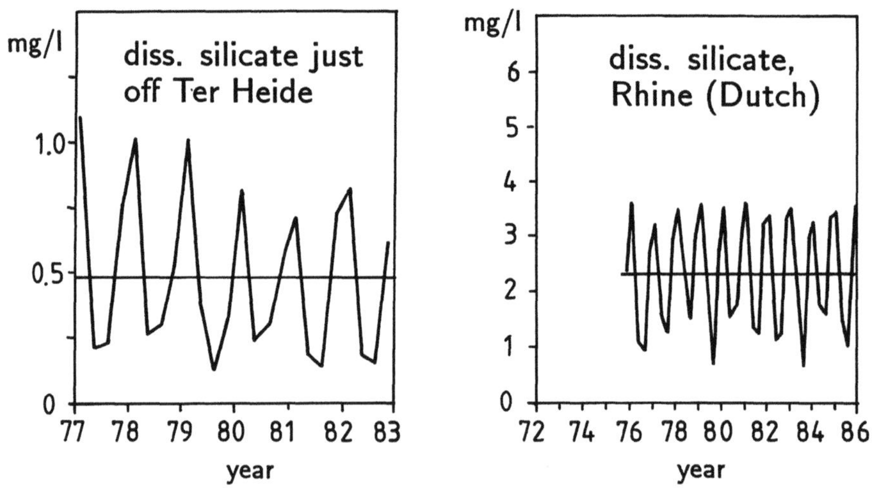

Figure 4: Silicate concentrations in the Rhine and in the Rhine river plume just off ter Heide (province of South Holland)

3. CORRELATION BETWEEN ANNUAL RIVER DISCHARGE AND ALGAL BIOMASS NEAR THE DUTCH COAST

The stations of the monitoring programme of Rijkswaterstaat in the North Sea are located in the area directly affected by the Rhine (cf. Fig. 1). Most Rhine water is discharged into this area; the Rhine tributary IJssel (cf. Fig. 1) contributes most of the freshwater that reaches the Wadden Sea (de Jonge, 1989).

It is of importance to realize that more water reaches the North Sea when there is high rainfall over northwestern Europe. The amount of water that flows through a river must be expected to be variable since each year the amount of rain and snow is different due to meteorological circumstances, that are, of course, different both seasonally and annually. The temporal variation in Rhine river discharge is shown in Fig. 5. The three years with the highest discharge are marked with an arrow. We will discuss the reason why we have marked these years in the last section.

One can expect that when a river is charged with rainwater above average levels, any pollutant will be diluted. This dilution effect is seen when the total P content near Lobith is compared with the annual discharge at Lobith, the station where the Rhine enters the Netherlands from Germany (Fig. 6). Why is this dilution effect not seen in the that of the Rhine downstream of Lobith ?

This Dutch part of the Rhine differs from the more upstream reaches. When a river is flushed with a high discharge, sediment on the bottom is resuspended so nutrient concentrations are raised, e.g. by the bottom mud's interstitial water. The lower course of the Rhine, in the Netherlands, is a sedimentation basin (Eisma et al., 1982). In times of slow river flow expecially, up to 60 % of the suspended matter supplied settles in the Netherlands. When the river runs fast after heavy rains in the catchment area, material is resuspended; at the same time the percentage that is retained in the sediment is lower because suspended material does not sink readily in a fast-flowing river. On the other hand, a slow river implies a long residence time of the water in the lower reaches, and possibly also a higher water temperature, so microbial activity affects nutrient concentrations more than before: nitrification, denitrification, and mineralization of organic matter supplied allochthonously and autochthonously (see above).

Another factor, the one that is probably of most importance in the nutrient budget of the Dutch part of the Rhine, is nutrient charging of the soil in the catchment area. Agriculture nowadays implies the use of fertilizers and excessive amounts of manure, which has boosted the economy through greatly increased productivity of crops, especially in the Netherlands. Soils have an enormous binding capacity for nutrients, in the case of phosphorus, for example, through adsorption and chemical reaction processes, but this binding capacity has a limit. When saturation is reached, nutrients go into the deep groundwater, and heavy rains are then followed by flushing of the "excess" nutrients into the rivers connected with the

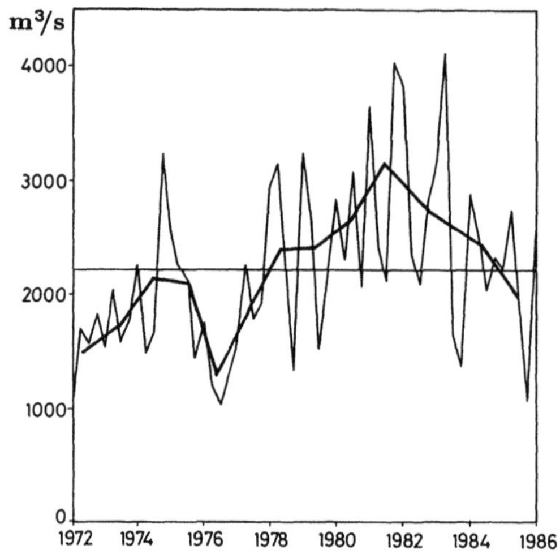

Figure 5: Discharge at Lobith station (cf. Fig. 1)

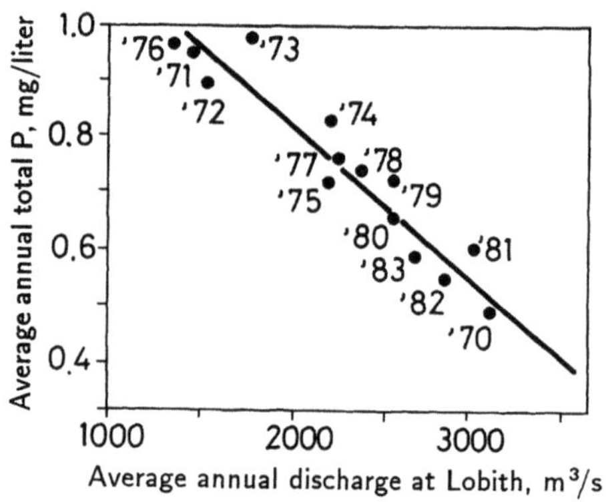

Figure 6: Average total P concentration and river discharge at Lobith (after Olsthoorn, 1985)

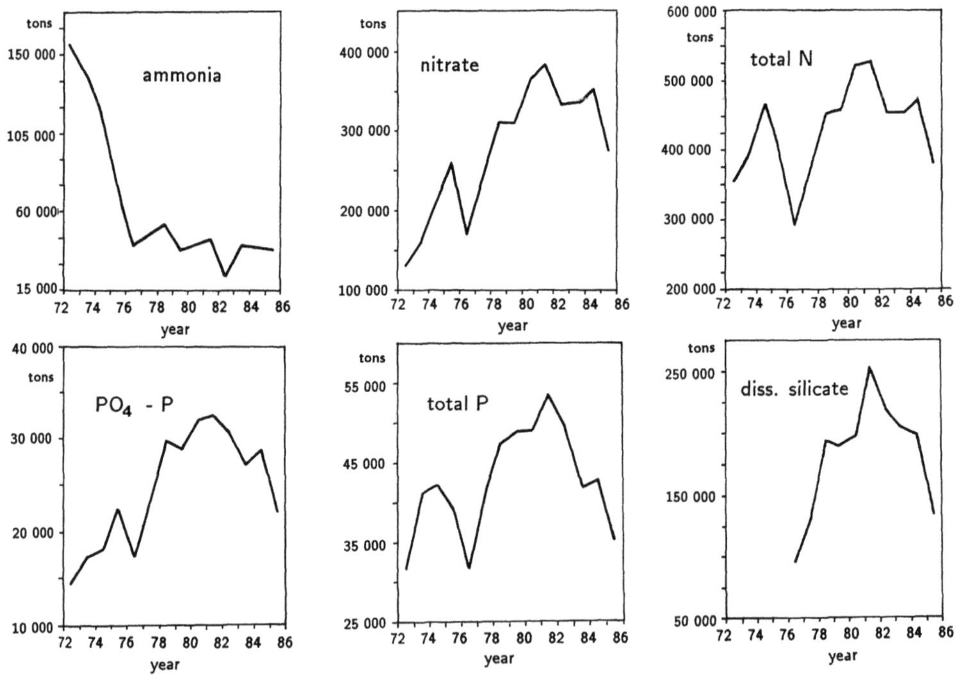

Figure 7: Calculated discharge of nutrients into the North Sea from the Rhine proper

Thus, soil erosion increases with increased rainfall. Hickel (1984) has pointed out that high nitrogen concentrations in the German Bight correlated with riverine N supplied through the Elbe. This nitrogen probably originated from the low-lying marshes near the mouth of the Elbe and was flushed into the estuary by heavy rains. Hagebro et al. (1983) presented another report of a positive correlation between freshwater discharge and nitrogen transport to the sea. The small Danish rivers are located in agricultural land and the groundwater appeared to be the most important nitrogen source here.

In the case of the Rhine we have seen (Fig. 6) that near Lobith, at the German-Dutch border, nutrients are still diluted by rain. However, we have also seen that downstream of Lobith, in the Dutch part of the Rhine, the concentration of most nutrient components remained roughly the same from year to year, no matter if the year was a wet or a dry one. Apparently, the discharge-related processes mentioned above counteracted dilution in the Netherlands. Of these processes, nutrient flushing from agriculture-enriched soils and resuspension of sediment have probably been of major importance: agriculture is far more developed in the Netherlands than along the Rhine in Germany; and the lower Rhine is a more important sedimentation basin than the course upstream (Eisma et al., 1982).

We have multiplied the seasonal figures of water discharge of the Rhine (Fig. 5) with the mean nutrient concentrations measured at the stations along its lower course, so we could calculate how much nutrient reached the sea in the years between 1972 and 1986. The results of this exercise are presented in Fig. 7. Notice that exactly in the wet years 1981, 1982 and 1983 (see Fig. 5) low oxygen values, down to depletion, were registered in the German Bight dúring summer (Gerlach, 1984). It is textbook knowledge that much freshwater flushed into a marine environment is followed by stratification, especially in summer. This may enhance algal growth - but the nutrients that fuel primary production are of course a "conditio sine qua non". Vertical stability may be greater in the German Bight after increased run-off, and possibly along the Dutch coast although tidal currents normally hinder the firm establishment of stratification here.

It is clear that in seasons with high river discharge, during "wet years", much nutrient reaches the sea, and favourable growth conditions for algae are created. However, the question remained whether or not these conditions actually induced more growth of phytoplankton in the eutrophied Dutch coastal zone. We tried to answer this question by putting all the chlorophyll measurements of the Rijkswaterstaat monitoring programme collected in the North Sea (between 1977 and 1982) in one figure. The results were surprising; they are shown in Fig. 8. In the phytoplankton growing season, which lasts from April to October (Fransz and Gieskes, 1984), the relation between river discharge and phytoplankton crop (as chlorophyll) was quite good. We admit that the correlation coefficient is not very high (Fig.8), but higher than one could reasonably expect given the sampling frequency which does not allow corrections for phytoplankton patchiness in space and time; moreover,

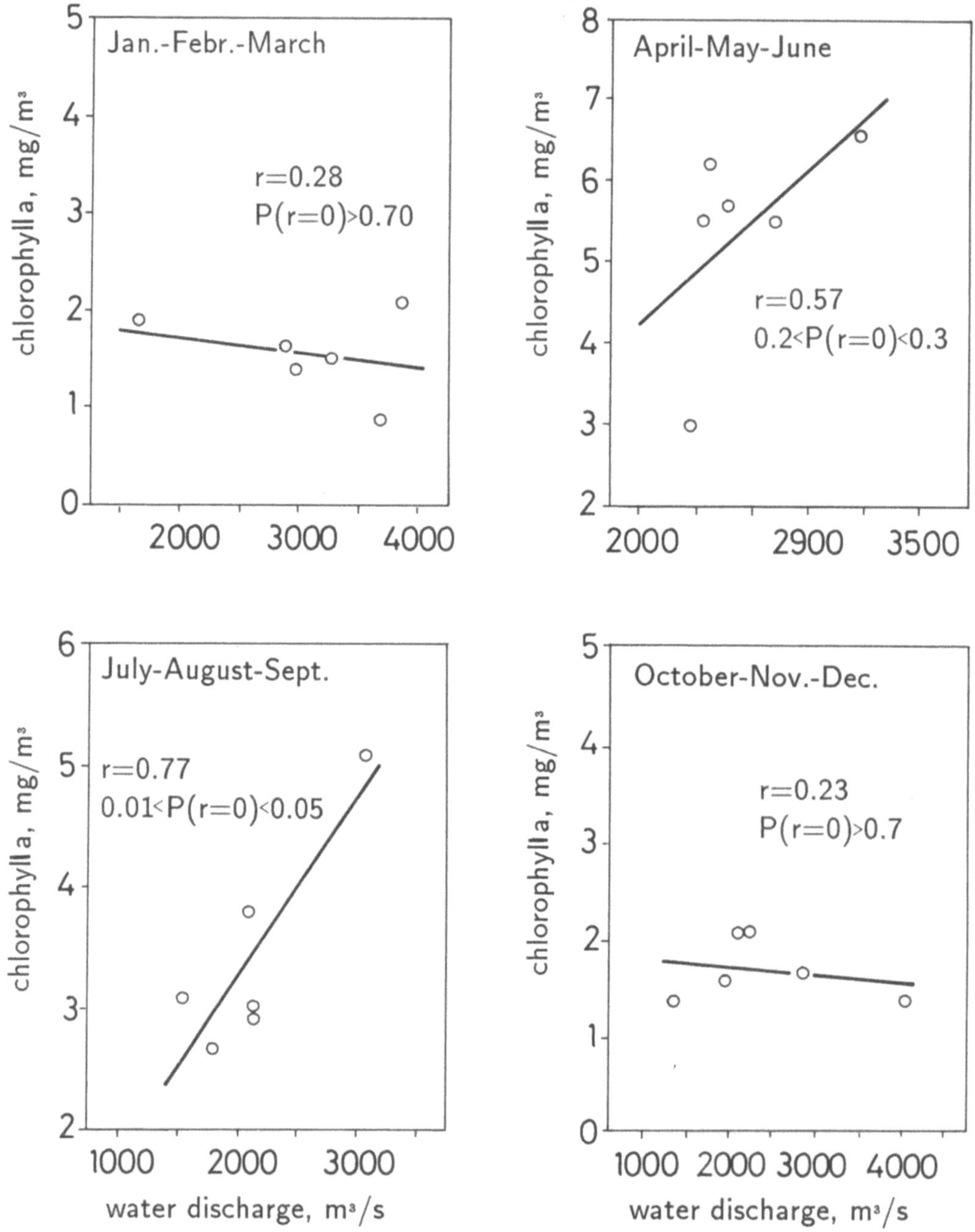

Figure 8: Correlation between Rhine discharge and mean chlorophyll concentration at the sampling stations in the North Sea (cf. Fig. 1)

chlorophyll is not the best biomass index. The good correlation between loading of nutrient and algal biomass in the eastern part of the North Sea's Southern Bight suggests that well over 50 % of the variance in primary production can be explained by nutrient availability. Neither light nor grazing, but the cocktail of nutrients from the Rhine seems to be the limiting factor for the amount of algal biomass in Dutch coastal waters. When enough nutrient is discharged into coastal waters, excess biomass may especially be produced during hot summers such as the one of 1989.

We may compare the situation in Dutch coastal North Sea waters with the one in a batch culture: the addition of nutrient is almost immediately followed by a response in the plankton. The measurements indicate that the time-lag between a discharge event and an algal population increase is shorter than a fortnight.

4. CONCLUDING REMARKS

The direct relation between algal crop and Rhine river discharge greatly facilitates the predictability of the Dutch coastal North Sea ecosystem. On the short time-scale of weeks the effect of high nutrient discharge is an increase in primary production, at least during the growing season (April to October). On longer time-scales of months, the effect of more algal growth is more difficult to predict, but such effects may be measurable as far downstream as the German Bight, the frontal regions north of the Netherlands, and in sedimentation areas of the central North Sea where the inflow of organic matter produced in the most eutrophic region, i.e. near Holland, may be considerable. On the time-scale of years nothing can be said with any degree of certainty; Sutcliffe (1972) has noted annual variations in primary production and algal biomass in Canadian marine bays close to land drainage. These variations were correlated with fish catches. However, the complex interactions in the ecosystem of the North Sea probably preclude such a direct relation between phytoplankton growth and fish landings one or more years later.

We have drawn attention to the rather neat relation between Rhine river discharge and algal crop: a wet period appears to be followed by rich phytoplankton growth - more often than not with negative consequences such as foam on beaches and low oxygen values when the biomass is degraded by heterotrophs.

We have presented evidence for our conclusion that as long as the saturation of the groundwater in the Netherlands with fertilizer nutrients is not diminished, heavy rainfall will not dilute river-borne nutrients (as is the case in the German part of the Rhine) but will enhance the diffuse source of groundwater leaching into the river, next to enrichment due to sediment resuspension. At the Lobith station, where the Rhine has not yet been in contact with Dutch point and diffuse nutrient sources, nutrients in the river are, as we have seen, proportionally diluted by rain; a situation probably typical only for the upstream, not the downstream Rhine. Over-fertilization of agricultural Dutch land should be reversed in order to escape the direct effect of a wet season on high algal crop abundance in the North Sea.

REFERENCES

Admiraal W, van Zanten B, and de Ruyter van Steveninck E (1989) Biological and chemical processes in communities of bacteria, phytoplankton and zooplankton in the lower river Rhine. Arch Hydrobiol (in press)

Anon. (1986) Resultaten van het waterkwaliteitsonderzoek in de Rijn in Nederland 1985. DBW/RIZA, Notanummer 86 - 21

Beukema JJ (1986) Eutrophication of the North Sea: reason for satisfaction or concern? In: Proc 2nd North Sea Seminar, Werkgroep Noordzee (Amsterdam): 27 - 38

Conley DJ, Kilham SS, and Theriot E (1989) Difference in silica content between marine and freshwater diatoms. Limnol Oceanogr 34: 205 - 213

Dijken G van (1989) Pigmentsamenstelling vom zoetwaterfytoplankton in de Rÿn en in de Rÿnpluim (Noordzee). University Groningen, Dept Marine Biology

Eisma D, Cadée GC, and Laane R (1982) Supply of suspended matter and particulate and dissolved organic carbon from the Rhine to the coastal North Sea. Mitt Geol-Paläontol Inst Univ Hamburg, SCOPE/UNEP Sonderband 52: 483 - 505

Fransz HG and Gieskes WWC (1984) The unbalance of phytoplankton and copepods in the North Sea. Rapp P-v Réun Cons int Explor Mer 183: 218 - 225

Gerlach SA (ed) (1984) Oxygen depletion 1980 - 1983 in coastal waters of the Federal Republic Germany. Ber Inst Meeresk Kiel no 130

Gieskes WWC and Kraay GW (1977) Continuous plankton records: Changes in the plankton of the North Sea and its euthrophic Southern Bight from 1948 to 1975 Neth J Sea Res 11: 334 - 364

Hagebro C, Bang S, and Sommer E (1983) Nitrate load - discharge relationships and nitrate load trends in Danish rivers. Proc Hamburg Symp, August 1983, IAHS Publ no 141

Hickel W (1984) Particulate nitrogen in the German Bight and its potential oxygen demand. In: Oxygen Depletion 1980 -1983 in Coastal Waters of the Federal Republic of Germany (ed SA Gerlach). Ber Inst Meeresk Kiel No. 130

Jonge V de (1989) Response of the Dutch Wadden Sea ecosystem to phosphorus discharges from the river Rhine. Hydrobiologia (in press)

Nelissen PHM and Stefels J (1988) Eutrophication in the North Sea. Neth Inst Sea Res Rept 1988 - 4: 1 - 100

Olsthoorn CSM (1985) Fosfor in Nederland 1970 - 1983. CBS, dept Natuurlyk Milieu. Staatsuitgeverij, The Hague

Owens NJP, Cook D, Colebrook M, Hunt H, and Reid PC (1989) Long term trends in the abundance of <u>Phaeocystis</u> sp in the north-east Atlantic. J mar biol Ass UK, in press

Sutcliffe WJ (1972) Some relations of land drainage, nutrients, particulate material, and fish catch in two eastern Canadian bays. J Fish Res Bd Canada 29: 357 - 362

NUTRIENT VARIABILITY IN A SHALLOW COASTAL LAGOON (Ria Formosa, Portugal)

M. Falcao and C. Vale
Instituto Nacional de Investigaçao das Pescas
C.R.I.P.A., Av 5 de Outubro 8700 Olhao, Portugal

ABSTRACT

Water samples have been collected at four stations in Ria Formosa every two weeks, at low and high tide, over one year. From an inner zone of the lagoon, water was sampled daily, at low and high tide, over three neap-spring tidal cycles. Results showed that nutrients in the lagoon water vary with both the season and the tide. While nitrates are higher in winter/spring, increasing with seawater influx, concentrations of phosphates and silicates tend to fluctuate mainly with the tide, displaying higher values in periods of low tide. Therefore, nitrates seem to be imported from the coastal waters, while benthic regeneration is probably the major factor controlling the levels of phosphates, silicates and ammonium in the lagoon.

INTRODUCTION

Factors determining the chemistry and hence the ecology of coastal lagoons are freshwater inputs, the rate of evaporation and circulation processes with the adjacent sea. On the other hand, circulation of metabolically active elements is strongly influenced by biochemical processes in the water column, sediments and interfaces (UNESCO, 1981).

Ria Formosa (Fig. 1) is a shallow (3.5 m depth) coastal lagoon (area 160 km²) with a permanent connection to the sea.

Freshwater inflow is very small, showing a significant influence only during sporadic run-off periods. Otherwise, the amplitude of the tide is relatively high (maximum 3.5 m) which causes an important tidal fluctuation of the water volume inside the lagoon. So, one major feature of the Ria Formosa is the extensive intertidal area (50 km²). In this paper we report on the temporal variability of nutrients, measured fortnightly over an annual cycle and daily over neap-spring tidal cycles, in a number of areas of the lagoon. Nutrient concentrations in the lagoonal water are greatly influenced by exchanges with the adjacent seawater and with the bottom sediment.

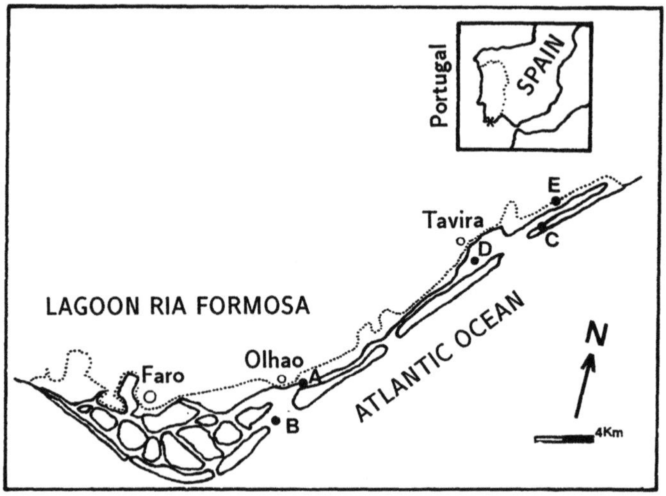

Figure 1: Map of lagoon Ria Formosa; locations of sampling stations A, B, C, D, E (•).

SAMPLING, METHODS

In order to study the nutrient variation in the Ria Formosa, water samples were collected fortnightly at low and high tide, including neap and spring tides, from September 1985 to September 1986 at four stations (Fig. 1, A-D). During January, February and March 1988, water samples were collected daily over three neap-spring cycles at low and high tide in a shallow flat area (station E). Silicates, nitrates, phosphates and ammonium were determined with a Chemlab autoanalyser, dissolved oxygen by the Winkler method, and salinity with a Beckman salinometer.

RESULTS

The sampling schedule followed in this study resulted in information on nutrient concentratrions in the lagoon water at different seasonal and tidal periods. Principal component analysis (PCA), applied to the gathered results, gave us a multivariate concept of all information and its variability in time (Legendre and Legendre, 1984). The multivariate vectors are viewed through a two-dimensional space, defined by two principal coordinate axes (A1, A2).

Seasonal variation of nutrients in the water

Salinity inside the lagoon varied within a narrow interval. Values were closest to those found in the coastal water, except during warmer periods when values were slightly higher (37 ‰) and at station D, in proximity to a freshwater input where salinity decreased to 33.8 ‰ at low tide.

Contrarily, nutrients changed seasonally within a broad range of values. Nitrates varied from 0.2 to 8.8 µM, phosphates from 0.05 to 5.6 µM and silicates from 0.2 to 14 µM. Principal component analysis was applied to the results from stations A, B, C and D. Figure 2 shows the dispersion of the parameters (concentrations of DO, NO_3, PO_4 and SiO_2 at each station) and the objects (low spring tide, high spring tide, low neap tide and high neap tide over the study year) in planes defined by two principal axes. Inertia as explained by two axes was always higher than 80 %.

The ordination of dissolved oxygen (DO) indicated that this parameter was closely associated with the winter/spring period (Fig. 2-I), that is, with the period of lower water temperature. At stations A and B located near a wide entrance channel, the highest DO levels were recorded at high spring tide (∇), when seawater influx was higher, but at stations C and D located at inner areas, the highest values were measured at neap tide (•,▼). The increase at neap tide may be due to photosynthesis. Indeed, DO at these stations ranged from 10.7 mg l^{-1} (180 % saturation) in March to 5 mg l^{-1} (98 % saturation) in September. Results of principal component analysis of nitrates are plotted in Figure 2-II. Axis A1 opposes samples collected in winter/spring to those obtained in other periods of the year. Nitrates in all the stations are closely associated with the winter/spring samples. Less clear seasonal patterns were found for phosphates and silicates. Although the highest PO_4 concentrations were measured in spring/summer (Fig. 2-III), closer association was obtained for those samples collected at spring tide (o,∇). For silicates, the fluctuation with tide is superimposed on seasonal variations. In fact, a rod including only samples collected at low tide of both neap and spring tide (o,•) may be delimited (Fig. 2-IV).

This means that the incoming seawater dilutes the silicate and phosphate-enriched sub-tidal volume. For these three nutrients, the axis A2 opposes station C to the other stations, suggesting also a spatial variability.

Fluctuation of nutrients with the tide

Nutrient concentrations at station E over the three neap-spring tidal cycles also varied within a broad range. Nitrates ranged from 0.1 to 4.6 µM, phosphates from 0.1 to 3.5 µM and silicates from 1 to 165 µM. The levels of dissolved oxygen, pH and temperature varied from 5.8 to 18.1 mg l^{-1}, from 7.5 to 8.9 and from 11 to 24 °C, respectively. The results obtained in this sampling allowed us to study the effect of the tide in a short period of the year (from January to March). We assume therefore, that a seasonal effect is not taken into account. The results of PCA for T, pH, DO, NO_3, PO_4, SiO_2 and NH_4 are presented in Figure 3. Inertia explained by the two principal axes was 59 %. Parameters may be agglomerated in three different groups. Temperature, DO and pH were associated mainly with low neap tide (•). This means that in winter DO (and pH), at inner zones, increased with the temperature, which indicated again the influence of photosynthesis. A second group formed by SiO_2, PO_4 and NH_4 was closely associated with situations of low spring tide (o).

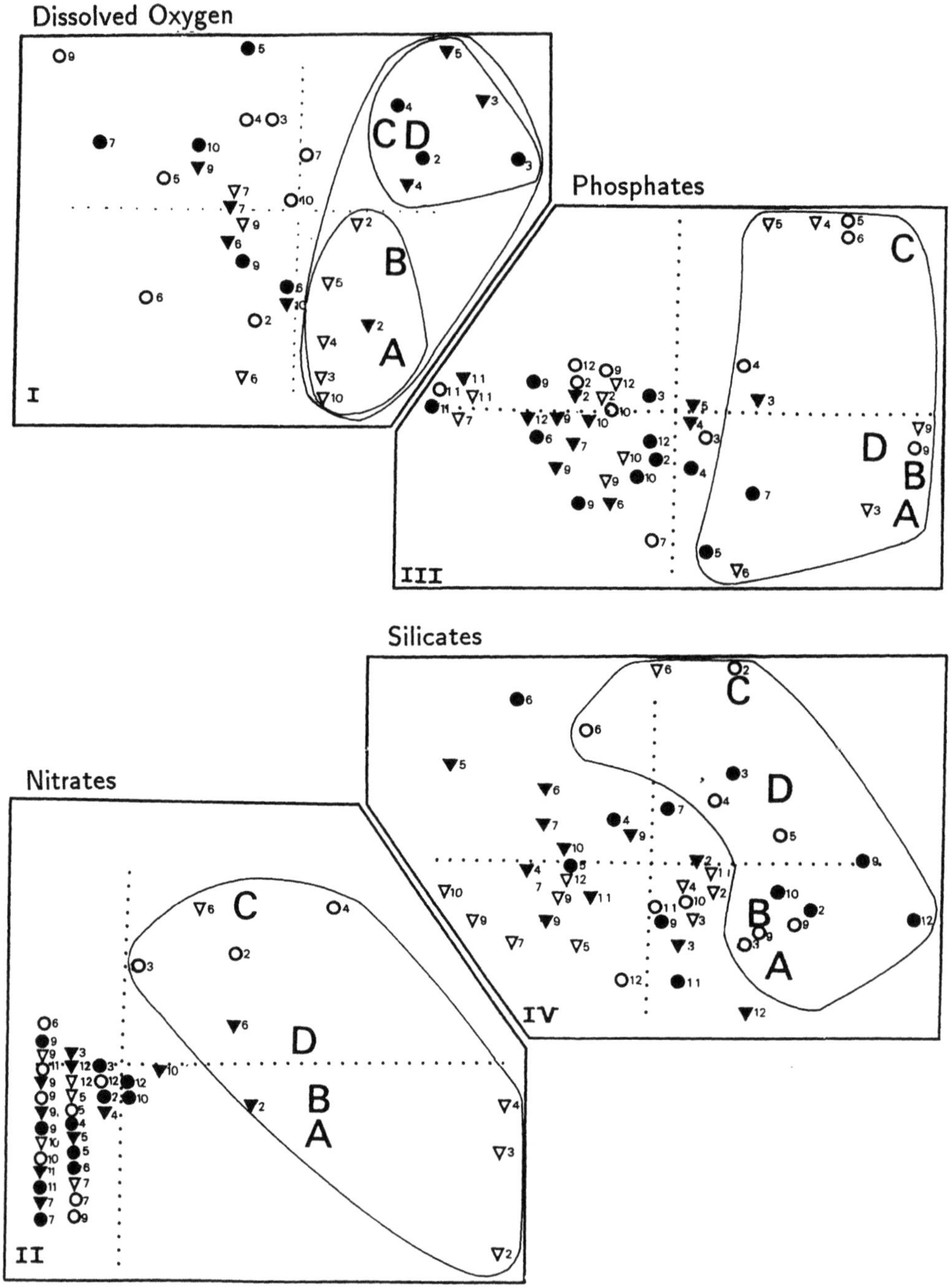

Figure 2: Dispersion of the parameters Dissolved Oxygen (I); Nitrates (II); Phosphates (III) and Silicates (IV) at stations A, B, C, D and the objects (low spring tide o; low neap tide •; high spring tide ∇ and high neap tide ▼ over the year, months marked by numbers) in planes defined by two principal axes A1 (horizontal) and A2 (vertical)

This means that these parameters reached maximum values at periods of low water volume. Otherwise, NO_3 values were higher at periods of high tide independently of the tidal amplitude (∇,▼). This means semi-diurnal fluctuations are superimposed on the fortnightly variations. As observed in the annual surveys, nitrates increased with the seawater influx.

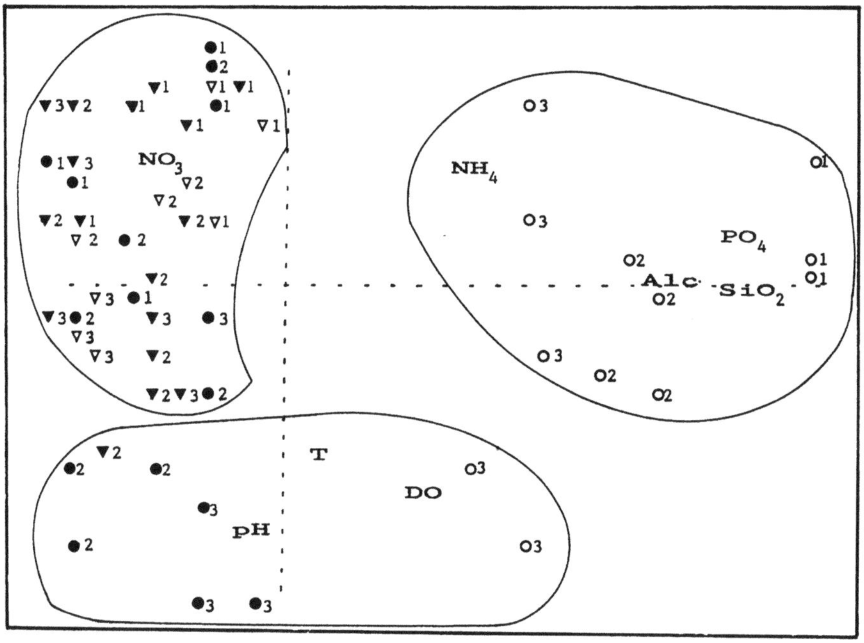

Figure 3: Dispersion of the parameters T, pH, DO, NO_3, PO_4, SiO_2, Alc. and NH_4 at station E and the objects (low spring tide o; low neap tide •; high spring tide ∇ and high neap tide ▼ in January (1), February (2) and March (3)) in a plane defined by two principal axes A1 (horizontal) and A2 (vertical)

DISCUSSION

The results obtained in these surveys showed that nutrients in the Ria Formosa varied in different time-scales: seasonally with the rise of temperature and all its induced processes; and with the tide, due to the large water volume exchanged with the sea in comparison with the subtidal volume of the lagoon.

Concentration of phosphate in the lagoon was higher in spring/summer, particularly at low spring tide. For silicates, a more conservative nutrient, the sediment-water exchange appears to be the major factor regulating their concentration in the water (except in short periods of run-off). Indeed, silicates were higher at low tide during the annual survey, that is, at situations of lower water volume in the lagoon, independent of the season. The increased concentrations of silicates and phosphates in these periods may be attributed to release from the bottom sediment (Aller, 1980). In microcosms, we have observed that phosphate and silicate fluxes are higher in summer, principally across the interface of mud sediments covered with plants and containing clams (Falcao and Vale, 1988). In the field, however, increases in phosphates and silicates are mainly observed in periods of higher physical perturbation of the bottom, and with lower water volume. An opposite pattern was found for nitrates, as maximum concentrations occurred in periods of maximum water volume. Semi-diurnal fluctuation indicates a rapid denitrification of the nitrates supplied by the coastal water. Our microcosms experiment also indicated a consumption of nitrates from the water column and an intense release of ammonium. In the Ria Formosa the levels of ammonium are around 50 times higher than the nitrate concentrations. Bacterial decomposition of organic matter (Helder, 1983) and digestion of organic matter by the abundant filter feeders of the bottom (Pacheco et al., 1988) may both contribute to the domination of ammonium. As nitrates increase in winter/spring and phosphates in warmer periods, the nitrate/phosphate ratio varies seasonally.

These results suggest that, as for other coastal environments (Nixon, 1980), Ria Formosa acts as a nitrogen transformer, importing dissolved oxidized forms and exporting nitrogen as reduced forms. Otherwise, increase of phosphates in warmer periods and situations of minimum water volume may suggest an export of phosphorus from the domestic sewage which has been temporarily retained in sediments.

ACKNOWLEDGEMENT

We thank M.L. Inacio, A.J. Figueiredo and A. Figueiredo for their help in the field and laboratory work. Support was provided by the PIDR-Ria Formosa on the behalf of CCRA.

REFERENCES

Aller RC (1980) Diagenetic processes near the sediment-water interface of Long Island Sound. Decomposition and nutrient element geochemistry (S, N, P). Advances in Geophysics 22: 237-350

Falcao MM, Vale C (1988) Fluxos de nutrientes em viveiros de ameijoa da Ria Formosa. Proceedings of VI Simposio Iberico de Estudio del Bentos Marino, Palma de Mallorca (in press)

Helder W, De Vries RTP (1983) Estuarine nitrite maxima and nitrifying bacteria (Ems-Dollard Estuary). Netherlands J Sea Research 17 (1): 1-18

Legendre L, Legrende P (1984) Ecologie numérique. 2. Le traitement multiple des données écologiques. Deuxieme edition française. Masson, Paris et les presses de l'Université du Québec, 335 p

Nixon SW (1980) Between coastal marshes and coastal waters - a revue of twenty years of speculation and research on the role of salt marshes in estuarine productivity and water chemistry. In: Hamilton P, Mc Donald KB (eds) Estuarine and Wetland Processes, pp 437-525

Pacheco L, Vieira A, Ravasco J (1988) Crescimento e reproduçao de ruditapes decussatus na Ria Formosa (Sul de Portugal). Proceedings of VI Simposio Iberico de Estudio del Bentos Marina, Palma de Mallorca (in press)

UNESCO (1981) Coastal lagoon research present and future. Seminary Duke University, Marine Laboratory Beaufort, NC, USA, August 1978, UNESCO Technical Paper 32, 98 p

SEASONAL CHANGES OF DISSOLVED AND PARTICULATE MATERIAL IN THE TURBIDITY ZONE OF THE RIVER ELBE

U.H. Brockmann and A. Pfeiffer
Institut für Biochemie und Lebensmittelchemie der Universität Hamburg, Martin-Luther-King Platz 6, D-2000 Hamburg 13, F.R.G.

ABSTRACT

During 1986 and 1987 four seasonal investigations on nutrients and particulate material were carried out in the lower Elbe estuary (Hamburg to the German Bight). Measurements made along transects portray the degree of summer nitrification, the extent of local sinks or sources, and the location of a turbidity zone. Gradients of total suspended material generally increased towards the sea. Frequent measurements at anchor stations within the salinity gradient and turbidity zone revealed persistent local nutrient remobilisation, nitrification and showed positive correlations between turbidity and total suspended material. The anchor station measurements indicated shifts in end-member concentrations, which were also seen along the transects. Imports from costal water were particularly evident in summer and autumn.

INTRODUCTION

Much of the dissolved and particulate load transported by rivers is converted in the estuaries by biological or physico-chemical processes before being discharged into the sea. The transport, transfer and transformation of chemical compounds are controlled by freshwater residence time and discharge rate. Seasonal effects on both nutrients and suspended material have been described in many studies (e.g. ARGE 1978-1987, Caffrey & Day 1986, Ward & Twilley 1986).

The river Elbe is the largest nutrient source (about 250.000 t N per year) of the German Bight (Anonymous 1980, Brockmann et al. 1988). During summer, nitrogen compounds are transformed by nitrification. This nitrification starts close to the river mouth and moves upstream with increasing temperatures (ARGE Elbe 1978-1987). Direct sources of ammonium are sewage inputs and the ammonification of organic material. A high nitrate load is characteristic of the river Elbe (Brockmann & Eberlein 1986). This results from agricultural fertilization, nitrification of ammonium upstream and weathering processes. In the Elbe estuary classified as between a partially and well mixed type B-D (Morris 1985), location and structure of gradients are mainly controlled by tidal action and river flow rates. In the oligohaline zone, a turbidity maximum is often found, indicating that formation, trapping, decomposition and export of fine particulate material are in a steady state situation.

METHODS

In 1986 and 1987 four cruises (winter, summer 1986, and spring, autumn 1987) were carried out on sections along the Elbe river between Hamburg and the German Bight (54°00'N, 08°00'E) (the first cruise was limited to 695 downstream km). Anchor stations, with frequent sampling during 4 - 5 days, were located at 695 downstream km within the oligohaline and turbidity zone (see map, Fig. 1). Measurements were performed at three depths. Samples were taken with Niskin bottles attached to a multi-sensor probe (ME Meerestechnik). Salinity was measured using a probe and in addition calculated from a chloride titration with $AgNO_3$. Oxygen content was measured by the Winkler method. Turbidity was estimated in terms of Nephelometric Turbidity Units (NTU) with a Turner Nephelometer. For the analysis of dissolved compounds, samples were filtered at a constant vacuum of 0.2 atm using glass fiber filters (GF/C Whatman). Nutrients were determined by AutoAnalyser methods, partly after dilution (Eberlein et al., 1983); dissolved organic nitrogen (DON) and phosphorus (DOP) after combustion by peroxidisulphate. Dried filters (60°C, 6 h) were used to estimate suspended matter (SM). Following hydrolysis of filter residues, particulate phosphorus (PP) and carbohydrates (PCH) were also

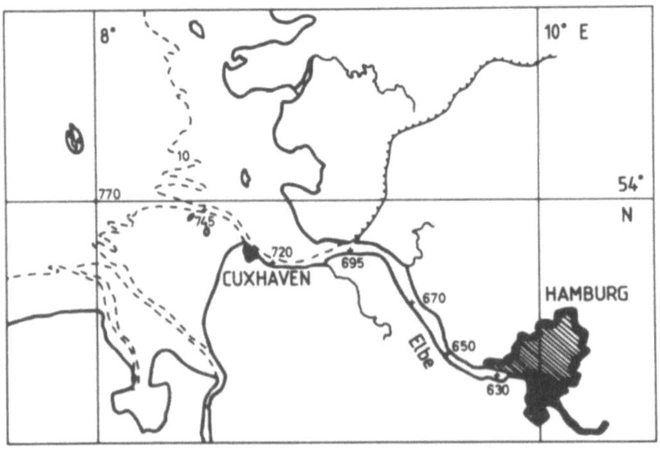

Fig. 1: Map of transect stations (+) in the Elbe. Numbers indicate km positions. Km 695 is the position of the anchor station.

determined by AutoAnalyzer methods. A Coulter Counter was used for analysis of particle concentration and particle size spectra. Using this method only statistical spectra of particle fragments are obtained.

RESULTS and DISCUSSION

The variability of dissolved and particulate parameters between Hamburg and the German Bight are presented (Tab. 1). Seasonal variability was most evident for the gradients of temperature, ammonium, silicate and oxygen. The ammonium gradients were steepest, due to

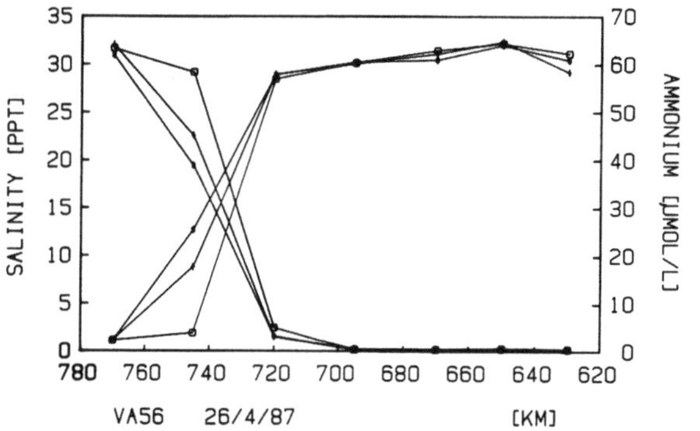

Fig. 2: Salinity (ppt) and ammonium concentration (µmol/l) along a transect (26/4/87) from Hamburg (km 620) to the German Bight (km 770; 54°N, 08°E)
Sampling depths:
* = surface
+ = mid depth
□ = above bottom

Table 1: Nutrients and suspended material in the Elbe estuary
Transects km 630-770 (German Bight) (VA 37: km 630-695)

Cruise Date	VA 37 20-26/1 1986	VA 56 26/4-2/5 1987	VA 47 4-9/7 1986	VA 64 26/9-2/10 1987
m^3/s*	1000 (+)	1800 (-)	625 (-)	700 (±)
°C	1.5- 2.6	4 - 13	11 - 24	14 - 16
Cl^- (g/l)	0.2- 0.5	0.1- 18	0.2- 18	0.2- 18
S (‰)	0.3- 0.9	0.2- 32	0.9- 32	0.4- 31
O_2 (mg/l)	10 - 12	8 - 13	3 - 8	2 - 8
DON(µgatN/l)	40 - 80	8 -160	12 -100	10 - 70
NH_4^+ (µM)	160 -220	2 - 65	2 - 28	2 - 11
NO_2^- (µM)	0.5- 3	0.5- 7	0.5- 2	0.5- 12
NO_3^- (µM)	250 -280	30 -380	10 -350	10 -370
DOP(µgatP/l)	0.5- 2	0 - 1.	0 - 2.5	0 - 1
PO_4^{3-} (µM)	5 - 6	0.3- 2.5	0.5- 6	1 - 7
SiO_2 (µM)	208 -220	10 -200	10 - 80	20 -140
PP (µgatP/l)	4 - 40	1 - 9	1 - 10	1 - 8
PCH (µM)	3 - 30	1 - 9	1 - 10	3 - 22
DW (mg/l)	30 -300	10 -240	20 -300	20 -300
NTU	8 - 70	3 - 35	2 - 35	5 - 35
Part.x10^{-3}/ml	-	1 - 15	6 - 25	2 - 10

* Discharge rates from ARGE (1978-1987) with indication of tendency (±)
PP = Particulate Phosphorus; PCH = Particulate Carbohydrates
DW = Dry Weight; NTU = Turbidity; Part. = Number of particles

enhanced nitrification during summer and autumn, causing local oxygen depletion. Silicate was incorporated into diatoms in the upstream freshwater regime and as a result of slow remineralisation probably remobilised on an annual scale. The freshwater discharge varies also seasonally, the largest varation occuring during spring. High discharge rates are often correlated with increasing nutrient concentrations, especially nitrate (Brockmann & Eberlein 1983).

In spring, end of April (1987), ammonium concentrations were still high downstream as far as the oligohaline zone (Fig. 2) where values decreased from 60 to 5 µM. Vertical gradients were observed at stations with significant vertical salinity gradients (km 745). In general, ammonium decreased (exept in winter) already at low salinities. A linear relationship between chlorinity and ammonium, measured along the Elbe transect during late spring, indicated conservative mixing (Fig. 3). However, frequent measurements lasting for longer than five days at the anchor station, show that this was not the case at low salinity. Besides the dominanting tidal influence on nutrient concentrations (Fig. 4), a persistent shift of endmember concentrations was observed, causing a consistent increase in chlorinity and a decrease of ammonium during all tidal stages. Compared to the nitrite and nitrate gradients, ammonium gradients were inverted. The convertion of ammonium via nitrite to nitrate at low chlorinity was indicated by the mixing curves of the nutrients (Fig. 5). There was a steep increase of about 40 µM for nitrate and nitrite. The corresponding decrease of ammonium was only 30 µM. The increase of the intermediate nitrite was less than 2 µM.

In summer, when the main load of ammonium had already been nitrified upstream, positive significant correlations were found between chlorinity and ammonium, silivate and phosphate at the anchor station. The ammonium increased up to 3 µM, silicate 24 µM and phosphate 0.5 µM. Nitrate was significantly negatively correlated with chlorini-

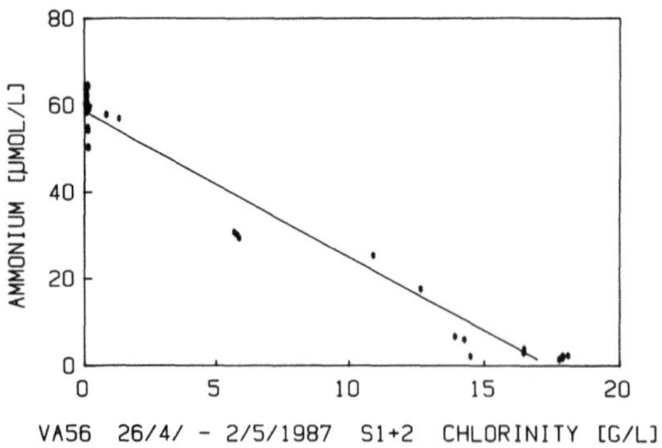

Fig. 3: Ammonium concentrations from two transects (Fig. 2) plotted against chlorinity

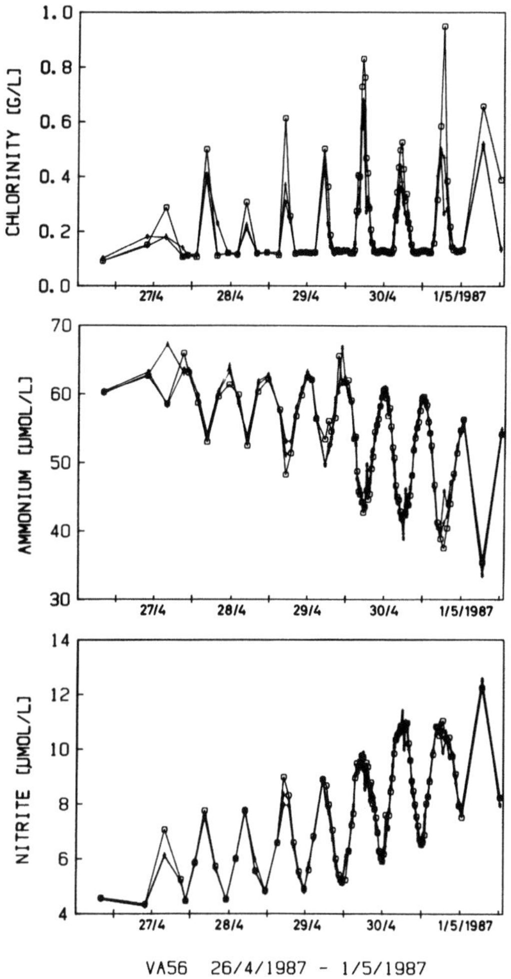

Fig. 4: Tidal variability and shift of endmember concentrations for ammonium, nitrite and chlorinity at the anchor station (km 695) in spring 1987

ty. However, locally non-conservative behaviour, shown by an intermediate maximum of 8 µM in the mixing curve, corresponded to a relative ammonium minimum of about 0.5 µM at the same chlorinity. This indicates that part of the ammonium, produced in the oligohaline zone,was immediately converted via nitrite to nitrate. Therefore the amount of released ammonium was probably higher. One probable source could have been the ammonification of DON compounds, which decreased by 7 µM. Remineralisation of DOP compounds might have caused the increase of phosphate, since DOP decreased by 0.5 µM. However, other sources must also be taken into consideration, such as the particulate fraction or remobilisation from the sediment. The high silicate increase showed the significance of the latter source.

Along the transects, which were not sampled during the same tidal phases, concentrations of particulate material may have been strongly affected by locally different current velocities, which may reach up to 4 knots (2 cm/s) at the surface. Nevertheless, turbidity maxima (Fig. 6) were detected, with high concentrations of organic material (phosphorus and carbohydrates). The total dry weight increased towards the sea, indicating that the turbidity maximum was mainly caused by organic material associated with smaller optically active particles (Campell & Spinrad 1987). These relationships were found for all cruises, extended to the Germen Bight, revealing that the suspended

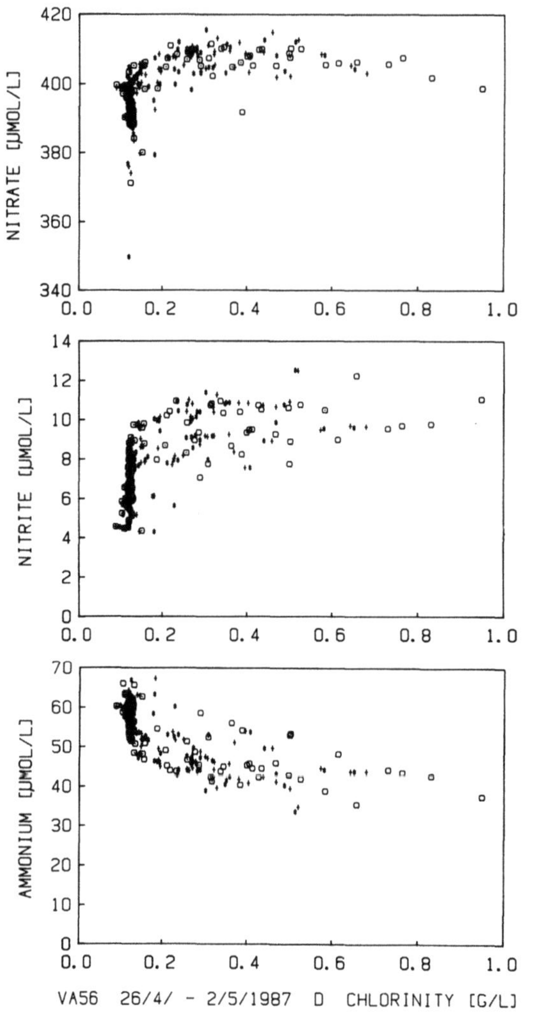

Fig. 5: Ammonium, nitrite and nitrate concentrations plotted against chlorinity. Measurements carried out at the anchor station. (compare Fig. 4)

material (SM) within the turbidity zone was not homogeneous. However, measurements at the anchor station show that locally there were highly significant positive correlations between turbidity and total particulate material, estimated as dry weight. Here tidal action influenced the distribution of suspended matter to a greater degree than regional differences.

The particle size spectra shifted during autumn simultaneously with the load of SM: Mean particle size increased with increasing concentration of SM at high current speeds. During spring, maximum particle size occurred at high water, independent of SM concentrations. Seasonal variations of SM were more evident as qualitative differences (Tab. 1) rather than as local quantitative changes since the latter were controlled mainly by tidal action.

Particularly in the turbidity zone, the frequent resuspension of different particulate material in the water column will enhance transfer processes, not only between the dissolved and particulate phase, but also between the water column and the sediment, giving the estuary filtering qualities (Schubel & Kennedy 1984).

ACKNOWLEDGEMENTS

This work was supported by the Deutsche Forschungsgemeinschaft within the SFB 327 "Tide-Elbe". For assistance we are grateful to the captains Klaaßen and Krull and crews of FS VALDIVIA. Thanks are given to the Elbe team, especially to Ilse Büns, Renate Lucht, Monika Schütt and Annette Schliephake.

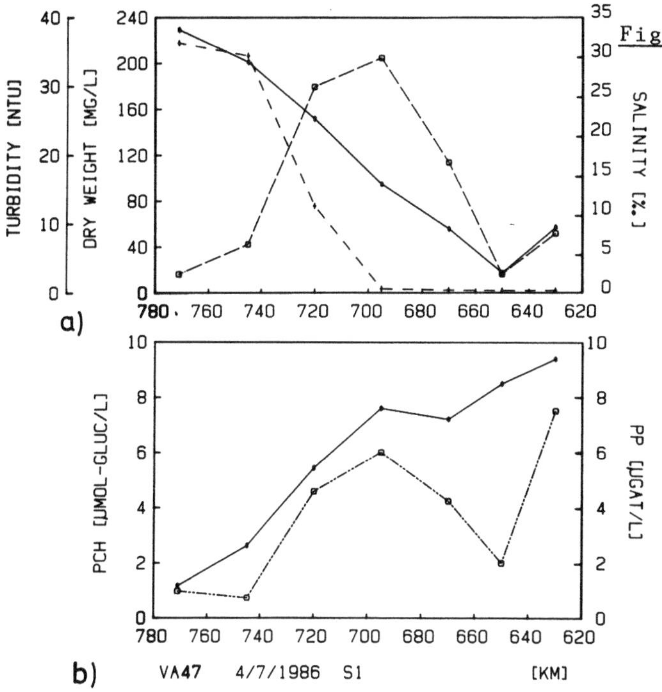

Fig. 6: Salinity (---) and a) concentrations of particulate material (means of three depths): turbidity (-.-), dry weight (——),

b) phosphorus (-..-..), carbohydrates (——) along a transect from Hamburg to the German Bight

LITERATURE

Anonymous (1980) Umweltprobleme der Nordsee. Sondergutachten Juni 1980, 503 pp

ARGE Elbe (1978-1987) Wassergütedaten der Elbe. Arbeitsgemeinschaft für die Reinhaltung der Elbe, Hamburg

Brockmann UH, Eberlein K (1986) River input of nutrients into the German Bight. In: Skreslet S (ed) The role of freshwater outflow in costal marine ecosystems. Springer, Berlin heidelberg New York Tokyo, p 231

Brockmann U, Billen G, Gieskes WWC (1988) North Sea nutrients and eutrophication. In: Salomons W, Bayne BL, Duursma EK, Förstner U (eds) Pollution of the North Sea: an assessment. Springer, p 348

Caffrey JM, Day jr JW (1986) Control of the variability of nutrients and suspended sediments in a gulf coast estuary by climatic forcing and spring discharge of the Atchafalaya river. Est 9: 295-300

Campbell DE, Spinrad RW (1987) The relationship between light attenuation and particle characteristics in a turbid estuary. Est Coast Shelf Sc 25: 53-65

Eberlein K, Brockmann UH, Hammer KD, Kattner G, Laake M (1983) Total dissolved carbohydrates in an enclosure experiment with unialgal Skeletonema costatum culture. Mar Ecol Prog Ser 14: 45-58

Morris AW (1985) Estuarine chemistry and general survey strategy. In: Head PC (ed) Practical estuarine chemistry. Cambridge Univ. Press, Cambridge London New York New Rochelle Melbourne Sydney, p 1

Schubel JR, Kennedy VS (1984) The estuary as a filter: an introduction. In: Kennedy VX (ed) The estuary as a filter. Academic Press, Orlando San Diego New York London Tronto Montreal Sydney Tokyo, p 1

Ward LG, Twilley RR (1986) Seasonal distributions of suspended particulate material and dissolved nutrients in a costal plain estuary. Est 9: 156-168

ANNUAL FLUX BUDGET OF DISSOLVED INORGANIC NUTRIENTS THROUGH A WELL-MIXED ESTUARY

Dan Baird and Deo Winter
Department of Zoology, University of Port Elizabeth, Box 1600
Port Elizabeth, South Africa

INTRODUCTION

The hypothesis that estuarine systems produce more material (e.g. inorganic nutrients, particulate carbon, dissolved organic carbon) that can be utilized or degraded within, and that excess material is being exported to the coastal marine environment where it may contribute to coastal ocean productivity has received considerable attention during the past few years. This "outwelling" hypothesis of Odum (1980) was examined for estuarine, saltmarsh and mangrove swamp systems by, for example, Valiela et al. (1978), Nixon (1980), Black et al. (1981), Boto and Bunt (1981), Daly and Mathieson (1981), Chrzanowski et al. (1982, 1983), Wolaver et al. (1983), Woodroffe (1985), Dame et al. (1986) and Baird et al. (1987).

There have been two basic approaches to testing the "outwelling" hypothesis (Dame et al., 1986). The indirect approach is concerned with the compilation of production and consumption budgets for the estuarine or marsh system and imbalances in the budgets are ascribed to either the import or export of material. The second approach involves the direct measurement of the fluxes of water and material across a cross-section of an inlet (Chrzanowski et al., 1982, Baird et al., 1987).

In this study the fluxes of dissolved inorganic nutrients (NH_4-N, NO_2-N, NO_3-N, PO_4-P and DIC) were directly measured and estimates are presented of spring and neap tidal transport as well as annual flux budgets for each of the constituents.

MATERIAL AND METHODS

The study was conducted in the constricted inlet of the Swartkops estuary (33°52S, 25°39E), near Port Elizabeth, South Africa. The estuary is shallow, well mixed, with extensive saltmarshes (Spartina maritima) and with a single, permanently open inlet to the ocean. The tidal prism of the estuary is ca 2.88 x 10^6 m^3 and the average river inflow per tidal cycle during the study period was ca 22.3 x 10^3 m^3. The flushing time is about 22 hours.

During the first spring and neap tide of each lunar cycle from June 1983 to May 1984 samples were collected at hourly intervals in 250 ml dark glass, acid washed bottles from 0.2 m below surface, mid-water and 0.5 m off the bottom. A total of 13 spring and 13 neap tides were sampled during this period. One sample from each depth per sampling time was used to determine a depth-averaged concentration.

The concentrations of all nutrient species were analysed by means of a Technicon Autoanalyzer (AAII) according to the analytical procedures described by Mosterd (1983a) for dissolved carbon and Mosterd (1983b) for other nutrients in sea water.

The volumes of water flowing respectively in (flood tide) and out (ebb tide) were calculated at hourly intervals, by means of a one-dimensional hydrodynamic model (Huizinga, 1985), which was calibrated for each sampling date as the product of the volume and the concentration. The flux of each nutrient was calculated and the total transport over the tidal cycle determined by summing the quantities imported (+) and exported (-). Appropriate statistical tests were applied to estimate confidence limits of these residual transports. An annual flux budget was calculated by multiplying the mean transport per tide with the number of tides.

RESULTS AND DISCUSSION

Maximum water volume transport coincided with maximum current velocities at approximately mid ebb and flood tides. Ebb volumes consistently exceeded flood volumes because of fresh water inflow. The mean spring and neap tidal amplitudes relative to mean sea level (MSL) at the mouth are 1.74 and 0.62 m, respectively (Baird et al., 1987). During most of the sampling period concentrations for the various nutrient species did not differ significantly ($p > 0.05$) with depth, justifying the use of a depth averaged mean value for each of the constituents. The homogeneity of concentrations in the water column at the inlet also indicates a relatively well-mixed and unstratified water body.

The concentrations of the inorganic nutrients varied over time within a tidal cycle as illustrated by a randomly selected data set (Fig. 1). Fig. 1b illustrates the concentrations of NH_4 over a tidal cycle, measured on January 10, 1984. Variations in concentrations of the individual nutrient species between tidal cycles were observed. Results have shown that the mean concentrations of NH_4, NO_2, NO_3 and PO_4 measured during ebbing tides of both spring and neap tides, were consistently and statistically significantly higher ($p<0.05$) than flood tide concentrations. Similar trends were also observed by Dame et al. (1986) for these inorganic nutrients in the mouth inlet.

The mean concentrations (per tide) of NH_4, NO_2 and NO_3 were, in general, higher during spring tides than during neap tides. This was, however, not observed for PO_4 and DIC whose concentrations during neap tides were, in general, higher than during spring tides. The emerging pattern is thus higher concentrations for nitrogen during the ebb of all tides, and higher concentrations during spring than neap tides.

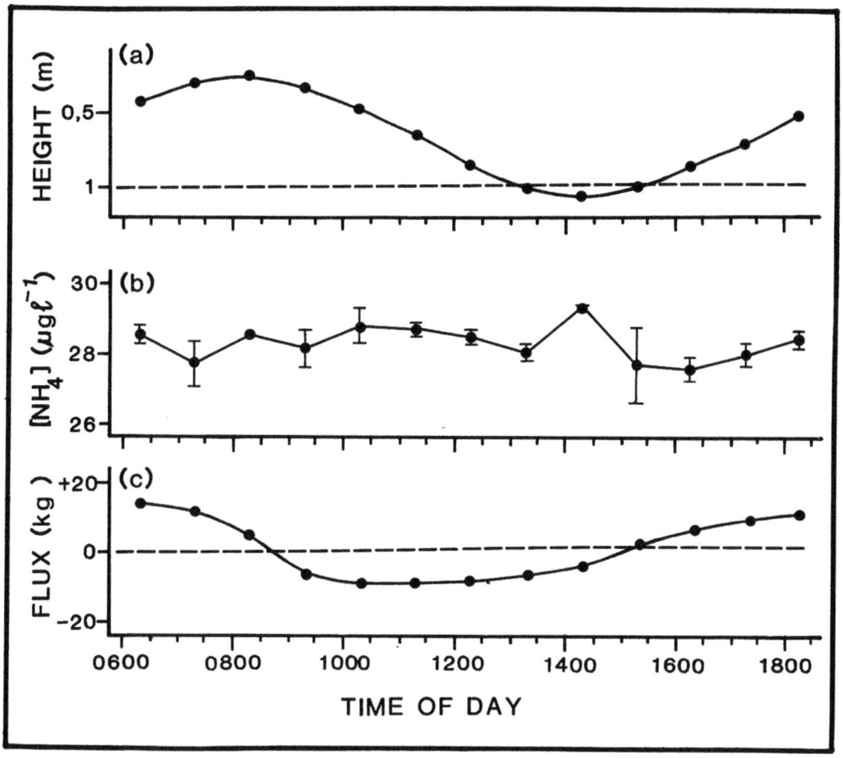

Figure 1: Concentration and computed flux of NH_4-N over one tidal cycle

There appears to be no clear seasonal trend in the mean concentrations of any of the nutrients (Fig. 2). The mean monthly concentrations of DIC remained remarkably constant throughout the study period (Fig. 2a). The others show some fluctuations with higher concentrations measured towards the end of winter and early summer, June - October (Fig. 2b, c, d and e).

The quantities of nutrients exchanged were significantly ($p > 0.05$) less during neap tides than during spring tides. Fig. 1c illustrates the flux of NH_4 in relation to the state of the tide. In this example, the flux yielded a net export of ca 5.25 kg NH_4. Table 1 shows the net flux for each month for each nutrient. Nutrients were exported from the estuary during most of the months with an annual budget indicating a net export from the estuary. No clear seasonal pattern for any one of the nutrients is in evidence and no clear explanation can be given for the inter-month variations.

Several studies have been conducted elsewhere on the import-export function of entire estuarine systems by, for example, Dame et al. (1986), Woodroffe (1985), Woodwell and Whitney (1977), Keizer and Gordon (1985), Baird et al. (1987).

All these studies reported on the export of particulate organic material and, in some cases, also of inorganic nutrients (Dame et al., 1986, Keizer and Gordon, 1985, Woodwell and Whitney, 1977).

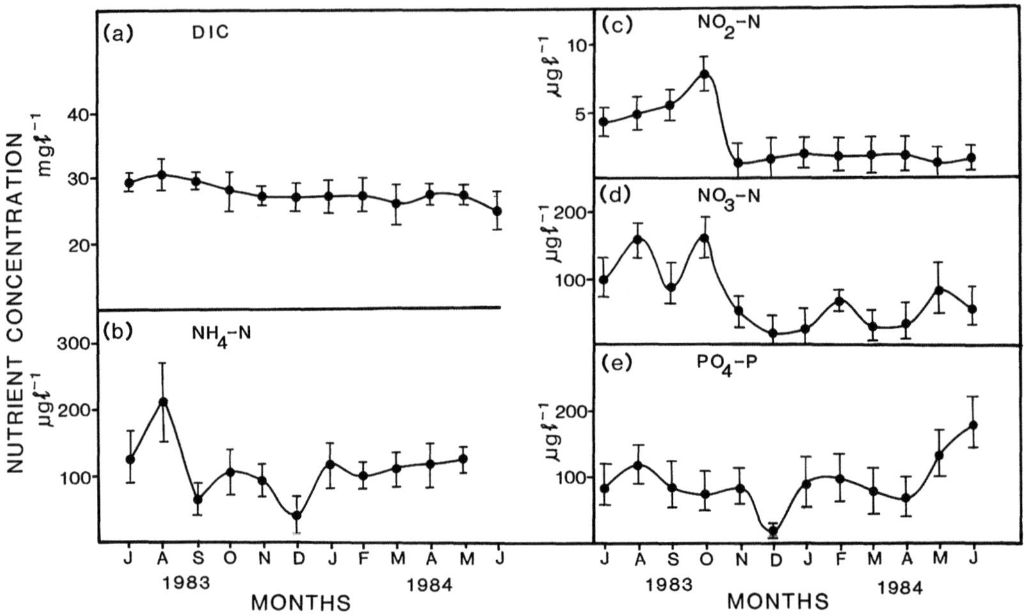

Figure 2: The mean monthly concentrations of inorganic nutrients for the period July 1983 - June 1984 (bars denote standard error, n ranged from 58 to 78 per nutrient per month)

Phosphorus was consistently exported from the estuary, and the magnitude of orthophosphate export was at the lower end of the range of exports reported by Nixon (1980) in his review of 7 marsh systems. Nixon (1980) suggested that exported phosphorus is most probably derived from the decomposition of Spartina in the marshes. Although the annual net exchange of phosphorus is negligible (Table 1), the forms and concentrations in which phosphorus occurs are important, especially in the saltmarshes.

The general explanation of the magnitude and direction of transport of inorganic nutrients between estuarine systems and the nearshore ocean is dependent on the dynamics of water movement and the mechanisms, biologically and physically, by which nutrients are removed from or added to tidal waters (Dame et al., 1986). The Swartkops estuary exports most types of materials to the ocean on an annual basis and thus has, in general, an "outwelling" function.

Table 1: Mean monthly flux of inorganic nutrients in the Swartkops estuary (+ = import, - = export, SE = standard error of mean, n.d. = no data).

					NUTRIENT					
Month	$[DIC]$ kg	SE	$[NH_4]$ kg	SE	$[NO_2]$ kg	SE	$[NO_3]$ kg	SE	$[PO_4]$ kg	SE
Jul 83	1698.36	1061.93	4.49	2.45	n.d.	-	5.51	3.67	4.20	2.85
Aug 83	-1343.35	1518.41	-35.35	16.91	n.d.	-	-44.73	27.86	- 9.09	5.31
Sep 83	2019.76	2617.60	1.48	5.69	- 0.01	0.45	1.26	2.18	- 0.98	1.14
Oct 83	161.31	1350.87	- 8.02	4.88	- 1.68	0.47	-34.29	9.68	- 8.97	3.83
Nov 83	418.35	1102.51	- 1.27	3.93	0.01	0.04	- 0.76	3.27	- 1.37	2.90
Dec 83	- 612.57	1118.10	0.79	2.00	- 0.02	0.04	- 0.51	0.30	- 2.08	0.94
Jan 84	- 278.50	1187.46	- 1.85	7.31	0.00	0.09	- 0.61	0.91	- 3.24	3.88
Feb 84	- 783.20	1245.02	- 0.54	5.74	- 0.04	0.10	0.31	3.04	- 2.57	4.12
Mar 84	632.02	1247.59	- 1.57	4.53	0.02	0.05	- 0.21	0.75	- 1.18	2.99
Apr 84	- 601.31	1026.83	- 3.03	3.87	- 0.02	0.04	- 1.15	0.36	- 3.57	2.06
May 84	- 815.01	1057.49	- 2.33	4.30	- 0.01	0.02	- 3.72	4.46	1.16	4.53
Jun 84	- 119.85	917.10	- 3.66	9.18	- 0.05	0.05	1.12	2.49	- 5.42	6.37

REFERENCES

Baird D, Winter PED, Wendt G (1987) The flux of particulate material through a well-mixed estuary. Continental Shelf Res 7:1399-1403

Black RE, Lukatelich RJ, McComb AJ, Rosher JE (1981) Exchange of water, salt, nutrients and phytoplankton between Peel Inlet, Western Australia, and the Indian Ocean. Aus J Mar Freshw Res 32: 709-720

Boto KG, Bunt JS (1981) Tidal export of particulate organic matter from a Northern Australian mangrove system. Estuar coast Shelf Sci 13: 247-245

Chrzanowski TH, Stevenson LH, Spurrier JD (1982) Transport of particulate organic carbon through the North Inlet ecosystem. Mar Ecol Prog Ser 7: 231-245

Chrzanowski TH, Stevenson LH, Spurrier JD (1983) Transport of dissolved organic carbon through a major creek of the North Inlet ecosystem. Mar Ecol Prog Ser 13: 167-174

Daly MA, Mathieson AC (1981) Nutrient fluxes within a small north temperate salt marsh. Mar Biol 61: 337-344

Dame R, Chrzanowski T, Bildstein K, Kjerfve B, McKellar H, Nelson D, Spurrier J, Stancyk S, Stevenson H, Vernberg J, Zingmark R (1986) The outwelling hypothesis and North Inlet, South Carolina. Mar Ecol Prog Ser 33: 217-229

Huizinga P (1985) A dynamic one-dimensional water quality model. CSIR Res Rep 562: 1-23

Keizer PD, Gordon DC (1985) Nutrient dynamics in Cumberland Basin-Chignecto Bay, a turbid macrotidal estuary in the Bay of Fundy, Canada. Neth J Sea Res 19: 193-205

Mosterd SA (1983a) Photochemical procedure used in South Africa for the photometric determination of dissolved carbon in seawater. S Afr J mar Sci 1: 57-60

Mosterd SA (1983b) Procedures used in South Africa for the automatic photometric determination of micronutrients in seawater. S Afr J mar Sci 1: 189-198

Nixon SW (1980) Between coastal marshes and coastal waters - a review of twenty years of speculation and research on the role of salt-marshes in estuarine productivity and water chemistry. In: Hamilton R, MacDonald KB (eds). Estuaries and Wetland Processes with Emphasis on Modelling. Plenum Publishing Company, New York, p. 437

Valiela I, Teal JM, Volkman S, Shafer D, Carpenter EJ (1978) Nutrient and particulate fluxes in a saltmarsh ecosystem: tidal exchanges and inputs by precipitation and groundwater. Limnol Oceanogr 23: 798-812

Wolaver TG, Zieman JC, Wetzel R, Webb KL (1983) Tidal exchange of nitrogen and phosphorus between a mesohaline vegetated marsh and the surrounding estuary in the Lower Chesapeake Bay. Estuar coast Shelf Sci 16: 321-332

Woodroffe CD (1985) Studies of a mangrove basin, Tuff Crater, New Zealand: III. The flux of organic and inorganic particulate matter. Estuar coast Shelf Sci 20: 447-461

Woodwell GM, Whitney DE (1977) Flux pond ecosystem study: Exchange of phosphorous between a salt marsh and the coastal water of Long Island Sound. Mar Biol 41: 1-6

STUDIES ON THE DISTRIBUTION OF OXYGEN AND NUTRIENTS IN THE WESER ESTUARY

I. Grabemann, H. Kühle, B. Kunze, A. Müller, L.J.R. Neumann *
Institut für Physik, GKSS Forschungszentrum Geesthacht GmbH
Max Planck Straße, D-2054 Geesthacht

1 INTRODUCTION

The Lower Weser river (Müller et al., chap. 7, this vol., Fig. 1) flows through a densely populated area in North Germany and is used as a navigational channel, harbour area, industrial location and receives water from both urban and industrial sewage. Together with other rivers the Weser carries pollutants to the North Sea. Tidal and ocean currents cause an accumulation of these pollutants in the German Bight (Hainbucher et al. 1987).

The already polluted river water coming from upstream passes a sewage inlet several times and is manifold loaded. Depending on the river discharge a water-body needs 2 to 50 days to travel from Bremen to Bremerhaven (Fig. 1). In the past, critical oxygen-situations have occured in the Lower Weser during both summer and autumn periods. Because of this, the capabilities of the sewage treatment plants have been greatly improved during the 1980's, e.g. the largest plant (Seehausen (Bremen)) was equipped with a biological stage at the end of 1985.

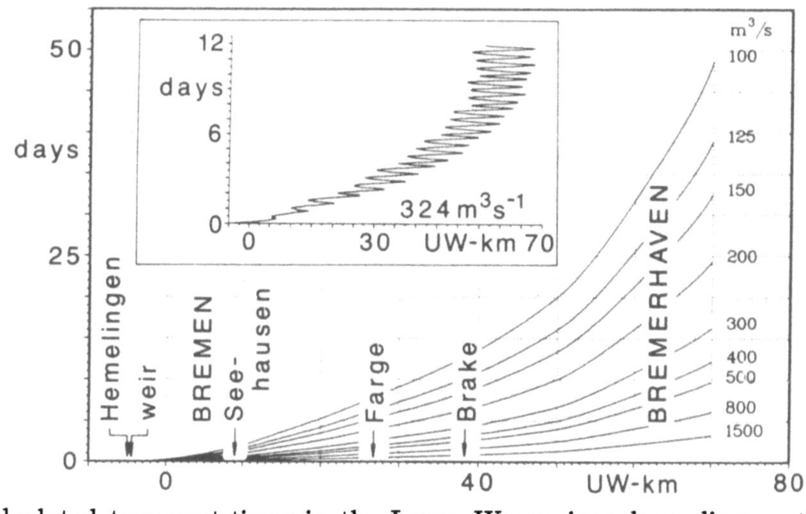

Fig. 1: Calculated transport-times in the Lower Weser river depending on the river discharge. UW-km: Lower-Weser-Kilometers.

This short paper deals with the oxygen budget affecting nutrients with regard to the plant Seehausen. Chlorinated hydrocarbons and heavy metal concentrations were also investigated and are described elsewhere (Knauth and Sturm, chap. 10, this vol.; Schirmer, chap. 9, this vol.).

* present affiliation: Salzgitter Elektronik GmbH, D-2302 Flintbek

2 WATER QUALITY MEASUREMENTS AND THEIR REPRESENTATIVITY

To document the actual water quality situation, to detect long-term changes and to get better information about sources and sinks of water constituents, about their distribution and influence on the environment (especially on the German Bight), the water quality of the Lower Weser has been monitored continuously at five onshore stations (three in the freshwater zone and two in the marine zone) since 1979. Additionally, since 1984, regular longitudinal profile measurements are being undertaken by ship. As such investigations require great efforts on measuring capacity they are combined with a one-dimensional numerical water quality and transport model to get a complete overview of the pollution situation (Müller et al., chap. 7, this vol.).

In order to check the representativity of the onshore stations an extensive field experiment was performed (Kühle et al. 1989). It shows that upstream of the turbidity zone the vertical and horizontal distribution of all parameters is nearly homogeneous; with some exceptions this is also the case as far as Bremerhaven, aside from the turbidity which becomes strongly inhomogeneous. For all dissolved substances, therefore, agreement exists between measurements at the banks and the mean value of the cross section. This differs from substances which are bound to the suspended matter in the region of the turbidity zone (Fig. 2). Here measurements at the banks are not representative.

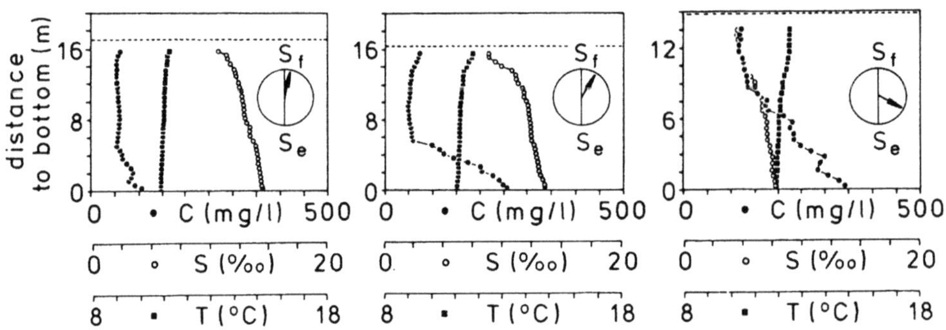

Fig. 2: Some typical vertical profiles for three different tidal elevations showing the vertical distribution of suspended matter concentration C, salinity S and temperature T in the turbidity zone (measured in the centre of a cross section at Bremerhaven). The arrows in the tide clocks show the respective tidal phase of measurement. S_f: flood slack, S_e: ebb slack.

3 NUTRIENT AND OXYGEN CONCENTRATIONS

Whereas the longitudinal sections of nitrate- and phosphate-content show no significant peak, nitrification usually causes an ammonium and nitrite maximum. Within the brackish water zone, the content of all nutrients decreases due to mixing with the intruding seawater. The nutrient concentrations depend on the river discharge in different ways. As nitrate is easily washed out of soil by rain, the concentration of nitrate increases with increasing river discharge. All other substances , however, are diluted and their concentration decreases.

Maximum ammonium and nitrite concentrations are measured between UW-km 10 and 40 and between UW-km 20 and 50 respectively, caused by inputs from the Seehausen sewage treatment plant (Fig. 3). Because nitrification depends on water temperature and season (Schirmer et al. 1983), in times of high temperatures the ammonium content is lower whereas the nitrite content is higher with more pronounced peaks. During periods of low temperatures the preload of ammonium is high and this large content is carried without nitrification and with an additional amount through the Lower Weser to the North Sea. As model calculations show (Fig. 4), the Seehausen plant considerably increases the ammonium content (typical input 1987: 39 mg/l equivalent to 66 g/s NH_4-N) over a distance of more than 20 km. Smaller inputs would result in smaller ammonium peaks. Because of biodegradation and nitrification, lower oxygen concentrations are usually measured in the Bremen/Brake region during periods of higher temperatures and lower river discharges (Fig. 3).

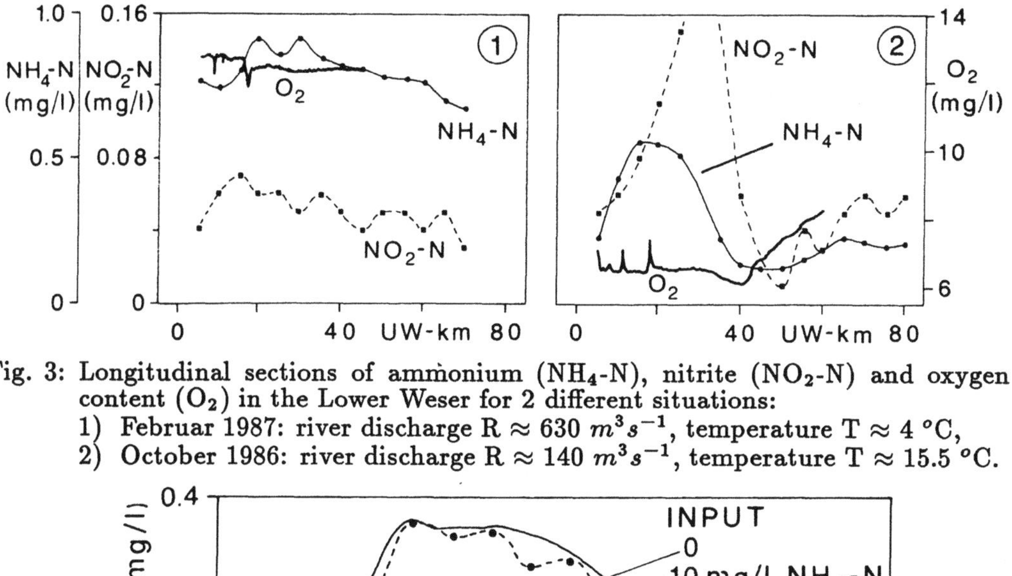

Fig. 3: Longitudinal sections of ammonium (NH_4-N), nitrite (NO_2-N) and oxygen content (O_2) in the Lower Weser for 2 different situations:
1) Februar 1987: river discharge R $\approx$ 630 m^3s^{-1}, temperature T $\approx$ 4 ^oC,
2) October 1986: river discharge R $\approx$ 140 m^3s^{-1}, temperature T $\approx$ 15.5 ^oC.

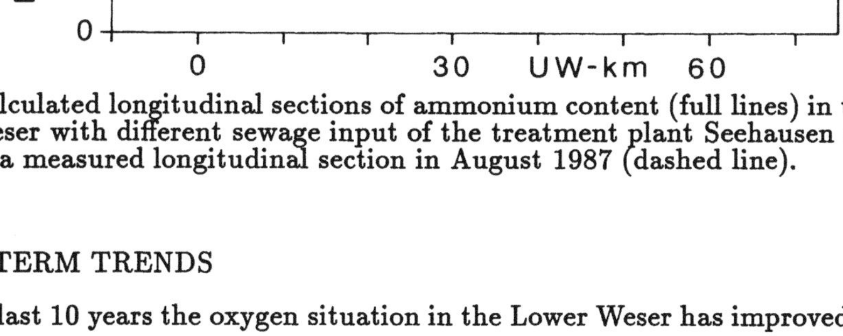

Fig. 4: Calculated longitudinal sections of ammonium content (full lines) in the Lower Weser with different sewage input of the treatment plant Seehausen compared to a measured longitudinal section in August 1987 (dashed line).

4 LONG-TERM TRENDS

During the last 10 years the oxygen situation in the Lower Weser has improved (Fig. 5). Between 1979 and 1982 the content in the Lower Weser is continuously lower than in the incoming water, whereas since 1983 there is nearly no difference. Model calculations (Müller et al., chap. 7, this vol.) show, that the significant lower input of organic bound nitrogen and C-BOD after installation of the biological stage (Seehausen plant) contributes to the improvement of the oxygen situation of the Lower Weser. The input of nutrients by the Seehausen plant, however, is nearly constant since 1979. Therefore the Weser carries today similar nutrient amounts to the North Sea as in the years before (Tab. 1). In Winter, especially, the effluents of the Lower Weser contribute greatly to the ammonium flux.

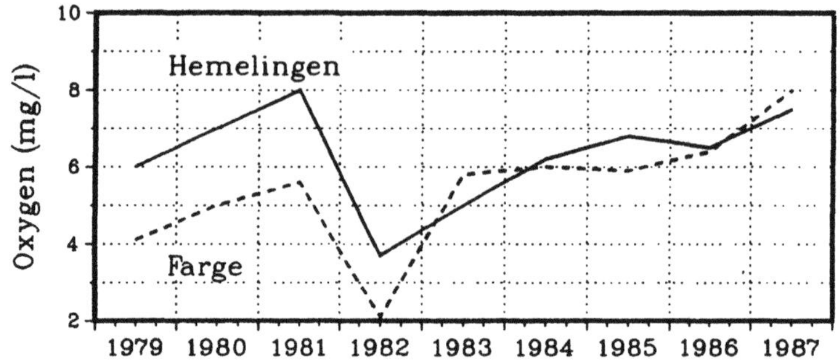

Fig. 5: Long-term trends of oxygen content between 1979 and 1987. The values are summer-minimum averages of each year.

Table 1: BOD_5- and nutrient fluxes (annual means, ARGE Weser 1979-1987) at Hemelingen and the additional fluxes (crude approximations, evaluated from the monitored discharge of effluent data) into the Lower Weser.

	BOD_5	NH_4-N	NO_3-N	O-PO_4-P	P_{sum}
Hemelingen	800-1900 g/s	100-390 g/s	1130-2830 g/s	80-110 g/s	190-380 g/s
Lower Weser	27 g/s	112 g/s	18 g/s	20 g/s	25 g/s

5 REFERENCES

ARGE Weser (Arbeitsgemeinschaft der Länder zur Reinhaltung der Weser) (1979 - 1987) Zahlentafeln der physikalisch-chemischen Untersuchungen.

Hainbucher D, Pohlmann T, Backhaus J (1987) Transport of conservative passive tracers in the North Sea: first results of a circulation and transport model. Cont Shelf Res 7:1161-1179

Knauth HD, Sturm R (1989) Use of nonvolatile chlorinated hydrocarbons in suspended particulate matter as anthropogenic tracers for estimating the contribution of the rivers Weser and Elbe on the pollution of the German Bight. This volume, chapter 10

Kühle H, Grabemann I, Kunze B, Müller A, Neumann LJR, Prange A (to be published) Field measurements for better understanding of transport processes in the Weser estuary.

Müller A, Grodd M, Weigel P (1989) Lower Weser monitoring and modelling. This volume, chapter 7

Schirmer M, Hackstein E, Liebsch H (1983) Kritische Belastung des O_2-Haushaltes der Unterweser durch kommunale und industrielle Abwässer. Verhandlungen der Gesellschaft für Ökologie 10:337-344

Schirmer M (1989) Monitoring the bioavailability of heavy metals in relation to the sediment pollution in the Weser Estuary. This volume, chapter 9

MANAGEMENT OF NITRATE LOADING IN THE CHAO PHRAYA ESTUARY

Prida Thimakorn
Regional Research and Development Center, Asian Institute of Technology
Bangkok, Thailand

ABSTRACT

Nitrate loading in the Chao Phraya estuary, derived primarily from agricultural waste, is considered to be the most significant factor in causing oxygen depletion. In order to maintain appropriate levels of dissolved oxygen in the estuary control of nitrate loading is required. This paper describes a simulation model of nitrate dispersion in the Chao Phraya estuary, which allows the normalized nitrate distribution over the estuary in the dry season to be determined. By introducing a number of fictitious nitrate loadings at the upper boundary, the distribution of dissolved oxygen in the estuary is obtained. Further by employing the limiting level of dissolved oxygen at 2 mg/l (the survival limit of biota), the pattern of allowable nitrate loading is elaborated.

1. INTRODUCTION

The Chao Phraya River in Thailand is one of the most important rivers in Asia. The estuary of this river is influenced by the densely populated city of Bangkok as well as the intensive mix of industry and agriculture. The river flows in the north-south direction having its origin from the Tennessarim mountain in Northern Thailand. With its four major tributaries, the Chao Phraya River winds through a length of 240 km. It drains into the northern boundary of the Gulf of Thailand. The overall drainage area of this river is 177,000 km^2 while the estuary itself is 120 km long. Within the boundary of the estuary the Chao Phraya River possesses an average depth of about 10 m and a width of about 200 m. Tidal height at the river mouth is about 1.5 to 3.5 m. Fresh water flow into the estuary is derived not only from the basin of the river but also from the adjacent Pasak River basin in the northeastern region of the country. Due to its location, inflow derived from rainfall follows the seasonal monsoon pattern, mainly from the southwest monsoon during the period May to October, having its peaks in June and September. Average inflow of the Chao Phraya River is about 400 m^3/s, with peak values at 4,000 m^3/s.

It was found that at relatively high nitrate concentrations (about 98 mg/l) at the mouth the level of dissolved oxygen was as low as 2.3 mg/l which indicates danger to the biota in the estuary (Chernbumroong, 1987). In addition, degradation of water quality was observed in the coastal region of the upper Gulf of Thailand resulting in high growth of phytoplankton (Hormchong, 1989). In the rainy season, it was found that nitrate content in the water body is relatively low while the amount of dissolved oxygen is rather high. On the other hand, due to low fresh water flow in the dry season concentration of nitrate is relatively high and as a result the dissolved oxygen drops to a low level. There is an urgent need to properly manage the Chao Phraya estuary to sustain its water quality for the conservation of valuable marine resources. This paper mainly describes a simulation model of the water quality in the Chao Phraya estuary and attempts to determine the levels of nitrogen loading that can be maintained for the survival of biota in the water body.

2. THE MODEL

The lower boundary of the estuary is designated at km 0 - the river mouth where full information on tide and salinity is available. The upper boundary of the estuary is at a location 120 km upstream from the river mouth where the fresh water inflow and stage variation are known. The nitrate content at the upper boundary of the estuary is assumed as being a single point source with a constant release. Lateral sources of nitrate input along the estuary are assumed to have a constant distribution along the total 120 km length of the estuary; they are combined together as a point source at the upper boundary. The fresh water flow along the whole length of the estuary is also assumed to be constant.

A one-dimensional flow field is assumed and the dynamic equations describing the flow comprise both the continuity and the momentum equation. There are two main sources of pollutants in the estuary, namely the salinity of water at the lower boundary and the nitrate loading at the upper boundary. These two pollutants will be dispersed along the axis following established dispersion equations.

The dynamic and dispersion equations are solved employing the finite difference scheme on the rectangular grids of size Δx by Δt on the x-t plane. Values of stage (H), discharge (Q), salinity (S) and nitrate (N), at all grid points are then calculated. With n = 0.026 (Torranin, 1969), the model is executed using the mean monthly discharge of the lower Chao Phraya River as the fresh water input into the upper boundary. At the lower boundary hourly tide data are used as the downstream input. Due to the fact that depletion of dissolved oxygen is found mostly in the dry season, the model is applied for the low flow with fresh water discharges of 40, 60 and 100 m^3/s. The dispersion coefficient, E, is obtained from

$$E(x,t) = K_1 \, nuR^{5/6} + K_2 \left|\frac{\partial S}{\partial x}\right| \quad \text{(Thatcher and Harleman, 1972)} \tag{1}$$

where

x	=	distance along the estuary	n	=	Manning's roughness coefficient
t	=	time	R	=	hydraulic radius
K_1, K_2	=	coefficients to be calibrated	$\frac{\partial S}{\partial x}$	=	salinity gradient
u	=	flow velocity			

The values of K_1 and K_2 are obtained from simulated salinity data. In the case of the Chao Phraya estuary the values are $K_1 = 600$ and $K_2 = 400$ $(m^2/s)/(ppt/km)$. The dispersion coefficient, E, must be approximately the same for all dissolved pollutants, therefore $K_1 = 600$ and $K_2 = 400$ $(m^2/s)/(ppt/km)$ are adopted for determining nitrate dispersion. Due to the fact that there is at present no reliable information on the nitrate content at the upstream boundary of the estuary as well as at other locations along the estuary, modelling is executed using an assumed value of $N_0 = 1$ and its variation is obtained in terms of the dimensionless value of the nitrate distribution, N/N_0. Further, the model is applied using the same values for fresh water discharge and vertical tide as that of the salinity estimation.

Results show that when nitrate input into the estuary is constant it exhibits less fluctuation within a tidal cycle as compared to salinity fluctuation; at the river mouth the salinity varies with tide.

3. DISSOLVED OXYGEN IN THE CHAO PHRAYA ESTUARY

The most crucial factor affecting the micro-organisms in an estuary is the amount of dissolved oxygen (DO) in the water body, especially at the lower region of the water column. A DO value lower than 2 to 3 mg/l can be harmful for the survival of the living biota. Measurements which had been conducted in many estuaries in Thailand produced a picture of the unique relationship between DO and NO_3 as shown in Fig. 1.

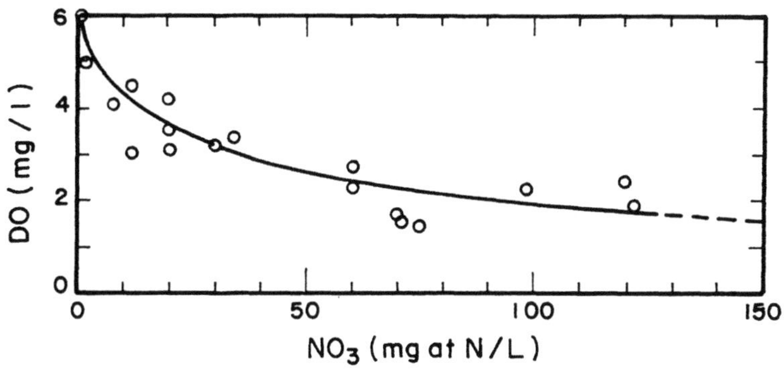

Figure 1: Relationship between DO and NO_3 from estuaries in Thailand (after NRCT)

Applying regression analysis to the data in Fig. 1, the DO-NO_3 relationship can be stated as:

$$DO = 6.13 - 2.013 \log NO_3 \qquad (2)$$

Fig. 2 shows the percentage distribution of N/N_0 at $Q = 40$, 60 and 100 m^3/s along the estuary within a tidal cycle.

Substituting equation (2) into the normalized NO_3 distribution, that of DO over a tidal cycle in the dry season is obtained as displayed in Fig. 3. It can be noted

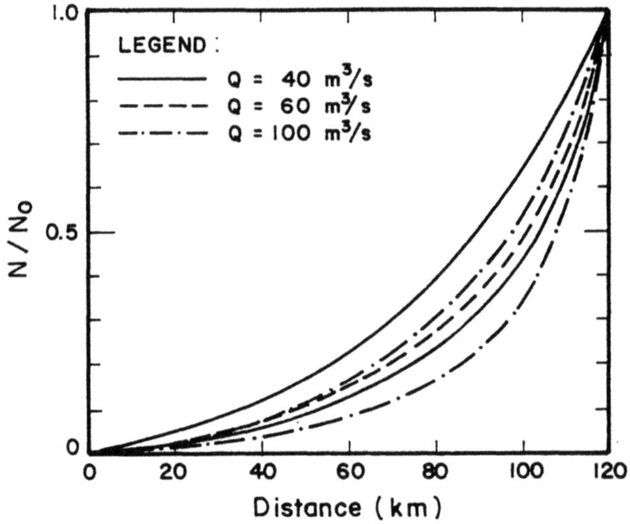

Figure 2: Distribution of normalized NO_3 in the dry season along the estuary

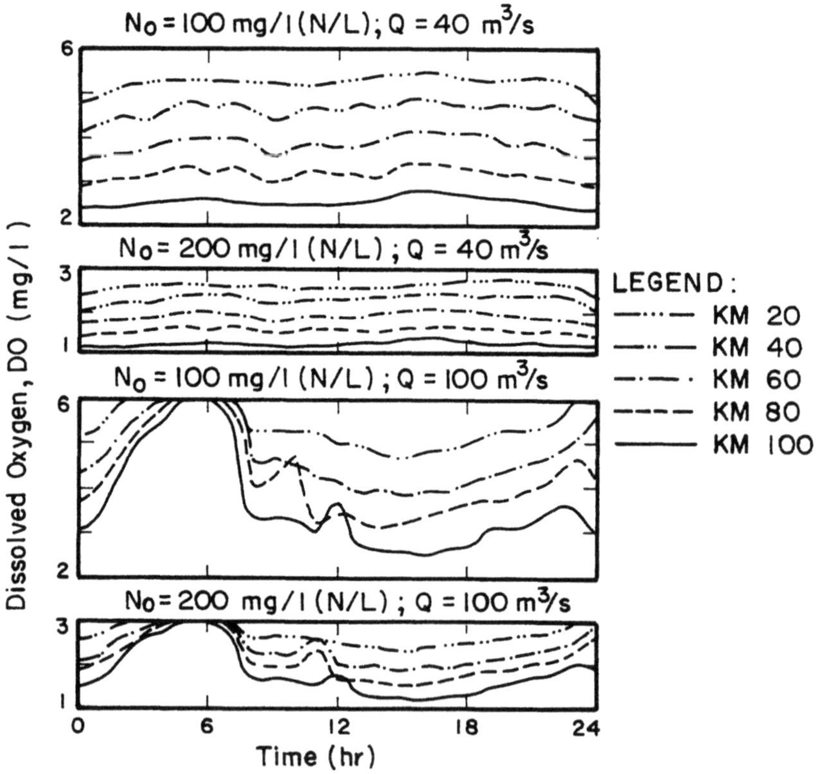

Figure 3: Examples of DO distributions over a tidal cycle at Q = 40 and 100 m³/s for N_0 = 100 mg/l (N/L) and N_0 = 200 mg/l (N/L)

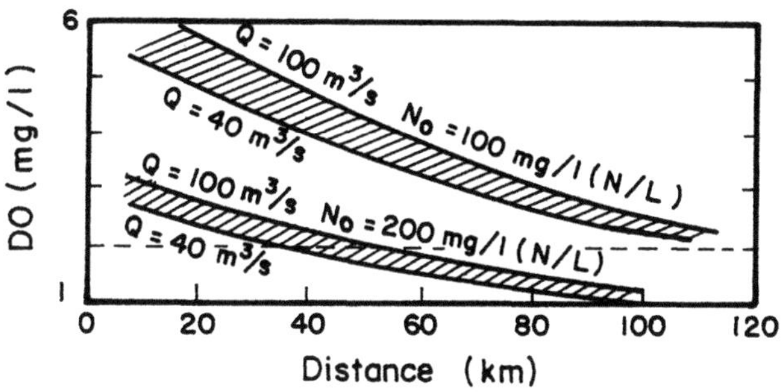

Figure 4: Distribution of DO along the estuary for N_0 = 100 mg/l (N/L); 200 mg/l (N/L) and Q between 40 and 100 m³/s

that the lower limit of DO at 2 mg/l would be imposed. Furthermore, by means of generating fictitious initial nitrate concentrations at various river flows throughout the year a full range of DO distributions is obtained as shown in Fig. 4.

4. CONTROL OF NITRATE INPUT

The occurrence of DO lower than 2 mg/l in the estuary, as shown in Fig. 4, can be found during the four months of low flow period in the dry season (January, February, March and April). Also, the low level of DO in the estuary occurs only about 60 percent of the time, which is during the half ebb flow. Therefore, in order to maintain the DO concentration at a level not lower than 2 mg/l the upper bound distribution of NO_3 can be suggested.

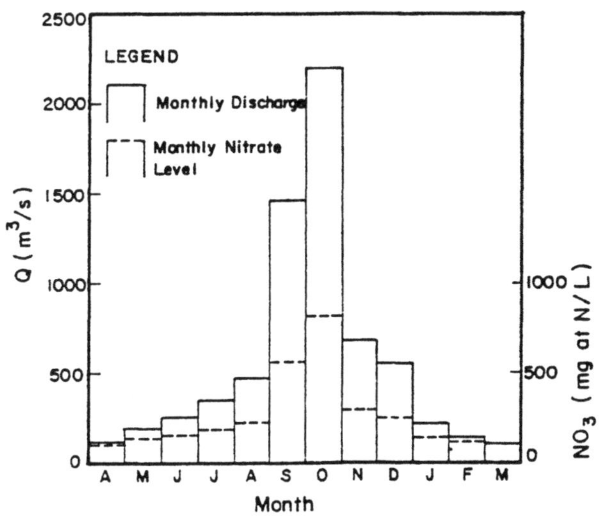

Figure 5: Upper limit of monthly distribution of nitrate concentration at the upper boundary of the estuary

When a reversible process utilizing the nitrate distribution model and the $DO-NO_3$ relation is tested together with the inflow hydrograph of the fresh water, the results obtained (Fig. 5) could be suggested for sustaining the DO concentration in the estuary and in turn the quality of water necessary for the survival of living resources in the water body.

5. CONCLUSION AND RECOMMENDATION

The exercise presented in this paper is aimed at formulating simplified control of nitrate loading in the Chao Phraya estuary in order to maintain the survival of marine habitats. The key issue in the survival is the limiting level of dissolved oxygen in the water column which exhibits a unique relationship with nitrate loading. In order to determine the dispersion coefficient, E, of the estuary, salinity modelling must firstly be executed since salinity data are available. Since information on nitrate loading is not available and since there is only limited information on water quality in the estuary, fictitious nitrate loadings are assumed. This exercise can be considerably improved if systematic monitoring of water quality and the dynamics of flow at strategic stations along the estuary is conducted. To be more accurate, the model should be upgraded into a two-dimensional flow with a substantial amount of field data to be gathered for supporting the model.

REFERENCES

Chernbumroong S (1987) Water quality of estuaries in the inner Gulf of Thailand, 1986. Proceedings of the Fourth Seminar on the Water Quality and the Quality of the Living Resources in Thai Waters, 7 - 9 July 1987. National Research Council of Thailand, Bangkok, Thailand

Hormchong T (1989) Population dynamic studies on phytoplankton and zooplankton in connection with Eastern Seaboard Development Project. First Symposium of the Association of Southeast Asian Marine Scientists (ASEAMS), United Nations Environment Programme, Manila, February 1989

Thatcher ML, Harleman DRF (1972) A Mathematical Model for the Prediction of the Unsteady Salinity Intrusion in Estuaries. Technical Report No. 144, Ralph M. Parsons Laboratory for Water Resources and Hydraulics, Department of Civil Engineering, MIT, USA

Torranin P (1969) A Tidal Mathematical Model of the Chao Phraya River. Thesis No. 247, Asian Institute of Technology, Bangkok, Thailand

CHAPTER IX

BIOLOGICAL PROCESSES, ENVIRONMENTAL IMPACT

BIOLOGICAL PROCESSES IN THE ESTUARINE ENVIRONMENT

H. Kausch
Institut für Hydrobiologie und Fischereiwissenschaft,
Hydrobiologische Abteilung, Universität Hamburg, Zeiseweg 9, D-2000 Hamburg, FRG

1. INTRODUCTION

Estuaries are complex systems which are governed by hydrographical factors, such as the tidal action and the mixing of freshwater and seawater, which produce complicated structural patterns that undergo continuous change in space and time. Ecological research in estuaries has to take into account that these changes are typical for the system and that variation in rather than the mean of the recorded values has to be considered. Furthermore, not only a structural analysis but also a detailed knowledge of the many different biological processes, their interactions and their interdependence with abiotic processes are the basis for a better understanding of the estuarine environment as a whole. In order to illustrate the complexity of the estuarine environment and the great importance of biological processes for the functioning of this type of ecosystem, a number of examples will be given, mostly from the turbid Elbe Estuary. These will confirm the need for more interdisciplinary research in this field.

2. THE ESTUARINE ENVIRONMENT

2.1 Hydrological implications

The environmental and biological structures of estuaries are substantially influenced by hydrological conditions, such as tidal movements and the salt gradient, which is formed by the mixing of the freshwater discharge from the river with seawater (Figure 1).

Modelling of currents and water transport has improved very much recently (Duwe, 1988). Different types of hydrological regimes produce different types of estuaries (Odum, 1973), some with and others without stratification. In most of the cases, however, there is, at least at certain times of the year, a brackish water zone, separating a freshwater zone upstream and a marine zone downstream. The brackish water zone oscillates with the tide, causing extreme salinity changes at the sediment surface and in the water column.

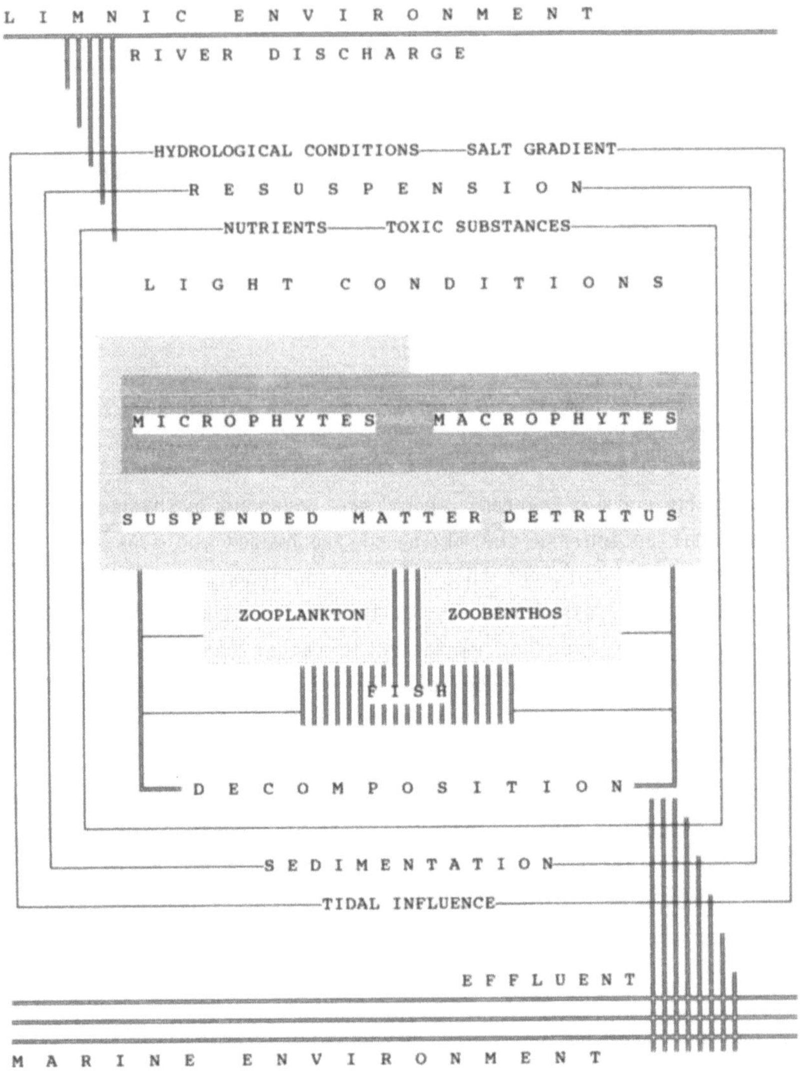

Figure 1: Simplified structural diagram showing the relative importance of various external factors, such as hydrographical features, resuspension and sedimentation, nutrients and toxic substances, and illumination to the internal biological functioning of an estuarine ecosystem

In the Elbe Estuary, the borders between the zones can move back and forth over a distance of about 20 km. At the same time the tidal change in the water level causes periodic immersion and exposure of littoral mudflats to the air. This is independent of salinity and brings about considerable variations in temperature, oxygen availability, and related factors. Thus, during one tidal cycle, benthic organisms are exposed to considerable changes in their environment, and planktonic organisms are transported over long distances in the estuary. In order to maintain their populations, these different organisms must have very different strategies at their disposal. Research in this field therefore must include physical, chemical, biological and physiological factors in order to understand what these strategies are and how they are put into practice.

2.2. Turbidity and primary production

A typical feature of an estuary is the large amount of suspended matter in the water. Much sediment from the catchment area is transported to the estuary. Furthermore, in the brackish water zone, both freshwater and marine plankton organisms die due to osmotic stress. In addition, material from the marshes along the estuary, such as detritus from reeds and other plants, decomposes and disintegrates into small pieces. Finally, there are strong erosional forces from the tidal movements, which change their direction up to four times a day, and relatively calm periods at high and low tide, which promote an alternation of resuspension and sedimentation. The hydrographical and morphological conditions produce distinct zones in each estuary, in which a maximum amount of suspended matter remains, and the water is extremely turbid.

Because of their important roles, resuspension and sedimentation were included in Figure 1 as an outer frame around the inner diagram. The suspended particles are important vehicles for nutrients, heavy metals, and halogenated hydrocarbons originating from pollutants. They are transported through the estuary across the salt gradients. There are many different physically, chemically and biologically driven interactions between the particulate and the dissolved phases which are very important for the fate of the substances involved (Schoer, 1985; Knauth et al., 1987). The state of speciation, which can be influenced by the salt gradient in different ways for different substances (Calmano et al., 1988), is of substantial importance for their biological availability. For heavy metals (Ahlf, 1985; Ahlf et al., 1986), this can determine whether they will be toxic or not. In Figure 1, these important processes are accounted for by a third frame around the inner diagram. Research in this field will require cooperation among chemists, mineralogists and toxicologists in order to clarify the nature of particle-water interactions, speciation of substances and their biological roles.

In turbid estuaries, light penetration of the water column is often very poor. In many turbidity zones, the euphotic layer has a depth of less than 1 m (Wassergütestelle Elbe, 1985). In well mixed, deep estuaries, this is only a small fraction of the mixing depth. Therefore, even when the nutrient content of the water is rich, primary production by the phytoplankton in deep, turbid and sufficiently well mixed estuaries is limited by light, and no algal blooms occur. This is true for the Elbe and the Weser estuaries (Kausch et al., in prep. 1990).

Thus, in turbid estuaries like the Elbe, the limitation of the illumination necessary for primary production by phytoplankton is effected by suspended matter from other sources rather than the phytoplankton itself. The epipelic microphytes on the mudflats and the diatoms that migrate during low tide onto the sediment surface, as well as the macrophytes, such as reeds (Phragmites australis), in the freshwater and oligo- or mesohaline zones can be exposed to the full intensity of photosynthetically active radiation at least during some of the daylight period. Therefore, macrophytes and epipelic algae are often more important for primary production in turbid estuaries than the phytoplankton. For primary production "there are large variations from one type of estuary to another but the order of magnitude may be summarized as: phytoplankton 50 - 200 g C m^{-2} yr^{-1} in shallow turbid estuaries and 100 - 500 g C m^{-2} yr^{-1} in clear deep estuaries; benthic microalgae 100 g C m^{-2} yr^{-1}; seagrasses about 500 g C m^{-2} yr^{-1}; salt marsh macrophytes (including mangroves) 500 - 1000 g C m^{-2} yr^{-1}" (Day & Grindley, 1981). For benthic microalgae, more recent measurements revealed higher production rates between 63 and 253 g C m^{-2} yr^{-1} for sand and sand/mud, respectively (Fielding et al., 1988).

The distribution of chlorophyll a usually shows a maximum upstream of the salt gradient (Anderson, 1986) and its abundance is often found to be inversely related to suspended matter and/or turbidity (Pennock, 1985; Fisher et al., 1988; cf. Figure 2).

Primary production in the water column, however, is closely related to the chlorophyll distribution. As data from the Elbe Estuary show, the chlorophyll content is not always proportional to the phytoplankton cell counts, the distribution of which on a longitudinal section can differ remarkably from the chlorophyll distribution (Figure 3). We do not yet know the reason for this. It will not be enough, however, to treat large algal groups as one taxon, such as "diatoms" or "green algae". Is is hypothesized that the discrepancy arises either because there are many photosynthetically inactive or dead shells of diatoms in the turbidity zone, which are resuspended by tidal action (DeJonge & Van den Bergs, 1987), and/or because there are great numbers of very small algae well above the turbidity zone, which are attached to suspended particles, and which normally are not included in the cell counts.

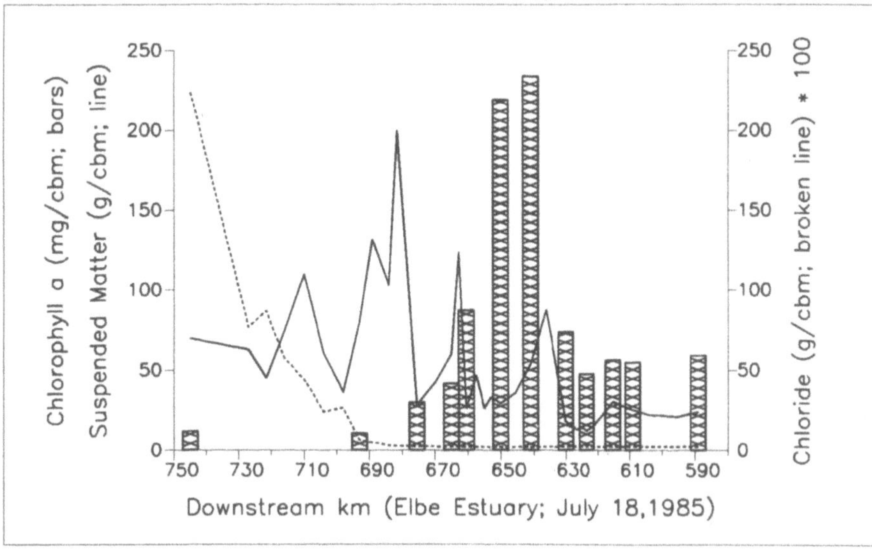

Figure 2: Chlorophyll a (bars), suspended matter (continuous line) and chloride (broken line) on a longitudinal section of the Elbe Estuary. Data from Wassergütestelle Elbe (1986); samples taken 2 hours after high tide at each position

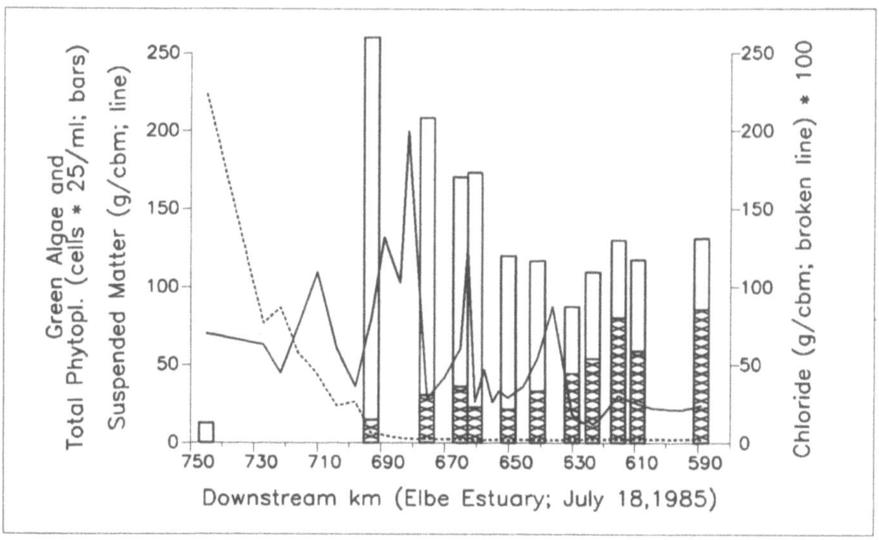

Figure 3: Total phytoplankton (bars) and green algae cell counts (shaded part of the bars) along a longitudinal section of the Elbe Estuary, together with the amounts of suspended matter and chloride concentrations in the water column. Data from Wassergütestelle Elbe (1986); samples taken 2 hours after high tide at each site

It will be an important task to establish quantitatively which species of phytoplankton algae are important primary producers, how they are transported through the estuary, and how they are influenced by the salt gradient and the grazing activity of herbivorous zooplankters.

2.3 Detritus

As already pointed out, many particles of organic suspended matter are flocs, which are composed of constituents from a number of different sources. Some are small mineral grains, but most of them are of plant origin. Not only freshwater and marine phytoplankton but also macrophytes from terrestrial, freshwater, and marine environments contribute a great deal to this detritus (Fenchel, 1970; Day & Grindley, 1981; Greiser, 1988). The flocs are colonized by bacteria, and mucous substances give them a firm consistency.

It is important to know how these particles are produced and transported along the estuary. This can only be answered by multidisciplinary research, required to determine the origin of the flocs, their changes in shape (Greiser, 1988), their sedimentation characteristics (Puls, 1986) and their role as substratum for microbes. Moreover, this detritus supplements phytoplankton and epipelic microphytes as an important source of food for brackish water zooplankters, like Eurytemora affinis and Acartia tonsa (Heinle et al., 1977; Roman, 1984) and a great number of benthic, suspension - and deposit-feeding species. Furthermore, Mann (1969) estimated that 30 to 60 % of the food for fish in the River Thames originated as detritus. We are far from understanding the quantitative relationships in this food web.

2.4 Oxygen depletion

Turbid estuaries undergo intensive nitrification processes during summer (Owens, 1986), which are sometimes followed by oxygen depletion (Caspers, 1984; Relexans et al., 1988). In the Elbe Estuary, this is intensified by the heavy pollution from the Elbe River. Each year, a large area of rapid oxygen depletion moves upstream from the region near the mouth of the estuary in May to just west of Hamburg or even to Hamburg itself in June or July, covering a distance of almost 100 km (Figure 4).

Intensive decomposition of organic substances is partly responsible for the development of this zone, but about 50 % of the oxygen is consumed by nitrification (Figure 5). These processes also occur in the Weser and Ems Estuaries, but there they are less intensive because of the smaller nutrient load and different morphological conditions.

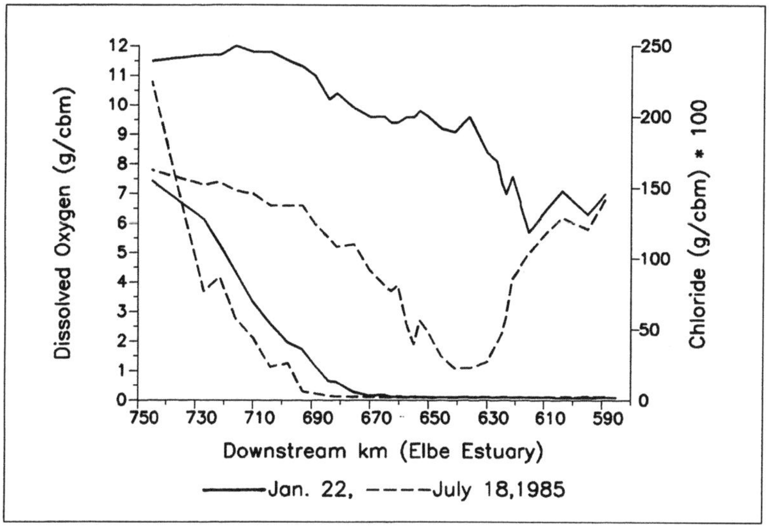

Figure 4: Dissolved oxygen in water at a longitudinal section of the Elbe Estuary in January (continuous line) and July 1985 (broken line) showing a large area of heavy oxygen depletion in the limnic zone west of Hamburg. The respective chloride curves for orientation. Data from Wassergütestelle Elbe (1986); samples taken 2 hours after high tide at each position

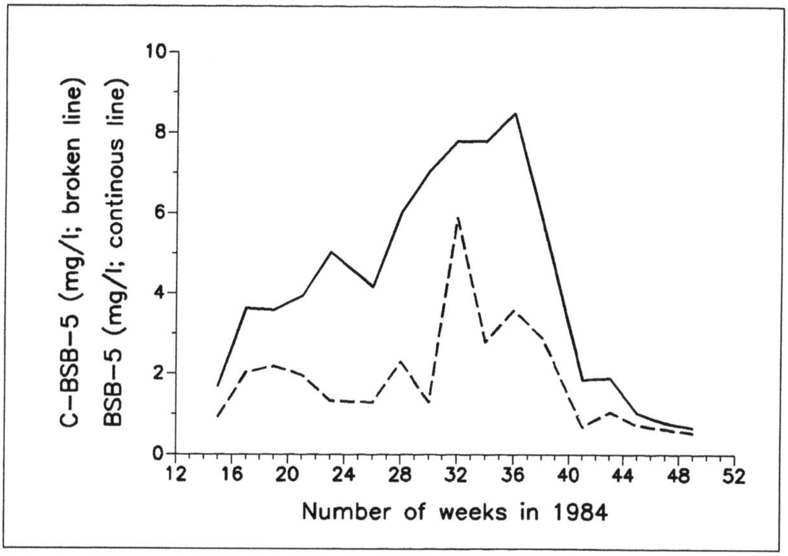

Figure 5: Total BOD-5 and C-BOD-5 at ambient water temperature in the freshwater zone of the Elbe Estuary at Hamburg Harbour (km 620). The difference between the two lines signifies the fraction of N-BOD-5, which accounts for 50 % or more of the total BOD-5 during the warm season (Kausch & Hinz, unpublished data)

Nitrification is only one of the decomposition processes occurring. Most of them have not yet been quantified. A great many of the organic substances undergoing microbial decomposition have not yet been carefully investigated.

3. NEAR FUTURE NEEDS

Research on biological processes in the estuarine environment is essentially a multidisciplinary task. In near future, it will be necessary to solve a number of problems that cannot be solved without a multidisciplinary approach. Examples are

- modelling the transport of suspended matter, phytoplankton and zooplankton, which is important for the calculation of population dynamics;
- modelling primary production in the water column, at the sediment surface, and by the macrophytes;
- accounting for the energy transfer from primary production to secondary production and fish, taking into account the allochthonous components of food from both ends, the river and the sea, and including the important role of detritus. For this, a quantification of the microbiological aspects of decomposition in the nutrient cycles is of great importance;
- balancing input and output values for nutrients, heavy metals and other toxic substances, taking into account the processes at the sediment-water interface and the interactions of particulate and dissolved substances.

The knowledge of the quantitative aspects of biological processes and their interactions will provide a better basis for understanding special features of different estuaries and improve their management and protection.

REFERENCES

Ahlf W (1985) Verhalten sedimentgebundener Schwermetalle in einem Algentestsystem, charakterisiert durch Bioakkumulation und Toxizität. Vom Wasser 65: 183-188

Ahlf W, Calmano W, Förstner U (1986) The effects of sediment-bound heavy metals on algae and importance of salinity. In: Sly PG (ed) Sediments and Water Interactions, Springer, New York, 319-324

Anderson GF (1986) Silica, diatoms and a freshwater productivity maximum in Atlantic Coastal Plain estuaries, Chesapeake Bay. Estuar Coast Shelf Sci 22: 183-197

Calmano W, Ahlf W, Förstner U (1988) Study of metal sorption/desorption processes on competing sediment components with a multichamber device. Environ Geol Water Sci 11: 77-84

Caspers H (1984) Seasonal effects on the nitrogen cycle in the freshwater section of the Elbe Estuary. Verh Internat Verein Limnol 21: 866-870

Day JH, Grindley JR (1981) The estuarine ecosystem and environmental constraints. In: Day JH (ed): Estuarine Ecology, A.A. Balkema, Rotterdam, 345-372

DeJonge VN, Van den Bergs J, (1987) Experiments on the resuspension of estuarine sediments containing benthic diatoms. Estuar Coast Shelf Sci 24: 725-740

Duwe K (1988) Numerische Simulation von Bewegungs- und Transportvorgängen in der Brackwasserzone eines Tideästuars am Beispiel der Unterelbe. Dissertation, Universität Hamburg

Fenchel T (1970) Studies on the decomposition of organic detritus derived from the turtle grass Thalassia testudinum. Limnol Oceanogr 15: 14-20

Fielding PJ, Damstra KS, Branch GM (1988) Benthic diatom biomass, production and sediment chlorophyll in Langebaan Lagoon. South Africa Estuar Coast Shelf Sci 27: 413-426

Fischer TR, Harding LW Jr, Stanley DW, Ward LG (1988) Phytoplankton, nutrients and turbidity in the Chesapeake, Delaware, and Hudson estuaries. Estuar Coast Shelf Sci 27: 61-63

Greiser N (1988) Zur Dynamik von Schwebstoffen und ihren biologischen Komponenten in der Elbe bei Hamburg. Hamburger Küstenforsch, Heft 45

Heinle DR, Harris RP, Ustach JF, Flemer DA (1977) Detritus as food for estuarine copepods. Mar Biol 40: 341-353

Kausch H, Flügge G, Gaumert T, Kies L, Nöthlich I, Schirmer M, Weigel HP, Breckling P (in prep. 1990): Tidegewässer.- In: Hauptausschuß Phosphate und Wasser in der Fachgruppe Wasserchemie der GDCH (Hrsg.): Wirkungsstudie Fließgewässer.

Knauth HD, Schwedhelm E, Sturm R, Weiler K, Salomons W (1987) The importance of physical processes on contaminant behaviour in estuaries. In: Coastal and Estuarine Pollution, IAWPRC/JSWPR Fukuoka, 236

Mann KH (1969) The dynamics of aquatic ecosystems. Adv Ecol Res 6: 1-81

Odum EP (1973) Fundamentals of ecology, 3rd edition WB Saunders Company, Philadelphia

Owens NJP (1986) Estuarine nitrification: A naturally occurring fluidized bed reaction? Estuar Coast Shelf Sci 22: 31-44

Pennock JR (1985) Clorophyll distributions in the Delaware Estuary: Regulation by light-limitation. Estuar Coast Shelf Sci 21: 711-725

Puls W (1986) Field measurements of the settling velocities of estuarine flocs. In: Wand SY, Shen HW, Ding LZ (eds), River Sedimentation, Vol III Proc 3rd Intern Symp River Sed, 525-536

Relexans JC, Meybeck M, Billen G, Brugeailly M, Etcheber H, Somville M (1988) Algal and microbial processes involved in particulate organic matter dynamics in the Loire Estuary. Estuar Coast Shelf Sci 27: 625-644

Roman MR (1984) Utilization of detritus by the copepod, Acartia tonsa. Limnol Oceanogr 29: 949-959

Schoer J (1985) Iron-oxo-hydroxides and their significance to the behaviour of heavy metals in estuaries. In: Heavy Metals in the Environment, Proc Int Conf, Athens, 384-388

Wassergütestelle Elbe (1984) Gewässerökologische Studie der Elbe von Schnackenburg bis zur See. Arbeitsgemeinschaft für die Reinhaltung der Elbe (ARGE Elbe)

Wassergütestelle Elbe (1986) Wassergütedaten der Elbe von Schnackenburg bis zur See. Arbeitsgemeinschaft für die Reinhaltung der Elbe (ARGE Elbe)

A MASSIVE FISH KILL CAUSED BY AN ANOXIA ON BRITTANY COAST (FRANCE)
- CAUSES AND MECHANISMS -

Michel Merceron
IFREMER - Centre de Brest, B.P. 70 29263, Plouzane, France

INTRODUCTION

In the end of July 1982, a massive fish kill occurred in the bay of Vilaine (Southern Brittany, France) (Fig. 1). Several tens of tons of benthic fish (Conger-eels, Sea-Bass, shrimps, crabs, etc.) were killed within some days. The suddenness of the event prevented the gathering of data on sea water quality. Only meteorological data and river discharges (Loire and Vilaine) are available for that period. Nevertheless, the mortality conditions which were observed led to the assumption that it was caused by an anoxia or a severe hypoxia. The dying fishes which were actually caught and immersed in aerated sea water, recovered quickly. Similar events occur more or less frequently in many coastal areas: Baltic Sea, Kattegat, German Bight, Northern Adiatic Sea, New York Bight, Louisiana coast, etc.).

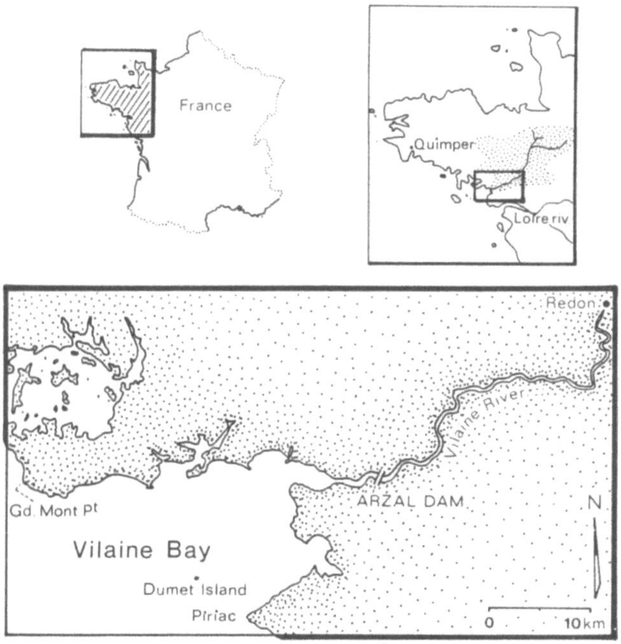

Figure 1: Brittany coast with the bay of Vilaine

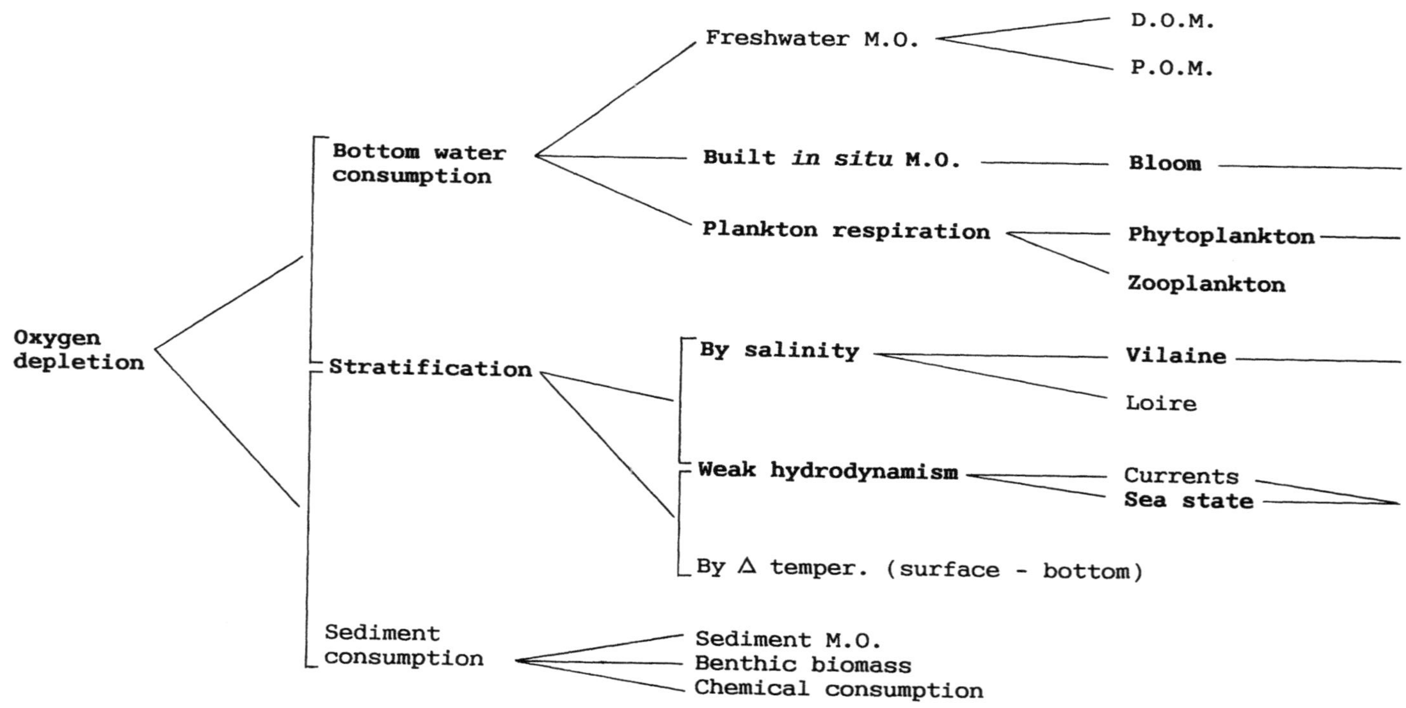

Figure 2: Anoxia in the Bay of Vilaine (July 1982): Causal diagram. The simple lines depict possible causal links; the words in bold type represent the most likely effective causes in the 1982 fish kill; the braces represent the interactions between two causes, which are necessary for inducing the effects.

ANALYSIS OF THE EVENT

Many field studies were undertaken in 1983, in order to understand the phenomenon and prevent it occurring again. These concerned the water quality of the coastal and estuarine zones and the drainage basin of the river Vilaine. They were carried out over a five-year period, and they provided the opportunity of finding out if chronic summer oxygen depletions existed in the bay.

To override the lack of sea water quality data during the event period, an inductive line of thinking was required. Starting from the anoxic event, we proceeded back from the effects, to the causes, by successive steps, and thence up to the primary causes. At each step, we selected, among the possible causes, the one or two whose effect was likely to predominate (Fig. 2). Each selection was supported by the data of the event period where possible. Often, however, the numerous hydrological data collected in the bay of Valaine from 1983 to 1988 were used. Thus, an absolute certainty about the causes could not be obtained, but the very large amount of data collected in the subsequent years, particularly during oxygen depletion periods, gives a high level of confidence in our outline of the causes and the mechanisms (Fig. 3).

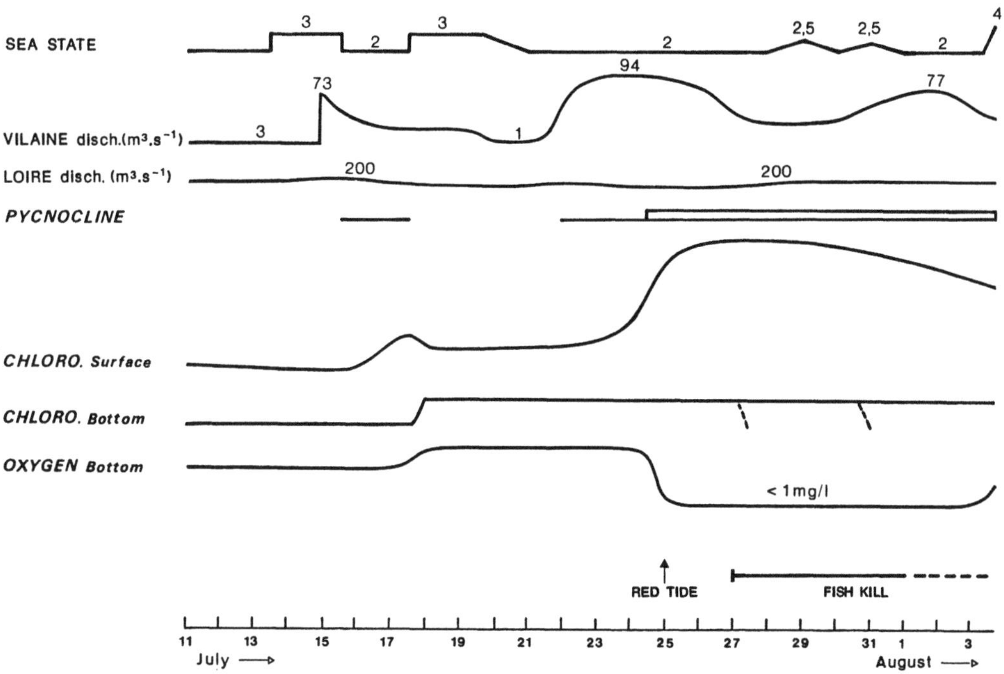

Figure 3: Anoxia in the bay of Vilaine (July 1982)

In 1982, the first mortalities were observed on July 27th. Twelve days before (July 15th) heavy precipitation on the drainage basin induced an increase of Vilaine river discharges from three to 73 $m^3 \cdot s^{-1}$ (yearly average: 68 $m^3 \cdot s^{-1}$). Subsequently discharges decreased to 25 $m^3 \cdot s^{-1}$ and stayed at this level for several days. During this first period, the sea remained, at first, smooth for two days, and the large freshwater supply should have generated a density stratification. The important irradiance and the surface temperature at that time, associated with the stratification and the nutrient supply (both caused by the high discharges), would most likely have induced a phytoplankton bloom in the upper layer. During this first period, the moderate sea which followed would probably have mixed the upper and the lower layers together. Thus the bottom phytoplankton content increased. Then, discharges decreased to one $m^3 \cdot s^{-1}$, while the sea got calm again.

After that, very important rainfalls on the river basin during three days caused a considerable increase in Vilaine discharges on and after July 22nd. Discharges were very high for that season, and lasted for five days (77 $m^3 \cdot s^{-1}$ on average). The sea remained smooth till August 4th. That should have allowed the development of a steep and lasting stratification, which should have prevented the supply of oxygen from the surface to the bottom waters. Moreover, the conditions were once more favourable for a phytoplankton bloom, but, on this one occasion, it was more important and more intense because the nutrient flux from the river and the phytoplankton inoculum were very large. Extensive discoloured red waters were observed in the bay on July 25th. That particularly important second superficial bloom could have drastically reduced the bottom irradiance, resulting in a rise of the compensation depth to a level situated above the pycnocline (Fig. 4). By this shading, the bottom phytoplankton, originating from the first bloom, could have become unable to photosynthesize and to produce oxygen in situ. It should have thus quickly consumed bottom oxygen by respiring. The oxygen consumption might also have been due to a beginning of phytoplankton cell decomposition. The whole process should have lasted for several days, and has been amplified by decomposition of dead animals. On August 4th, the sea became moderate, and the stratification should have been broken. So, oxygen should have been once again supplied to the bottom layers.

During the summer of 1984, we observed such a pattern though less pronounced than in 1982. The illustration of this phenomenon is in the temporal evolution of dissolved oxygen which is opposite between the upper and the lower layers (Fig. 5). The trend pattern is found in the expression of a more or less abundant surface phytoplankton and of a bottom irradiance which is or isn't important enough to allow phytoplankton photosynthesis.

Several primary causes of oxygen depletion stand out in the bay of Vilaine. Firstly, this natural environment is rather sensitive. This is due to the enclosed nature of the bay and its very weak dispersive capacities compared with the large size of the drainage basin. Here, the current velocities are reduced

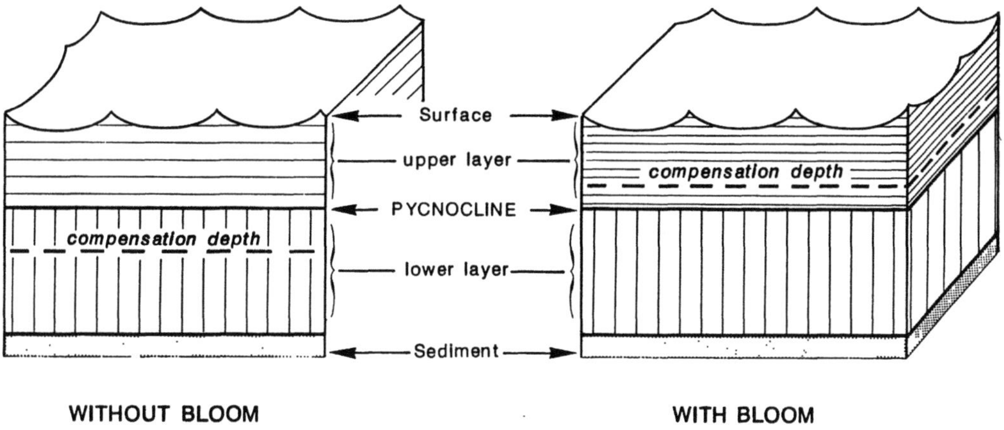

Figure 4: The rise of compensation depth by superficial planctonic bloom

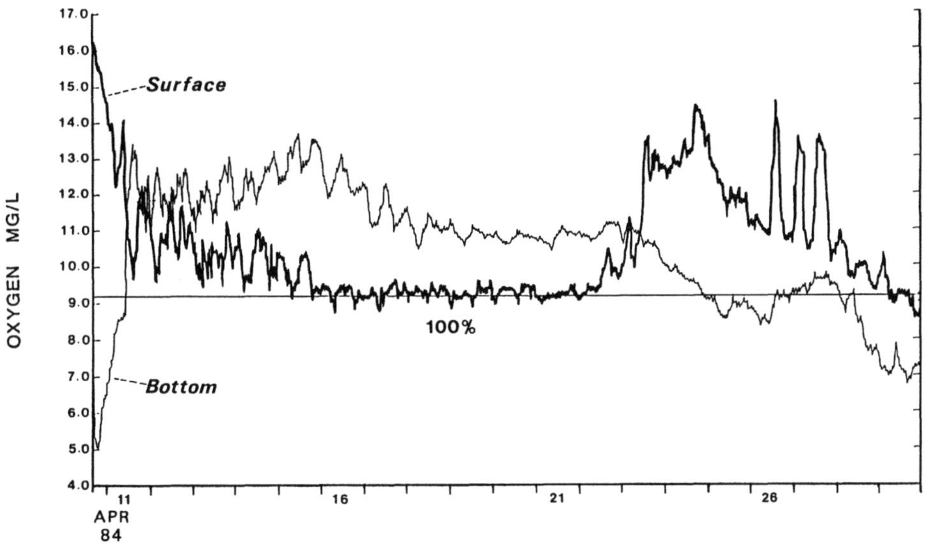

Figure 5: Dissolved oxygen in the bay of Vilaine

(maximum = 0.5 m $\cdot$ s^{-1}); residual currents are weak, and depend mostly on winds. Flushing times are at least in the bay of some days in duration and they may be much longer. Moreover, the islands and the shallows located seawards act as hydrodynamic shelters causing stratified situations of longer duration. The schistose nature of the Valaine river watershed tends to increase the river level more quickly and thus the stratification in the estuary and in the bay.

Several human activities on the river basin enhance this natural fragility. The intensification of nitrogen supply to soil and the raising of surface flow as a result of the drainage of damp meadows and destruction of bocage embankments, contribute to an increase of both the primary production and bay stratification. The existence of an estuarine dam further aggravates the situation. This dam cancelled out forty kilometers of the old estuary, which was fifty long. It shifted the estuarine functions which consume dissolved oxygen (M.O. mineralization, nitrification) into the bay; it also contributes to reinforce the stratified character of the estuary and the superficial distribution of phytoplankton blooms in the bay. Further, the increase of urbanization over several decades is likely to have contributed an increase of phosphorus supplies (evolution of hygienic uses, increase of the number of washing machines and depuration plants).

In 1982, meteorological factors played a vital role for the water quality: high summer precipitation, weak winds, and, secondarily, small tide coefficients. The triggering of the anoxic event of July 1982 was quite likely due to a very unusual meteorological sequence, e.g. very heavy precipitation associated with and followed by a period of calm sea.

PHYTOPLANKTON AND EUTROPHICATION MODELLING IN THE VILAINE BAY

A. Chapelle
IFREMER-Centre de Brest, B.P. 70 29263 Plouzané, France

1. INTRODUCTION

The Vilaine Bay (South Brittany) has been showing evidence of increasing ecological problems for several years, e.g. red tides, fish mortality caused by oxygen depletion and shell toxicity induced by Dinophysis sacculus. Therefore, scientific studies were initiated in order to improve the ecological knowledge of the bay and to work out solutions leading to the restoration of the water quality. Taking into account the results to date a box model of the ecosystem is used to provide a dynamic simulation of the eutrophication, to discover its major causes and to test possible restoration schemes.

2. MODEL CONSTRUCTION

2.1 The biological system

The model is based on seven state variables X_i (Fig. 1): Five nitrogen variables (µM) describing the nitrogen cycle, salinity (‰), a conservative variable used to calibrate the physical fluxes, and dissolved oxygen (mg ℓ^{-1}), a key-variable for water quality.

Furthermore, the system is driven by temperature, light and phosphates. Silicon, which is not limiting the phytoplankton growth (Queguiner, 1986), is not included in the model.

The evolution of variables is described by the following set of equations:

dX_1/dt = mineralization + zoo-excretion - phyto-uptake

dX_2/dt = phyto-growth - phyto-mortality - phyto-sedimentation - zoo-grazing

dX_3/dt = zoo-growth - zoo-mortality - zoo-excretion

$dX_4/dt = dX_1/dt - dX_2/dt - dX_3/dt - dX_5/dt$

dX_5/dt = phyto-sedimentation + detritic sedimentation - resuspension

$dX_6/dt = 0$

dX_7/dt = aeration + photosynthesis - phyto-respiration - zoo-respiration - benthic demand - mineralization.

The parameter values used are given in the Appendix.

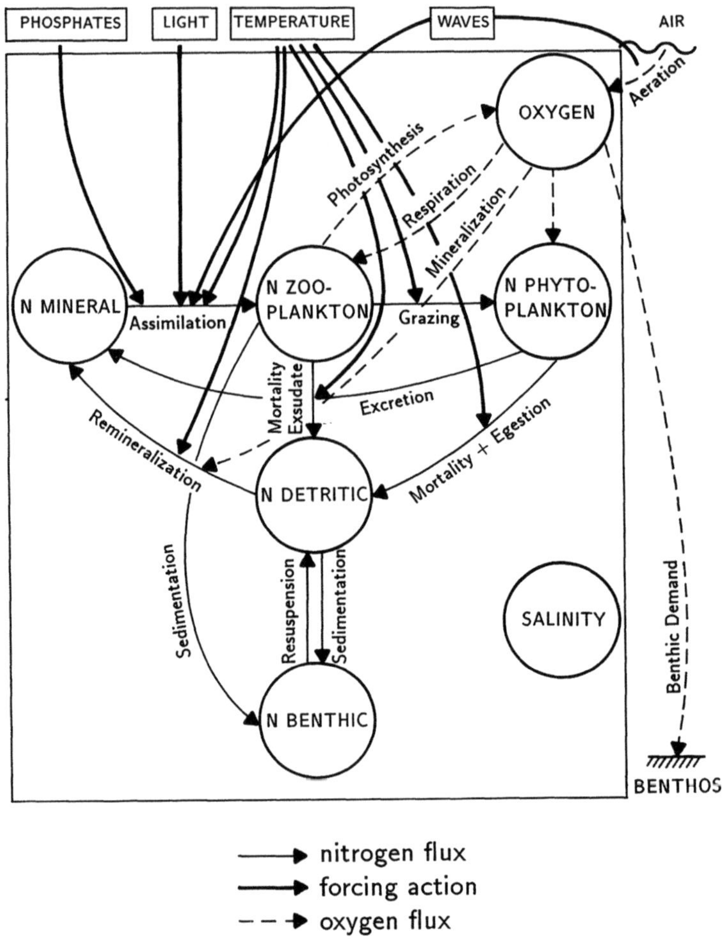

Figure 1: Sketch of the biological model

Phytoplankton uptake

$$\mu_m \exp(\alpha T) f_n \ f_I$$

is accelerated by temperature T and is dependent on light intensity I (Steele function f_I) as well as on nutrient concentrations (nitrogen N and phosphorus P; Michaelis function f_n).

Here,

$$f_I = \frac{I(z,t)}{I_s} \exp\left(1 - \frac{I(z,t)}{I_s}\right) dz\, dt$$

and

$$f_n = \min\left(\frac{N}{N+K_N}, \frac{P}{P+K_P}\right).$$

The irradiance at depth z is given by the relation

$$I(z,t) = I(0,t) \exp(-kz),$$

where I(0,t) is the irradiance at the surface calculated by the Brock method (Brock, 1981). The light extinction coefficient k is determined by background extinction and attenuation by chlorophyll (Riley et al., 1956). The calibration for background extinction depends on the detritic matter concentration.

Grazing by zooplankton is also accelerated by temperature and is a function of phytoplankton concentration according to the Ivlev formula:

$$0.17 \exp(\alpha T) [1 - \exp(K_I \cdot \max(0, X_2 - B_0))].$$

Mortality, excretion and mineralization are dependent on temperature; the sedimentation, resuspension and egestion rates are assumed to be constant.

The formulation of the oxygen terms is more arduous. The first step is to calculate the saturated oxygen concentration (O_{2S}) from the Weiss equation (cf. Aminot et al., 1983):

$$\log O_{2S} = A_1 + A_2 \frac{100}{T} + A_3 \log \frac{T}{100} + A_4 \frac{T}{100} + S [B_1 + B_2 \frac{T}{100} + B_3 (\frac{T}{100})^2]$$

Here, S is the salinity (‰). The constants A_i, B_i are given in the Appendix.

The aeration of water, k_a, which is correlated to the sea-surface agitation (Wilson and MacLeod, 1974) is complex to simulate. Therefore, a constant value is chosen using a mean current velocity and water depth. Then, the aeration is given by the expression $k_a (O_{2S} - DO)$.

The oxygen produced by photosynthesis is related to growth by an O/N ratio deduced from the equation: $CO_2 + H_2O \rightarrow CHOH + O_2$. Eight per cent of the maximum photosynthesis are consumed by phytoplankton respiration. Zooplankton respiration is a simple constant.

The combustion of detritic matter is related to mineralization by the O/N ratio:

$$N_{org} \rightarrow NH_4^+ \xrightarrow{2\,O_2,\ OH^-} NO_3^- + 2\,H_2O + H^+.$$

$$C_{org} \xrightarrow{O_2} CO_2 \text{ with } C/N = 7.$$

The benthic demand is the result of mineralization, benthic respiration and the chemical demand from sediment. It is, however, too difficult to evaluate all these processes. Therefore, a temperature dependent constant is used.

2.2 The physical system

The ecological model is coupled to a box-model of the bay (Fig. 2) in accordance with the horizontal dilution gradient (cost → ocean) and the pycnocline (depth). Advection corresponds to the Valaine river flow, discharging into the surface boxes. The influence of the Loire river and of the tidal residual drift (Salomon and Lazure, 1988) is weak and is ignored.

However, the tidal excursion is included in the dispersion term. Wind effects are taken into account as well. All the fluxes are calibrated by comparing simulated salinity and data.

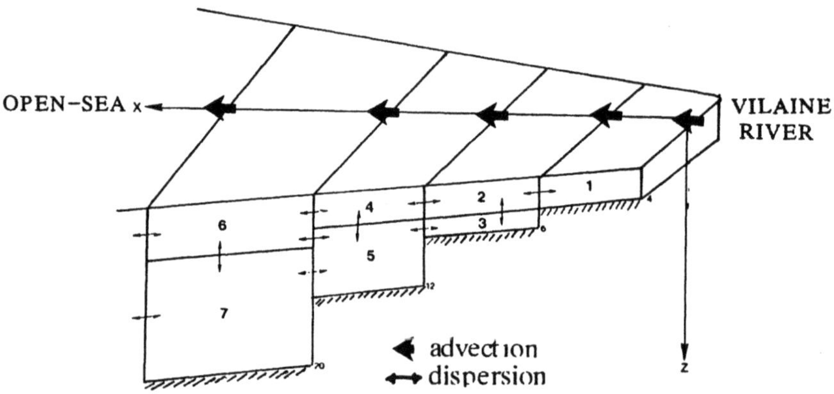

Figure 2: The box model

3. SIMULATIONS

3.1 Standard run

1984 was chosen to test the model, simulating the entire year. Fertilization by the Vilaine river has a strong, quasi-permanent effect as shows the great amount of nutrients in the bay. In response to this enrichment, phytoplankton biomass increases in spring and summer attaining more than 10 µg Chlo ℓ^{-1} (Fig. 3) and oxygen produced by photosynthesis reaches high levels in the surface boxes (180 % oversaturation was observed by Maggi et al., 1984).

On the other hand, bottom boxes are oxygen depleted in summer (40 % observed by Maggi et al., 1984) after blooms; when the phytoplankton sinks down to the aphotic zone, respiration and degradation become dominant. This may cause fish mortalities, as it appeared in July 1982 (Merceron, 1987, 1989).

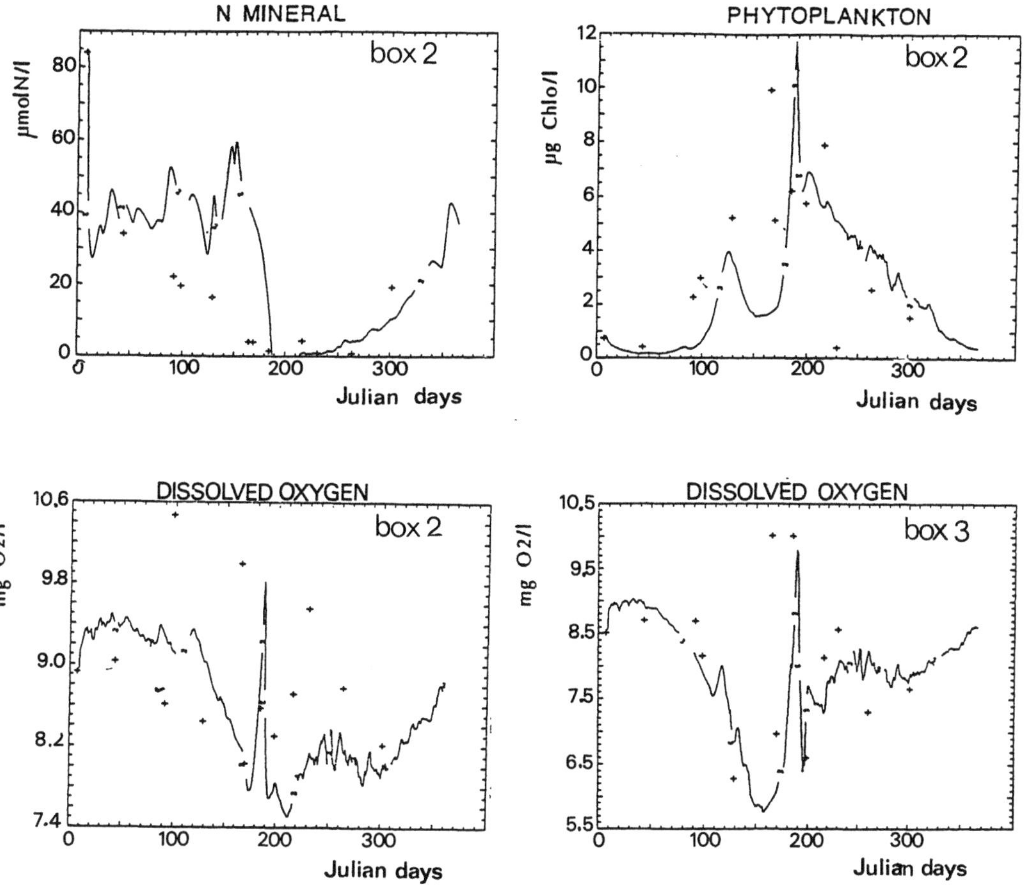

Figure 3: Simulations and field data for 1984

3.2 Model Applications

- Simulation of the 1982 anoxia

This dramatic and singular event occured in conjunction with a smooth weather and a sudden Vilaine flood. On the basis of meteorological data and discharges only, the model was used to reproduce the anoxia (Fig. 4). The two factors responsible for this drastic oxygen drop are phytoplankton bloom and strongly stratified water. Confirmation of their role is achieved by a sensitivity test.

- Sensitivity analysis of the oxygen terms

Simulations have been made neglecting one oxygen process in each case. Quantitative differences for the oxygen depletion in bottom boxes between test and reference are shown in Fig. 5. The most important conclusion of this study is the considerable importance of phytoplankton respiration as an oxygen sink.

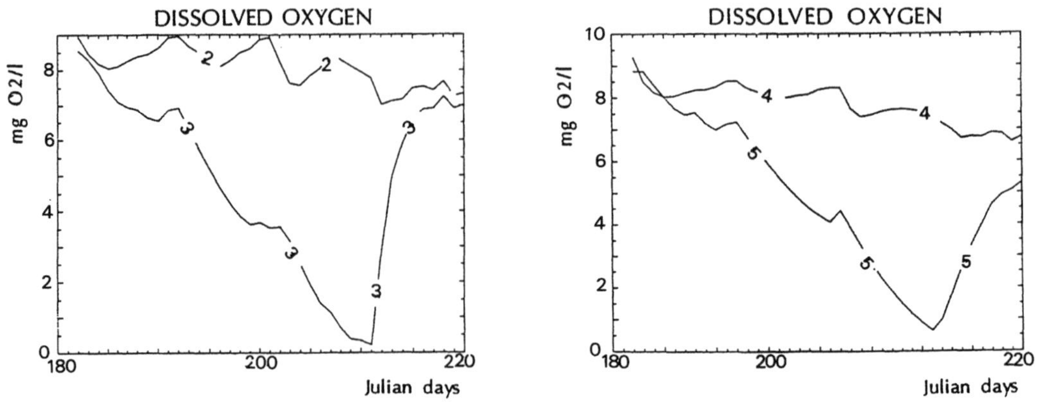

Figure 4: 1982 oxygen simulation

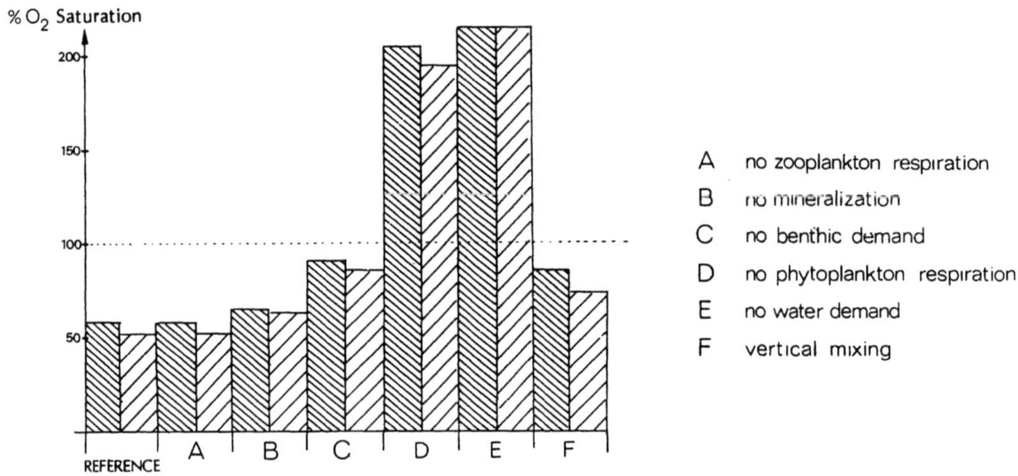

Figure 5: Sensitivity test

4. DISCUSSION

These first results lead to the conclusion that managing the Vilaine water quality must end with reducing phytoplankton biomass. A further development of the model with inclusion of the phosphorus cycle simulation will help to test the best ways of limiting primary production.

APPENDIX
Ecological parameters

a assimilation rate for zooplankton: 0.6

α augmentation rate with temperature (1/°C): 0.07

B_o limit of predation (μmol N/ℓ): 0.1

I_s optimum light intensity (W/m²): 80

k light extinction coefficient

k_a reaeration rate (1/d): 0.7

K_I Ivlev constant (1/μmol N): 0.1

K_N Michaelis constant for nitrogen (μmol N/ℓ): 4

K_P Michaelis constant for phosphorus (μmol P/ℓ): 0.2

μ_m maximum phytoplankton growth rate at 0 °C (1/d): 0.5

Oxygen equation constants:

$A_1 = -173.4292$	$B_1 = -0.033096$
$A_2 = 249.6339$	$B_2 = 0.014259$
$A_3 = 143.3483$	$B_3 = 0.0017$
$A_4 = -21.8492$	

assimilation for phytoplankton (%): 0.8 (Falkowski et al., 1980)
benthic respiration (g O/m² d): 1 (Merceron, 1987; Thouvenin, 1984)
benthic resuspension rate (1/d): 0.07
detritic sedimentation rate (m/d): 1
nitrogen remineralization rate at 0 °C (1/d): 0.01
non-chlorophyllic extinction (1/m): 0.3 - 0.8
N/chlo a ratio (μmol N/mg): 1
O/N ratio for remineralization (mg O/μmol N): 0.288
O/N ratio for photosynthesis or phyto-respiration (mg O/μmol N): 0.5
phytoplankton mortality rate at 0 °C (1/d): 0.005
phytoplankton respiration: 0.8
phytoplankton sedimentation rate (m/d): 0.5
zooplankton excretion rate at 0 °C (1/d): 0.025 (Mayzaud, 1973)
zooplankton mortality rate at 0 °C (1/d): 0.025
zooplankton respiration rate (mg O/d μmol N): 0.0084 (Mayzaud, 1973)

REFERENCES

Aminot A et al (1983) Manuel des méthodes d'analyses chimiques (RNO)

Brock TD (1981) Calculating solar radiation for ecological studies. Ecol Model 14: 1-19

Falkowski PG et al (1980) An analysis of factors affecting oxygen depletion in the New York Bight. J Mar Res 38: 479-505

Maggi P et al (1980) Facteurs hydroclimatiques et apparitions d'eaux colorées en baie de Vilaine durant l'année 1984. Rapport IFREMER DERO-86.06-MR

Mayzaud P (1973) Respiration and nitrogen excretion of zooplankton. II Studies of the metabolic characteristics of starved animals. Mar Biol 21: 19-28

Merceron M (1987) Mortalités de poissons en baie de Vilaine (Juillet 1982). Causes, mécanismes, propositions d'action. Rapport IFREMER DERO-87.14-EL, 100 p

Merceron M (1989) A massive fish kill caused by an anoxia on Brittany coast (France) - causes and mechanisms. This volume

Queguiner B (1986) Programme PROBRAS: Mise en évidence des facteurs limitants la production de phytoplancton dans le MOR-BRAS (UBO). Cahiers du MOR-BRAS 17

Riley GA et al. (1956) Bulletin of the bingham oceanographic collection peabody Museum of National History. Yale University, Vol XV: 15-46. Ozeanography of Long Island Sound 1952-1954

Salomon JC, Lazure P (1988) Etude par modèle mathématique de quelques aspects de la circulation marine entre Quiberon et Noirmoutier. Rapport IFREMER DERO-88.26-EL

Thouvenin B (1984) Baie de Vilaine - Oxymor 1983. Modèle mathématique vertical. Evolution de l'oxygène dissous au fond. Rapport IFREMER: 56 p

Wilson GT, MacLeod N (1974) A critical appraisal of empirical equations and models for the prediction of the coefficient of re-aeration of deoxygenated water. Wat Res 8: 341-366

MONITORING THE BIOAVAILABILITY OF HEAVY METALS IN RELATION TO THE SEDIMENT POLLUTION IN THE WESER ESTUARY (FRG)

Michael Schirmer
Universität Bremen, Fachbereich 2
Postfach 330 440, D-2800 Bremen 33

1. INTRODUCTION

The geochemistry of heavy metals in aquatic ecosystems is dominated by solution, complexation, oxidizing and reducing chemical reactions, precipitation and remobilization processes. Their ecotoxicological properties, especially their bioavailability, change markedly in accordance with the respective chemical status. When heavy metals reach the tidal and/or brackish water region they are subjected to pronounced changes in hydrographic and chemical conditions. The accumulation of suspended solids, an increase of fine-grained and organic material in the sediments and growing salinity, are the predominating factors affecting their geochemical status within the estuarine environment. The usual heavy metal monitoring practices, however, do not regard these dynamics. The methods in use, i.e. total digestion of unfiltered water samples and occasionally of the silt fraction of the sediment, yield only gross values without any differentiation of the geochemical speciation. Thus, processes of geo- and bioaccumulation can hardly be described nor predicted, or controlled with regard to the available information. We made an attempt to improve the data base by following a highly integrated two-year-survey program in the Weser estuary. The River Weser has a well documented anthropogenic heavy metal pollution with particularly high cadmium levels (Schirmer et al., 1989). The Weser estuary (Fig. 1) comprises an inner part (km 0 - 50) which is slightly brackish due to effluents from the potash industry, a middle reach (km 50 - 70), naturally mixomesohaline with high salinity variations, and the outer estuary leading through the Wadden Sea into the German Bight. With average headwater discharge of 323 m^3/sec and a tidal range of about 4 m at km 0 a suspended particle needs about eight days to reach the turbidity zone at km 60.

2. MATERIALS AND METHODS

Samples were collected from four locations along the estuary (km 0, 20, 40, 60; see Fig. 1) every 2 weeks between March 1984 and 1986. Suspended solids were prepared by filtration of about 250 ml through 0.45 µm, sediment was collected from the aerated sediment surface near the low-water line. Sediments were analysed for their heavy metal content in the total sample. They were in addition treated with oxalic acid/ammonia oxalate at pH 3 and the leachate was also analysed ("moderately reducible fraction" according to Förstner and Calmano, 1982). Organisms were collected from the same locations including benthic Green Algae (Enteromorpha) and various mobile Crustacean species. All samples were digested with nitric acid (suprapure) and analysed by atomic absorption spectrometry for Cd, Pb, Ni and Zn. Correlation of metal concentrations in organisms and solids was calculated by linear regression analysis.

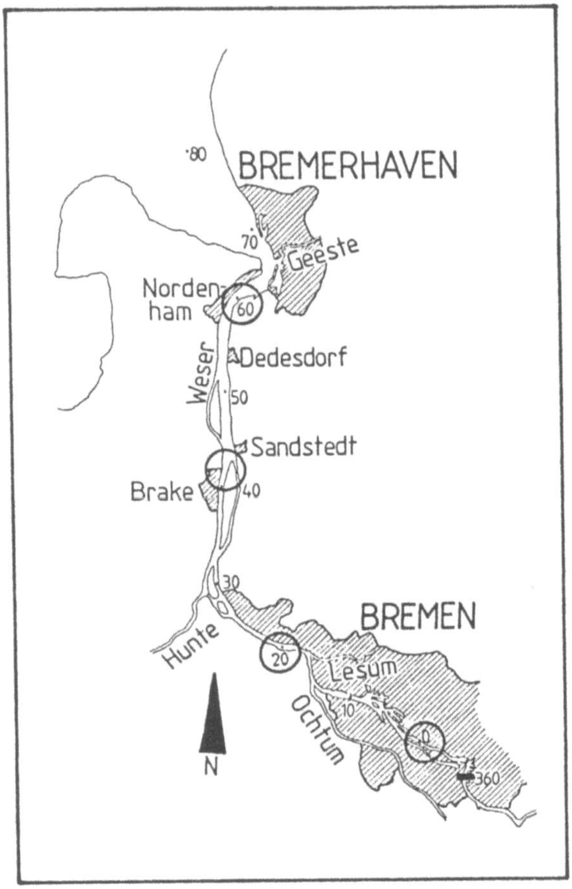

Figure 1: The Weser estuary, FRG

3. RESULTS

3.1 Contamination level

The mean concentrations of Cd, Pb, Ni and Zn in suspended solids, sediments, sediment leachate, algae (Enteromorpha spec.) and Crustacea (Gammarus spec.) on the four locations are given in Table 1.

Concentrations of Cd and Zn in suspended solids and total sediment samples tend to be highest in the freshwater region at km 0 and 20 and lower at km 60, within the turbidity maximum, where salinity S rises to approximately 10 ‰ and less contaminated marine material is imported. Pb and Ni show a more even distribution, probably due to industrial effluents entering the estuary. The leachate indicates a relative and absolute seaward decrease in the "moderately reducible" phase of the sediment, i.e. Cd: 18 to 4 %, Pb: 10 to 5 %, Ni: 63 to 46 % and Zn between 62 and 74 %. Förstner and Wittmann (1983) suggested that a weak bonding of heavy metals in sediments could be used as an indication of anthropogenic pollution, which is certainly true for the River Weser. A different approach in the estimation of man-made sediment pollution is the comparison of actual heavy metal concentrations in the silt fraction to those in pre-industrial sediments as published by Turekian and Wedepohl (1961). In relation to their geochemical background values the sediments of the Weser estuary contain up to 70-fold more Cd, 18-fold more Pb, 16-fold more Zn and 2,5-fold more Ni. The high standard deviations in Table 1 reflect the environmental variations during two complete seasonal cycles. Statistical analysis could not verify an influence of river discharge or water temperature.

Table 1: Concentrations of Cd, Pb, Ni and Zn in various matrices from the Weser estuary; locations given as km downstream of the tidal weir; mean concentrations in mg/kg dry matter, number of samples n and standard deviation; leaching with oxalic acid/ammonia oxalate at pH 3

Location km:		0			20			40			60		
		n	$\bar{x}$	± s	n	$\bar{x}$	± s	n	$\bar{x}$	± s	n	$\bar{x}$	± s
Sus-pended solids	Cd	28	6,5	5,7	27	5,7	4,1	25	4,5	3,1	28	2,2	1,6
	Pb	28	79	73	27	82	47	25	80	44	28	84	50
	Ni	28	36	40	27	52	32	25	47	36	27	42	26
	Zn	27	770	500	30	680	300	28	650	360	28	470	240
Total sedi-ments	Cd	55	7,5	3,7	43	4,0	3,6	78	2,9	1,2	82	1,9	0,8
	Pb	55	109	53	43	67	47	78	74	19	82	77	34
	Ni	55	40	14	43	37	28	78	33	11	82	29	9
	Zn	55	540	180	43	350	260	79	340	80	82	370	190
Sedi-ment lea-chate	Cd	52	1,4	0,5	34	0,6	0,7	69	0,1	0,1	71	0,1	0,1
	Pb	50	11	6	34	7	5	70	4	2	70	4	3
	Ni	52	25	10	34	22	15	69	18	6	72	13	5
	Zn	46	380	190	29	220	190	60	250	110	58	250	160
Green Algae	Cd	15	3,2	0,7	25	1,9	0,8	32	1,1	0,4	51	0,9	0,6
	Pb	16	47	15	25	44	12	32	43	15	52	28	16
	Ni	15	27	13	24	33	10	32	24	10	52	17	6
	Zn	16	330	70	25	300	110	32	220	100	52	150	100
Crusta-cea (Gam.)	Cd	160	1,3	0,9	144	0,4	0,4	145	0,4	0,3	167	0,4	0,3
	Pb	159	1,3	1,1	149	1,4	1,7	141	1,1	1,3	155	0,9	0,9
	Ni	160	3,3	3,2	143	2,2	1,6	144	2,2	1,5	169	1,8	1,2
	Zn	158	83	19	149	82	18	147	84	23	169	92	31

Benthic Green Algae show a trend of decreasing contamination by Cd, Pb and Zn in a seaward direction, only Ni-values peak at km 20, probably due to the emission of highly bioavailable Ni from shipyards. As the uptake and excretion of heavy metals by crustaceans are mediated by other biochemical processes than in algae, their bioaccumulation shows a different pattern: again the peak concentrations of Cd, Pb and Ni occur in the inner part of the estuary, but Zn shows a stable low level since its excretion can be regulated by these animals. The standard deviations as given in Table 1 for heavy metals in organisms can in part be correlated to the weight of the animal, e.g. in Gammaridae, and to seasonal cycles as indicated by the water temperature. With regard to crustaceans and algae, our own data and those published by Zauke et al. (1987) show a similar or even higher degree of contamination in organisms from the Weser estuary compared to those collected from the Elbe estuary (Schirmer et al., 1989).

3.2 Geoaccumulation, bioaccumulation

For a calculation of the accumulative capacity of solids and organisms in the Weser estuary an estimation of the amount of heavy metals in solution must be made. Böddeker et al. (1988) analysed heavy metals in suspended sediments and filtrates ("dissolved heavy metals") collected from the inner estuary under normal conditions with an average of about 26 mg/l dry suspended matter. These data allow an estimation of the partition of the heavy metals analysed in unfiltered samples between

Table 2: Heavy metals in various compartments; samples collected between March 1984 and 1986 from the locations Weser-km 0 and 20

	Cd	Pb	Ni	Zn
unfiltered water sample (µg/l)	0.5	8.8	8.5	56
dissolved portion (%)	24	6	50	23
dissolved concentration (µg/l)	0.1	0.5	4.3	13
conc. in suspended solids (mg/kg)	6.1	81	44	723
geoaccumulation factor SUS	0.6×10^5	1.6×10^5	1.0×10^4	5.6×10^4
conc. i. sediment < 63 µm (mg/kg)	15	234	85	1100
geoaccumulation factor SED	1.5×10^5	4.7×10^5	2.0×10^4	8.5×10^4
Green Algae (Enteromorpha)(mg/kg)	2.4	45	31	312
bioaccumulation factor ENT	2.4×10^4	9.0×10^4	7.2×10^3	2.4×10^4
Crustacea (Gammarus)(mg/kg)	0.9	1.3	2.8	83
bioaccumulation factor GAM	0.9×10^4	0.3×10^4	0.7×10^3	0.6×10^4

March 1984 and 1986. Table 2 gives the concentrations in solution, solids and organisms and the respective geo- and bioaccumulation factors, calculated for samples from locations km 0 and 20.

3.3 Analysis of variance

Statistical analysis of covariation of heavy metal concentrations compared one algal and five crustacean species to suspended solids, sediments and sediment leachate. The variables considered were metal concentration, animal weight, location and water temperature. Crustacea included Gammarus tigrinus, Neomysis integer, Palaemon longirostris, Crangon crangon and Eriocheir sinensis. Considering the suspended solids, their Cd content corresponded significantly ($P \leq 5$ %) to the concentrations found in algae and the crustacea (except Palaemon), whereas Zn in suspended solids was only correlated with Zn in algae, and Pb and Ni varied independently (Table 3). A closer correspondence was found between heavy metals in biota and in the surface-sediment from the respective location. Again Cd is highly correlated, Ni in the case of two crustacean species and algae, Pb and Zn, however, only in the case of gammarids and algae, respectively.

Table 3: Significant simple and/or multiple correlations between heavy metal concentration in organisms and in suspended solids (SUS), sediments (SED) and sediment-leachate (LEA); * = P < 5 %, ** = P < 1 %, *** = P < 0.1 %.

		GAM	NEO	PAL	CRA	ERI	ENT
SUS	Cd	**	*	-	*	*	*
	Pb	-	-	-	-	-	-
	Ni	-	-	-	-	-	-
	Zn	-	-	-	-	-	*
SED	Cd	***	*	***	***	***	***
	Pb	*	-	-	-	-	-
	Ni	*	-	*	-	-	**
	Zn	-	-	-	-	-	*
LEA	Cd	***	***	**	***	***	***
	Pb	***	*	-	**	***	***
	Ni	***	*	-	-	**	***
	Zn	-	-	-	-	-	**

When the amounts of moderately reducible heavy metals within the sediment are considered even more significant correlations are obtained. Zn, however, is reflected only in algae, due to the ability of Crustacea to control their Zn budget.

Considering that the algae and most of the Crustaceans live in little or no contact with the sediments and that the main route of heavy metal uptake is the resorption of ions from solution via gills and skin rather than from contaminated food, the high degree of correlation between the bioavailability of heavy metals and their presence in the moderately reducible fraction in the sediments should not be interpreted as a direct interaction. We must instead assume that the concentrations found in these two compartments are the result of a varying supply of heavy metals in solution delivered by the headwater and polluters. However, this prime source of heavy metals for aquatic ecosystems can hardly be controlled directly due to analytical problems and high variability. But as shown above, algae and certain crustacean species reflect the pollution of the sediments with a high degree of probability, especially the amount of moderately reducible heavy metals that can easily be remobilized by changing environmental conditions.

3.4 Application scheme

We provisionally propose to establish a low-cost biomonitoring program using monthly samples of littoral benthic algae and gammarids, which are frequently found in the estuaries of Weser and Elbe, in addition to the routine of monitoring heavy metals in solution and suspended solids. The information gained comprises the following:

(i) The level of heavy metal contamination in estuarine key species and its temporal and spatial variation as it develops under chronic exposition of all life-stages and as a result of synergisms and antagonisms that cannot be simulated in any laboratory; such data are indispensable for the assessment of estuarine pollution and water quality standards.

(ii) Variation and trends in bioavailability of heavy metals, especially regarding changes in pollution quality such as improved retention of suspended solids in sewage treatment plants; remobilization of heavy metals by increasing emission of complexing agents such as EDTA and NTA or resuspension of sediments by dredging activities and shipping.

(iii) The level of and changes in the pollution of sediments; the tight correlation between bioavailable heavy metals and the moderately reducible sediment-bound fraction allows the detection of changes in the chemical state and tightness of bonding to sediment particles that occur due to changes in sediment composition or redox-potential and that are not detected by plain sediment analysis.

ACKNOWLEDGEMENT

This work was funded by Bundesminister für Forschung und Technologie, FKZ 525-3891-MFU 0533.

REFERENCES

Böddeker H, Jablonski R, Kramer K, Prange A, Schnier C, Wellhausen K (1988) Schwermetalluntersuchungen im Weserwasser mit Hilfe der TRFA zur Weser-Kampagne 21. Mai - 3. Juni 1987. GKSS Geesthacht, Januar 1988

Förstner U, Calmano W (1982) Bindungsformen von Schwermetallen in Baggerschlämmen. Vom Wasser 59: 83-92

Förstner U, Wittmann GTW (1983) Metal Pollution in the Aquatic Environment. Springer Verlag, Berlin Heidelberg Tokio

Schirmer M, Jathe B, Schuchardt B, Busch D (1989) Gutachten zur Beurteilung des Gewässergütezustandes der Unterweser. Teilgutachten: Belastung der Unterweser mit Schwermetallen. I Auftr d WWA Bremen

Turekian KK, Wedepohl KH (1961) Distribution of the elements in some major units of the earths crust. Bull Geol Soc Am 72: 175-192

Zauke G-P, Meurs H-P, Todeskino D, Kunze S, Bäumer H-P, Butte W (1987) Untersuchungen zur Verwendung von Bioindikatoren für die Überwachung im Ästuarbereich der Elbe, Weser und Ems. Teil 3: Zum Monitoring von Cadmium, Blei, Nickel, Kupfer und Zink in Balaniden (Cirripedia: Crustacea), Gammariden (Amphipoda: Crustacea) und Enteromorpha (Ulvales: Chlorophyta). Bundesministerium für Umwelt, Naturschutz und Reaktorsicherheit. Forschungsbericht Wasser, FKZ 10 20 52 09

SEASONAL AND SPATIAL PATTERNS OF THE DIATOM ACTINOCYCLUS NORMANII IN THE WESER ESTUARY (NW-GERMANY) IN RELATION TO ENVIRONMENTAL FACTORS

Bastian Schuchardt and Michael Schirmer
University of Bremen, FB 2, PO-Box 330440
D-2800 Bremen 33

1. INTRODUCTION

During late summer and fall the tidal freshwater reaches of the estuaries of Elbe and Weser are dominated by the autochthonous Actinocyclus normanii, a large centric diatom which forms up to 95 % of phytoplankton biomass (Schuchardt, to be published). In spite of much work that has been done on phytoplankton dynamics in the outer reaches of estuaries, seaward of high suspended matter (SPM) concentrations (Cloern, 1987), only few studies have focussed on phytoplankton dynamics and the controlling processes within the tidal freshwater reaches.

2. MATERIALS AND METHODS

The Weser Estuary is a coastal plain estuary opening into the German Wadden Sea (North Sea). Descriptions have been given by Lüneburg et al. (1975) and Wellershaus (1981). Sampling was carried out between June 1986 and November 1988 on monthly cruises in the inner part of the estuary from the upstream border of the estuary, a tidal weir (Lower Weser-km 4.7), to the mesohaline reaches (Lower Weser-km 65) at 10 kilometer intervals. The master variables were recorded continuously on board ship and samples were taken from 1.3 m below surface for phytoplankton cell counts, determination of SPM, nutrients, chlorophyll a and others.

3. Results and Discussion

A. normanii shows a consistent seasonal and spatial pattern of abundance in the estuary (Figs. 1 and 2). Peak abundance occurs between August and October. At lower cell numbers the species persists throughout the year. Abundance (N) in the seasonal cycle (Fig. 3) can be described by: $\ln N = 0.14\ Temp - 1.77 \ln Q_5 - 10.85$ ($r^2 = 0.85$) where Q_5 is the average discharge 5 days before sampling. There is no significant correlation with global irradiance. The negative correlation with river discharge supports our suggestion that longer residence time of the species within its habitat inforces the abundance (see below).

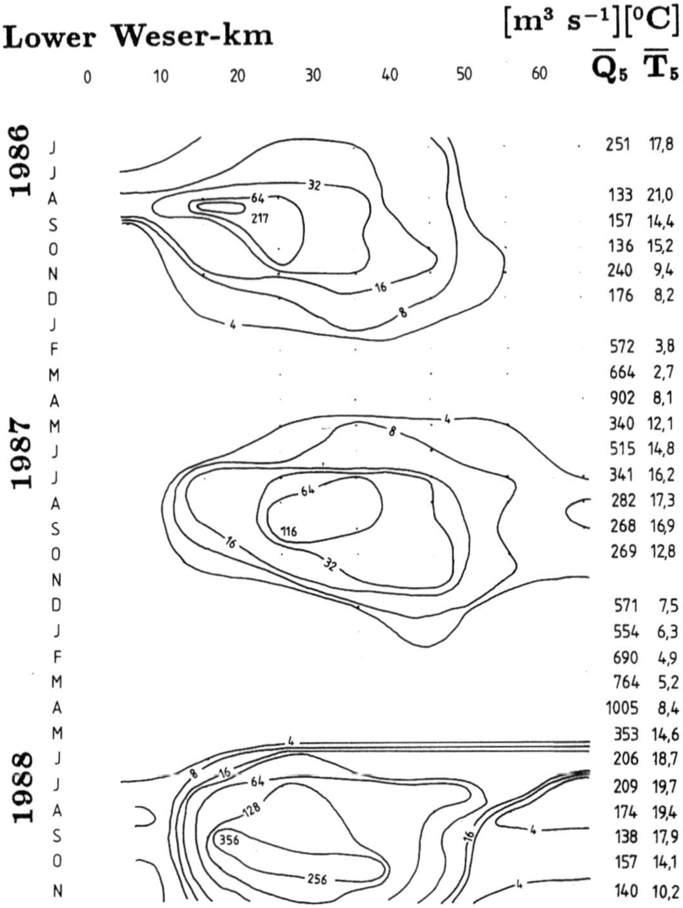

Figure 1: Temporal and spatial pattern of A. normanii cell numbers (cells ml^{-1}). Q_5 = average river discharge 5 days before sampling date

A. normanii is restricted to the freshwater reaches of the estuary. The upstream border of the population coincides with the upstream limit of flood currents (approximately 4 km downstream of the tidal weir); the species is not introduced from upstream. In the longitudinal axis the population shows steep gradients of abundance, especially during bloom conditions. During all cruises peak abundance occurred clearly upstream of the salinity intrusion (Fig. 2). Cell numbers are strongly depressed before salinity increases. So we can exclude osmotic stress as the most important factor limiting the seaward extension of the population.

Peak abundance occurred upstream of the SPM-maxima of the turbidity zone (Fig. 2), which forms at the tip of the salt wedge and is a persistent feature of the estuary. A qualitatively inverse relationship exists between SPM and cell numbers along the longitudinal axis. Calculating photic depth (Z_p) from SPM

(Cloern, 1987) and using a tidally averaged total depth of 11 m (Z_m, assuming the inner part of the estuary as vertically well mixed), the ratio of Z_p to Z_m varies between 0.46 for lowest observed SPM (about 15 mg l^{-1} upstream) and 0.03 for highest concentrations (about 250 mg l^{-1}). Thus, photic depth is a very small fraction of the well mixed water column. As high anthropogenic nutrient inputs provide non-limiting nutrient concentrations in the Lower Weser (data not presented) regulation of phytoplankton growth is considered to be controlled by light limitation due to the steep gradient of SPM concentration.

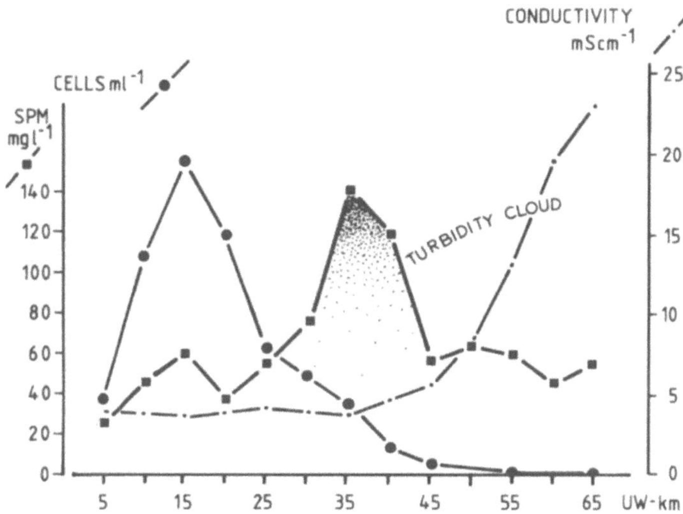

Figure 2: Longitudinal section of A. normanii cell numbers, SPM, and electrical conductivity in the Weser Estuary (August 13, 1986, Q_5 = 133 m^3 s^{-1})

The position of peak abundance (P) within the estuary varies throughout the year mainly due to changes in river discharge and temperature and can be described by: P (LW-km) = 11.58 ln Q_5 -0.66 Temp -19.83 (r^2 = 0.71).

The habitat of A. normanii is strongly influenced in two ways by river discharge. During periods of low discharge in summer and fall (when the species is abundant), the length of the habitat is reduced due to the upstream movement of the turbidity zone, which is the seaward border of the population. This is, however, counteracted by a strongly prolonged flushing time following reduced river discharge. The residence time of the water within the habitat (here assumed to be between LW-km 5 and 45) increases from only 1 to 2 d at 1180 m^3 s^{-1} (mean high discharge) to 3 d at 324 m^3 s^{-1} (mean discharge). During summer conditions with 119 m^3 s^{-1} (mean low discharge) the residence time increases to approximately 12 d (data calculated by Müller, GKSS, using the model FLUSS). So we can assume that autoch-

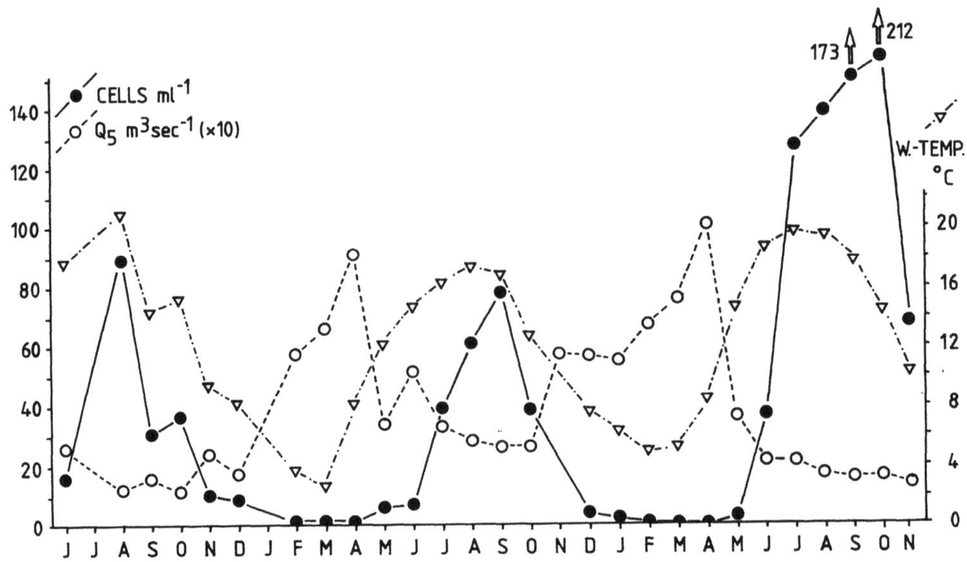

Figure 3: Seasonal cycles of A. normanii cell numbers, water temperature, and river discharge (Q_5). Cell numbers are averaged values from Lower Weser-km 15 - 45, the main habitat of the species

thonous plankton blooms within the freshwater tidal region of the Weser Estuary are restricted to periods of low river discharge which result in increased residence time and therefore decreased exchange ratio.

REFERENCES

Cloern JE (1987) Turbidity as a control on phytoplankton biomass and productivity in estuaries. Cont Shelf Res 7 (11/12): 1367-1381

Lüneburg H, Schaumann K, Wellershaus S (1975) Physiographie des Weser-Ästuars (Deutsche Bucht). Veröff Inst Meeresforsch Bremerhaven 15: 195-226

Schuchardt B (to be published) Phytoplankton maxima in the tidal freshwater reaches of the estuaries of Weser and Elbe

Wellershaus S (1981) Turbidity maximum and mud shoaling in the Weser Estuary. Arch Hydrobiol 92: 161-198

THE IMPORTANCE OF STRATIFICATION FOR THE DEVELOPMENT OF PHYTOPLANKTON BLOOMS - A SIMULATION STUDY

Günther Radach and Andreas Moll
Institut für Meereskunde, Universität Hamburg
Troplowitzstr. 7, D-2000 Hamburg 54

ABSTRACT

Physical processes are important for plankton dynamics on many space and time scales. Based on model simulations, the effect of stratification for the formation of plankton blooms is discussed. Annual phytoplankton simulations incorporate the effect of stratification on algal blooms by a generalized time dependent Sverdrup approach. The simulations show that the main features of the plankton blooms are caused by stratification effects.

INTRODUCTION

From the beginning of plankton modelling the outstanding role of the mixed-layer dynamics for the formation of plankton blooms was recognized. Usually a two-layer system was assumed, the mixed-layer itself and the underlying layer. Thus, stratification was used to avoid the determination of the vertical structure of physical, chemical and biological processes.

Riley, Stommel & Bumpus (1949) were the first to introduce vertical eddy diffusion as a forcing mechanism for plankton dynamics. Sverdrup (1953) introduced the "critical depth", as that depth down to which the total photosynthesis of the algae in the water column is equivalent to their total respiration. For effective production the critical depth must exceed the thickness of the mixed-layer. Thus, the outburst of spring growth is mainly dependent on a stabilizaton of the water. The strong vertical dependence of phytoplankton and light intensity and the complicated exchange of material in the vertical makes a vertical resolution necessary.

During the 1970's physical upper-layer models were designed to calculate time-dependent profiles of temperature and turbulent diffusion coefficients or the mixed-layer depth as a function of time (then assuming instantaneous mixing within the mixed-layer). An overview of modelling the physical upper-layer dynamics is given by Price et al. (1987). The simulated quantities may be used as physical forcing in the plankton models. The first authors to introduce vertical eddy diffusion profiles as a forcing mechanism for plankton dynamics seemingly did so at the same time (Winter, Banse & Anderson, 1975 and Radach & Maier-Reimer, 1975).

Most work deals with the spring phytoplankton bloom during the formation of the seasonal thermocline. Little work has been done so far on the annual cycle of phytoplankton development, including the vertical structure in relation to the development and erosion of the seasonal thermocline (Kiefer & Kremer, 1981; Wolf & Woods, 1988). We present here, as an example, two simulations of annual cycles of plankton dynamics, performed with our 4-component time-dependent water column model under actual meteorological forcing.

THE MODEL

The model we use is a modification of the model from Radach (1983). It is presented in detail by Moll (1988). The original simulation model, which was designed to cover the spring phytoplankton bloom, was extended to the full annual cycle by including the regeneration part of the nutrient dynamics. The model is now able to produce full annual cycles of nutrient levels, yielding, amongst others, the same concentration at the end of the year as at the beginning.

The numerical experiments described here are based upon two second order, non-linearly coupled partial differential equations for the concentrations of phytoplankton and a limiting nutrient (phosphate), with both variables depending on time and depth. In addition, two first order differential equations, one for light intensity and one for detritus, are solved simultaneously.

The plankton model includes the physical processes of turbulent diffusion, sinking, light penetration, and the biological processes of photosynthesis with light limitation (self-shading) and nútrient limitation, respiration, mortality, grazing, nutrient uptake and regeneration.

We assume that dead organic material is collected in a detritus pool at the bottom. The detritus will be regenerated with a rate of about 1/(60 days), thus providing regenerated phosphate in the lowest layers. Physical vertical mixing then serves for the redistribution of phosphate in the water column, e.g. by storms, which mix the water column down to the bottom.

The dead organic material consists of dead phytoplankton, fecal pellets and dead zooplankton. We assume that the time-scales of sinking are short compared to the time-scale of regeneration. Therefore, we let the dead organic material fall immediately out of the water column, thus avoiding a further partial differential equation for detritus in the water column. The regenerated nutrient is diffused upwards by its coupling to the boundary condition for phosphate.

The physical upper-layer model by Kochergin et al. (1976) was driven by 3-hourly wind, radiation and heat flux data, derived from meteorological standard observations at lightvessel ELBE 1, to generate the time-dependent profiles of temperature and turbulent diffusion coefficients, which then drive the phytoplankton simulations. Further forcing fields are underwater light and herbivorous zooplankton biomass.

RESULTS

Two scenarios are presented, which are characterized by different weather situations: a "windless winter and long summer" year (1965) and a "stormy winter and short summer" year (1967). We derived these characterizations from the time series of wind velocity, air and sea surface temperature and calculated net surface energy balance. We assumed an initial phosphate concentration of 1.2 µmol/l, which is comparable to inner German Bight conditions at the time of outgoing winter. Thus, the scenarios are suited to investigate the influence of the development and decay of stratification under the assumption that "everything else" remains constant, i.e. the dynamical mechanism as well as the zooplankton biomass.

The results of the simulations are sequences of profiles. Figs. 1 and 2 show the annual cycles of turbulent diffusion coefficients, phytoplankton and the corresponding phosphate concentration for 1965 and 1967, respectively.

The distribution of the different turbulent diffusion coefficients exhibits for 1965 a relatively long period with weaker near-surface mixing (March to November, 258 days), but for 1967 a considerably shorter period (April to October, 202 days). The depth of the turbocline is in both cases at 30 m and characterizes the upper boundary of the thermocline.

The resulting annual cycles of phytoplankton are totally different: 1965 is dominated by a spring bloom, while 1967 shows a strong summer bloom. Self-shading restricts primary production in 1965 to the upper 30 m. The outburst of the spring bloom occurs two weaks later in 1967 and reaches down to a depth of 50 m, and the fall bloom stops three weaks earlier.

The corresponding phosphate development shows differing phases of depletion. In 1965 depletion exists from July to October, while in 1967 the depletion starts two weaks later and ends two weaks earlier (see concentration of 0.6 µmol/l). Thus, the depletion time in 1967 is shorter, but more intensive. For both years the bottom phosphate accumulates due to the regeneration process in the depletion phase, until storms mix the water column fully. To our surprise the different spring and summer blooms, caused by totally different stratification scenarios, yield in both cases similar integrated annual gross primary production values of about 150 gC/y.

CONCLUSIONS

Very different annual cycles of phytoplankton may be brought about by different meteorological and oceanographic cycles, although total annual primary production may remain unaltered. Thus, the optimum biomass will be accumulated in every year, but its distribution in time evolves in relation to the actual environmental conditions. The sequence of bloom events within each year is related to the sequence of meteorological events in a nonlinear manner.

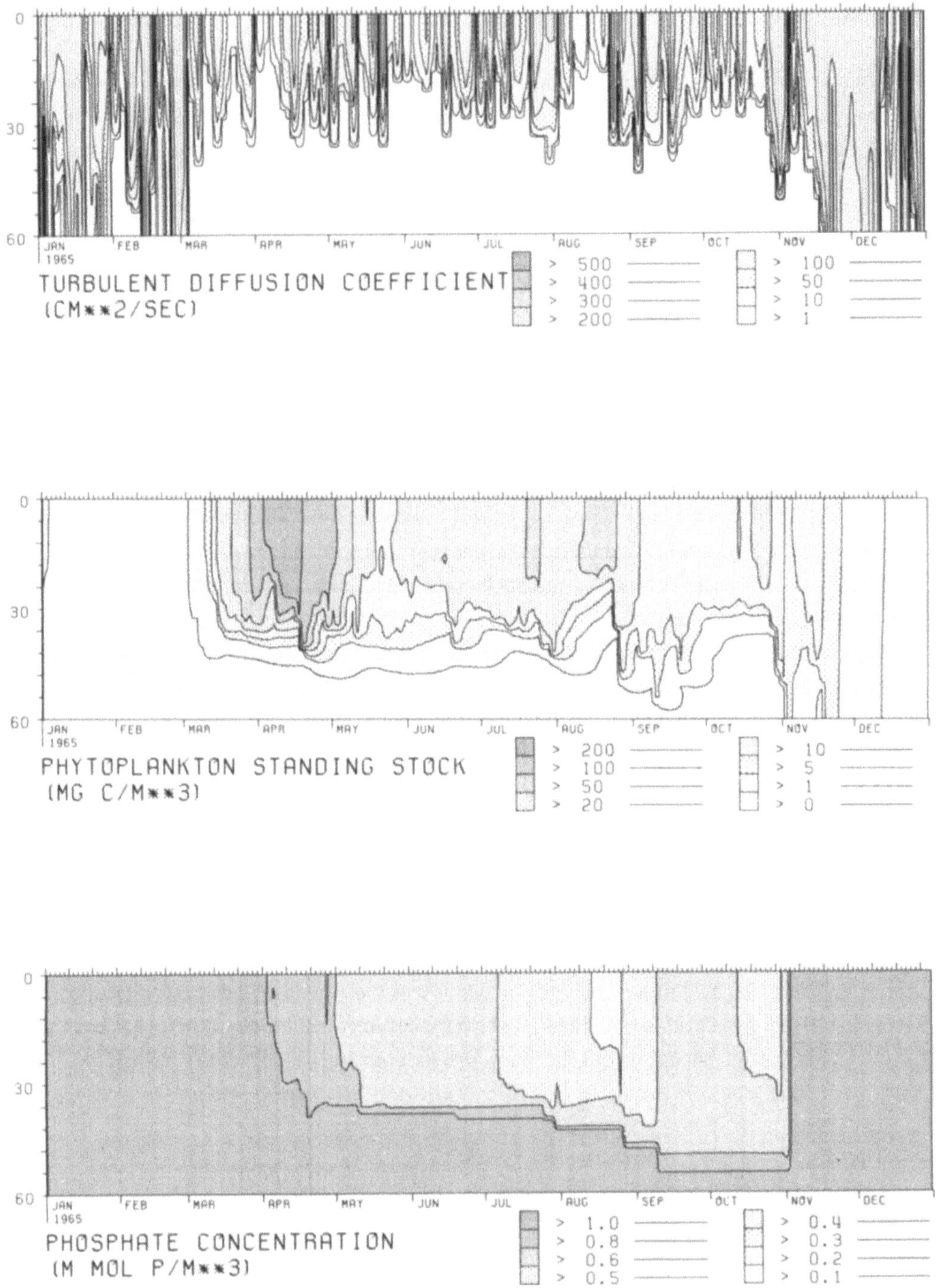

Figure 1: Simulated profiles for the annual cycle of 1965

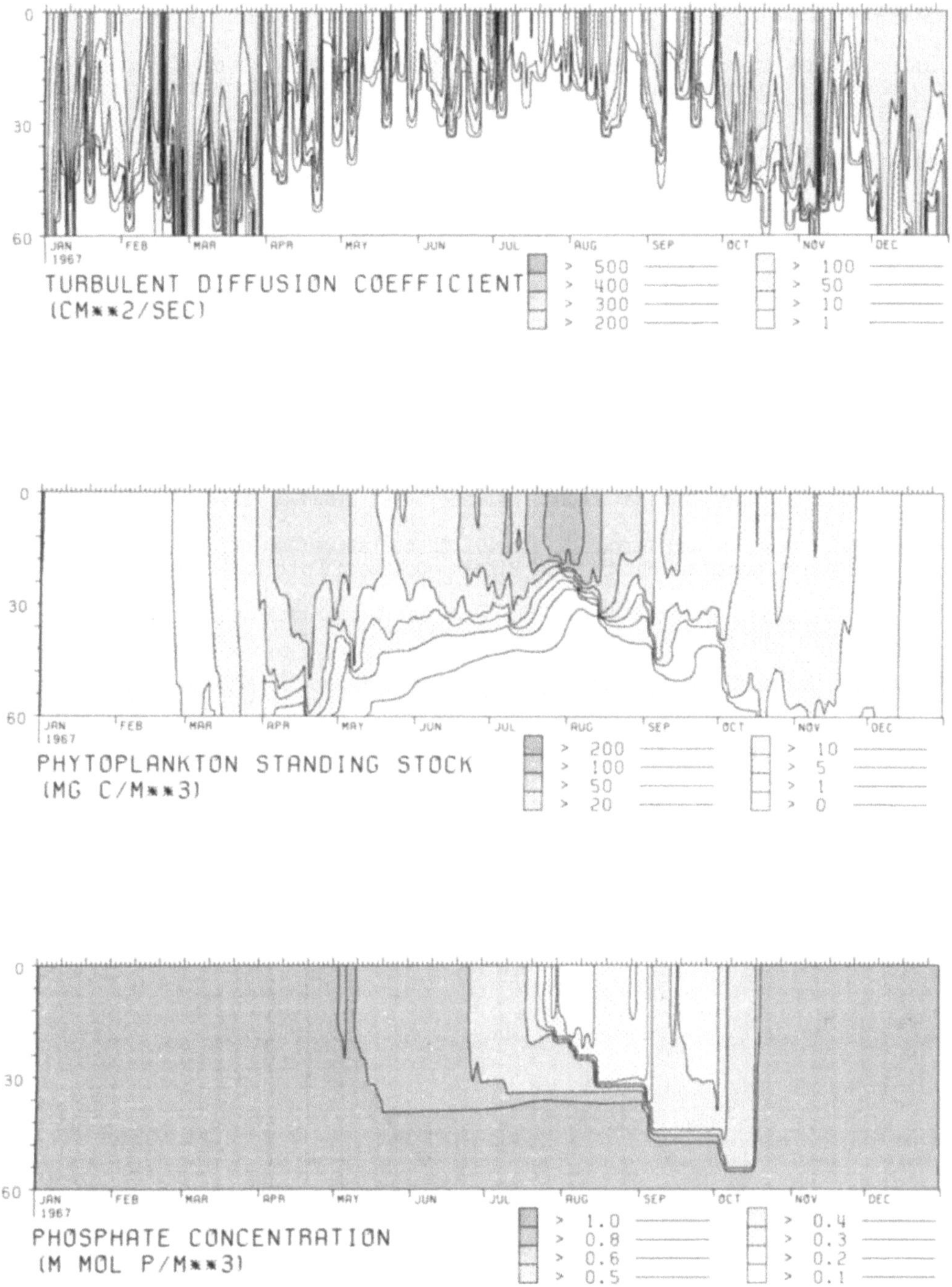

Figure 2: Simulated profiles for the annual cycle of 1967

REFERENCES

Kiefer DA and Kremer JN (1981) Origins of vertical patterns of phytoplankton and nutrients in the temperate, open ocean: a stratigraphic hypothesis. Deep-Sea Res 28: 1087-1105

Kochergin VP, Klimok VA and Sukhorukov VA (1976) A turbulent model of the ocean Ekman layer (in Russian). Sb Chisl Metody Mekhan Sploshnoi Sredy 7(1): 72-84

Moll A (1988) Simulation der Phytoplanktondynamik für die zentrale Nordsee im Jahresverlauf - Diplomarbeit im Fachbereich Geowissenschaften der Universität Hamburg, 141 S

Price JF, Terray EA and Weller RA (1987) Upper ocean dynamics. Rev Geophys 25(2): 193-203

Radach G (1983) Simulations of phytoplankton dynamics and their interactions with other components during FLEX'76. In: Sündermann J and Lenz W (eds) North Sea Dynamics, Springer Verlag, Berlin, pp 584-610

Radach G and Maier-Reimer E (1975) The vertical structure of phytplankton growth dynamics - a mathematical model. Mem Soc R Sci Liege 6: 113-146

Riley GA, Stommel H and Bumpus DF (1949) Quantitative ecology of the plankton of the western North Atlantic. Bull Bingham Oceanogr Coll 7: 1-169

Sverdrup HU (1953) On conditions for the vernal blooming of phytoplankton. Journal du Conseil 18(3): 287-295

Winter DF, Banse K and Anderson GC (1975) The dynamics of phytoplankton blooms in Puget Sound, a fjord in the northwestern United States. Mar Biol 29: 139-176

Wolf KU and Woods JD (1988) Lagrangian simulation of primary production in the physical environment - the deep chlorophyll maximum and the nutricline. In: BJ Rothschild (ed) Towards a Theory of Biological-Physical Interactions in the World Ocean, D Reidel, Dordrecht (in press)

BIOMASS AND PRODUCTION OF PHYTOPLANKTON IN THE ELBE ESTUARY

T. Fast and L. Kies
Institute for General Botany, University of Hamburg,
Ohnhorststraße 18, D-2000 Hamburg 52, Federal Republic of Germany

INTRODUCTION

The brackish water section of the Elbe Estuary is characterized by strong salinity gradients and high turbidity, both acting as stress factors for autotrophic planctonic algae.

METHODS

Phytoplankton biomass was calculated from chlorophyll-a measurements according to Nusch and Palme (1975). Phytoplankton production was measured under in-situ condition using the ^{14}C-technique.

RESULTS AND DISCUSSION

In the Elbe Estuary the salinity rises from about 0.4 ‰ in the freshwater reach (Neumühlen) to about 15 ‰ in the mesohalinic section (Altenbruch). Within this gradient the chlorophyll-a content decreased from about 180 $\mu g \cdot l^{-1}$ to about 6 $\mu g \cdot l^{-1}$ (Fig. 1). Primary production decreased from about 260 $\mu gC \cdot l^{-1} \cdot h^{-1}$ (Neumühlen) to about 40 $\mu gC \cdot l^{-1} \cdot h^{-1}$, whereas the quotient production/biomass (P/B) increased from 1.2 to 5 $\mu gC(\mu gChl \cdot h)^{-1}$ (Fig. 2). When freshwater plankton enters the brackish water section of the estuary, photosynthesis and growth is adversely affected and freshwater phytoplankton finally disintegrates. Decrease of production due to salt stress can be simulated by slowly adding concentrated seawater to freshwater samples. When the salinity is thus raised from 0.7 ‰ to 3.4 ‰ S within 5 hours, primary production decreases to 30 % of the control values (Fig. 3). This effect was also observed in an experiment conducted in a similar way with cyanobacteria from the Potomac River Estuary (Sellner et al., 1988), where the production decreased to 8 % of the initial values. In the measohalinic section of the estuary the freshwater phytoplankton is replaced by a brackish water community, which is adapted to higher and fluctuating salinities. This phytoplankton cummunity might be

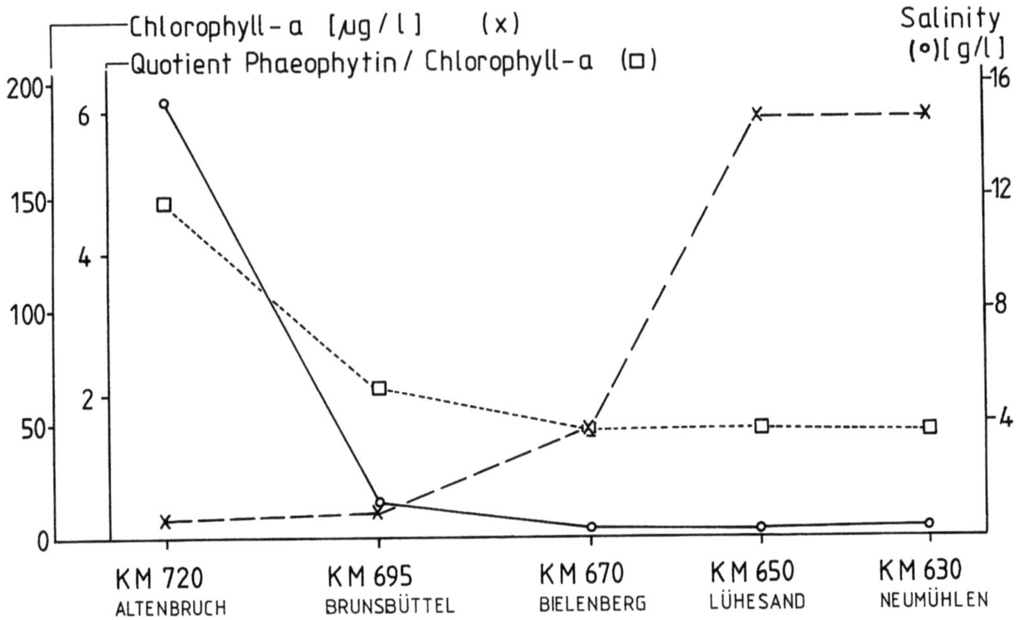

Figure 1: Chlorophyll-a content in a longitudinal profile of the Elbe Estuary. September 27, 1987

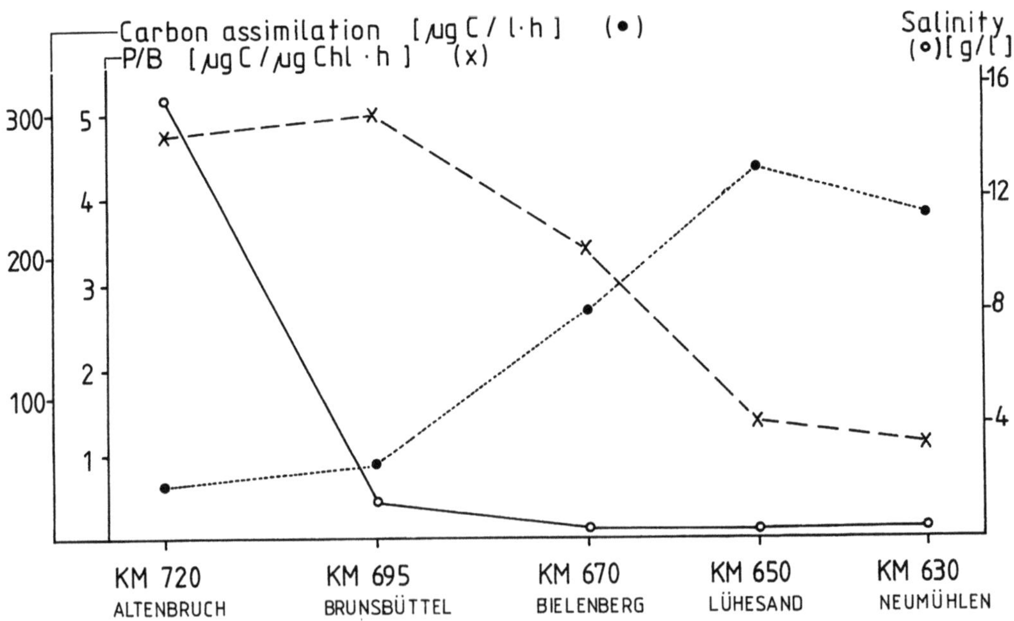

Figure 2: Primary production and P/B in a longitudinal profile of the Elbe Estuary. September 27, 1987

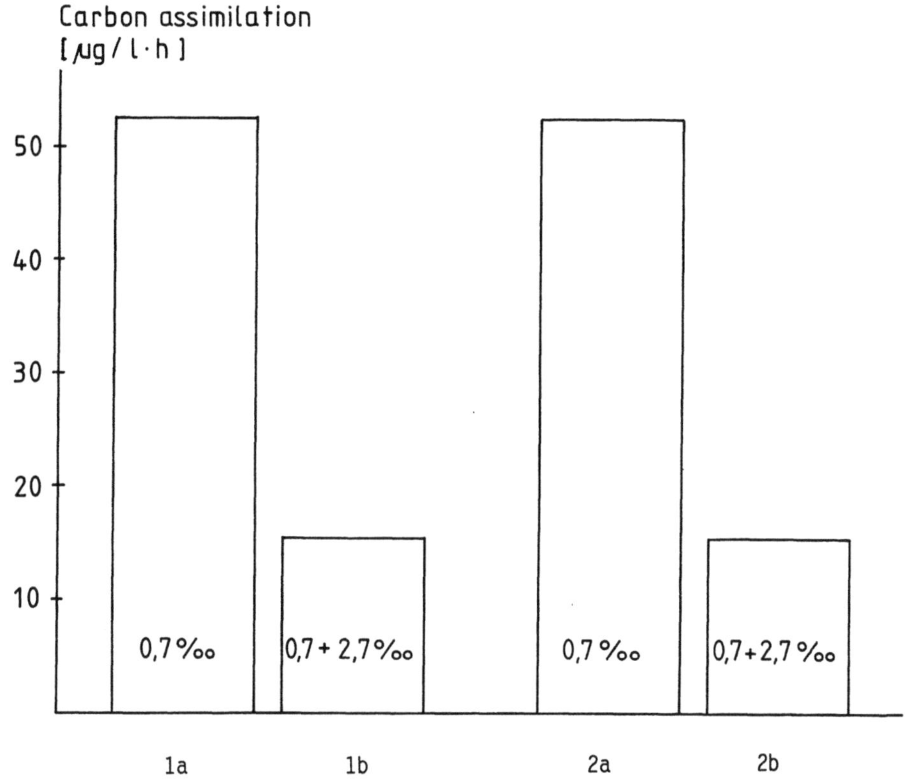

Figure 3: Effect of salinity increase on phytoplankton samples from the oligohalinic section of the Elbe Estuary. September 30, 1987
1a, 2a : controls, 0.7 ‰ salinity
1b : incubation 3 hours after final salinity is reached
2b : incubation 5.5 hours after final salinity is reached

Table 1: Carbon-contribution to the amount of total organic carbon

Location	Salinity [$g \cdot l^{-1}$]	Total organic carbon [$mg \cdot l^{-1}$]	Phytoplankton carbon [$mg \cdot l^{-1}$]	Contribution [%]
Bielenberg	0.40	13	2.4	18.4
Brunsbüttel	1.21	11	0.4	3.6
Cuxhaven	26.01	8	0.3	3.5

responsible for the higher P/B values found in this area. The contribution of the phytoplankton biomass to total organic carbon (values from ARGE ELBE, 1988) drops from 18.4 % in the freshwater section to 3.5 % in the mesohalinic section (Table 1). This is due not only to dilution of freshwater plankton by inflowing marine water poor in phytoplankton or to grazing, but also to disintegration after osmotic stress. This can be deducted from the quotient phaeophytin/chlorophyll-a which increases towards the open sea.

ACKNOWLEDGEMENTS

This work from Sonderforschungsbereich 327 "Tide-Elbe", subproject B2, was supported by Deutsche Forschungsgemeinschaft. We thank Dr. U. Brockmann for providing the salinity data.

REFERENCES

ARGE ELBE (Arbeitsgemeinschaft für die Reinhaltung der Elbe) (1988) Wassergütedaten der Elbe von Schnackenburg bis zur See. Abflußjahr 1987

Nusch EA, Palme G (1975) Biologische Methoden für die Praxis der Gewässeruntersuchung. Bestimmung des Chlorophyll-a- und Phaeopigmentgehaltes in Oberflächenwasser. GWF-Wasser/Abwasser 116: 562-565

Sellner KG, Lacouture RV, Parris CR (1988) Effects of increasing salinity on a cyanobacteria bloom in the Potomac River estuary. J Plankton Res 10: 49-61

MICROPHYTOBENTHOS IN THE ELBE ESTUARY: BIOMASS, SPECIES COMPOSITION AND PRIMARY PRODUCTION MEASUREMENTS WITH OXYGEN MICROELECTRODES

C. Gätje and L. Kies
Institute for General Botany, University of Hamburg
Ohnhorststraße 18, D-2000 Hamburg 52

In 1986 and 1987 sediment samples were taken every four weeks (except during ice periods) from the upper and lower eulittoral zone of eight stations in the Elbe estuary. The sampling area comprised the tidal flats bordering the river between the city of Hamburg and the North Sea including the freshwater-, oligo- and mesohaline section.

Chlorophyll-a/phaeopigment concentrations, algal cell numbers and diatom species composition were determined as well as physical parameters (temperature, pH, electrical conductivity) and sediment properties (water content, loss on ignition).

The microphytobenthos in the Elbe estuary was dominated by pennate diatoms. In summer, they were occasionally accompanied by blue-green algae (Oscillatoria, Agmenellum) or by flagellates (Euglena, nanoflagellates) which temporarily occurred in high cell numbers. The diatom flora consisted of species living on mud which are able to migrate within the upper few millimeters of the sediment (Navicula, Nitzschia, Cylindrotheca) or species attached to sand grains (Opephora, Achnanthes, Rhaphoneis). A total number of 145 diatom species, belonging to 42 genera, was identified. Twenty obviously extremely euryhaline species occurred over the whole salinity gradient.

The chlorophyll-a concentration in the upper eulittoral zone generally exceeded that of the low level sediment twofold. During the investigation period the annual mean of chlorophyll-a concentration in the 0.5 cm top layer of the sediment was highest at the stations in the freshwater section (120-220 mg chl-a/m^2) and lowest in the oligohaline section (27-56 mg chl-a/m^2). Medium values were reached in the mesohaline section (41-114 mg chl-a/m^2) (fig. 1).
The proportion of phaeopigments in relation to native chlorophyll-a showed a linear correlation with water content and loss on ignition of the sediment except at the stations with mud containing particular high percentages of water and organic matter.

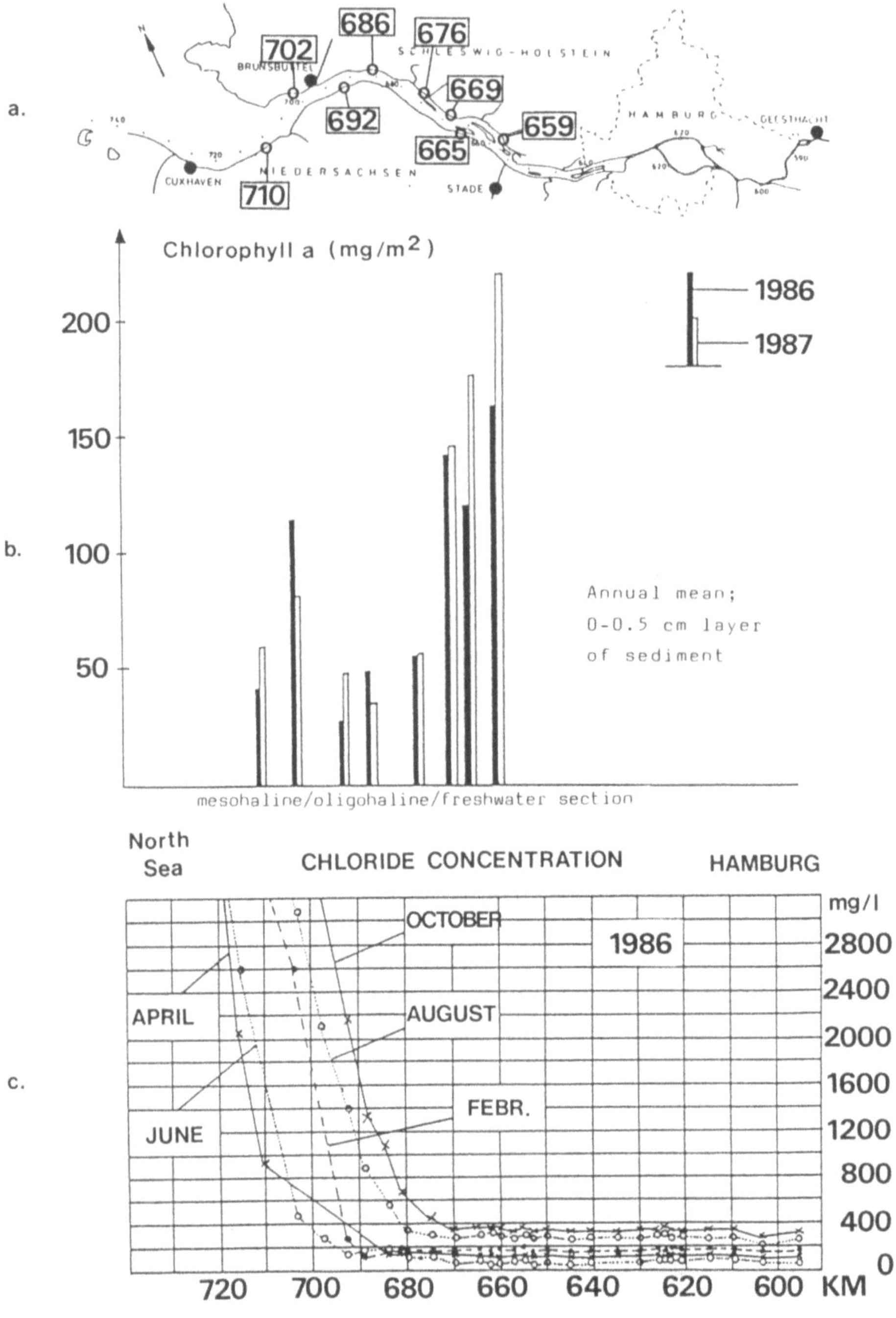

Fig. 1: a: Locations of sampling sites (characterized as km downstream)
b: Annual mean of chlorophyll-a concentration (0-0.5 cm layer of intertidal sediment)
c: Longitudinal profiles of chloride concentration (water column). Chloride values: ARGE Elbe (1987)

Primary production measurements were carried out in 1987 and 1988 at one sampling site in the freshwater section by the oxygen microprofile method according to REVSBECH & JÖRGENSEN (1983). This technique allows measurements in the sediment with high spatial and temporal resolution using oxygen microelectrodes.

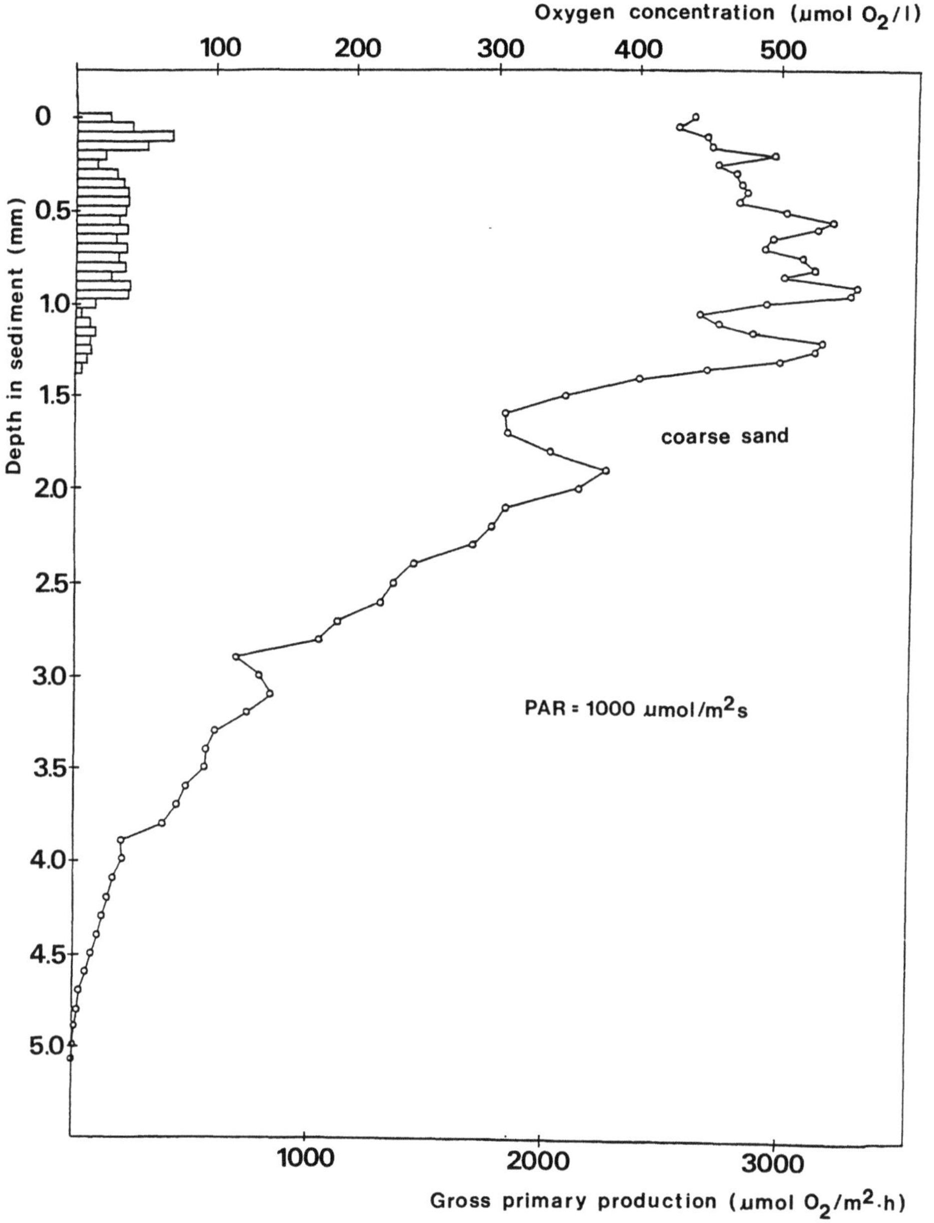

Fig. 2: Vertical microprofiles of oxygen concentration (open circles) and oxygen production (bars) measured with an oxygen microelectrode in coarse sand.

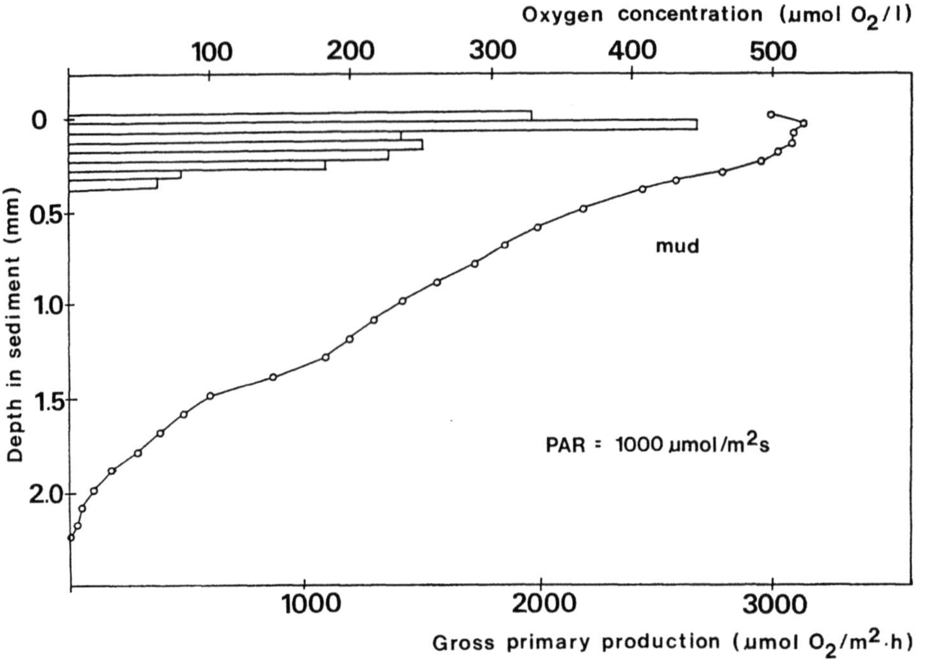

Fig. 3: Vertical microprofile of oxygen concentration (open circles) and oxygen production (bars) measured with an oxygen microelectrode in mud.

Figs. 2 and 3 show the marked differences between sandy and muddy sediment concerning thickness of the oxygenated layer and the photic zone. Due to deeper penetration of light in coarse sand an apparent photosynthesis could be measured up to a depth of 2.0 mm (mean: 1.2 mm). The thickness of the photic zone in mud and fine sand amounted to an average of 0.5 mm and 0.8 mm, respectively.
Gross primary production was remarkably higher in 1988 than in 1987 with maximum values in spring (up to 6.0 g O_2/m^2d) and in Aug./Sept. (up to 3.8 g O_2/m^2d) coinciding with considerable biomasses (104-517 mg chl-a/m^2). Production was relatively low in summer (max. 3.4 g O_2/m^2d) although medium chlorophyll-a concentrations (100-276 mg/m^2) occurred.

Acknowledgements: The present study is part of the research program SFB 327 "Interactions between abiotic and biotic processes in the Elbe estuary" supported by the Deutsche Forschungsgemeinschaft.

References

ARGE Elbe (Arbeitsgemeinschaft für die Reinhaltung der Elbe) (1987) Wassergütedaten der Elbe von Schnackenburg bis zur See. Zahlentafel 1986

REVSBECH NP, JÖRGENSEN BB (1983) Photosynthesis of benthic microflora measured with high spatial resolution by the oxygen microprofile method. Limnol Oceanogr 28:749-756

CHAPTER X

WATER CHEMISTRY AND SPECIATION

ADSORPTION AND RELEASE PROCESSES IN ESTUARIES

W. Salomons and J. Bril

Institute for Soil Fertility. P.O.Box 30003

9750 RA Haren (Gn). The Netherlands

Abstract

Experimental laboratory studies simulating changing estuarine conditions show that copper contaminated sediments from the Fly river are able to release copper. A comparison between calculated and experimental release curves shows that kinetics play an important role.

An estuarine chemical model, which includes settling of particles, salt wedge and increase in surface sites, showed the importance of adsorption, desorption and dilution processes for zinc, cadmium and copper; and the differences in behavior between these three elements in the estuarine environment.

Introduction

Understanding and predicting dissolved metal concentrations in estuaries is complex. A large number of chemical and physical factors have to be considered. In an effort to predict dissolved copper concentrations for a tropical estuary we experienced a number of problems which will be highlighted in this article. Predictive modelling of metals concentrations in estuaries is complex because of the sheer number of physical, chemical and hydrodynamic factors involved. As a first step towards this kind of predictive modelling we will discuss results of a model which takes into account the major processes in the surface waters of a stratified estuary. This model uses an earlier conceptual description of estuarine adsorption/desorption processes (Salomons, 1980).

Predicting metal concentrations in estuaries: Laboratory approach for sediments from the Fly river estuary in Papua New Guinea

Introduction

The Fly river and its tributaries rise in the central mountain range of New Guinea, flows through the southern plains and lowlands of the island and finally, with many other rivers, in the Gulf of Papua (figure 1). Its total catchment area is about 68.000 km^2.

The sediment load of the Middle Fly is estimated to be 16 million tons/year and that of the Lower Fly (which contains material from the Strickland) about 80 million tonnes/year (Salomons et al. 1988).

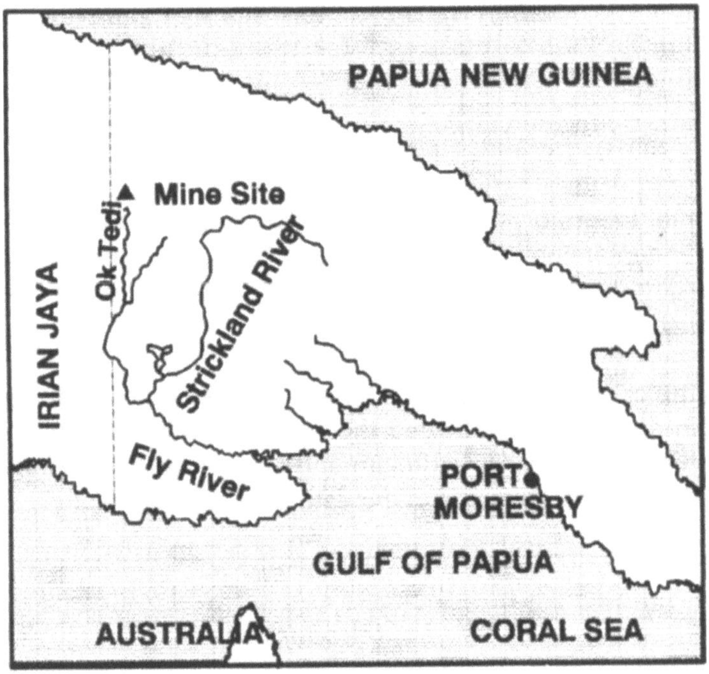

Figure 1. Papua New Guinea with the Fly River basin and the location of the mine site.

The source for elevated metal (copper) levels in the sediments from the river Fly is mining activities in its drainage area. Production of copper started in May 1987. During mine life an estimated 1000 million tonnes of material comprising 400 million tonnes of mill ore residue and 600 million tonnes of overburden will be generated.

This material contains elevated copper levels which, when no tailings dam or waste retention scheme is constructed, enters the river system (Salomons and Eagle, 1989). One of the questions which needs to be answered is the copper concentrations in the estuary for various mining scenarios.

To date (1989) copper levels in the estuary are still at the base line level. Therefore, it is not yet possible to study any changes in the behavior of copper in this estuary. A series of laboratory simulations were carried out to get a first approximation of likely effects.

Laboratory study to simulate estuarine processes

The experimental setup for the laboratory simulation of estuarine processes is similar to the one used earlier for riverine processes (Salomons et al. 1988). A schematic diagram is given in figure 2A .

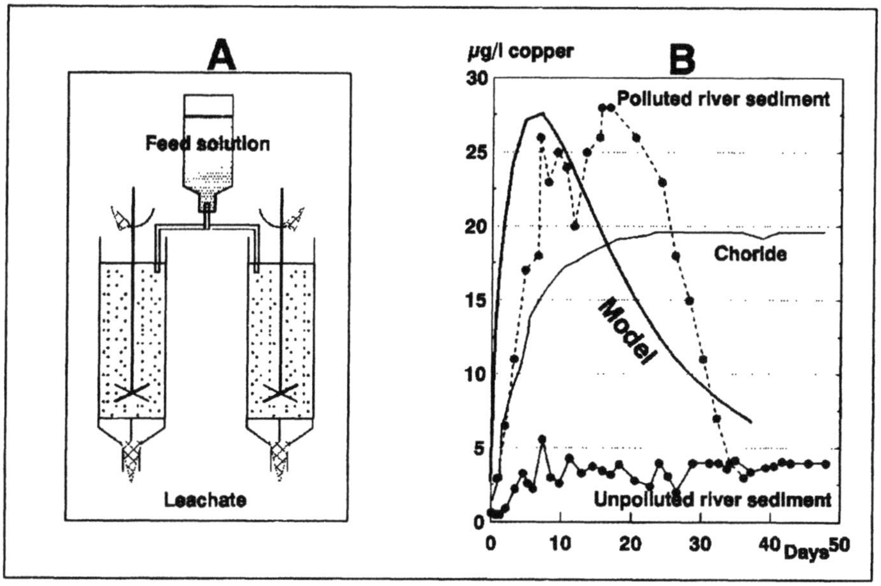

Figure 2. A. The experimental setup for studying possible release of copper from sediments in artificial Fly water. B. Experimental and modelling results for a 40 day experiment with sediments from the Middle Fly. For comparison the release of copper from an unpolluted sediment sample (Strickland) is also given.

The sediment sample is suspended in 6 l of artificial Fly water (major cation and anion composition equal to that of the Fly). The addition of seawater from the feeding flask compensates for the flow of water from the column. As a result the chloride concentration in the column increases gradually. The particles are subject to an increasing chloride concentration in the same way as happens during transport in an estuary. After about 25 days the freshwater in the column is completely replaced by seawater. The experiment lasted for about 50 days. For comparison an unpolluted sediment sample from the Strickland river (figure 1) was studied in a separate column.

The results (figure 2B) show a strong increase in the concentration of copper for the Fly river sediment which peaks between 10 and 20 days and then gradually decreases. After 35 days the concentration of copper becomes similar to that of the unpolluted Strickland river sediment. A balance calculation, including experiments with other sediment samples from the Fly and mine residue tested, showed that between 15 and 20 % of the total copper is amenable to release in the estuarine environment.

A simple explanation for the copper release from these sediments is ion exchange processes. The high cation concentrations of the estuarine waters (especially magnesium) cause a replacement of copper held at ion exchange positions on the particulates by cations prevalent in estuarine waters. Simulating the experiment with the CHARON model and using instantaneous ion-exchange as the hypothesis was able to predict the general shape of the curve. However, the model overpredicts the release at low salinities (figure 2B). Apparently some other, not yet identified kinetic process, is going on.

The rate of release (e.g, no chemical equilibrium between the copper on the particles and in the estuarine waters) has some rather serious implications for predicted copper levels in the estuary. Dissolved copper concentrations are expected to increase as the salinity of the estuarine waters increase, and will reach maximum concentrations at the delta front. Upon reaching the delta front however, the processes of dispersion in the coastal waters will reduce the dissolved copper concentrations. The chemical kinetics of the estuarine system are an important mechanism in assessing the likely magnitude of copper release. For example, if the release mechanisms are slow (as in the experiment), copper enriched particulates entering the Fly estuary will be subject to a high rate of dispersion before significant copper mobilization takes place. A low rate of release will prevent the occurrence of high dissolved copper concentrations in the inner estuary. In order to predict actual concentrations more information is needed on rates of dispersion in the Fly estuary; data which are now lacking. In addition more quantitative information is needed on the nature (speciation) of the "chloride-releasable" fraction of the copper.

One may conclude that further study into the rate of release of copper for this tropical estuary versus the rate of dispersion not only satisfies the scientific curiosity, but has also important implications for the environmental management of mine waste.

Predicting metal concentrations in estuaries: a model approach of sorption/desorption processes

Apart from desorption processes, which occur when sediments are subject to an increase in chlorinity, the reverse (adsorption) also occurs. Adsorption processes are possible when the total number of surface sites in the water column increases. However, even with an increase in surface sites (e.g, a turbidity maximum) the increase in chlorinity may counteract this effect and desorption may occur. Important is the location of the turbidity maximum in relation to the chlorinity gradient (Salomons, 1980).

To show the effects of an increase in chlorinity and turbidity maximum on heavy metal behavior in a stratified estuary, a mathematical model was constructed. The coupled transport-chemical equilibrium model CHARON (de Rooy 1988) was used to construct a model for estuarine adsorption-desorption processes.

The CHARON model differs from more common K_D models in several ways:

- all dissolved and adsorbed molecular species are accounted for.

- K_D is not constant but a function of the estuarine environment (e.g, changes in pH (calculated by the model) and changes in relative proportions of the major cations influence the value of K_D)

- because of its rigorous use of mass balance equations, effects of kinetic processes and transport on chemical equilibrium can be included.

The model as defined for this application simulates the salt wedge, the turbidity maximum and the flocculation of organic matter as basic parameters determining adsorption/desorption processes. Flocculation of organic matter was introduced in order to increase the amount of sites for adsorption. There are other processes which cause an increase in adsorption sites, however the net effect for the model is the same (e.g. an increase in the cation exchange capacity). Sedimentation in the model is a function of the mean flow velocity. Interaction with bottom sediments (e.g, erosion) is not included in the model.

Earlier studies (Salomons, 1980) provided adsorption date for the model. The estuarine schematisation (figure 3) consists of three layers of 6 boxes each. Between the layers and at the freshwater/estuarine interface dispersion takes place. It is realized that this schematisation and the hydrodynamics involved is an oversimplification of reality. However, the calculated salinity versus suspended matter concentrations for the three layers tend to confirm field observations (figure 4A).

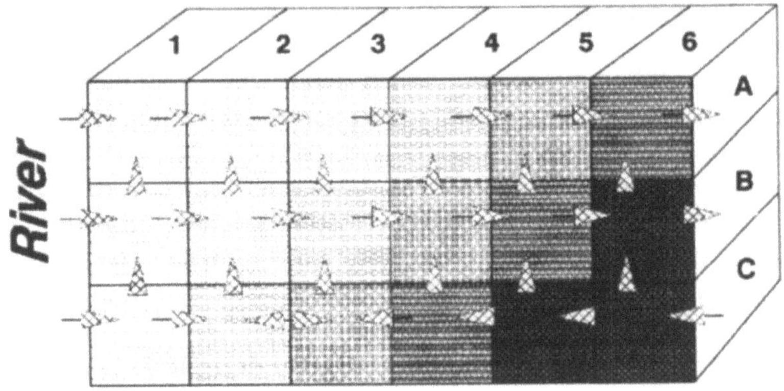

Figure 3. Schematization of a stratified estuary in the model. The arrows incidcate dispersion processes. The shading gives a schematic view of the calculated salinity gradient.

The chemical part of the coupled transport-chemical equilibrium model includes ion exchange, complexation, ionic strength and kinetic effects.

Inputs into the model, for this exercise, are the major ionic composition of estuarine water (Na, K, Mg, Ca, C-inorg, sulfate, chloride) and dissolved organic carbon. The model was run for cadmium, zinc and copper.

The calculated dissolved concentrations of copper, cadmium and zinc are plotted as a function of chlorinity for the three layers in the hypothetical estuary (figure 4B). In all cases a removal of copper is simulated because we allowed for the removal of dissolved organic matter from solution in the estuary. The formation of particulate organic matter enhances the adsorption of the suspended matter for copper. Cadmium and zinc desorb from the particulates in the bottom and middle layer at intermediate chlorinities. Chloride complexation and ion exchange are responsible. The release is followed by dilution.

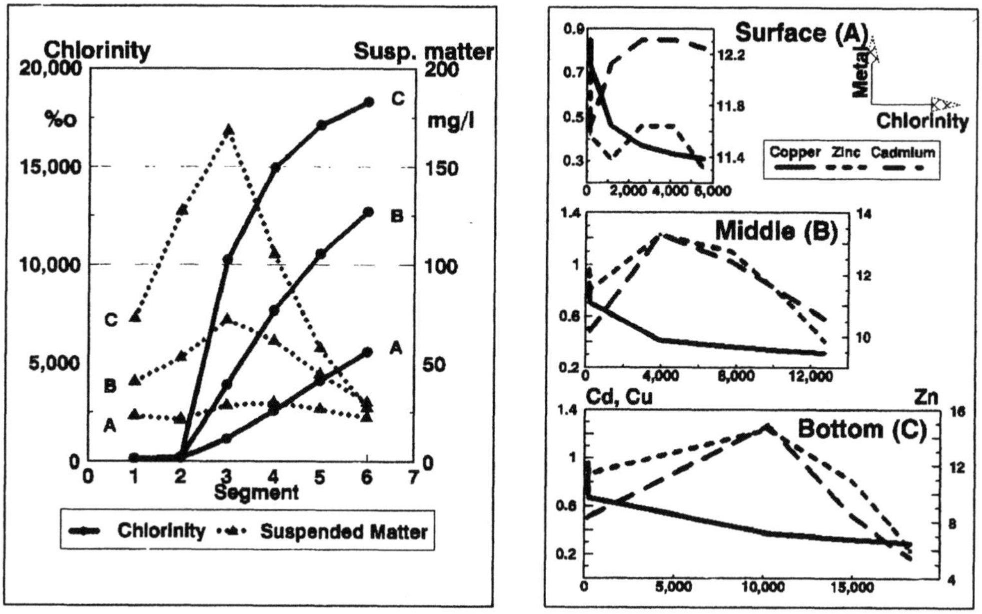

A B

Figure 4. A. Calculated salinity and suspended matter concentrations for the surface (A), middle (B) and bottom layer (C) in the estuary. B. Calculated cadmium, zinc and copper concentrations for the three layers in the estuary.

The behavior of cadmium and zinc in the surface layer differs because of the particular combination of an increase in chlorinity and in suspended matter concentration. It should be noted however, that the resolution of our model (18 boxes) is rather coarse. Effects found for the surface layer may also occur in the bottom and middle layer over the low salinity range for which no data were calculated. Zinc is first removed from solution, then a release takes place followed by dilution. This type of curve has been found in the Scheldt estuary (see Zwolsman and van Eck, this volume). Cadmium on the other hand exhibits a strong release as can be expected from its strong tendency to form cadmium-chloride complexes.

Although the model at present is an approximation of the "real world" situation, it is possible to simulate adsorption-desorption behavior of heavy metals. Furthermore the model makes it easily possible to verify hypotheses by changing input data.

REFERENCES

Rooy N M de (1988) Mathematical simulation of bio-chemical processes in natural waters by the model CHARON. Delft Hydraulics Report T68

Salomons W (1980) Adsorption processes and hydrodynamic conditions in estuaries. Environmental Technology Letters 1:356-365

Salomons W, M Eagle, E Schwedhelm, E allersma, J Bril and W G Mook (1988) Copper in the Fly River System (Papua New Guinea) as influenced by discharges of mine residue: Overview of the study and preliminary findings. Environmental Technology Letters 9: 931-940

Salomons W and M Eagle (1989) Hydrology, sedimentology and the fate and distribution of copper and mine related discharges in the Fly River System - Papua New Guinea. In Press

THE BEHAVIOUR OF DISSOLVED Cd, Cu AND Zn IN THE SCHELDT ESTUARY

J. J. G. Zwolsman[1] & G. T. M. van Eck[2]

[1] Department of Geochemistry, Institute of Earth Sciences, University of Utrecht
P. O. Box 80.021, 3508 TA Utrecht, The Netherlands
[2] Ministry of Transport and Public Works, Tidal Waters Division,
P. O. Box 8039, 4330 EA Middelburg, The Netherlands

ABSTRACT

In 1987, the Scheldt estuary was sampled during winter and spring to study the behaviour of Cd, Cu and Zn. The results show that seasonal mechanisms control the behaviour of trace metals.

During spring the river water entering the estuary is anoxic. The low concentration of dissolved trace metals in the fresh water is caused by precipitation of metal sulphides. After estuarine mixing, oxidation of these compounds leads to apparent desorption of Cd, Cu and Zn. In the lower estuary, phytoplankton blooms cause subsequent removal of dissolved Cd and Zn.

During winter, dissolved Cu and Zn concentrations in the fresh water entering the estuary are high since the river water is not completely anoxic. The existence of a turbidity maximum in the low salinity region causes rapid removal of dissolved Cu and Zn during the very first stages of estuarine mixing. Several mechanisms (oxidation of metal sulphides, desorption induced by chloride complexation, resuspension of reduced bottom sediments) can explain the subsequent increase of dissolved metal concentrations in the lower estuary.

1. INTRODUCTION

The behaviour of trace metals in estuaries is governed by various processes, which differ from one estuary to another (Salomons and Förstner, 1984). For example, extensive removal of dissolved Cd, Cu and Zn in the Rine estuary has been observed (Duinker, 1985), whereas the behaviour of these metals in the Scheldt estuary is dominated by remobilization from suspended matter (Wollast, 1976; Salomons et al., 1981; Duinker et al., 1982). The situation is even more complicated because the estuarine behaviour of trace metals may also depend on the season. Therefore, seasonal sampling of estuaries is essential to obtain a basic understanding of the relevant processes. Such a monitoring programme has been carried out in the Scheldt estuary, which was sampled eight times in 1987. In this paper the results of a winter and a spring cruise will be presented. Before dealing with the behaviour of dissolved Cd, Cu and Zn, the general geochemistry of the Scheldt estuary will be discussed.

2. THE SCHELDT ESTUARY

The Scheldt estuary is a coastal plain estuary, situated in South-West Holland and North-West Belgium (Fig. 1). According to its hydrodynamical characteristics, the estuary can be divided into three zones (Wollast and Peters, 1978); Wollast and Duinker, 1982). The lower estuary, commonly called the Western Scheldt, extends from the North Sea (km 0) to the Dutch-Belgian border (km 55). The upper estuary is situated between the border and the mouth of the river Rupel (km 93). The fluvial estuary, which contains only fresh water, is situated between the Rupel mouth and Ghent (km 160), where a tidal range of two meters still exists.

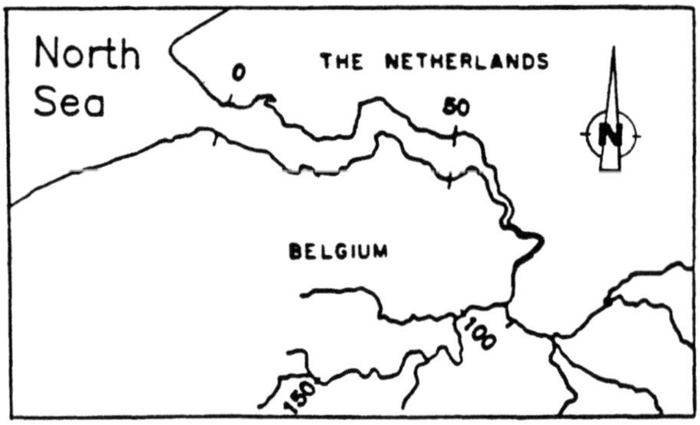

Figure 1. The Scheldt estuary (distance towards the mouth in km)

During average freshwater discharges the Scheldt estuary is well-mixed. The residence time of the water is about two to three months. In the upper estuary, a turbidity maximum is created because the residual upstream current in the upper estuary and the residual downstream current in the fluvial estuary cancel each other out. Flocculation of dissolved and colloidal matter during the first stages of estuarine mixing also contributes to the formation of the turbidity maximum (Wollast and Duinker, 1982).

Both the fluvial and the upper Scheldt estuary are heavily polluted as a result of large domestic, industrial and agricultural waste-water discharges (Billen et al., 1985). Since most of the waste-waters are discharged without prior treatment, intensive degradation of organic matter occurs in the fluvial and upper estuary. In this zone a permanent oxygen undersaturation and often complete depletion of dissolved oxygen in the water column is common. During dry summers an anoxic zone extends from the mouth of the river Rupel up to the Dutch-Belgian border, covering a range of 40 km. During winter the anoxic zone is much smaller or non-existent; in the upper estuary, however, a pronounced undersaturation of dissolved oxygen is still present (Somville and De Pauw, 1982).

3. SAMPLING SCHEME AND METHODS

The Scheldt estuary was sampled on 18-2-1987 and 13-5-1987. The February cruise is typical of a winter situation, because the river discharge was high (200 m^3s^{-1}) and the water temperature was low (4° C). The May cruise resembles the spring situation which is characterized by a decreasing freshwater discharge (100 m^3s^{-1}) and increasing water temperature (14° C).

Salinity, pH and dissolved oxygen were continuously monitored during each cruise. Surface water samples for heavy metal analyses were taken at twelve fixed salinities, covering the entire estuary. Special care was taken to prevent oxidation of the anoxic water from the upper estuary during sampling and filtration. Filtration of the water samples was carried out immediately in an all-teflon apparatus, which was placed in a laminar-flow box. The water samples were filtered under nitrogen pressure through 0.45 µm cellulose nitrate filters. The filtrate was collected in a polypropylene flask, acidified with HNO_3 and stored in a closed polyethylene bag until analysis. All materials were acid-cleaned prior to use.

In the laboratory, Cd, Cu and Zn were determined following standard procedures (Dornemann and Kleist, 1979). Briefly, the method involves pre-concentration in HMDC-DIPK and subsequent analysis by flame AAS (Zn) or graphite furnace AAS with Zeeman correction (Cd and Cu). According to a recent ICES inter-comparison exercise, this method yields reliable data on dissolved Cd, Cu and Zn concentrations in coastal sea waters (ICES-MCWG, 1988).

4. RESULTS AND DISCUSSION

The data on dissolved oxygen, suspended matter, chlorophyll and pH as a function of salinity are shown in Figure 2. This figure shows that the estuarine chemistry of the Scheldt is highly dependent on the season.

The distribution of dissolved oxygen in the Scheldt estuary depends both on the water temperature and the river discharge (Somville and De Pauw, 1982). Low levels of dissolved oxygen in the fluvial and upper estuary are common during winter when the river discharge is high and the water temperature is low. During spring the situation is reversed, leading to the development of an anoxic zone in the fluvial and upper estuary. On both occasions, dissolved oxygen levels rapidly increase during estuarine mixing due to re-aeration and dilution with sea water. Supersaturation of dissolved oxygen in the lower estuary occurs during phytoplankton blooms, which can be seen from the spring cruise results.

The suspended matter profiles are somewhat difficult to interpret since sampling was performed irrespective of the state of the tide. Duinker et al. (1982) have shown that suspended matter concentrations in the upper estuary are extremely variable during a tidal cycle as a result of alternating sedimentation and resuspension processes. Nevertheless, it is clear that suspended matter concentrations in the upper estuary are elevated. It can also be seen that the suspended matter concentration in the lower estuary decreases from winter to spring. The resulting increase in light penetration enables a phytoplankton bloom to develop, which dramatically increases the pH in the lower estuary.

The spring pH profile agrees qualitatively with previously published results (Mook and Koene, 1975; Salomons et al., 1981; Duinker et al., 1982). Mook and Koene (1975) attribute the minimum in the pH vs. salinity plot to the effect of increasing ionic strength on the apparent dissociation constant of carbonic acid. However, during four winter cruises we did not find a pH minimum in the Scheldt estuary, and neither did Kramer and Duinker (1984). Therefore, the influence of ionic strength on the estuarine pH distribution seems to be less important than generally assumed.

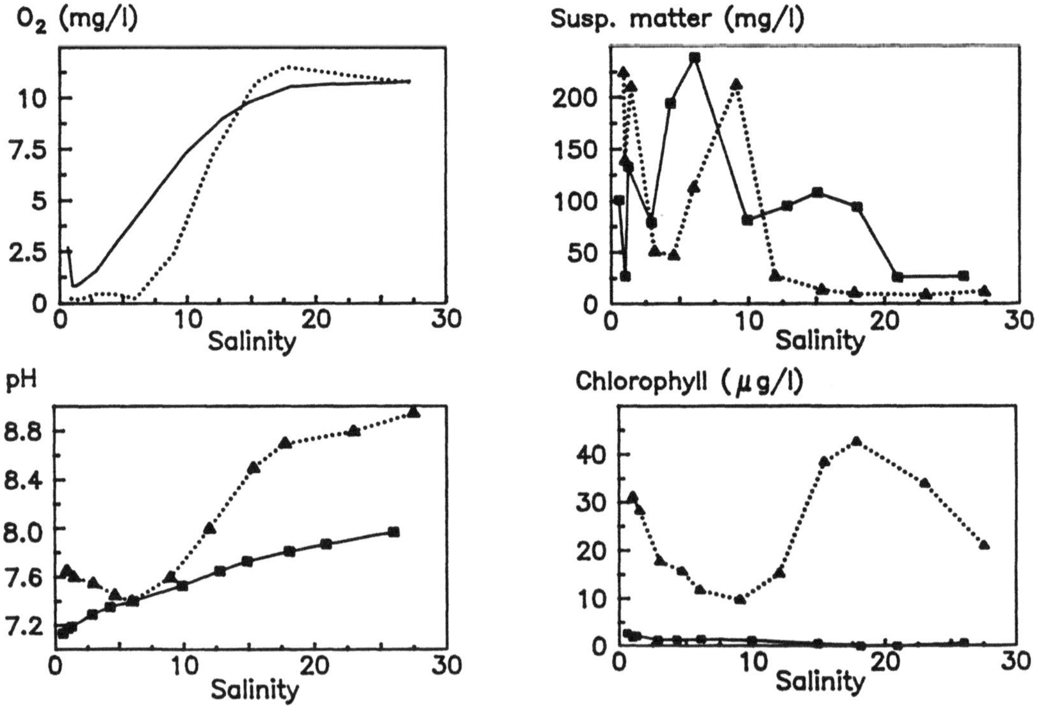

Figure 2. Distribution of dissolved oxygen, suspended matter, chlorophyll-a and pH in the Scheldt estuary during February (—) and May (...) 1987.

The distribution of dissolved Cd, Cu and Zn in the Scheldt estuary is shown in Figure 3. Again, a clear seasonal influence is observed. The dissolved Cu and Zn concentrations in the fresh water entering the estuary are elevated during winter, but very low during spring. However, a rapid decrease of dissolved Cu and Zn concentrations in the very low salinity region is observed during winter. Both the spring and the winter profiles show mid-estuarine maxima, occurring in the salinity range 7-10 (Zn), 12-18 (Cu) and 18-21 (Cd). The dissolved Cd and Zn concentrations in the lower estuary are much lower during spring than during winter.

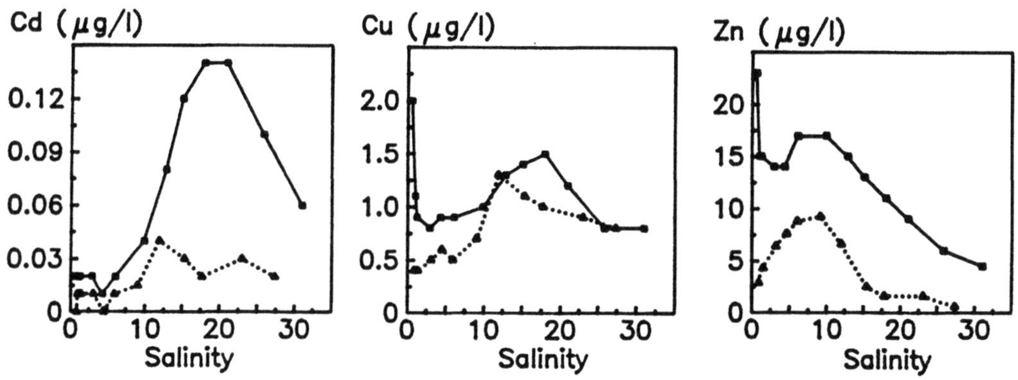

Figure 3. Distribution of dissolved Cd, Cu and Zn in the Scheldt estuary during February (—) and May (...) 1987).

The spring cruise results agree with previous studies in the Scheldt estuary as far as the remarkably low concentrations of dissolved Cd, Cu and Zn in the fresh water region are concerned (Wollast, 1976; Salomons et al., 1981; Duinker et al., 1982). All these studies have in common the fact that the river water was anoxic, suggesting the presence of dissolved sulphide in the water column. This hypothesis has been confirmed by the spring cruise, during which we observed a dissolved sulphide concentration of 30 $\mu g l^{-1}$ in the river water entering the estuary. Thus, the low concentrations of dissolved Cd, Cu and Zn which enter the estuary during spring are caused by precipitation of insoluble metal sulphides in the anoxic river water. Direct evidence for sulphide precipitation is given by the presence of CuS and ZnS in the suspended matter of the upper estuary, which has been demonstrated qualitatively by electron microprobe analysis (Bernard, unpubl.). The presence of metal sulphides in suspended matter is characteristic of heavily polluted estuaries like the Scheldt estuary (Luther et al., 1986).

With increasing oxygen concentrations in the upper estuary, the particulate metal sulphides will lose their stability and dissolve (Khalid et al., 1978; Delaune and Smith, 1985). Thus, the apparent "desorption" of Cd, Cu and Zn in the

Scheldt estuary during spring is due to oxidative dissolution of metal sulphides. However, in spite of this "desorption", low concentrations of dissolved Cd and Zn in the lower estuary are observed during spring (i.e. with respect to the winter situation). Moreover, the rapid decrease of dissolved Zn concentrations in the salinity range 10-15, is far too large to be explained by dilution with sea water. Therefore we conclude that, during spring, "desorption" of Cd and Zn is counteracted by their subsequent removal in the lower estuary, which is most likely caused by the phytoplankton bloom occurring in this region (Fig. 2). Direct uptake of Cd by phytoplankton may be involved, similar to recent findings in the Delaware estuary (Sharp et al., 1984). The phytoplankton bloom may also affect the behaviour of Cd and Zn by causing a sudden pH increase in the lower estuary, which strongly favours adsorption of these metals by suspended matter (Salomons, 1980; Bourg, 1983). Furthermore, the pH increase will lead to surface precipitation of Cd and Zn onto freshly formed calcite (Comans and Middelburg, 1987; Papadopoulos and Rowell, 1988; Zachara et al., 1988).

Contrary to Cd and Zn, dissolved Cu levels in the lower estuary are hardly affected during phytoplankton blooms, which may be because of differences in speciation. Theoretical calculations predict significant complexation of dissolved Cu by humic acid in estuaries. Organic complexation of dissolved Cd and Zn is less significant because the stability constants of their respective humic acid-metal complexes are some four orders of magnitude lower (Turner et al., 1981). The presence of organically complexed Cu in the Scheldt estuary has also been demonstrated experimentally (Kramer and Duinker, 1984; Van den Berg et al., 1987). The strong association between Cu and humic acid not only prevents uptake by phytoplankton, but also suppresses adsorption onto suspended matter at high pH values (Bourg, 1983).

During winter, low levels of dissolved oxygen are present in the water column of the fluvial estuary (Fig. 2). In this oxidizing environment formation of metal sulphides is impossible. On the contrary, any metal sulphides present in the suspended matter (e.g. by resuspension of reduced bottom sediments) will dissolve in the fluvial estuary. Therefore, high concentrations of dissolved trace metals enter the estuary during winter (Fig. 3). The rapid decrease of dissolved Cu and Zn concentrations in the very low salinity region may be attributed to their adsorption onto suspended matter which comprises the estuarine turbidity maximum. A similar removal mechanism has been found in the Tamar estuary (Morris, 1986; Ackroyd et al., 1986) and the Rhine estuary (Duinker, 1985).

After removal in the very low salinity region, desorption of Cd, Cu and Zn is apparent from the winter profiles. Several mechanisms may be involved in the desorption process. Firstly, oxidation of particulate metal sulphides which have escaped oxidation in the fresh water may be possible, and this is similar to the spring situation. Desorption from tidally resuspended sediments may also play a significant role (Morris et al., 1986; Ackroyd et al., 1986). Formation of dissolved metal complexes as a result of estuarine mixing will also be important, at least for Cd. Recent investigations show that the estuarine chemistry of Cd in oxygenated waters is dominated by chloride complexation (Elbaz-Poulichet et al., 1987; Comans and Van Dijk, 1988), leading to significant desorption.

ACKNOWLEDGEMENTS

This study was carried out under contract DGW-989 as part of the SAWES project (System Analysis Western Scheldt) of the Tidal Waters Division. The support given by A. Holland, C. Buurman and T. Siderius throughout this study is gratefully acknowledged. Professor C. H. Van Der Weijden (Institute of Earth Sciences) and R.W.P.M. Laane (Tidal Waters Division) critically read the manuscript. We express our gratitude to Professor R. Wollast (Université Libre de Bruxelles) for stimulating discussions and P. Bernard (University of Antwerp) for carrying out the electron microprobe analyses.

REFERENCES

Ackroyd DR, Bale AJ, Howland RJM, Knox S, Millward GE, Morris AW (1986) Distributions and behaviour of dissolved Cu, Zn and Mn in the Tamar estuary. Estuar Coast Shelf Sci 23:621-640

Billen G, Somville M, de Becker E, Servais P (1985) A nitrogen budget of the Scheldt hydrographical basin. Neth J Sea Res 19:223-230

Bourg, ACM (1983) Role of fresh water/sea water mixing on trace metal adsorption. In: Wong CS, Burton JD, Boyle E, Bruland K, Goldberg ED (eds) Trace Metals in Sea Water. Plenum Press, New York, 195-208

Comans RNJ, Middelburg JJ (1987) Sorption of trace metals on calcite: applicability of the surface precipitation model. Geochim Cosmochim Acta 51:2587-2591

Comans RNJ, Van Dijk CPJ (1988) Role of complexation processes in cadmium mobilization during estuarine mixing. Nature 336:151-154

Delaune RD, Smith CJ (1985) Release of nutrients and metals following oxidation of freshwater and saline sediment. J Environ Qual 14:164-168

Dornemann A, Kleist H (1979) Extraction of nanogram amounts of cadmium and other metals from aqeous solution using hexamethyleneammonium and hexamethylenedithiocarbamate as the chelating agent. Analyst 104:1030-1036

Duinker JC, Nolting RF, Michel D (1982) Effects of salinity, pH and redox conditions on the behaviour of Cd, Zn, Ni and Mn in the Scheldt estuary. Thalassia Jugosl 18:191-202

Duinker JC (1985) Estuarine processes and riverborne pollutants. In: Nürnberg HW (ed) Pollutants and their Ecotoxicological Significance. Wiley, New York, 227-238

Elbaz-Poulichet F, Martin JM, Huang WW, Zhu JX (1987) Dissolved Cd behaviour in some selected French and Chinese estuaries. Consequences on Cd supply to the ocean. Mar Chem 22:125-136

ICES-MCWG (1988) A preliminary assessment of 1985 baseline data on trace metals in coastal and shelf sea waters. Report MCWG 1988/7.1.2/1. ICES, Copenhagen

Khalid RA, Patrick Jr. WH, Gambrell RP (1978) Effect of dissolved oxygen on chemical transformations of heavy metals, phosphorus, and nitrogen in an estuarine sediment. Estuar Coast Mar Sci 6:21-35

Kramer CJM, Duinker JC (1984) Complexation capacity and conditional stability constants for copper of sea and estuarine waters, sediment extracts and colloids. In: Kramer CJM, Duinker JC (eds) Complexation of Trace Metals in Natural Waters. Nijhoff/Junk, The Hague, 217-228

Luther III GW, Wilk Z, Ryans RA, Meyerson AL (1986) On the speciation of metals in the water column of a polluted estuary. Mar Pollut Bull 17:535-542

Mook WG, Koene BKS (1975) Chemistry of dissolved inorganic carbon in estuarine and coastal brackish waters. Estuar Coast Mar Sci 3:325-336

Morris AW (1986) Removal of trace metals in the very low salinity region of the Tamar estuary, England. Sci Total Environ 49:297-304

Morris AW, Bale AJ, Howland RJM, Millward GE, Ackroyd DR, Loring DH, Rantala RTT (1986) Sediment mobility and its contribution to trace metal cycling and retention in a macrotidal estuary. Wat Sci Tech 18:111-119

Papadopoulos P, Rowell DL (1988) The reactions of cadmium with calcium carbonate surfaces. J Soil Sci 39:23-36

Salomons W (1980) Adsorption processes and hydrodynamic conditions in estuaries. Environ Technol Lett 1:356-365

Salomons W, Eysink WD, Kerdijk HN (1981) Inventarisation and geochemical behavior of heavy metals in the Scheldt and Western Scheldt (in Dutch). Report M1640/M1736 Delft Hydraulics Laboratory, Delft, The Netherlands

Salomons W, Förstner F (1984) Trace metals in estuaries: field investigations. In: Metals in the Hydrocycle. Springer Verlag, Berlin, 223-233

Sharp JH, Pennock JR, Church TM, Tramontano JM, Cifuentes LA (1984) The estuarine interaction of nutrients, organics and metals: a case study in the Delaware estuary. In: Kennedy VS (ed) The Estuary as a Filter. Academic Press, New York, 241-258

Somville M, de Pauw N (1982) Influence of temperature and river discharge on water quality of the Western Scheldt estuary. Water Res 16:1349-1356

Turner DR, Whitfield M, Dickson AG (1981) The equilibrium speciation of dissolved components in freshwater and seawater at 25° C and 1 atm pressure. Geochim Cosmochim Acta 45:855-881

Van den Berg CMG, Merks AGA, Duursma EK (1987) Organic complexation and its control of the dissolved concentrations of copper and zinc in the Scheldt estuary. Estuar Coast Shelf Sci 24:785-797

Wollast R (1976) Transport et accumulation de polluants dans l'estuaire de l'Escaut. In: Nihoul JCJ, Wollast R (eds) L'Estuaire de l'Escaut (Projet Mer, rapport final, vol 10). Services du Premier Ministre, Programmation de la Politique Scientifique, Bruxelles, 191-218

Wollast R, Peters JJ (1978) Biogeochemical properties of an estuarine system: the river Scheldt. In: Goldberg ED (ed) Biogeochemistry of Estuarine Sediments. UNESCO, Paris, 279-293

Wollast R, Duinker JC (1982) General methodology and sampling strategy for studies on the behaviour of chemicals in estuaries. Thalassia Jugosl 18:471-491

Zachara JM, Kittrick JA, Harsh JB (1988) The mechanism of Zn^{2+} adsorption on calcite. Geochim Cosmochim Acta 52:2281-2291

USE OF NONVOLATILE CHLORINATED HYDROCARBONS AS ANTHROPOGENIC TRACERS FOR ESTIMATING THE CONTRIBUTION OF THE RIVERS WESER AND ELBE TO THE POLLUTION OF THE GERMAN BIGHT

H.-D.Knauth, R.Sturm
GKSS Research Centre, Institute of Chemistry
Max-Planck Str.1, D-2054 Geesthacht, FRG

Introduction

The determination of different organochlorines in estuaries, coastal waters and in the North Sea exhibited concentration gradients, pointing to the fact that rivers are a major source for these compounds [1-4]. Model calculations for the spread of dissolved substances with conservative behaviour, introduced into the mouths of different rivers, show a distribution pattern with comparatively high densities close to the shore [5,6]. Improved models must take into account the transport of dissolved and particle bound substances. A better understanding of the transport and mixing behaviour of pollutants in the Elbe estuary was obtained from recent studies on water, suspended particulate matter (SPM) and sediments. It could be shown that water discharge and, to a lesser extent, windstress situations typically affect the transfer rates of fine-grained polluted solids into the outer estuary [7-10]. A quantitative approach must allow for the determination of net transport rates of contaminants through selected river sections [11]. The objective of this investigation was to assess the degree of influence of Weser and Elbe on water bodies in the German Bight, using nonvolatile chlorinated hydrocarbons as anthropogenic tracers.

Sampling Scheme and Methods

In May 1987 and in May 1988 water samples were taken at 17 stations, S1-S17, along an Elbe-Weser profile from Hamburg to Bremen and at four additional stations S18-S21 west of lightship Weser during '88 (see figure 1). After sampling at flood tide, SPM was separated from the water phase by continuous flow centrifugation ("Labofuge 15000", Heraeus Christ), at 11000 r.p.m. on board the research vessel "Ludwig Prandtl". The degree of separation was from approximately 95% at river stations down to about 50% at SPM-concentrations as small as 1 mg/l. The extraction of organic pollutants from SPM was carried out using a steam distillation-extraction method. Organochlorines from the water phase were preconcentrated by adsorption onto reversed-phase material. Concentrations of selected chlorinated compounds were determined by means of capillary gaschromatography using SE-54-phases and an electron capture detector. A description of analytical procedures for organochlorines is given in [7], for other compounds including heavy metals in [12].

Results and Discussion

The hydrographic conditions of the study area were influenced in both sampling periods by preceding high river discharges with maximum discharge rates at the end of March : Weser 1700; 1800 m^3/s, Elbe 2600; 3500 m^3/s in ‘87 and in ‘88, respectively. However, while the Weser discharge changed to values of between 400 and 350 m^3/s in both years, the Elbe discharge decreased to different values: approximately 1200 m^3/s in May ‘87 and from 1000m^3/s to 625m^3/s in May ‘88 (longtime annual mean discharges: Weser 320 m^3/s, Elbe ~720 m^3/s). Wind conditions were also different. Before and during the first phase of the ‘87 sampling period, the wind came from a West to Northwest direction with velocities as low as 2-3, increasing later to 4-5 on the Beaufort scale (Bft). In May ‘88 wind came preferentially from East with four up to seven Bft.

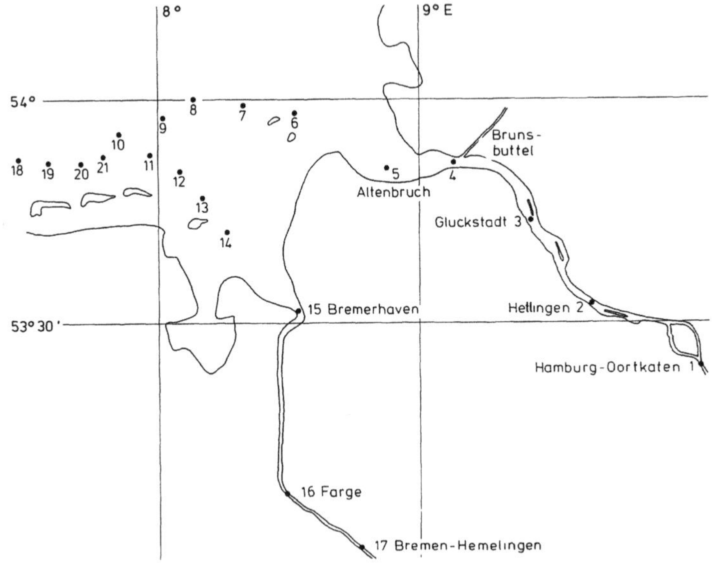

Figure 1
Sampling stations

1-17: profile Elbe-Weser from Hamburg to Bremen; (8 and 10: lightships Elbe1 and Weser)

18-21: additional stations west of lightship Weser

Freshwater content and SPM-concentrations

In accordance with the high freshwater discharges, the mixing zones started in the Elbe estuary at positions close to S5 (May ‘87) and S4 (May ‘88), and in the Weser estuary near S15 (see figure 2). The Elbe-Weser profile data show that the freshwater content decreased in ‘87 more or less continuously from the beginning of the mixing zones to a minimum at S11 with steeper gradients between S6, S7 and S14, S13. In comparision, the freshwater contents in ‘88 were found to be significantly lower at the same positions, showing comparatively high fluctuations at the coastal positions S8 to S12 and S18 to S21. The SPM-concentrations presented in figure 3 show maxima at those positions, where the freshwater content started to decrease (increase of salinity); this is in agreement with the results of previous investigations [2,13]. Special attention should be paid to the SPM material from the coastal stations. In ‘87 the concentration of SPM was found to be 9 to 13 mg/l for the stations 7 to 12 with particulate organic carbon (POC) contents of 10% to 44% (see figure 12) indicating the presence of large proportions of planktonic material. In ‘88 the SPM concentrations ranged from 1 to 13 mg/l for stations 7 to 12 and 18 to 21. The related POC values showed -when measured- large variations for different positions.

Hexachlorocyclohexanes (HCH)

In accordance with the low distribution coefficient (K_D) for the partition of γ-HCH (lindane) and α-HCH between SPM and the solvent phase (log [K_D/(L/kg)]~3-4, depending on SPM and water quality), about 90% and more of these isomers were in solution [2,3,7]. The measured lindane concentrations (see figure 4) show a pattern comparable to the freshwater proportion present in the Elbe-Weser transect. In the freshwater sections of the estuaries the γ-HCH concentrations at different stations were high although variable, caused by input variations of HCH into the river systems.

The γ-HCH concentrations decreased continuously in '87 from the mixing zones of the Elbe and Weser to station 10, whereas in '88 the situation was obviously characterised by steeper gradients between stations 5 and 6 for the Elbe, and between 15 and 14 for the Weser estuary. At these positions a sharp decrease in the freshwater contents was also observed. In '88, the γ-HCH concentrations of the coastal stations 8 to 12 were a factor of two to four smaller than in '87, and undiscernible from those of stations 18 to 21. Samples from the latter stations show γ-HCH levels characteristic for East-Frisian coastal waters [2,4] which are transported into the German Bight by residual currents [5,6]. Relations between γ-HCH and freshwater contents in the water bodies of the Elbe and Weser estuaries, as depicted in figures 6a and 6b, demonstrate the possibility of characterizing water bodies in coastal areas using lindane as an anthropogenic tracer compound.

When differences in the concentration ratios of HCH-isomers exist, they are an additional means of identifying water bodies from different sources [2]. The high loading of γ-HCH compared to that of the α-HCH isomer in the river Weser in particular allows an estimation of Weser water contributions to the water masses of the southern German Bight (see figure 5). It can clearly be seen from this ratio alone, that the Weser water contribution is as far as stations 13 ('87) and 14 ('88). Taking into account the absolute HCH-levels, this contribution may be found as far as stations 11 ('87) and 13 ('88). While values for the ratio determined in samples from the Elbe estuary were close to two (mean values for stations 1 to 6: 1.7 ('87), 2.1 ('88)), thus being in agreement with former data [2,3,4], the values for stations 18 to 21 obtained in May '88 were in the range 1 to 4.8. Of these stations three showed higher values than those cited in the literature, and these were similar to those of the Elbe estuary. A contribution of Weser water due to high water discharge four weeks before and strong easterly winds at the time of investigation could be one possible reason for the enlarged values.

Hexachlorobenzene and Polychlorinated biphenyl PCB-138

Those organochlorines HCB and PCB-138 which are much less soluble in water show a high affinity for adsorbtion onto SPM. The K_D-values discussed in the literature are in the range of $2 \cdot 10^5$ (Elbe/Weser) up to $5 \cdot 10^6$ (North Sea) [2,3,14]. This means that in the Elbe and Weser estuaries, with SPM concentrations of ≥20mg/l, at least 80% and in the coastal region with SPM concentrations of ≥1mg/l (K_D assumed to be $1 \cdot 10^6$) about 50% or more of

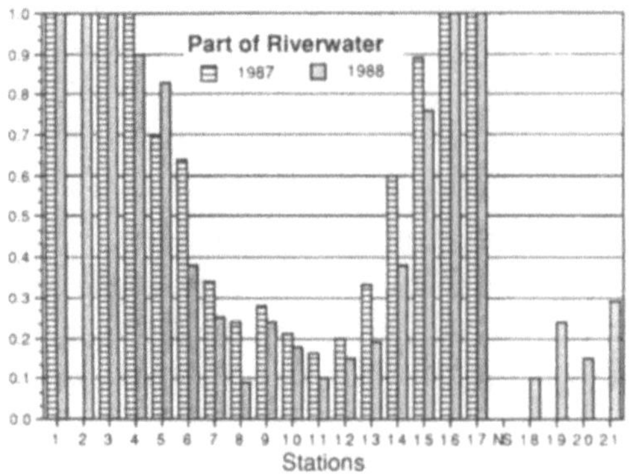

Fig. 2:Freshwater part derived from chlorinity

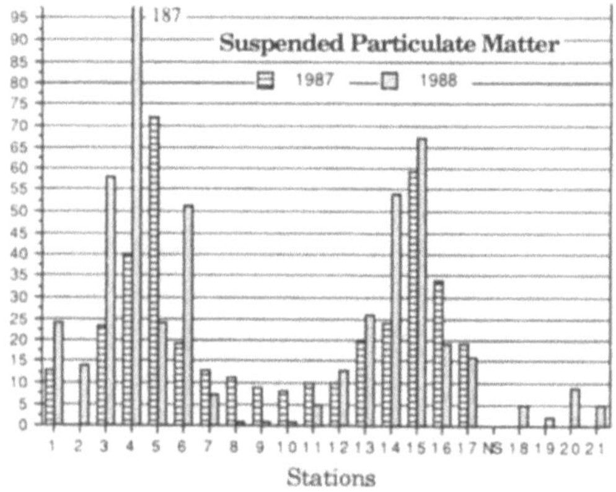

Fig.3: SPM- concentrations in mg/L

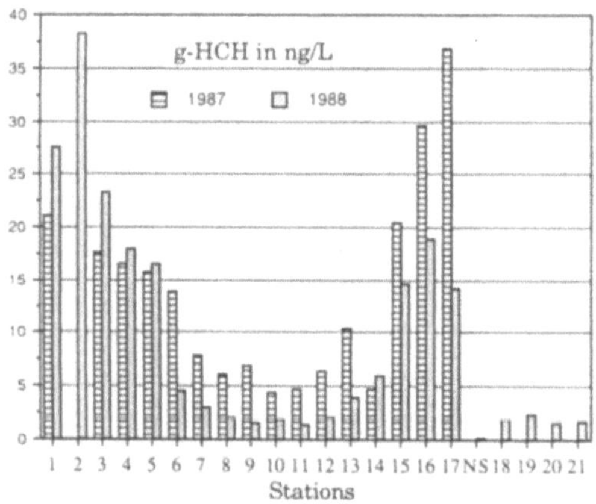

Fig.4: g-HCH concentration in "solution"

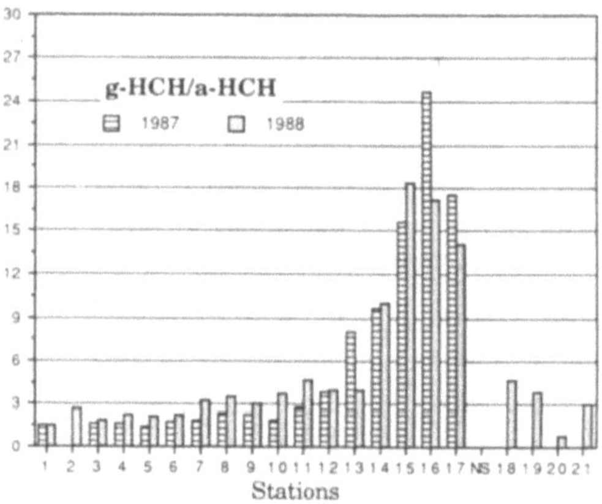

Fig.5: Ratio g-HCH/a-HCH in "solution"

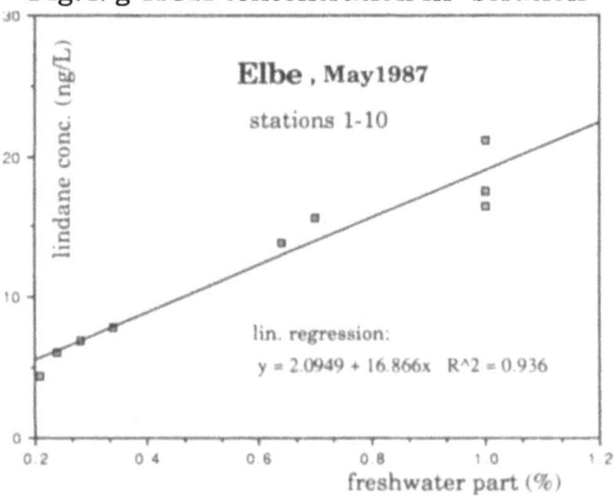

Fig.6a: Relation g-HCH ,Freshwater(Elbe)

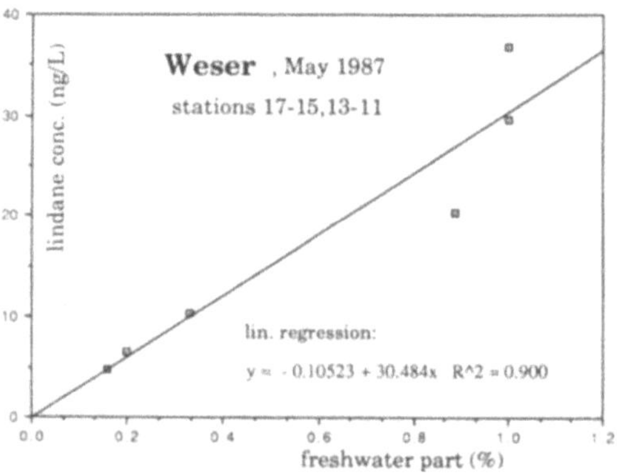

Fig.6b:Relation g-HCH,Freshwater(Weser)

Figures 2-13: Concentrations and contents of different substances in samples of surface water - separated in SPM and "solution" by centrifugation or filtration - taken in May ‘87 and in May ‘88 from stations given in fig.1

each compound is particle bound.

Measured data of seston bound HCB and PCB-138 concentrations are presented in figures 7 and 8. In accordance with comparatively high differences found for the SPM concentration and in part for the SPM composition (POC), the estuarine stations 1 to 6 and 17 to 13 can be distinguished from the coastal stations 7 to 12 and 18 to 21. The features of HCB profile data in the Elbe estuary (S1-S6), similar with those of PCB-138, p,p'DDD and mercury (figure 13), exhibit high concentrations in the upper part of the estuary and a sharp or less sharp decrease in seaward direction, depending on e.g. differences in discharge rates. At high freshwater discharges (1200 m^3/s) in '87, the decline in concentration was not so evident when compared with the '88 situation, where a discharge rate of 625 m^3/s was prevalent at the time of sampling. This behaviour - predominantly caused by differences in the transport rates and mixing efficiencies of fluvial and marine sediments- was observed and discussed in previous investigations [7-10].

It should be noted that, mainly as a consequence of the different discharge rates, the seston bound pollutant concentrations from the outer estuary station, S6, were significantly higher in May '87 than in May '88, i.e. for HCB, PCB-138, p,p'DDD and Hg by factors of 10; 5; 4 and 3.4, respectively.

While HCB concentrations were comparatively low in the Weser estuary and in samples of the coastal stations 18-21, PCB-138 concentrations in SPM exhibited high values in samples of the stations 17 and 16 (Weser) decreasing in seaward direction. But because of the high background level of PCB-138 in samples of S18 to S21 (a result consistent with literature data for an area which is normally influenced by water masses from the Rhine and the Channel [5,6]) SPM from stations downstream S16 cannot be identified unequivocally by PCB-138 load determinations only.

Another possible means of identifying SPM from different sources, may be by comparing the concentration ratios of simultaneously measured, strongly adsorbed chlorinated hydrocarbons. Profile data for the ratio PCB-138/HCB, as given in figure 9, show low values for stations 1 to 6 followed by an increase as far as S15. From these data Elbe and Weser originating SPM may be derived for samples of stations 1-6 and 17-15, respectively.

HCB/PCB ratios, given in figure 10, by their net decreases from S1 to S6 demonstrate, that the PCB decrease in the Elbe estuary is less rapid than that of HCB. This could in part be due to comparatively high background PCB concentrations in SPM and sediments from the outer estuary and coastal areas.

A simultaneous comparision of both ratios HCB/PCB and γ-HCH/α-HCH (figures 11a and 11b) which characterize the SPM material and the water bodies, respectively, clearly indicate the presence of zones which can unequivocally be related to different water bodies. In '87 and '88 samples from stations 1 to 6 show Elbe typical high HCB/PCB and low γ-HCH/α-HCH ratios while Weser typical low HCB/PCB ratios combined with high γ-HCH/α-HCH ratios are to be seen for stations 17 to 13 ('87) and 17 to 14 ('88).

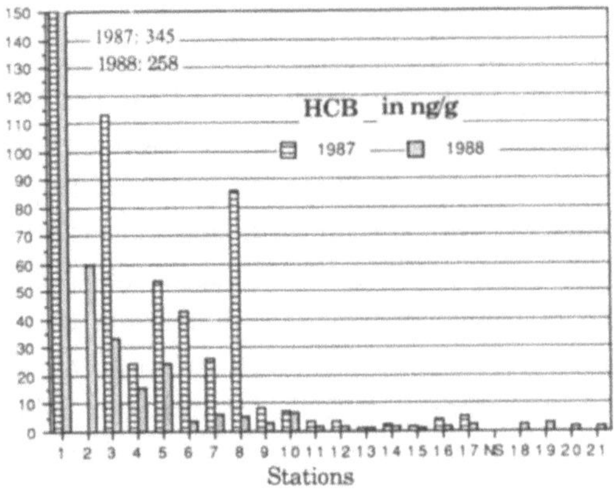

Fig.7: Sestonbound HCB in ng/g

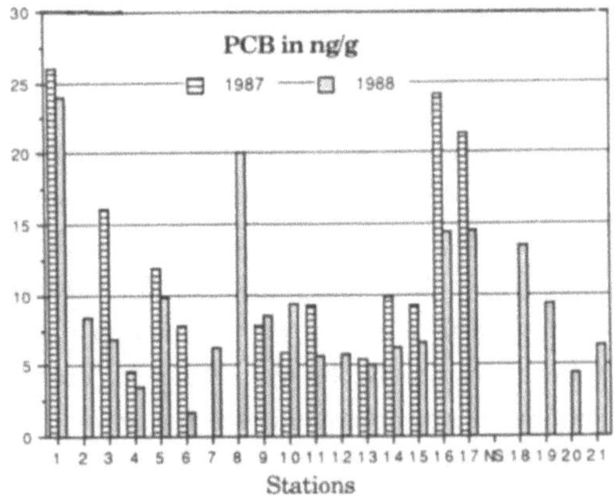

Fig.8: Sestonbound PCB in ng/g

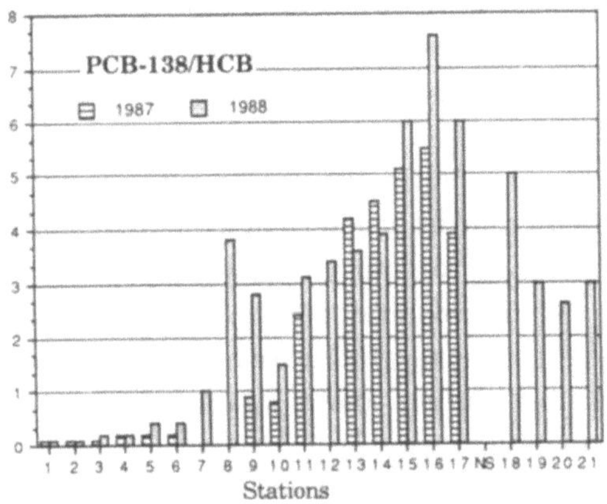

Fig.9: Ratio PCB-138/HCB in SPM

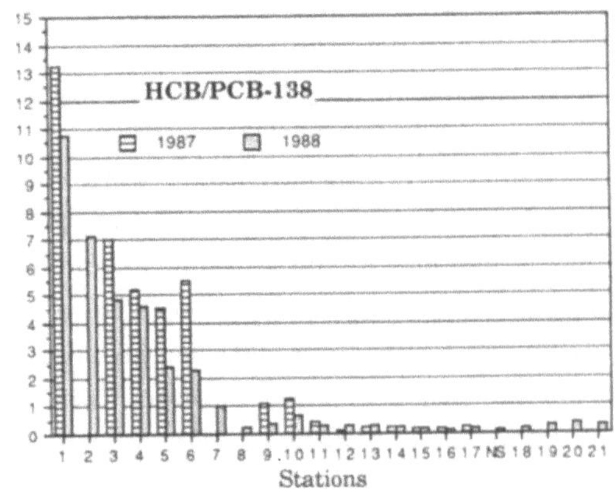

Fig.10: Ratio HCB/PCB-138 in SPM

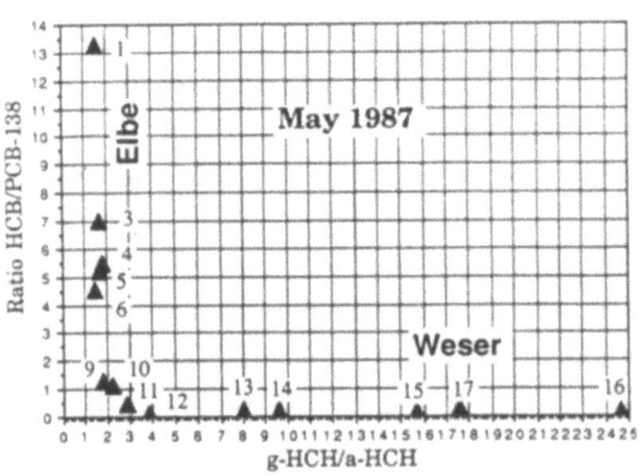

Fig.11a:Relation HCB/PCB to g-HCH/a-HCH

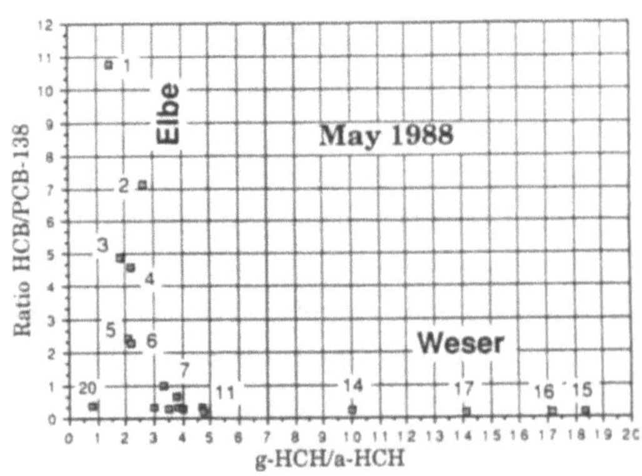

Fig.11b:Relation HCB/PCB to g-HCH/a-HCH

Particulate organic carbon and seston bound pollutants

As mentioned above, the quantity and, to a varying extent, the quality of SPM are different in the estuaries and neighbouring coastal regions. Whereas in the river reaches the SPM concentration range is relatively high, SPM concentrations in coastal environments are usually not higher than approximately 10 mg/l. A rough indicator of the SPM quality is its POC content, as given in fig.12, for the measured samples. The features of the POC profiles show high concentrations in the upper parts of the estuaries (~27% at S1 , 10 to 18% at S17 and 16) declining in seaward direction down to values between 5 and 2% at the river mouths (S6 (87,88); S14 (87), S12 (88)). This decrease is due to two effects. The first is a mixing of SPM with bottom sediments resulting in a seaward directed dilution, similar to seston bound pollutants [7,8]. The second effect is caused by diagenetic degradation,occuring to a large degree in surface sediments.

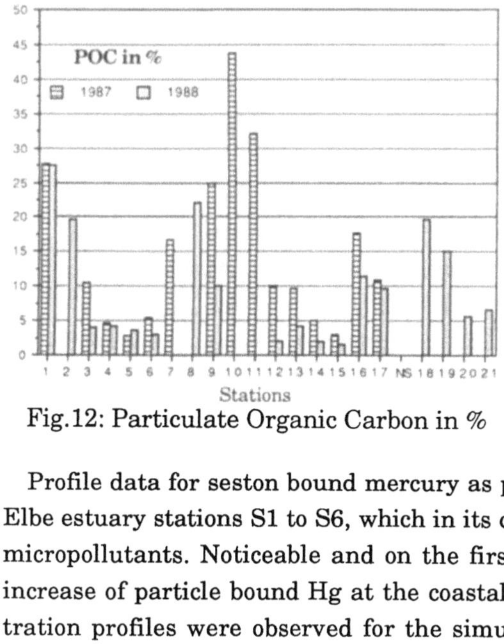

Fig.12: Particulate Organic Carbon in %

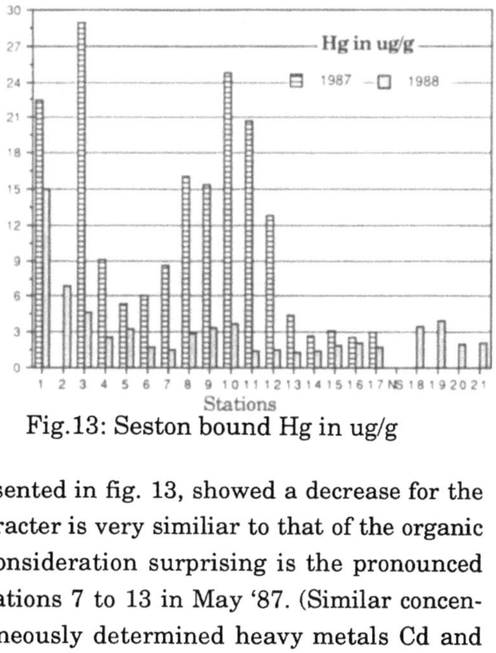

Fig.13: Seston bound Hg in ug/g

Profile data for seston bound mercury as presented in fig. 13, showed a decrease for the Elbe estuary stations S1 to S6, which in its character is very similiar to that of the organic micropollutants. Noticeable and on the first consideration surprising is the pronounced increase of particle bound Hg at the coastal stations 7 to 13 in May '87. (Similar concentration profiles were observed for the simultaneously determined heavy metals Cd and Cu .) A comparision of the Hg-concentrations with the relevant POC contents shows that for the coastal stations both parameters are highly correlated ($\sqrt{r^2}=0.95$,n=7 (S6-S13)).

This assists a possible interpretation of the unusual phenomenon as follows:The recently formed planktonic material, present extensively in the investigated area, facilitated a new and more effective adsorbance of quasi-dissolved pollutants which were transported into the coastal region by the river water.The Hg amounts,"filtered out" of the water phase by planktonic material, must be in the range 30 to 60 ng Hg/l which seems to agree in the light of Hg-concentrations, cited for waterbodies of this area [15].

On second observation the organic micropollutant concentrations reveal that in this coastal area enlarged values are connected with high POC contents in the SPM, demonstrating an accumulation of contaminants in the planktonic biomass.

Acknowledgement

The investigations, part of a study aiming to assess the water quality of the Weser estuary, were financially supported by the Water Authority of Bremen. Our special thank is directed to the participants for their effective and engaged help in planning and performance of the work: Gandraß, Kock, Mündner-Kuhr, Schnell (org. micropollutants, POC); Felsen, Weiler (heavy metals); Blöcker, Milde (planning, sampling); Alisch, Matheisel (navigation).

References

[1] Duinker JC, Hillebrand MTJ, Nolting RF, Wellershaus S (1982) The river Elbe: processes affecting the behaviour of metals and organochlorines during estuarine mixing. Neth J Sea Res 15, 141-169

[2] Ernst W, Boon JP, Weber K (1988) Occurence and fate of organic micropollutants in the North Sea. In: Salomons W, Bayne BL, Duursma EK, Förstner U(eds) Pollution of the North Sea. Springer, Berlin Heidelberg New York,pp284-299

[3] Ernst W, Gaul H (1984) Organische Schadstoffe in Wasser, Sediment und Organismen der Nordsee. In: Meereskundliche Beobachtungen und Ergebnisse Nr.55. Dtsch Hydrogr Inst, Hamburg, pp 53-88

[4] Anonymus (1985,1986,1987) Halogenierte Kohlenwasserstoffe. In: Überwachung des Meeres. Jahresberichte, Dtsch Hydrogr Inst, Hamburg

[5] Backhaus JO, Maier-Reimer E (1983) On seasonal circulation patterns in the North Sea. In: Sündermann J, Lenz W (eds) North Sea dynamics. Springer, Berlin Heidelberg New York pp 64-84

[6] Müller-Navarra S, Mittelstaedt E (1985) Schadstoffausbreitung und Schadstoffbelastung in der Nordsee. Dtsch Hydrogr Inst, Hamburg

[7] Sturm R, Gandraß J (1988) Verhalten von schwerflüchtigen Chlorkohlenwasserstoffen an Schwebstoffen des Elbe-Ästuars. In: Vom Wasser. Bd 70, VCH Verlagsgesellschaft Weinheim pp 265-280

[8] Knauth H-D, Schwedhelm E, Sturm R, Weiler K, Salomons W (1989) The importance of physical processes on contaminant behaviour in estuaries. GKSS Report (in preparation)

[9] Salomons W, Schwedhelm E, Schoer J, Knauth H-D (1987) Natural tracers to determine the origin of sediments and suspended matter from the Elbe estuary. In: Awaya J, Kusuda T (eds) Proc Spec Conf on coastal and estuarine pollution. pp119-132

[10] Förstner U, Schoer J, Knauth H-D (1989) Metal pollution in the tidal Elbe river . The Science of the Total Environment (in press)

[11] Fanger H-U, Kuhn H, Michaelis W, Müller A, Riethmüller R (1986) Investigation of material transport and load in tidal rivers. In: Moulder DS, Williamson P (eds) Estuarine and coastal pollution: detection, research and control.Water Sci Tech 18, pp 101-110

[12] Irmer U, Knauth H-D, Weiler K (1985) Einfluß der Schwebstoffbildung auf Bindung und Verteilung ökotoxikologischer Schwermetalle in der Tideelbe. In: Vom Wasser. Bd 65, loc cit [7], pp 37-61

[13] Eisma D, Irion G (1988) Suspended matter and sediment transport. In: Salomons et al (eds), loc cit [2], pp 20-35

[14] Duinker JC, Boon JP (1986) PCB congeners in the marine environment - a review. In: Björseth A, Angeletti D (eds) Organic micropollutants in the aquatic environment. Proc 4th Eur Symp, Vienna, Austria, October 1985. Reidel, Dortrecht , The netherlands, pp187-205

[15] Anonymus, loc cit [4] , Schwermetalle

MULTIELEMENT DETERMINATION OF TRACE ELEMENTS IN ESTUARINE WATERS BY TXRF AND INAA

A. Prange, R. Niedergesäß and C. Schnier
Institut für Physik, GKSS-Forschungszentrum Geesthacht GmbH,
Postfach 1160, D-2054 Geesthacht, Federal Republic of Germany

INTRODUCTION

According to the BILEX concept [1], numerous trace elements have to be determined, down to very low concentrations, and - because of different transport properties - both in the dissolved and the particulate phase. Moreover, inflow-outflow balancing requires good precision and accuracy of the analytical data. These challenges demand the availability of complementary analytical methods which are economical and characterized by a high detection power, a large dynamic range (mg/kg down to ng/kg) and few sources of systematic errors. Furthermore, the capacity for multielement detection is desired since it facilitates the detection of many ecologically relevant trace elements.

In this respect an analytical procedure for the determination of trace elements in estuarine waters has been developed, including all steps from sampling to the evaluation of the data. Two multielement methods are used. These are Total-Reflection X-Ray Fluorescence Spectrometry (TXRF) and Instrumental Neutron Activation Analysis (INAA). TXRF has proved to be an economical method for routine analysis of filtered river water as well as of suspended particulate matter (SPM) [2]. INAA is used as a reference method, for checking the accuracy of the results and also for the determination of further elements [3]. The present paper describes the analytical procedure as applied to estuarine research and gives some examples of application and evaluation of the trace element data for the Elbe and Weser rivers.

EXPERIMENTAL

Sampling and storage

River water samples are collected according to the sampling strategy of the BILEX concept. A 6 liter Niskin sampler, which is mounted horizontally, is used. Immediately after the samples are taken, 1 liter of the sample is filtered through weighed 0.4-µm Nuclepore filters by means of N_2 pressure filtration. The filtrates are acidified by conc. HNO_3 (suprapure®, Merck) to a final pH of around 1.7 and stored in precleaned polyethylene bottles in a refrigerator until trace element analysis in the shore laboratory.

Analytical procedures

TXRF is a multielement technique with substantially improved detection limits compared to conventional energy dispersive XRF [4]. Characteristic is the small angle of incidence of the exciting X-ray radiation, allowing total reflection from a highly polished quartz sample carrier. The sample is brought onto the quartz carrier by evaporating the solvent of solutions or fine grained suspensions resulting in a thin film-like residue. Figure 1 shows a scheme of the TXRF principle. This method has an excellent peak to

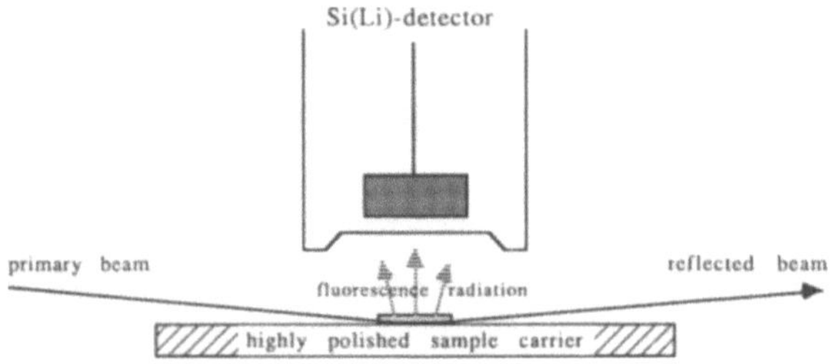

Figure 1: Schematic view of TXRF

background ratio resulting in instrumental detection limits of a few pg. TXRF spectrometers with molybdenum and tungsten excitation have been used. Details of the method and a review of applications are described elsewhere [5].

INAA is well known and is described in detail in several monographs [6]. At GKSS, research reactors of 5 and 15 MW with neutron fluxes of 10^{14} $cm^{-2}s^{-1}$ are available with tube- and core irradiation facilities.

For the element determination in filtrates and SPM, different sample preparation techniques are used. Figure 2 gives a survey of the sample preparation steps applied for TXRF and INAA.

Filtrates and SPM (TXRF): For element determination in the filtrates by TXRF, two techniques have been used, a direct measurement, and an acid digestion with subsequent matrix separation and enrichment of the trace elements.

For direct measurements, aliquots of the filtrates are diluted with ultrapure water to 1:3. After addition of an internal Co-standard (50 ng/g end concentration) 10 µl are then prepared as a thin film-like sample by evaporating the solvent and measured for 3000 sec. Thus the elements S, K, Ca, Cr, Mn, Fe, Ni, Cu, Zn, As, Rb, Sr and Ba are detectable. In order to take full advantage of the high sensitivity of TXRF even in the case of high salt matrix and organic content, it is necessary to separate the elements of interest from the bulk. The procedure used is shown in fig. 2 and described in more detail in [2]. Thus the elements V, Mn, Fe, Ni, Cu, Zn, Pb, Cd, Mo and U are determinable.

A simple and convenient sample preparation is used for SPM, shown in the middle of fig. 2. Thus the elements S, K, Ca, Ti, V, Cr, Mn, Fe, Ni, Cu, Zn, Ga, As, Rb, Sr, Y, Ba, Mo and Pb are determined.

Filtrates and SPM (INAA): For INAA 250 ml of the filtrate samples are freeze-dried in a polyethylene bag. The residue of about 0.2-0.3 g is transferred into an irradiation vial. The SPM laden filters are ready for exposure after freeze-drying.

Different exposure-, cooling- and measuring times are necessary for the determination of short lived and long lived radionuclides. A 5 minutes exposure time is used for isotopes with half-lives of 1 min. to several hours. Three measurements of the γ-ray activity with Ge-detectors follow after decay times of 5 min., 50 min. and 15 h. Thus the elements Na, Mg, Cl, K, Mn and As are determined in the filtrates, the elements Na, Mg, Al, Cl, K, Ti, V, Mn and As in SPM. An exposure time of 3 days is used for elements

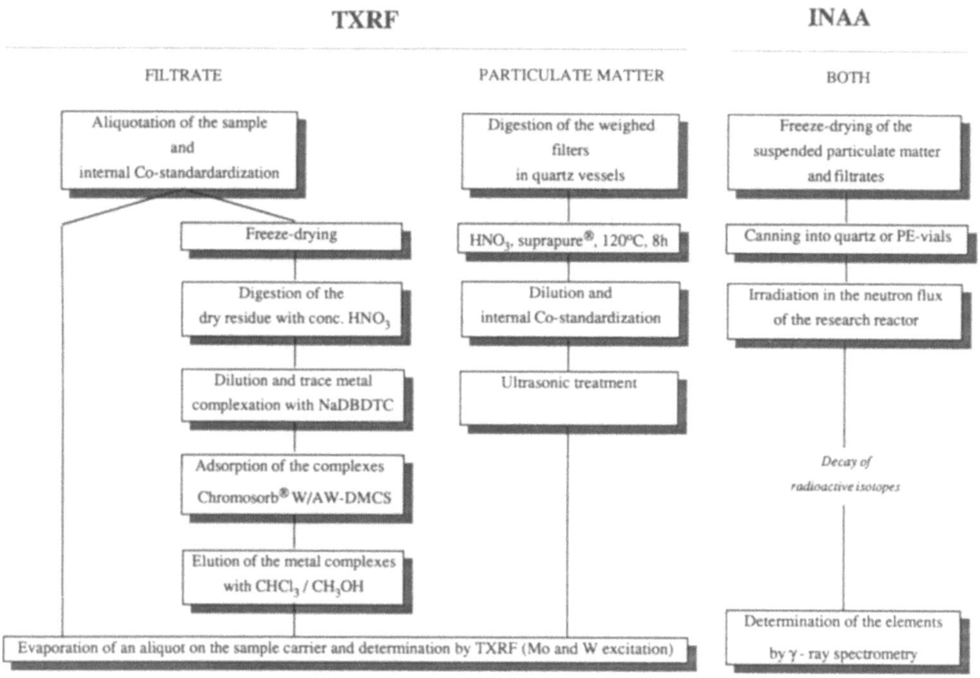

Figure 2: Scheme for the sample preparation steps applied for filtrates and SPM using TXRF and INAA

with longer lived activation products. After decay times of 7d, 15d and 30d the γ-radiation is measured. Thus in the filtrates the elements Na, K, Ca, Sc, Cr, Fe, Co, Ni, Zn, As, Se, Br, Rb, Sr, Mo, Ag, Sb, Cs, Ba, Eu, Hf, Hg, Th and U, in SPM the elements Na, K, Ca, Sc, Cr, Fe, Co, Ni, Zn, As, Se, Br, Rb, Sr, Zr, Ag, Cd, Sb, Cs, Ba, La, Ce, Nd, Sm, Eu, Tb, Yb, Hf, Ta, Au, Hg, Th and U are determined. This procedure is purely instrumental, since no chemical treatment is applied.

Quality assessment of the analytical procedures

To assess the quality of the analytical procedures, blank values, detection limits, precision and accuracy are investigated thoroughly on riverwater samples differentiated for filtrates and SPM.

Blank values. For the filtrates analyzed by TXRF, blanks of Fe, Ni, Cu, Zn and Pb must be taken into account. For SPM the elements Cr, Ni and Zn, originating from the filters, must be considered. For INAA the elements Cr, Br and Sb for SPM and Sc and Au for the filtrates have to be taken into account.

Detection limits. For direct measurement of the filtrates by TXRF, detection limits of 2-5 ng/g are achieved for freshwater samples. The method used for marine samples, results in detection limits of about 0.03-0.1 ng/g. For SPM detection limits of about 3-25 μg/g are achieved. The detection limits for INAA in filtrates and SPM vary over more than five orders of magnitude depending mainly on the nuclear data and the blank values. For the filtrates they range from 0.1-1 pg/g up to 100 ng/g, whereas in SPM 1-10 ng/g up to 100 μg/g are determined.

Precision (Reproducibility). For the filtrates analyzed by TXRF, reproducibilities of about 2-10 % are achieved for most of the elements. 20-25 % reproducibilities are determined for the elements As, Cd and Pb, because their concentrations are near the detection limits. The reproducibilities for SPM are in the range 1-15 %. For INAA reproducibilities of 1.5 to 5 % are determined for most of the elements. 10 % precision is found for Mg and Cl in SPM, 10-30 % for Ag, Eu, Hf and Th in the filtrates.

Accuracy. The accuracy of the results has been checked routinely by comparison of TXRF and INAA data, participation in various intercalibration tests (e.g. organized by the ARGE-Elbe and the International Council for the Exploration of the Sea (ICES)) and by analyzing standard reference material.

In the course of an ICES- intercomparison test for trace metals in estuarine water (filtrates from the Schelde river) [7] some TXRF-results exemplified by the elements Cd and Pb are shown in figure 3. Of

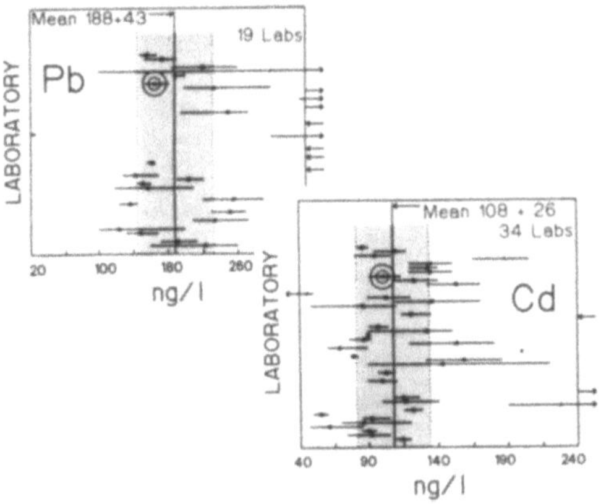

Figure 3: Pb and Cd data from the ICES intercalibration JMG 6/TMSW [7]

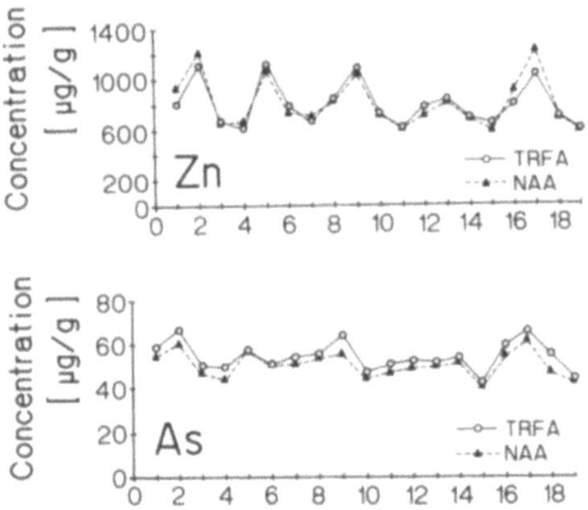

Figure 4: Comparison of Zn and As results in SPM of several Elbe samples determined by TXRF and INAA

45 participants, only 19 labs are used, for instance, for the evaluation of the Pb results, the others are out of the considered range. Our results are marked by a circle. Figure 4 shows a comparison of TXRF and INAA data exemplified by Zn and As in SPM of the Elbe river. There is an excellent agreement. In general the agreement of the results for the heavy metals is in the range 5 to 30 %.

RESULTS AND DISCUSSION

Analytical data for ten selected heavy metals are summarized in tables 1 and 2. Table 1 presents the ranges of the specific element concentrations in SPM for 3 sampling positions along the Elbe and Weser river respectively. The following tendencies can be recognized:

- For the elements Cr, Ni, Cu, Zn and Cd, a decrease of the concentrations is ascertainable in the downstream direction for the Elbe as well as for the Weser. For Mn this is valid for the Weser only.
- The element concentrations of As, Hg and Pb also decrease in the Elbe towards the marine environment, whereas no clear gradient for these elements is noticeable for the Weser.
- Comparing the concentration ranges between the Elbe and the Weser, high concentrations of As and Hg are conspicuous for the Elbe. However one should notice that sampling on the Elbe river was done in September whilst sampling on the Weser was carried out in May.

Table 1: Specific concentration ranges in SPM (µg/g) for a selected number of elements for the rivers Elbe and Weser

	ELBE			WESER		
Element	Oortkaten	Nienstedten	Brunsbüttel	Osterort	Brake	Bremerhaven
SPMµg/l	18700-28700	20300-299500	44700-674200	13100-40300	21500-67900	32700-346300
Cr µg/g	300-380	84-300	75-113	81-800	87-210	72-137
Mn	2200-2600	2300-6700	1600-3400	2900-4000	2100-3500	1300-2500
Fe	36000-44000	21000-48000	29000-43000	41000-58000	34000-51000	30000-46000
Ni	77-110	34-78	40-66	68-470	61-100	39-69
Cu	240-340	62-220	30-76	84-110	50-80	30-68
Zn	1500-2000	460-1300	290-450	860-1000	470-720	260-430
As	68-75	33-67	25-41	17-28	14-25	14-26
Cd	8,0-14	1,4-7,8	0,9-2,0	8,7-14	4,8-8,2	1,6-3,8
Hg	21-27	3,3-18	1,0-3,4	0,2-1,6	0,7-2,5	1,0-2,1
Pb	190-220	81-190	58-87	140-180	89-120	72-130

Table 2: Comparison of concentration ranges (µg/l) in filtrates and SPM from the Weser river

	Osterort		Brake		Bremerhaven	
Element	Filtrate	SPM	Filtrate	SPM	Filtrate	SPM
SPM		13100-40300		21500-67900		32700-346300
Cr	6,4-12	1,4-10	7,7-16	1,9-12	<5	3,5-34
Mn	30-54	49-117	20-79	75-170	7-88	53-546
Fe	10-20	600-1800	11-37	730-2700	2,0-5,1	1300-13900
Ni	3,3-3,8	1,1-6,1	3,6-4,2	1,3-5,8	2,3-5,2	1,9-16
Cu	2,2-3,3	1,4-3,6	2,4-4,5	1,0-4,0	2,6-10	2,0-13
Zn	6,6-11	11-38	6,3-9,2	10-42	4,4-8,9	12-110
As	1,6-2,6	0,23-0,97	1,7-2,5	0,3-1,4	<3	0,6-8,3
Cd	0,03-0,08	0,08-0,55	0,03-0,07	0,10-0,48	0,05-0,18	0,06-0,80
Hg	n.d.	0,003-0,032	n.d.	0,015-0,17	n.d.	0,05-0,47
Pb	0,16-0,36	2,0-5,7	0,10-0,28	2,2-6,8	0,09-0,30	3,6-33

In table 2 element concentrations in the filtrates and SPM for the Weser are given in µg per liter river water to compare the partitioning of the element masses in the dissolved and the SPM phases. The following tendencies can be stated:

- Especially the elements Fe and Pb are bound to particulate matter. To a lesser degree this is also valid for the elements Mn, Zn and Cd.
- The elements Cr and Cu are more or less equally distributed in both phases.
- Only As is found mainly in the dissolved phase especially at Osterort and Brake.

Hg has been determined only in SPM, because of adsorption- and loss problems during the filtration procedure, and contamination problems during storage.

In the following some results are presented for the river Elbe at positions upstream and downstream from the Hamburg harbour (Oortkaten - Nienstedten).

Filtrates. For the filtrates, constant concentrations of the dissolved elements are found at Oortkaten, showing no clear tidal profiles. In general, similar values have been determined at Nienstedten, which are constant over the tidal periods and also along the cross section. However, a small group of elements, Cr, Mn, Fe, Co and Ag, show pronounced variations of their concentrations over the tide at Nienstedten. Figure 5 demonstrates these conditions for the example of Co. According to the sampling strategy, two

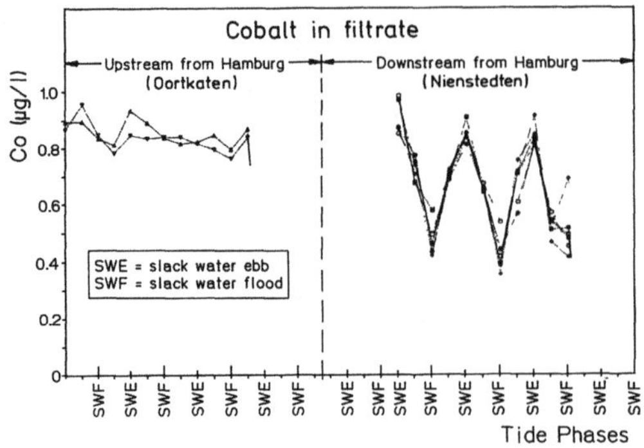

Figure 5: Time series of Co-concentrations in the filtrates upstream and downstream of Hamburg

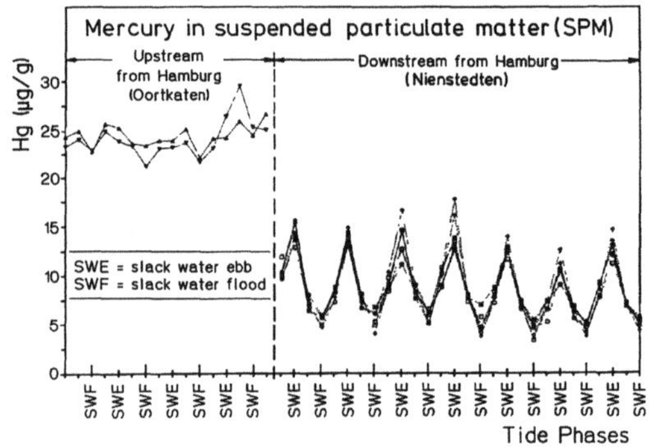

Figure 6: Hg in SPM over several tidal periods upstream and downstream of Hamburg

curves are produced for Oortkaten and six curves for Nienstedten. More or less pronounced maxima, which have a similar concentration level as at Oortkaten, are recognizable at the slack water point ebb (SWE), and clear minima at the slack water point flood (SWF). Thus these elements are less concentrated in the dissolved phase downstream from Nienstedten.

Suspended particulate matter. At Oortkaten, constant specific element concentrations were determined, showing no clear tidal profiles. However, at Nienstedten pronounced tidal variations can be seen. These profiles are reproduced over several periods. Fig. 6 describes the facts, showing Hg as a typical example.

Looking more closely at the variations in the different elements, the elements can be comprised into groups according to their characteristic behavior. This is supported by statistical cluster analysis which is performed for specific element concentrations in SPM for the example of surface samples from Oortkaten and Nienstedten. For the calculations, the correlation coefficient between the different elements is taken as the basis for a similarity measure. Clustering is carried out by the single linkage method ("nearest neighbor"), an agglomerative hierarchical procedure [8]. Figure 7 shows the resulting dendrogram. At a

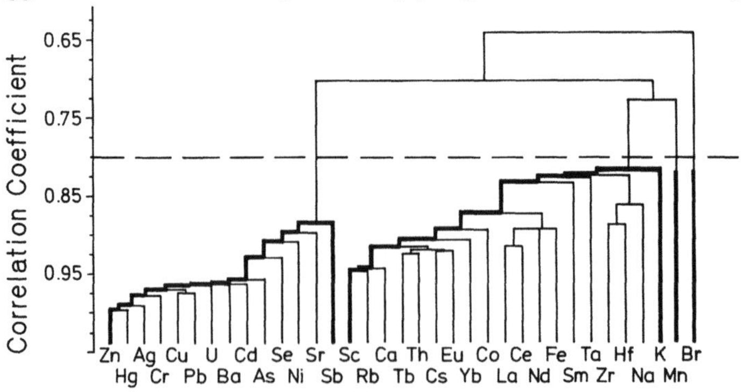

Figure 7: Dendrogram of the elements in SPM of surface samples from Oortkaten and Nienstedten

similarity level of about 0.8 (broken line), one gets four clusters which can be characterized as follows:

Cr, Ni, Cu, Zn, As, Se, Sr, Ag, Cd, Sb, Ba, Hg, Pb and U. These elements have high specific concentrations upstream from Hamburg, and downstream show pronounced smaller average amounts which, however, vary strongly with the tide. Along the cross section no remarkable differences in the concentration patterns are discernible. Maximum heavy metal contents in SPM are found at SWE, minima at SWF. These facts show that there is a considerable longitudinal gradient, which might be due to the admixture of less contaminated SPM at the downstream end of the Hamburg harbour. Most of these elements are enriched compared to average concentration values given for the earth crust [9]. This is especially true for Zn, As, Se, Ag, Cd, Hg and Pb. This finding indicates anthropogenic influences.

Na, K, Ca, Sc, Fe, Co, Rb, Zr, Cs, lanthanoids, Hf, Ta and Th. This group of elements shows upstream from Hamburg similar or slightly decreased average values as downstream, however, with tidal variations at Nienstedten. The concentrations of these elements are similar compared to average values of the earth crust. Thus anthropogenic influences are not detectable. These elements, designated as geogenic, are obviously supplied by natural sources from the Elbe river terretorium. Clear minima are found at SWF, and small minima at SWE especially for the surface samples. This might be due to the different settling velocities for the different components of SPM. During slack water the fast settling component disappears from the surface and the remaining SPM contains more fine grained and low density particles which consist probably of more organic material [3].

Mn and Br. Each of these elements forms a separate cluster. Conspicuous for both elements are their increased concentration levels at Nienstedten compared to Oortkaten. For Mn, clear minima at the concentration level of Oortkaten are recognizable for both slack water points, the maximum concentrations at ebb and flood are increased up to 3 times. For Br pronounced maxima appear at SWF. This might be due to a source downstream from Nienstedten.

For a more extensive interpretation of the analytical data with regard to inflow-outflow balances of the trace elements, transport calculations including hydrographical data are required. This has been discussed already elsewhere in these proceedings [10,11].

REFERENCES

[1] Michaelis, W. (1989) The BILEX concept: research supporting public water quality surveillance in estuaries. This volume, chapter III

[2] Prange, A., Knoth, J., Stößel, R.-P., Böddeker, H. and Kramer, K. (1987) Determination of trace elements in the water cycle by Total-reflection X-ray fluorescence spectrometry. Anal. Chim. Acta 195: 275-287

[3] Niedergesäß, R., Racky, B. and Schnier, C. (1987) Instrumental neutron activation analysis of Elbe river suspended particulate matter separated according to the settling velocities. J. Radioanal. Nucl. Chem. 114: 57-68

[4] Knoth, J. , Schwenke, H. (1980) A new totally reflecting X-ray fluorescence spectrometer with detection limits below 10^{-11} g. Fresenius Z. Anal. Chem. 301: 7-9

[5] Prange, A. (1989) Total reflection X-ray spectrometry: method and applications. Spectrochim. Acta 44B: 437-452

[6] Erdtmann, G. and Petri, H. (1986) Nuclear activation analysis: Fundamentals and Techniques. In: P. J. Elving (ed.) Treatise on Analytical Chemistry. Part I, Vol. 14. John Wiley and Sons, Inc., New York

[7] ICES (1988) ICES sixth round intercalibration for trace metals in estuarine water, JMG 6/TM/SW. Cooperative research report No. 152

[8] Steinhausen, D. and Langer, K. (1977) Clusteranalyse. De Gruyter, Berlin New York

[9] Bowen, H. J. M. (1979) Environmental chemistry of the elements. Academic Press, London, p. 36

[10] Kappenberg, J., Fanger, H.-U., Männing, V. and Prange, A. (1989) Suspended matter and heavy metal transport in the lower Elbe river under different flow conditions. This volume, chapter IV

[11] Fanger, H.-U., Kappenberg, J. and Männing, V. (1989) A study on the transport of dissolved and particulate matter through the Hamburg harbour. This volume, chapter IV

HIGH-SENSITIVE DETERMINATION OF ELEMENT TRACES BY MEANS OF X-RAY FLUORESCENCE ANALYSIS FOR AUTOMATIC ROUTINE MEASUREMENTS

H. Bruchertseifer, J.W. Leonhardt
Academy of Science of the G.D.R., Central Institute of Isotope and Radiation Research, Leipzig, G.D.R.

1. INTRODUCTION

Energy-dispersive X-ray fluorescence has become a well established analytical technique by means of which the elemental content of geological, marine, biological or other samples can be determined down to concentrations in the ppm range. However, there are many analytical problems with respect to environmental pollution control and industrial process control which demand the determination of ppb-concentrations. Therefore, a new special type of X-ray fluorescence analyzer, the RFAA-30, was developed (Leonhardt et al., 1986; Bruchertseifer et al., 1987).

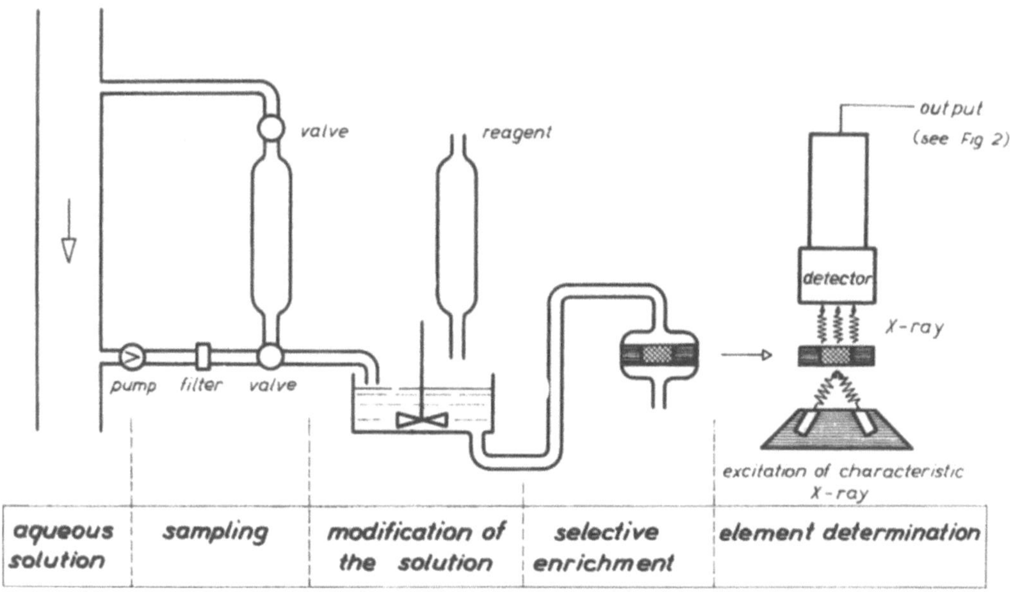

Figure 1: Scheme of X-Ray-Analyzer

2. APPARATUS AND ANALYTICAL PROCEDURES

The device consists of three main parts (Fig. 1):

- an electro-pneumatic system for sampling and solution modification
- a selective chemical enrichment unit, and the
- X-ray fluorescence analyzer including a sealed proportional scintillation counter (PSC), and isotope excitation sources (Fig. 2).

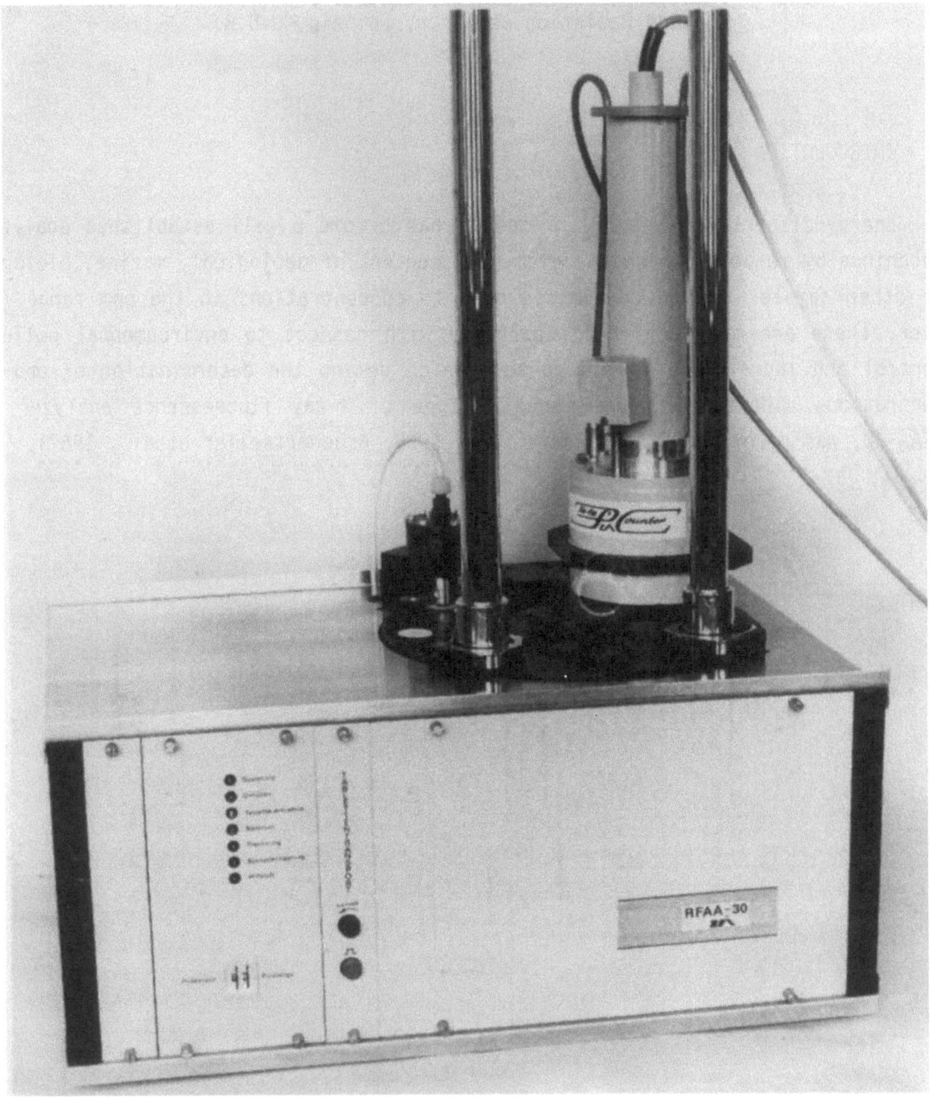

Figure 2: Sample changer with PSC

A modular assembly for the processing of analogue pulses is connected to the proportional scintillation counter. Its components are: high voltage supply units for 2 and 5 kV, a spectrometric amplifier, an electronic peak height stabilizer, and a 256 channel Wilkinson ADC integrated spectrum buffer. For the control of measuring data processing and of spectrum evaluation a software package is offered which can be run on any MS-DOS compatible computer system. The implementation of user-specific evaluation algorithms can be done on the basis of the programming language Turbo-PASCAL.

The proportional scintillation counter PSC (a novel X-ray detector) allows the detection of secondary X-ray quanta in a maintenance-free operation as it does not require any cooling, unlike most semiconductor devices. On the other hand, its spectrometric parameters as well as energy resolution, and thus the separation of neighbouring X-ray lines, are clearly superior to those obtained with conventional proportional counters. Due to the physical principle of the PSC (inert gas luminescence production resulting from the absorption of X-ray quanta) it is possible to detect X-ray fluxes up to nearly 10^5 s^{-1} without a deterioration of spectrometric properties, ensuring at the same time high counting efficiencies.

For the selective enrichment of the element(s) to be determined from liquid samples, ion exchange, liquid-solid extraction or other sorption processes whose distribution constants should be 10^2 may be employed. Extractants, such as tri-n-butylphosphate (TBP), di(2-ethylhexyl)-phosphoric acid (HDEHP), tri-n-octylphosphine oxide (TOPO), amines sorbed on inert support material (silica gel, PTFE) can completely separate elements, such as Au, Tl, platinum metals, Y and the rare earths from aqueous solutions. Chelating cellulose and ion exchangers have proved suitable for quantitatively sorbing trace amounts of Pb, Mn, Co, Ni, Fe, Cu and Zn. Depending on the concrete analytical task it is necessary to modify the starting solution by adding a reagent solution (e.g. setting a particular pH-value). The processes of sampling, the selective element enrichment and fixation, and transportation to the measuring position in the detector unit are automated.

3. APPLICATIONS

The analyzer was used to determine uranium (Fig. 3) and lead trace concentrations in the flow of aqueous process solutions and natural waters. Detection limits of 50 ppb and 10 ppb, respectively, could be achieved (Kießig et al., 1988a,1988b).

An automatic analyzer for element determination of this type can be integrated into industrial control systems, in automatic water control stations, but also used in laboratories. This opens up important applications in mining, in the metallurgical and chemical industries to process control and the monitoring of waste solutions (effluents), in energetics and power plants to the monitoring of polluting effluents, in environmental protection to the monitoring of effluents, river and drinking water, and in the microelectronic and galvanic industries to the recovery of valuable metals and bath control, to mention just the most important ones.

RFA spectrum measured with a Si(Li) semiconductor detector

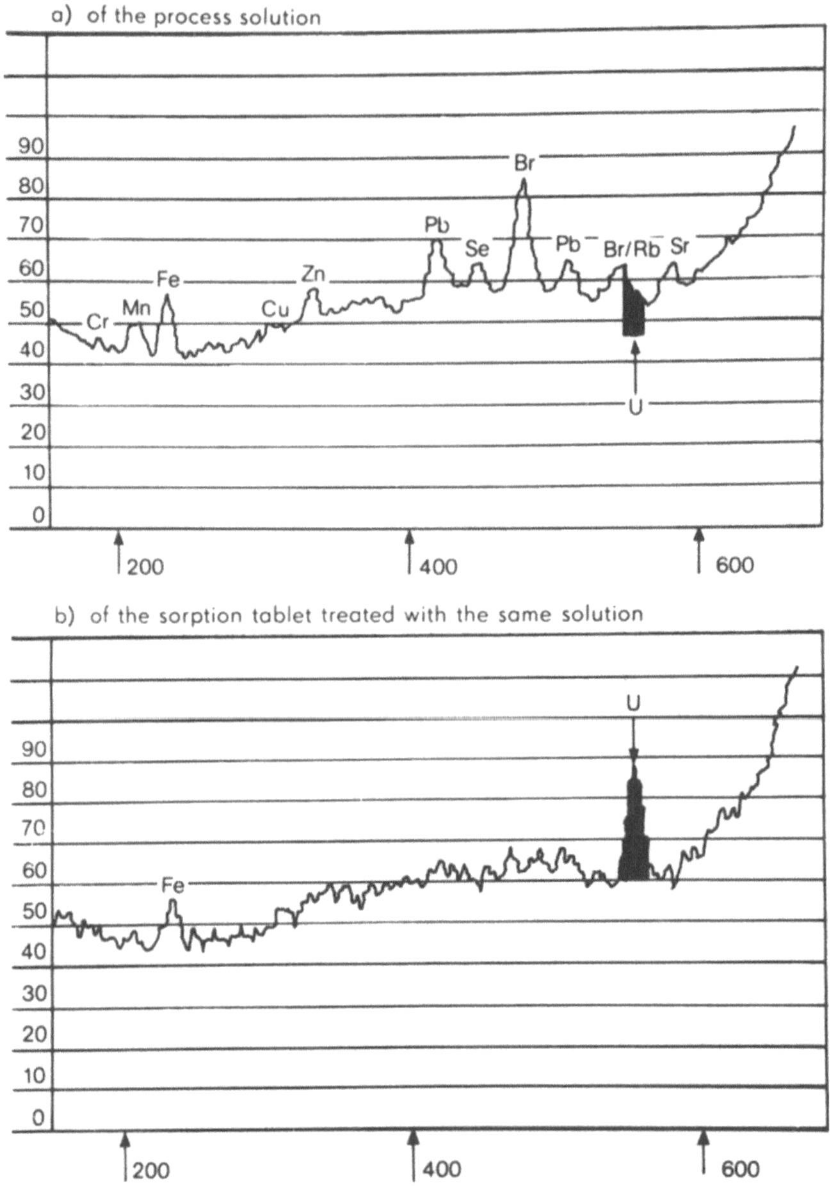

Figure 3: Automatic X-ray fluorescence analyzer - RFAA-30 - application example: determination of low uranium contents in technological solutions with high contents of salts

The benefits of the X-ray fluorescence analyzer RFAA-30 are:

- a robust construction which is largely protected so as to withstand industrial operating conditions which are often tough,
- fully automatic measurement, real time processing of measurement data and the provision of process control signals using compatible personal computers,
- easy operation by menu-driven evaluation programs,
- the detection of even very low element concentrations,
- the option of simultaneous measurement of other important parameters, such as temperature, CO content, pH value, conductivity, etc. by integrating further measuring sensors.

Only analyzers of the RFAA-30 type, using both the physical and chemical properties of elements and their compounds, are suitable for the simultaneous determination of elements in insoluble and soluble forms, such as sediments, precipitations, colloids and any really soluble chemical compounds. The RFAA-30 allows various chemical forms of the elements of interest in water to be distinguished. In the past few years it has become very well known that only some chemical compounds of the heavy metals are toxic.

Such a type of analyzer is expected to gain in importance because other heavy elements, ranging from V and Cr via Ni and Cu, the rare earths, Ag, Cd, the platinum metals, Au, Hg, Tl to U can be determined.

REFERENCES

Bruchertseifer H, Eckert B, Eife K-H, Kießig G, Leonhardt J, Morgenstern P, Riedel W, Süß R, Teller M (1987) Automatische Röntgenfluoreszenzanalyse von Schwermetallspuren in wäßrigen Lösungen. In: 4th Meeting on Nuclear Analytical Methods, May 4-8, 1987, Dresden, Proceedings Vol 2, p 489

Kießig G, Barborowski M, Bruchertseifer H (1988a) Möglichkeiten der röntgenfluoreszenzanalytischen Bestimmung von Schwermetallen in Wasser, Vortrag zum Magdeburger Gewässerschutzseminar, 27.10.1988, Magdeburg

Kießig G, Eife K-H, Eschrich B, Neumann U, Teller M, Bruchertseifer H, Eckert B, Leonhardt J, Morgenstern P, Müller D, Riedel W (1988b) Determination of U in industrial process solutions by means of XRA. In: 4th Conference on Radioisotope Application and Radiation Processing in Industry, Sept 17-23, 1988, Leipzig, Abstract of Papers, p 162

Leonhardt J, Kießig G, Bruchertseifer H, Eckert B, Eschrich B, Friedrich A, Hähnel W, Wünschmann G, Morgenstern P, Müller D (1986) Patent (DDR) WP G 01 N/2717366

GRADIENTS OF TRACE HEAVY METAL CONCENTRATIONS IN THE ELBE ESTUARY

Diether Schmidt
Deutsches Hydrographisches Institut, Laboratorium Sülldorf
Wüstland 2, 2000 Hamburg 55, Federal Republic of Germany

ABSTRACT

Concentration changes of Hg, Cd, Cu, Fe, Mn, and Ni in the water phase of the Elbe river were investigated on two cruises in August 1983 and in June 1985. Stations covered the full range of the estuary, from riverine water (Port of Hamburg) through the lower Elbe river course to the open North Sea (German Bight). From Hamburg to the sea, concentrations go down by 2 to 3 orders of magnitude. Plotted against salinity, all metals show an exponential decrease.

1. INTRODUCTION

The German Bight is located in a relatively remote corner of the North Sea. It is less accessible to the two openings for flushing with fresh Atlantic Ocean water, the entrance to the north and the English Channel. The bight itself is very shallow, mostly around 30 m deep. It is separated from the coastline by the Wadden Sea, a unique environment that normally falls dry twice a day. The coastal areas are subject to severe stress from tourism, fishery, maritime traffic, industrial and municipal outflows, and other human uses of the sea. Additionally, four major rivers empty dircetly into the German Bight: the Ems, Weser, Elbe, and Eider, with the Elbe being by far the most important.

Evaluation of the large pool of data from the monitoring programmes soon made it obvious that the distributions of the metals Hg, Fe, and Mn throughout the whole German Bight are influenced to a large extent by the inflow of waters from the major rivers, mainly the Elbe (Schmidt, 1976, 1980). Especially interesting was the case of Hg, which showed a pronounced input from the river Elbe (Schmidt and Freimann, 1983, 1984; Schmidt and Wendlandt, 1987).

The Elbe has the highest water flow among the rivers entering the German Bight. Its large drainage basin covers highly industrialized countries such as Czechoslovakia and the German Democratic Republic; on its lowermost course, it passes the Hamburg metropolitan area and some newly industrialized regions between this city and the sea.

The intention of the present study was to obtain some insight into the concentration changes of the six trace metals Hg, Cd, Cu, Fe, Mn and Ni determined regularly on samples covering the whole course from fully riverine water to the monitoring network stations in the open German Bight, using the same methods for sampling and analysis for both areas and thereby rendering data directly comparable.

2. METHODS

All samples were taken using the sampler system "MERCOS", a very simple technique developed for ultratrace metal analysis in coastal and surface sea water to a maximum depth of 100 m (Freimann, Schmidt, and Schomaker, 1983). To prevent contamination, exchangeable 500-ml teflon bottles were used as sampling and also as storage vessels. The sampler passes the contaminated surface layer of the sea in a closed configuration and is released at the desired depth by a conventional PTFE messenger. The water samples were stabilized in situ by pre-acidifying the sampling bottles with 10 ml ultrapure HNO_3 for Hg, and by immediately acidifying with ultrapure HCl for the other metals. All cleaning and handling of samplers and samples, aboard the ship and in the laboratories on shore, was done within clean benches.

Temperature and salinity profiles were measured, immediately prior to metal sampling, by means of a Niel Brown probe. Samples for Hg on the one hand, and for the five metals Cd, Cu, Fe, Mn, and Ni on the other were taken in two separate hydrocasts.

The acidified samples for Hg analysis were immediately irradiated on board for two hours by ultraviolet light to release any organically bound mercury. They were then stored at room temperature. Analysis was performed in the land-based clean laboratory by an improved method (Freimann and Schmidt, 1982). It employs cold vapour atomic absorption spectrophotometry, using nitrogen aeration, $SnCl_2$ reduction, and pre-enrichment/purification by amalgamation on finely dispersed gold.

Two sets of samples were taken: one directly into the teflon bottles of "MERCOS" for unfiltered water, and the other with pressure filtration on board inside a clean bench through 0.4 µm Nuclepore membrane filters in a Sartorius all-polycarbonate-plastic apparatus. For the other five metals, after acidification, all samples were deep frozen and stored at -28 °C for analysis. After complexation with APDC and liquid/liquid extraction with MIBK, determination was by graphite furnace atomic absorption spectrometry.

3. RESULTS AND DISCUSSION

The first series of samples were taken on a cruise with RV "GAUSS" out of Hamburg to the German Bight from 23 to 26 August 1983. Six stations spanned the lower course and estuary of the tidal Elbe river from the Port of Hamburg to the sea near Cuxhaven. Six other stations were selected as appropriate from the network of stations occupied regularly for the different monitoring programmes, to render a section across the coastal waters of the German Bight. Two stations described the outer estuary of the Elbe, and three stations were located around the island of Helgoland. Here, according to the prevailing wind direction and velocity, and the tidal phase, the outer river plume of the Elbe normally tends northwards east of Helgoland and can be discriminated from surrounding sea water, often as far north as the island of Sylt. The outermost station was located just northwest of the research platform "NORDSEE" which rests on the sea floor about 70 km northwest of Helgoland. This station is intended to display open sea conditions in the German Bight, in comparison with riverine, estuarine, and near coastal chemistry.

Salinity ranges were found to vary from 32 to almost zero upstream from Glückstadt. Samples 2 m below the water surface and about 2 m above the river or sea bottom were taken.

Some results from this first investigation have been published (Schmidt, 1985). The trace metal concentrations analyzed in unfiltered and filtered water samples have been displayed in a series of diagrams; a few selected graphs appeared in a previous publication (Schmidt, Freiman, and Zehle, 1986).

The range of metal concentrations found from river to sea water is generally more than 2 orders of magnitude, for some metals even more. For this reason a logarithmic scale has been chosen for the Y-axis, where commonly a linear scale is used. Generally the total Hg concentrations are higher in the lower sample than in the upper one, since the particle density increases down to the river or sea bottom. For Mn, unfiltered and filtered samples from the "surface sampler" show a decrease over more than 3 orders of magnitude from Hamburg to the open German Bight.

In another approach, metal concentrations (again on a logarithmic scale) were plotted against salinity. For this purpose, values from all sampler depths and stations were pooled. From the distributions, regression curves were calculated. For each metal, the Pearson correlation coefficient r was computed for different possible correlation functions. The pure mixing model, which is often used to describe the dilution of saline water with fresh water when going upstream in an estuary, should result in a linear decrease of metal concentrations with increasing salinity if mere dilution would occur without chemical reactions. For all our examples here, it could be shown that the correlation coefficients (r) for this linear relationship are so low that a significant correlation cannot be postulated. The same is true for other conceivable types of function; however, for all six metals the best fit is an exponential decrease in metal concentrations with increasing salinity of the form: $Y = A \exp(BX)$.

The best correlation exists for Hg; almost as highly correlated are Fe, Mn and Ni concentrations in filtered samples. These r values are from 0.92 to 0.94. All r values (with one exception) are higher than 0.80. For unfiltered Mn and Fe, the r values are the lowest, and the differences between them for exponential and linear decrease are smallest in these two cases.

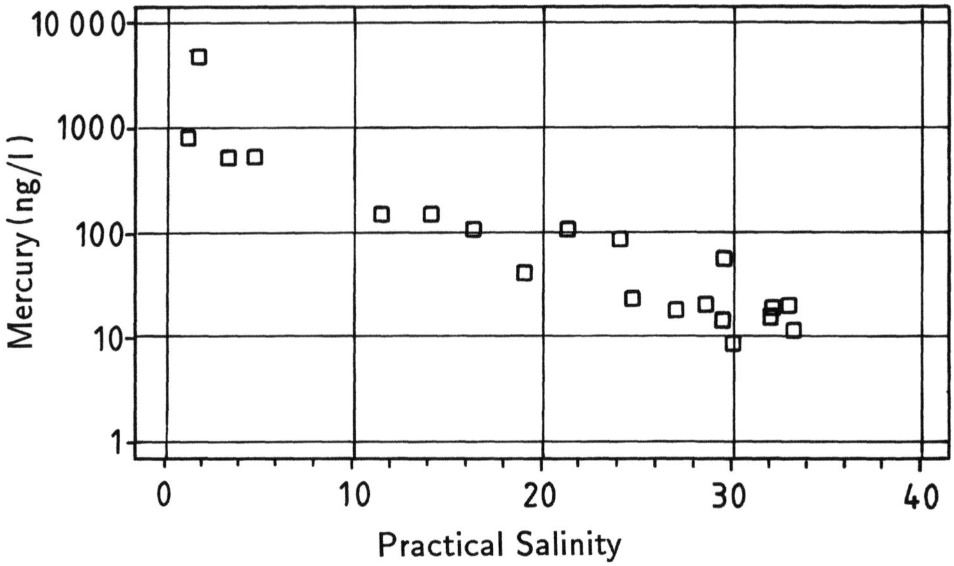

Figure 1: Diagram of mercury concentrations (in unfiltered samples) as a function of salinity

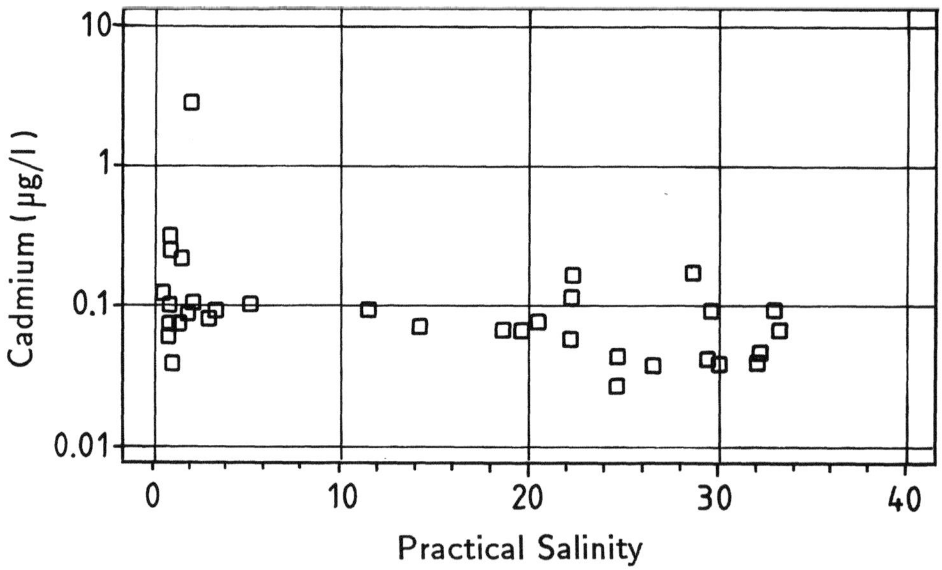

Figure 2: Cadmium concentrations in filtered samples

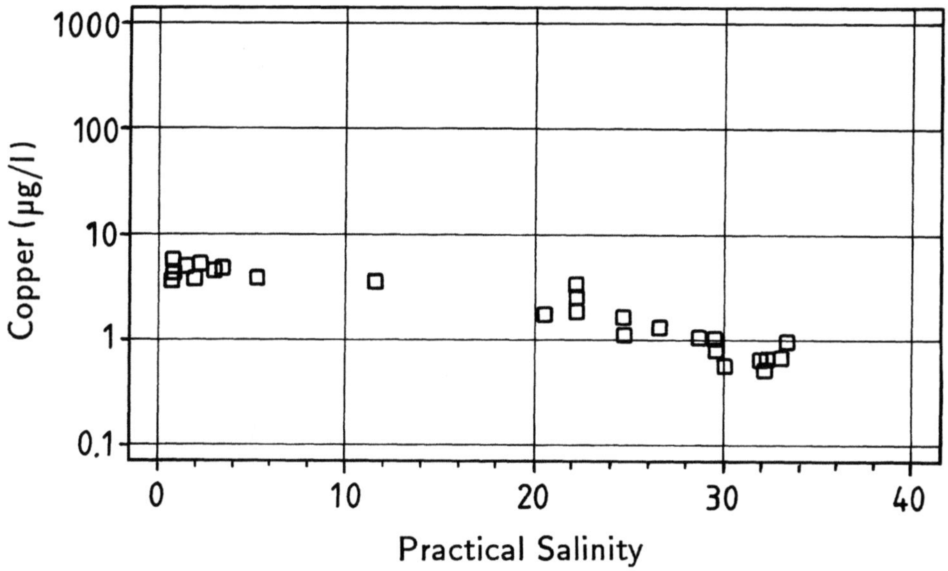

Figure 3: Copper concentrations in filtered samples

From the diagrams of metal concentration versus salinity it could be seen that most values accumulate in the low and high salinity regions, i.e. in river and open sea water. Relatively few data were acquired for brackish water because in the first investigation the river stations were selected according to geographical considerations. In order to enhance the representation of the intermediate brackish water, where the mixing occurs, additional sampling was obtained during a repetition of the cruise in June 1985.

A preliminary evaluation of the data from the second investigation yielded very similar results in the graphical displays. Some selected examples are given in Figures 1 to 3. Again, Hg in unfiltered samples shows a pronounced exponential decrease over more than two orders of magnitude, from fresh to sea water. Cd and Cu in filtered samples, in contrast, are more evenly distributed, with a much smaller salinity effect.

A more detailed discussion, including all data and diagrams from both cruises, will soon be given in a final publication.

REFERENCES

Freimann P, Schmidt D (1982) Determination of mercury in seawater by cold vapour atomic absorption spectrophotometry. Fresenius' Z Anal Chem 313: 200-202

Freimann P, Schmidt D, Schomaker K (1983) MERCOS - a simple teflon sampler for ultratrace metal analysis in seawater, Marine Chem 14: 43-48

Schmidt D (1976) Distribution of seven trace metals in sea water of the inner German Bight. ICES/CM 1976/C:10

Schmidt D (1980) Comparison of trace heavy-metal levels from monitoring in the German Bight and the southwestern Baltic Sea. Helgoländer Meeresunters 33: 576-586

Schmidt D (1985) Input of heavy metals from the river Elbe into the German Bight: a section from Hamburg harbour to the North Sea. Proceedings, International Conf "Heavy Metals in the Environment", Athens, September 1985, Vol 1: 530-532, Edinburgh

Schmidt D, Freimann P (1983) Determination of very low levels of mercury in sea water (North Sea, Baltic Sea, northern North Atlantic Ocean). Proceedings, International Conf "Heavy metals in the environment", Heidelberg, September 1983, Vol 1: 233-236, Edinburgh

Schmidt D, Freimann P (1984) AAS-Ultraspurenbestimmung von Quecksilber im Meerwasser der Nordsee, der Ostsee und des Nordmeers. Fresenius' Z Anal Chem 317: 385-387

Schmidt D, Freimann P, Zehle H (1986) Changes in trace metal levels in the coastal zone of the German Bight. Rapp P-v Réun Cons int Explor Mer 186: 321-328

Schmidt D, Wendlandt U (1987) Spurenbestimmung von Quecksilber im Meerwasser der Nordsee, in: Welz B (ed) 4. Colloquium Atomspektrometrische Spurenanalytik, Konstanz, April 1987, 617-628

CHLOROPHENOLICS IN THE FRASER RIVER AND ESTUARY

R.J. Allan
National Water Research Institute
Canada Centre for Inland Waters
Burlington, Ontario, Canada L7R 4A6

INTRODUCTION

In 1808, Simon Fraser found that the Fraser River was a route to the Pacific Ocean. Today, the river is no longer the pristine waterway of 1808. The river enters the sea just south of Vancouver, British Columbia (B.C.); has a length of some 1,253 km; a drainage basin of some 230,400 km², the largest in B.C. (Figure 1); and a mean annual discharge of 2,700 m³/sec. Sources of organic chemical contamination to the river include forest product industries. The river is the location of pulp and paper mills and wood processing operations. The objective of this paper is to summarize recent results on contamination of the river by chemicals from these forest products industries (Allan, 1989).

SOURCES AND CONCENTRATIONS OF CHLOROPHENOLIC COMPOUNDS

Approximately 680 tonnes of phenolic compounds are used as wood preservatives in B.C. annually. Many wood treatmant plants are located along the lower Fraser River and in the estuary. In the upper Fraser River basin, six plants pressure treat lumber and one uses the pentachlorophenol (PCP) process. Five pulp and paper mills are located upstream from Hope and four are bleached kraft mills. There are 16 sawmills using PCP and tetrachlorophenol (TeCP) for protection of cut lumber from sapstain fungi. Shavings from PCP treated lumber at mills may be used in pulp mills and thus the effluents from pulp mills may contain chlorophenols.

Krahn and Shrimpton (1988a) found that leaching of 2,3,4,6-TeCP and PCP from treated lumber began after this lumber was subjected to 1 to 1.5 mm of continuous rainfall. Krahn and Shrimpton (1988b) estimated that storm water runoff from surface treated lumber storage areas in the lower Fraser River totalled between 489,000 to 776,000 m³ per annum, resulting in an estimate of total chlorophenol input of from 226 to 916 kg/year.

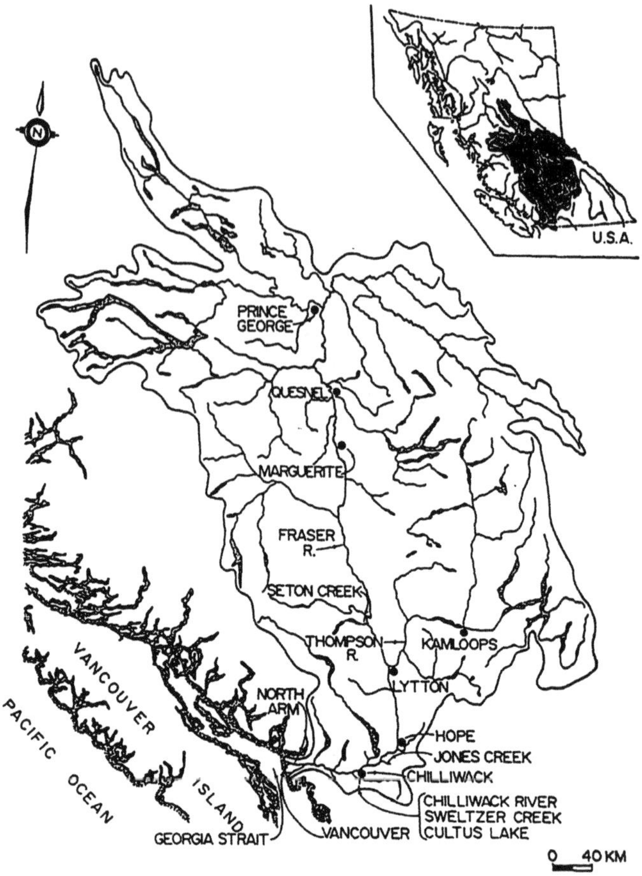

Figure 1: Drainage basin of the Fraser River

Water concentrations of total chlorophenols have been reported as 40 ppt in the east end of the north arm (Carey et al., 1988); and 90 ppt (Rogers et al., 1988) and 190 ppt in the upper river (Voss and Yunker, 1983); at various periods of the year. Carey and Hart (1988) investigated the effects of low and high flow conditions on the concentrations of chlorinated phenolic compounds in the Fraser River estuary (Figure 2). The major chlorophenolics present under both conditions were 2,4,6-trichlorophenol (TCP), 2,3,4,6-TeCP, PCP, tetrachloroguaiacol (TeCG), and a compound which they tentatively identified as 3,4,5-trichloroguaiacol (TCG). The concentrations of the guaiacols were higher than the chlorophenols under high flow conditions (Figure 3a). Concentrations of 2,3,4,6-TeCP and PCP were higher than the chloroguaiacols in some samples in the north arm of the estuary under low flow conditions (Figure 3b). However, chloroguaiacols still dominated in the site upstream of the estuary and in the main channel in low flow conditions. The conclusion was that pulp mills upstream of the estuary were major sources of chloroguaiacols under

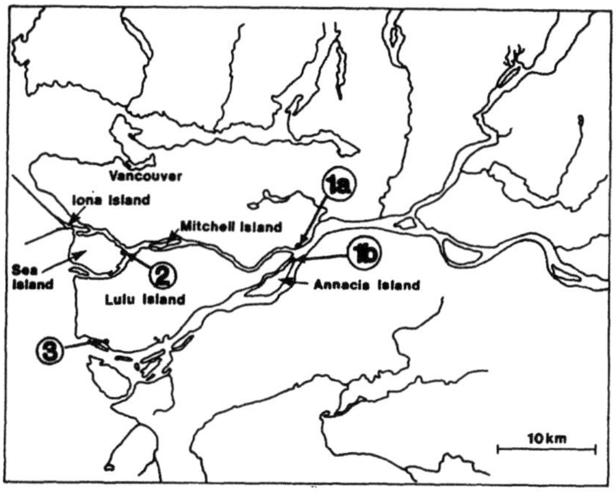

Figure 2: Location of sample sites in the Fraser River estuary - referred to in Figure 3

all flow conditions but that episodic inputs of chlorophenols into the estuary, particularly the north arm, can result in concentrations exceeding the chloroguaiacol concentrations during low flow.

BIOACCUMULATION OF CHLOROPHENOLIC COMPOUNDS

Jacob and Hall (1985) used leeches as biomonitors of chlorophenols in the Fraser River estuary. Bioconcentration was rapid and chlorophenol uptake after one-week exposures in the river varied over 1 to 2 orders of magnitude. The concentrations of TeCP in leeches in the estuary ranged from 28 to 2,946 ppb in the fall and 124 to 235 ppb in spring. For PCP the ranges were 28 to 846 ppb and 55 to 57 ppb, respectively, for the same time periods.

Millions of juvenile salmon use the Fraser River and its tributaries. All salmon pass through the estuary as juveniles and as returning adults. The highest chlorophenol in emergent pink salmon fry from the Fraser River was some 58 ppb (Servizi et al., 1988). Fingerling chinook from the Fraser River contained some 38 ppb total chlorophenols. Freshly spawned pink salmon from Sweltzeer Creek (Figure 1) contained 0.2 ppb PCP but no TeCP, TCP or dichlorophenol (DCP). Pink fry contained 0.8 ppb. Seton Creek and Jones Creek fry contained 2.9 and 2.4 ppb PCP, respectively. Pink salmon fry from the Fraser and Thompson River contained 5.4 and 5.8 ppb PCP, respectively, and significant concentrations of TeCP. The mean total chlorophenol content of overwintering feral chinook juvenile salmon from the upper Fraser River near Prince Geoge and Quesnel (Figure 1) was some 34 ppb in March 1987 (Rogers et al., 1988).

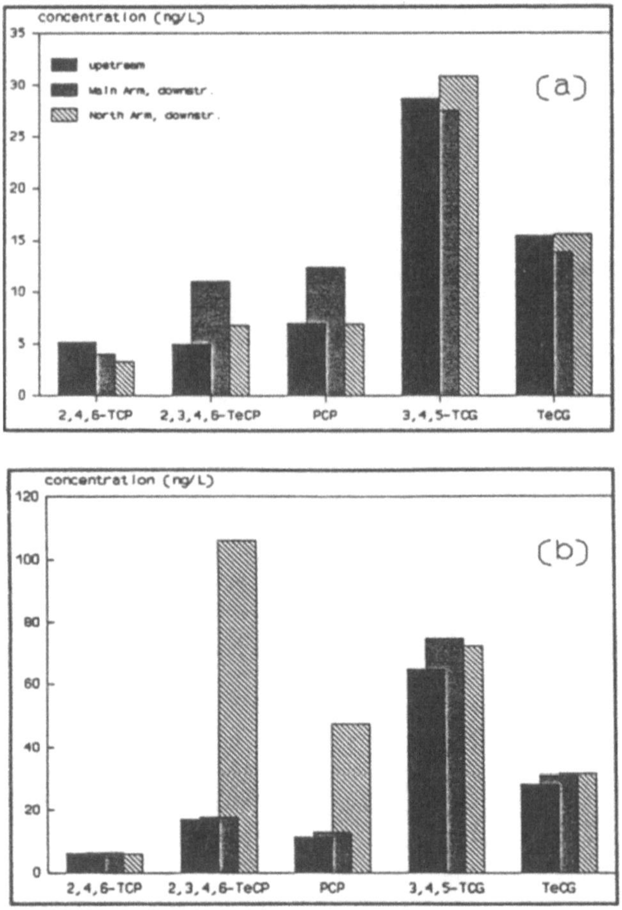

Figure 3: Comparison of average concentrations of five chlorophenolics sampled over four-day periods at three sites in the Fraser River estuary under (a) high and (b) low flow conditions (Carey and Hart, 1988)

MODELLING THE FATE OF CHLOROPHENOLIC COMPOUNDS

A one dimensional mathematical model to simulate 2,3,4,6-tetrachlorophenol (TeCP) fate in the Fraser River estuary (Lam et al., 1988) was verified using measured concentrations. The mean relative errors were within 10 to 20 % of observed values. During high flow, TeCP movement was essentially downstream. During low flow, both upstream and downstream movements were possible. In decreasing order of sensitivity, chemical loading, flow, sorption, photolysis, and particle settling affected the predicted concentrations.

No attempts have been made to model the fate of chlorophenols from the estuary out into the Strait of Georgia (Figure 1). Chlorophenols have been detected in marine waters bordering the Strait (Voss and Yunker, 1983). A strongly tidal estuarine system exists between the mainland of B.C. and Vancouver Island. In the summer, the shallow plume of the Fraser River is about 3 to 5 m thick and covers the bulk of the Strait. The main axis of the river jet curves north up along the mainland coast (Figure 4). During low flow conditions the plume can flow back up the river. Toxic organic chemical fate in the plume in the Strait of Georgia needs to be coupled with modelling of the salt wedge with a view to assessing movement of chemicals both out into the strait but also back into the Fraser River estuary.

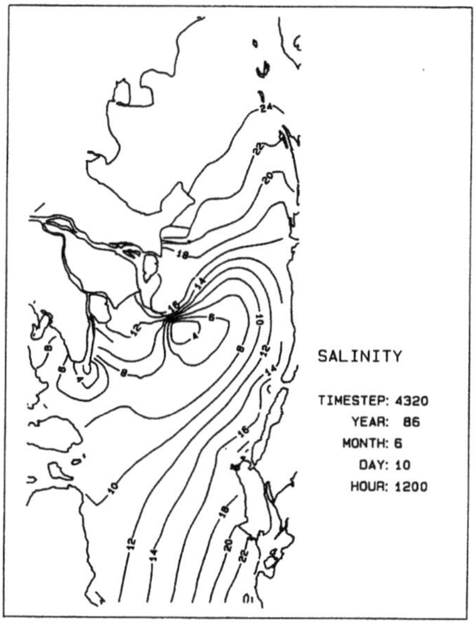

Figure 4: Modelled upper layer salinities in the Strait of Georgia. Estuary mouth at center left, Vancouver Island on right, top is south (Stronach et al., 1988)

REFERENCES

Allan RJ (1989) An overview of pollution of the Fraser River by chlorinated phenolic compounds. National Water Res Inst Contribution 89-30, 17 pp.

Carey JH and Hart JH (1988) Sources of chlorophenolic compounds to the Fraser River estuary. Water Poll Research J Can 23(1): 55-68

Carey JH, Fox ME and Hart JH (1988) Identity and distribution of chlorophenols in the north arm of the Fraser River estuary. Water Poll Research J Can 23(1): 31-44

Jacob C and Hall (1985). Use of bioconcentration capability of Hirudinea (leeches) to evaluate chlorophenol contamination of water. Proceedings of the International Conference on New Directions and Research in Waste Treatment and Residuals Management, Vol 2: 773-788, University of British Columbia, Vancouver, B.C.

Krahn PK and Shrimpton JA (1988a) Stormwater related chlorophenol releases from seven wood protection facilities in British Columbia. Water Poll Research J Can 23(1): 45-54

Krahn PK and Shrimpton JA (1988b) Modelling stormwater related chlorophenol releases to the Fraser River using the Motz-Benedict (Moben) model. Water Poll Research J Can 23(1): 114-121

Lam DCL, McCrimmon RC, Carey JH and Murthy CR (1988) Modelling the transport and pathways of tetrachlorophenol in the Fraser River. Water Poll Research J Can 23(1): 141-159

Rogers IH Servizi JA and Levings CD (1988) Bioconcentration of chlorophenols by juvenile chinook salmon (Oncorhynchus tshawytscha) overwintering in the upper Fraser River: field and laboratory tests. Water Poll Research J Can 23(1): 100-113

Servizi JA, Gordon RW and Carey JH (1988) Bioconcentration of chlorophenols by early life stages of Fraser River pink and chinook salmon (Oncorhynchus gorbuscha, O. tshawytscha). Water Poll Research J Can 23(1): 88-99

Stronach JA, Crean PB and Murty TS (1988) Mathematical modelling of the Fraser River plume. Water Poll Research J Can 23(1): 179-212

Voss RH and Yunker MB (1983) A study of chlorinated phenolics discharged into kraft mill receiving waters. Contract Rep. to Council of Forest Industries Technical Advisory Committee, 113 pp.

AN INTERCOMPARISON OF PARTICULATE TRACE METALS FROM FOUR LARGE ESTUARIES

A. Turner and G.E. Millward
Institute of Marine Studies, Polytechnic South West, Plymouth, PL4 8AA, UK

L. Karbe and M. Dembinski
Institut für Hydrobiologie und Fischereiwissenschaft
Universität Hamburg, D-2000 Hamburg 50, FRG

ABSTRACT

Suspended sediment samples have been taken during axial profiles of four major estuaries (viz. Mersey, Humber, Elbe and Thames) and analysed in a consistent manner for trace metals and specific BET surface area. The results indicate that a great deal of inter-estuarine variability exists.

INTRODUCTION

The Mersey, Thames, Humber (UK) and Elbe (FRG) are large, partially/well mixed macrotidal estuaries. Furthermore, all are industrialized with long histories of pollution problems. The former estuary discharges into Liverpool Bay (Irish Sea), whereas the remainder discharge into the North Sea, and both are shelf sea regions of growing concern regarding levels of toxic materials.

One form of pollution that has received much attention over the past two decades is that of trace metals (Salomons and Förstner, 1984). Trace metals occur in the water column in solution and associated with the suspended solid material.

It was the aim of this investigation to determine the levels of leachable (i.e. perhaps representative of bioavailable or geochemically reactive) particulate trace metals in each of the aforementioned estuaries in order to gain a basic insight into inter-estuarine variability, and relate such levels to physical properties of the particles as determined by BET analysis. The latter analyses are discussed in detail in a separate paper (Millward et al., 1989). However, combining the results of both trace metal and BET analysis may allow a fuller assessment of inter-estuarine geochemical variability.

SAMPLING AND ANALYTICAL METHODS

Between 10 and 20 samples of 1-10 l were taken during axial profiles of the Mersey, Humber and Thames and filtered through 0.45 µm Millipore filters (142 mm diameter) under N_2 pressure. The residue material was washed with about 20 ml distilled water and transferred into plastic pots for freeze-drying. During the

Elbe profile and for the Mersey samples for Hg analysis, a continuous flow centrifuge (10^4 rpm) was employed to collect material on teflon strips lining the interior of the rotating bowl. Samples were subsequently washed with distilled water and freeze-dried.

The salinity was monitored with an MC-5 T-S bridge and subsamples were taken to determine the suspended solids concentration (mgl^{-1}) by filtration through Whatman GF/F papers.

Table 1. Hydrodynamic conditions during sampling

	Salinity Range ‰	Suspended Solids Range mgl^{-1}
Mersey (Nov. 1987)	0.4 - 31.9	50 - 155
Mersey (July 1988; Hg only)	27.7 - 33.5	1 - 47
Humber (Jan. 1988)	< 0.5 - 25.9	35 - 248
Elbe (June 1988)	< 0.5 - 31.3	13 - 289
Thames (Feb. 1989)	4.7 - 27.4	35 - 424
Thames (March 1989)	0.4 - 4.5	17 - 70

The surface areas of samples were determined using a gravimetric BET N_2 adsorption technique (Crosby et al., 1983) and trace metals were determined using conventional flame AA techniques following leaching with 25 % acetic acid + 0.05 M hydroxylamine hydrochloride (Chester and Hughes, 1967; Tessier et al., 1980). The particulate Hg levels were determined by cold vapour AAS following leaching by dilute HNO_3.

RESULTS AND DISCUSSION

The results are expressed as means, standard deviations, ranges and coefficients of variation in Table 2. Although sampling was undertaken during different seasons and under different hydrodynamic conditions (Table 1), the significance of which to trace metal levels and distributions has previously been demonstrated (e.g. Morris et al., 1986), some useful comparisons can be made. Trace metal levels are generally highest in the Humber Estuary (especially for Cu and Cr; Fig. 1a, Table 2) reflecting the large industrial inputs, as is the Fe content (Fig. 1b). The latter is largely derived from two titanium dioxide manufacturing plants and, as Fe is very BET reactive (Crosby et al., 1983; Marsh et al., 1984), this leads to large surface areas and enhances scavenging properties of Humber particulates. The surface areas of Elbe material are equally high. However, the particles are relatively Fe poor (Fig. 1b, Table 2) and the high Mn contents (Table 2) suggest that this element is more significant to the surface properties.

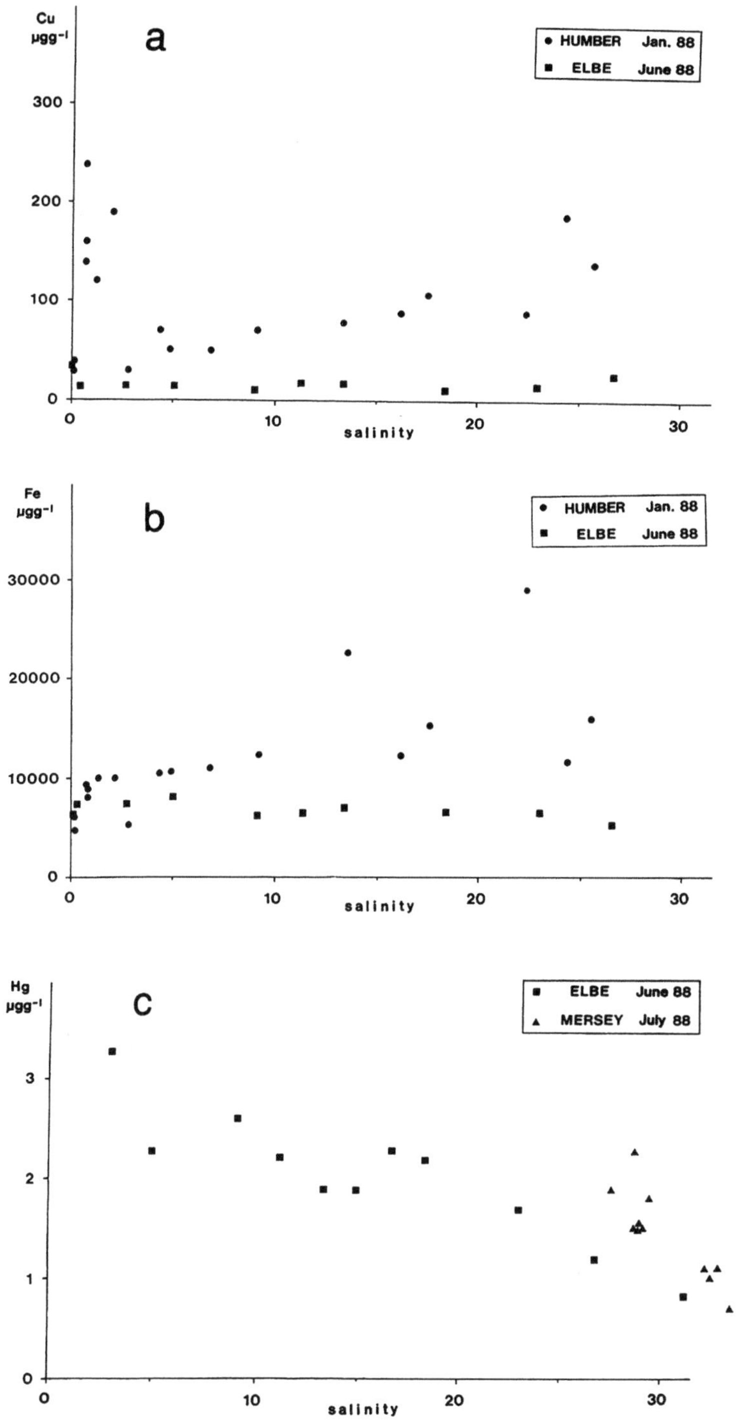

Figure 1. Profiles of the particulate trace metal levels. a) Cu, Humber and Elbe; b) Fe, Humber and Elbe; c) Hg, Elbe and Mersey

One suggestion regarding the origin of such high Mn levels is the mobilization of reduced Mn from the anoxic sediments of Hamburg Harbour, and subsequent uptake onto particles downstream as the O_2 levels in the water column improve and the turbidity increases. Trace metal levels and coefficients of variation are lower than in the Humber, possibly because of the absence of such large industrial discharges to the lower tidal waters. In fact much of the trace metal burden of the Elbe may have been derived from the industrialized regions of the eastern European countries in the upper catchment, and subsequently diluted by the large volume of water downstream.

Table 2. Particulate trace metal levels in the four estuaries (n = number of samples; R = range; CV = coefficient of variation).

	Statistical parameter	Fe $mg g^{-1}$	Mn $mg g^{-1}$	Zn $mg g^{-1}$	Cu $\mu g g^{-1}$	Pb $\mu g g^{-1}$	Cr $\mu g g^{-1}$	Hg $\mu g g^{-1}$	SSA $m^2 g^{-1}$
Mersey	$\bar{x} \pm \sigma$	4.8±0.8	1.2±0.3	0.31±0.08	30±13	-	-	1.4±0.4	9.6±2.2
	n	9	10	10	10	-	-	11	10
	R	3.5-6.5	0.7-1.6	0.23-0.50	20-60	-	-	0.7-2.3	6-14.9
	CV, %	17	25	26	43	-	-	28	23
Humber	$\bar{x} \pm \sigma$	12.5±5.7	0.99±0.17	0.53±0.43	110±60	55	37±144	-	24.0±6.3
	n	19	19	19	19	6	14	-	18
	R	4.8-29.1	0.64-1.29	0.25-2.22	30-240	-	17-74	-	9.8-38.3
	CV, %	48	17	81	55	-	38	-	26
Elbe	$\bar{x} \pm \sigma$	6.6±0.9	3.08±0.33	0.47±0.10	17±6	58	22	2.0±0.6	25.6±8.1
	n	10	9	10	10	4	4	11	10
	R	5.0-8.3	2.58-3.62	0.36-0.70	10-31	-	-	0.8-3.3	7.2-35.5
	CV, %	14	11	21	35	-	-	30	32
Thames	$\bar{x} \pm \sigma$	4.9±1.1	0.70±0.13	0.29±0.17	27±10	73±17	19±3	-	12.3±13.9
	n	14	14	14	14	14	13	-	14
	R	3.7-8.3	0.42-0.94	0.13-0.73	16-45	51-113	16-24	-	5.0-19.2
	CV, %	22	19	59	37	23	16	-	32

The surface area and Fe contents of the Mersey Estuary are low compared with the Humber and Elbe, and trace metal levels are somewhat lower than anticipated bearing in mind the abundance of industrial discharges to the tidal waters. This may be interpreted as coarse material of low adsorptive capacity being carried in suspension. A similar situation exists for the Thames Estuary, viz. low surface areas and Fe contents. The relatively low levels of trace metals such as Zn and Cr may be a result of the low BET reactivity as in the Mersey, and associated with the lack of large industrial effluents. The relatively high levels of Pb on the other hand, may be expected in an estuary receiving vast quantities of urban runoff.

Regarding the Hg results, although less comprehensive, it is worth mentioning that an almost conservative seaward decrease in the Elbe was noted (Fig. 1c). The Hg levels in the Mersey (Fig. 1c) also decrease seawards. Although levels in this estuary are lower than in the Elbe, it should be borne in mind that sampling was undertaken at higher salinities and extrapolation back to freshwater would yield a significant Hg burden in the Mersey, which would be consistent with previous observations (Airey and Jones, 1982; Campbell et al., 1986). Much of this Hg load is derived from anthropogenic discharges, notably the chlor-alkali industry, as well as effective scavenging of Hg from solution (Airey and Jones, 1982; Campbell et al., 1986).

Using previously published estimates of sediment discharges (Veenstra, 1970), the possible impact on the North Sea particulate trace metal budget may be assessed for the Elbe, Humber and Thames (Table 3). Because of its far greater water and sediment discharge, the Elbe provides the largest contribution of particulate trace metals to the North Sea through its input to the German Bight. Material derived from the Humber is discharged in a distinct but poorly dispersing plume (Lewis, undated) which is likely to have more of an impact on the local nearshore environment.

Table 3. Approximate discharge of particulate trace metals from the Humber, Elbe and Thames estuaries.

	Approximate sediment discharge* (tons/annum)	Cr	Cu	Fe (tons/annum)	Mn	Pb	Zn
Humber	10^5	4	10	1200	100	5	50
Elbe	$7 \cdot 10^5$	15	10	4600	2220	40	300
Thames	$2 \cdot 10^5$	4	5	1000	150	15	60

* Veenstra, 1970

CONCLUSIONS

Particulate trace metals studies in four large estuaries have shown that levels in the Humber are much higher than those in the Thames and Mersey, probably as a result of the abundance of large industrial discharges and the high (BET) reactivity of particles in the former estuary. The levels of particulate trace metals in the Elbe are generally intermediate, although its impact on the North Sea particulate trace metal budget is larger than that of the Humber through its far greater sediment discharge.

REFERENCES

Airey D, Jones PD (1982) Mercury in the River Mersey, its estuary and tributaries during 1973 and 1974. Wat Res 16: 565-577

Campbell JA, Chan EYL, Riley JP, Head PC, Jones PD (1986) The distribution of mercury in the Mersey Estuary. Mar Poll Bull 17(1): 36-40

Chester R, Hughes MJ (1967) A chemical technique for the separation of ferro-manganese minerals, carbonate minerals and adsorbed trace metals from pelagic sediments. Chem Geol 2: 249-262

Crosby SA, Glasson DR, Cuttler AH, Butler I, Turner DR, Whitfield M, Millward GE (1983) Surface areas and porosities of Fe(III)- and Fe(II)- derived oxyhydroxides. Env Sci Tech 17: 709-713

Lewis R (undated) Hydrography. In: The Effects of Four Industrialised Estuaries upon the Coastal Waters of North Eastern England. Brixham Laboratory. Report BL/A/1698

Marsh JG, Crosby SA, Glasson DR, Millward GE (1984) BET nitrogen adsorption studies of iron oxides from natural and synthetic sources. Thermochim Acta 82: 221-229

Millward GE, Turner A, Glasson DR, Glegg GA (1989) Intra- and inter-estuarine variablity of particle microstructure. Sci Tot Env Submitted

Morris AW, Bale AJ, Howland RJM, Millward GE, Ackroyd DA, Loring DH, Rantala RTT (1986) Sediment mobility and its contribution to trace metal cycling and retention in a macrotidal estuary. Wat Sci Tech 18: 111-119

Salomons W, Förstner U (1984) Metals in the Hydrosphere. Springer-Verlag, Berlin

Tessier A, Campbell PGC, Bisson M (1980) Trace metal speciation in the Yamaska and St. Francois Rivers (Quebec). Can J Earth Sci 17: 90-105

Veenstra HJ (1970) Sediments of the southern North Sea. In: Delany FM (ed) The Geology of the East Atlantic Continental Margin. Vol. 3, Europe. Institute of Geological Sciences, London, p 10-23

DISSOLVED, PARTICULATE AND SEDIMENTARY MERCURY IN THE COCHIN ESTUARY SOUTHWEST COAST OF INDIA

P.P Ouseph
Centre for Earth Science Studies
Cochin - 682018, India

ABSTRACT

The Cochin estuary is subjected to the discharge of effluents from the major industrial units situated on the banks of the rivers Periyar and Chitrapuzha. Since mercury and its compounds are highly toxic, a study has been initiated for understanding its distribution and seasonal variations. The concentration of mercury in the sediments varied with sediment texture and showed good correlation with organic carbon. The bottom waters are found to have higher concentrations compared to the surface. Total mercury for particulate matter was high in relation to the sediment concentrations.

1. INTRODUCTION

The southwest coast of India bordering the state of Kerala is connected by a chain of backwaters and rivers. The region surrounding Cochin accounts for more than 70 % of the chemical industries situated near the banks of the rivers Periyar and Chitrapuzha, and the effluents containing a heavy load of metals are discharged into the Cochin estuarine systems. The particulate matter of water in Cochin backwaters indicated that this environment contains high contents of Zn, Cr, and other metals (Sankaranarayanan and Rosamma Stephen, 1978). High concentrations observed in Crassostrea madrasensis is considered to be due to industrial and domestic pollution (Sankaranarayanan et al., 1978). A series of environmental-geochemical studies have been undertaken to determine the levels, behaviour and dynamics of heavy metals in the Cochin estuarine system. This paper reports on the dissolved, particulate and sedimentary mercury determined during the surveys undertaken in July and November 1985, and April 1986.

2. MATERIALS AND METHODS

Samples were collected from 18 locations as shown in Fig. 1. Sediment samples were taken using a van veen grab, surface water samples were collected with a plastic bucket and bottom water samples with a van dorn sampler. The analytical method for total mercury in sediment employs oxidative digestion (Thomson et al., 1980) and cold vapour atomic absorption (Hatch and Ott, 1968). Filtered and unfiltered water samples were used for the estimation of dissolved and particulate mercury (Ouseph, 1987). The method was further modified by a dithizone/CCl_4 extraction technique using 800 ml samples. The organic layer was extracted back by means

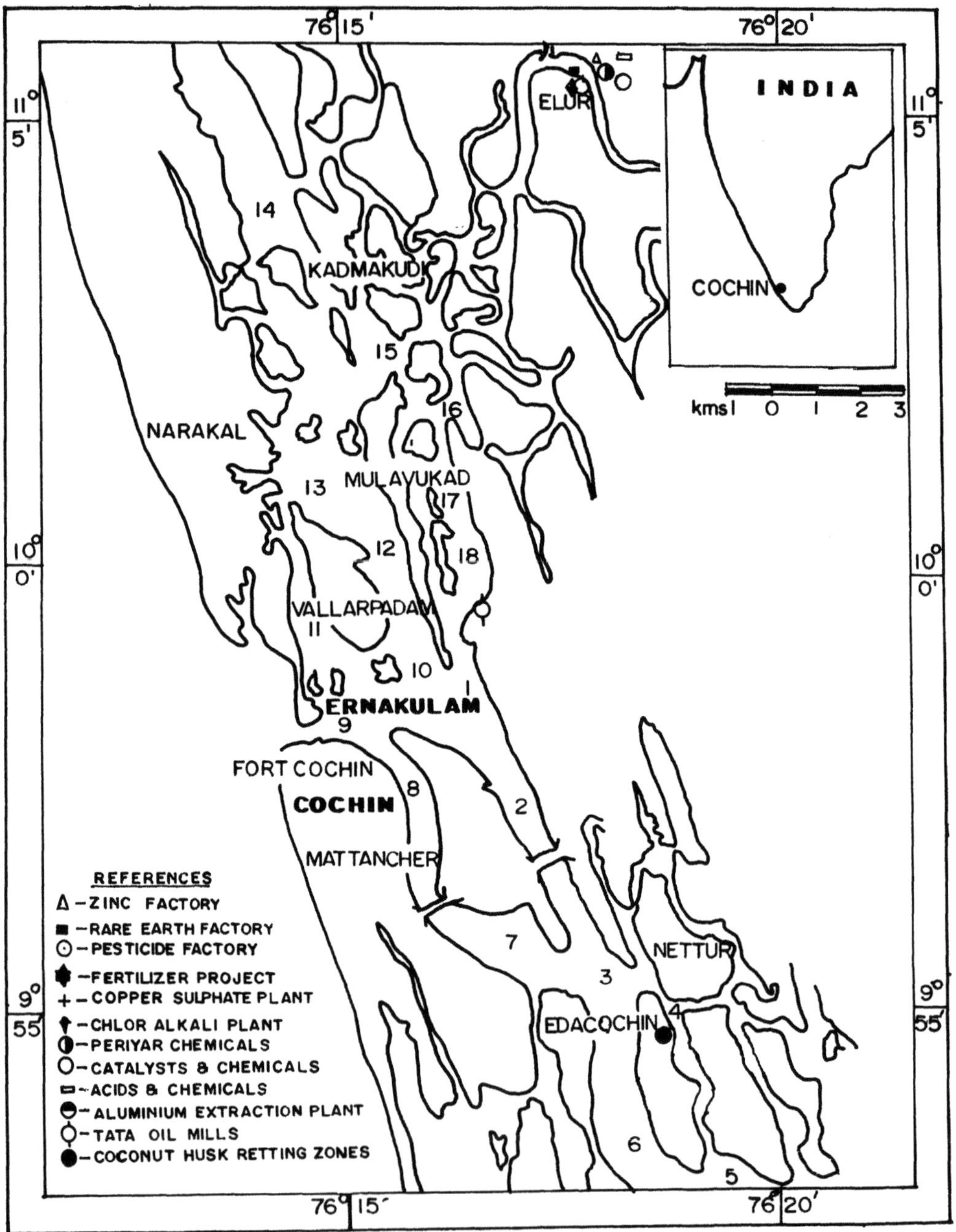

Figure 1: Location map with details of sampling stations and major industries

of sodium nitrite and hydroxylamine hydrochloride. The extracts served to estimate the mercury concentration by cold vapour AAS. In each case the blank values were determined. Five replicate determinations, done for one sample from each set of samples collected seasonally, gave a variation coefficient of 3.8 to 4.2 % for water samples at concentrations between 5 to 100 ng/l and 5.2 to 6.8 % for sediment samples. Silt is defined as the fraction < 0.062 mm.

3. RESULTS AND DISCUSSION

The mercury concentration in sediments ranged from 0.12 to 0.95 ppm and varied regionally with sediment texture. The background level of mercury was estimated on the basis of a 90 cm core sample collected at station 13 and found to be 0.08 ppm. Hence, the observed sediment concentrations are 1.5 to 11.8 times higher compared to the background levels, indicating an anthropogenic orgin. In order to find out whether mercury is held up in the silicate or non-silicate extractable fraction, the sediment samples were subjected to 8N HNO_3 attack (Cosma et al., 1983) and the extracts were used to estimate the mercury concentration. It was found that 80 to 90 % of the mercury is held in the non-silicate extractable fraction.

The total iron concentration ranged from 0.4 to 6.8 % and organic carbon from 0.26 to 3.2 %. Results of linear regression analysis are presented in Figs. 2, 3 and 4. The good interrelationship of organic carbon with mercury indicates that organic carbon plays a major role in the incorporation of mercury in the sediment compared to silt and iron. The correlation studies confirm that mercury appears to be trapped in the finer sediment. Reported values from other marine environments range, for example, from 0.06 to 2.57 ppm (Thomson et al., 1980) and from 0.2 to 4.4 ppm (Bartlett, 1978).

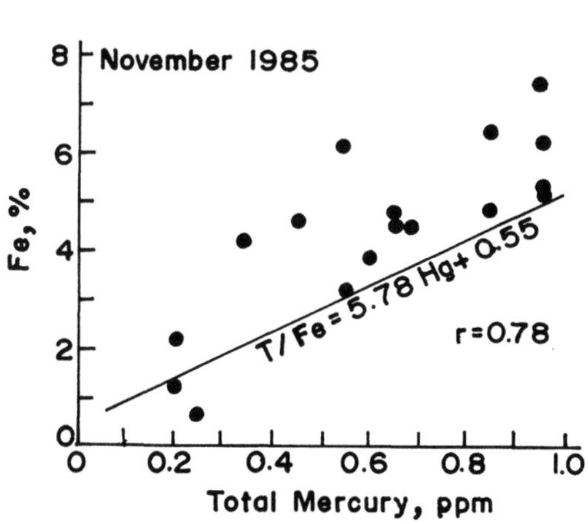

Figure 2: Total Fe and Hg in sediments

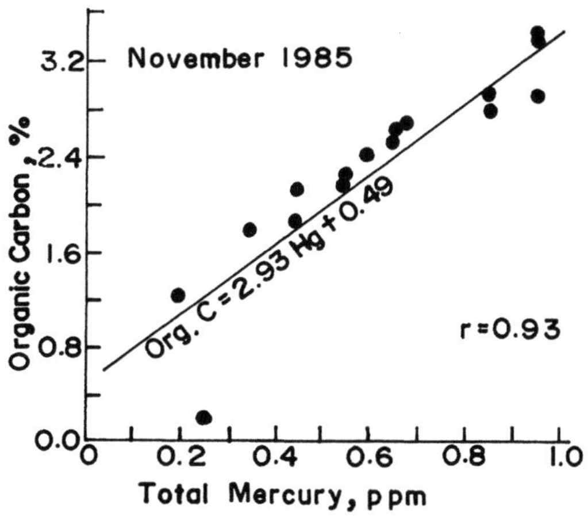

Figure 3: Organic C and Hg in sediments

The seasonal distribution of dissolved mercury ranged from 40 to 180 ng/l during southwest monsoon, 50 to 240 ng/l in the post-monsoon and 80 to 280 ng/l in the pre-monsoon season. For all seasons the concentration in the surface samples was found to be lower than in the bottom samples. This may indicate that mercury is released from the sediment or suspended particulate matter to the bottom water. To examine the possible contamination during sampling and analysis, water samples from unpolluted reaches of the upper Periyar river were analysed. The results were in the range between 15 to 25 ng/l. These values are approximately 2 to 10 times higher compared to the natural background levels. The relation between salinity and mercury concentration is shown in Fig. 5. During the SW monsoon season there is a strong input of fresh water and the mercury concentrations are low. This might be the result of dilution by fresh water.

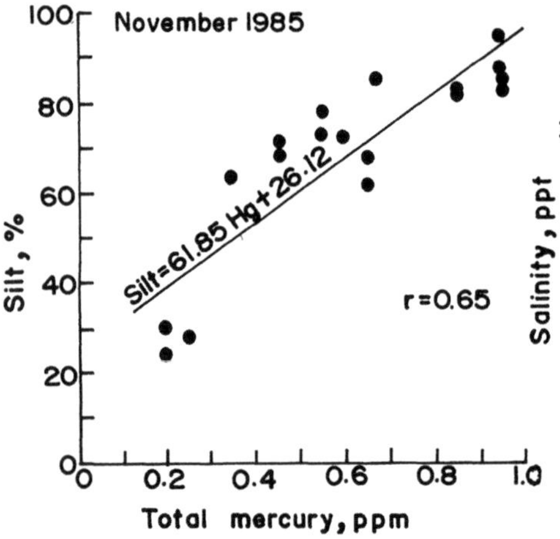

Figure 4: Silt and Hg in sediments

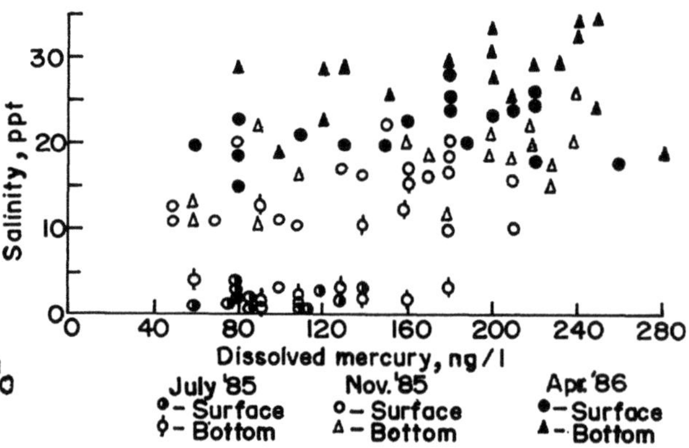

Figure 5: Salinity and dissolved Hg

The particulate mercury concentrations ranged from 2 to 20 ppm during SW monsoon, 2 to 25 ppm during post-monsoon, 6 to 36 ppm during pre-monsoon seasons. Particulate mercury is almost uniform for bottom samples except at a few stations. Bottom particulate matter has a higher content compared to the surface which is similar to dissolved mercury. Salinity has a profound influence on the precipitation of particulate matter. The relation between salinity and particulate matter indicates that the precipitation takes place with salinity > 10 ppt (Ouseph, 1987).

ACKNOWLEDGEMENT

The author is thankful to Dr. M. Ramakrishnan, Director in-charge, Centre for Earth Science Studies, for providing facilities, to Dr. M. Baba, Head RC, for his encouragement, to Dr. K. Premchand for useful discussions, and to Shri D. Raju for the preparation of the drawings.

REFERENCES

Bartlett PD (1978) Total mercury and methyl mercury levels in British estuarine and marine sediment. The Science of the Total Environment 10: 245-251

Cosma B, Contradi V, Zanicehi G, Capelli R (1983) Heavy metals in superficial sediment from the Ligurian Sea, Italy. Chemist in Ecol 1: 331-344

Hatch NR, Ott WL (1968) Determination of submicrogram quantity of mercury by atomic absorption spectrophotometry. Anal Chem 40: 2085-2087

Ouseph PP (1987) Status report on marine pollution along Kerala coast. Proc of the Interdepartmental Seminar on Status of Marine Pollution in India, Dept of Ocean Development, New Delhi, pp 86-105

Sankaranarayanan VN, Purushan KS, Rao TSS (1978) Concentrations of some of the heavy metals in the oyster Crassostrea madrasensis (Preston) from the Cochin region. Ind J Mar Sci 7: 130-131

Sankaranarayanan VN, Rosamma Stephen (1978) Particulate iron, manganese, copper and zinc in water of the Cochin backwater. Ind J Mar Sci 7: 201-203

Thomson JAJ, Macdonald RW, Wong CS (1980) Mercury geochemistry in sediments of a contaminated fjord of coastal British Columbia. Geochem J 14: 71-82

SUBJECT INDEX

AUTHOR INDEX

Coastal and Estuarine Studies

(formerly Lecture Notes on Coastal and Estuarine Studies)

MIX
Papier aus verantwortungsvollen Quellen
Paper from responsible sources
FSC® C105338

If you have any concerns about our products,
you can contact us on
ProductSafety@springernature.com

In case Publisher is established outside the EU,
the EU authorized representative is:
Springer Nature Customer Service Center GmbH
Europaplatz 3, 69115 Heidelberg, Germany

Printed by Libri Plureos GmbH
in Hamburg, Germany